Springer Series in Optical Sciences

Volume 184

Founded by

H. K. V. Lotsch

For further volumes:
http://www.springer.com/series/624

Springer Series in Optical Sciences

The Springer Series in Optical Sciences, under the leadership of Editor-in-Chief William T. Rhodes, Georgia Institute of Technology, USA, provides an expanding selection of research monographs in all major areas of optics: lasers and quantum optics, ultrafast phenomena, optical spectroscopy techniques, optoelectronics, quantum information, information optics, applied laser technology, industrial applications, and other topics of contemporary interest.

With this broad coverage of topics, the series is of use to all research scientists and engineers who need up-to-date reference books.

The editors encourage prospective authors to correspond with them in advance of submitting a manuscript. Submission of manuscripts should be made to the Editor-in-Chief or one of the Editors. See also www.springer.com/series/624

Vladimir Protopopov

Practical Opto-Electronics

An Illustrated Guide for the Laboratory

Springer

Vladimir Protopopov
Suwon
Korea, Republic of South Korea

ISSN 0342-4111 ISSN 1556-1534 (electronic)
ISBN 978-3-319-04512-2 ISBN 978-3-319-04513-9 (eBook)
DOI 10.1007/978-3-319-04513-9
Springer Cham Heidelberg New York Dordrecht London

Library of Congress Control Number: 2014933532

Printed on acid-free paper

Springer is part of Springer Science+Business Media (www.springer.com)

Preface

During the past 10 years I had a good opportunity to watch the work of university graduates and young PhDs in optical laboratories. The strongest observation is how deep the gorge is, dividing theoretical knowledge acquired at university from the ability to make the simplest opto-electronic systems work. For example, a freshman knows perfectly well that if a laser beam is directed onto a photodetector, the latter must generate a signal. But often, it is far from the scope of the engineer that the flux may be too strong, jeopardizing the very life of an expensive device, or how to correctly extract that signal from a «black box» named photodetector. It is even more dangerous for the company when an inexperienced opto-electronic engineer is granted the power to purchase equipment for the project. How to make the right choice? Do I need any additional equipment to put the system into operation? Are my requirements professional? What exactly to request from the vendor? These and many others are the questions to be answered. Finally, after having been explaining such simple but numerous things to my colleagues for years, it occurred to me that a book could be the best solution. Initially, the book was conceived as a practical guide to the laboratory. However, it immediately occurred to me that the questions to be answered are not solely of the «How?» type, but are immediately followed by the questions «Why?». Therefore, theoretical background, on the scale of higher mathematics, also proved to be necessary.

The result of such a fusion of practice and theory is the present book, teaching how to create opto-electronic systems in the most efficient way, avoiding typical mistakes, and explaining the theoretical background necessary to realize practical limitations. It covers light detection techniques, imaging, interferometry, spectroscopy, modulation–demodulation, heterodyning, beam steering, and many other topics common to laboratory applications. The focus is on self-explanatory figures rather than on words. Therefore, the density of figures per page far exceeds the standards for technical publications. The book will guide the reader through the entire process of creating problem-specific opto-electronic system, starting from optical source, through beam transportation optical arrangement, to photodetector, and data acquisition system. But not only this: relevant basics of beam propagation and computer-based raytracing routines are also explained, and sample codes listed. It will teach important know-how and practical tricks that are never disclosed in scientific publications, and protect from typical mistakes listed in the end of each chapter. I hope that the book will be the reader's personal advisor in the world of opto-electronics and navigator in the ocean called the market of optical components and systems.

The work on this book was intense, and I always received support from the Springer Senior Editor Dr. Claus Ascheron who represents the best traditions of this renowned publisher.

Suwon, South Korea Vladimir Protopopov

Contents

Chapter 1
Optical Elements

Optical manufacturers offer a wide variety of optical elements, covering almost all possible laboratory applications. The problem is not how to design what you want but how to choose what you need.

Abstract Great variety of optical elements for visible domain available on the market is summarized in seven structural sections: mirrors, simple lenses, plates and prisms, retroreflectors, beamsplitters, imaging lenses, and microscope objectives. The first section presents mirrors: types of reflecting surfaces, broadband and narrowband coatings, mounting options, typical mounting mistakes. Polarization rotation after several reflections may be unexpected. Basic functionality of simple lenses is described, beginning from the lens maker equation for the thick lens. Its derivation, however, is postponed until the Chap. 12 where it is obtained from the matrix formalism (ABCD law). Spherical aberration for basic single lens geometries is presented graphically and its minimum for the plano-convex configuration is explained. Best practical lens mountings are summarized and consequences of birefringence are emphasized in context with magneto-optics (Chap. 8). Spectral transparency of various optical materials should be considered as the first priority for the ultra-violet domain. High power applications require some simple but important know-how in order to avoid breakdowns. In aspheric lenses, spherical aberration may be almost completely removed by optimizing conic coefficients. Ray tracing is presented in the form of a picture, while detailed mathematics and computer codes are left to Chap. 12. Achromatic doublets is the better choice for wide spectrum, and the front surface must be chosen wisely to obtain the design performance. Steinheil and Hastings triplets are considered as the most popular choice for relay optics. Optical flat plates and prisms are discussed starting from the defocusing they produce, being installed in the focused rays. Simple practical formulas are given for estimating deviation of rays, passing through tilted flats. Along with an ordinary dispersive prism that is both chromatic and deflecting, some special types of combined prisms may be helpful to reduce either chromaticity or deflection. Functionality of reflecting prisms like penta, Amici, Porro, Dove, and Littrow is explained. The concept of the corner cube reflector can be easily understood geometrically, using three-dimensional vector presentation. Its imaging and polarizing peculiarities are also explained in context with interferometers (Chap. 6). The section devoted to beamsplitters summarizes performance

V. Protopopov, *Practical Opto-Electronics*, Springer Series in Optical Sciences 184,
DOI: 10.1007/978-3-319-04513-9_1, © Springer International Publishing Switzerland 2014

of the basic types: plates and cubes, including rather rare type—the energy separator. Among the imaging lenses, the C-mount TV lens is most frequently used. Design, formats, and recommended mounting techniques are carefully explained. Physical idea of the telecentric lens—a very popular element in machine vision—is presented in succinct form with minimum mathematics, and the design of multi-element commercial products is demystified. The last section provides practically indispensable but not widely published specifications for microscope objectives, like mounting thread diameters, tube lens focal distances, marking legend, etc. Finally, the so-called inspection objective is introduced—a handy tool to quickly assemble an electronic imaging system.

1.1 Mirrors

Mirror is a simplest optical element. Mirrors may be flat, concave or convex. The second two types are used for focusing or defocusing. As a focusing element, the concave mirror is inferior to a lens because aberrations cannot be minimized. Therefore, concave or convex mirrors are rarely used in laboratories. On the contrary, flat mirror (Fig. 1.1) is the most frequently used optical element.

The second-surface mirrors are more protected against scratches, but produce ghost reflections from the front surface. Another disadvantage is absorption inside the glass, which may be considerable in ultraviolet or infrared domains. Therefore, it is always better to use the first-surface mirrors. Although the substrate of such mirrors is practically always made of glass for the reason of better polishing, in applications with high thermal loads it may be a metal, for example aluminium or copper. For a glass substrate, the ratio of thickness to diameter must be around 5 in order to maintain flatness of the reflecting surface. When high thermal stability is needed, materials like Pyrex® or Zerodur® may be used. Pyrex® has low thermal expansion and is an excellent substrate for most applications. Zerodur® is a glass–ceramic material with thermal expansion being nominally zero. This extra stability

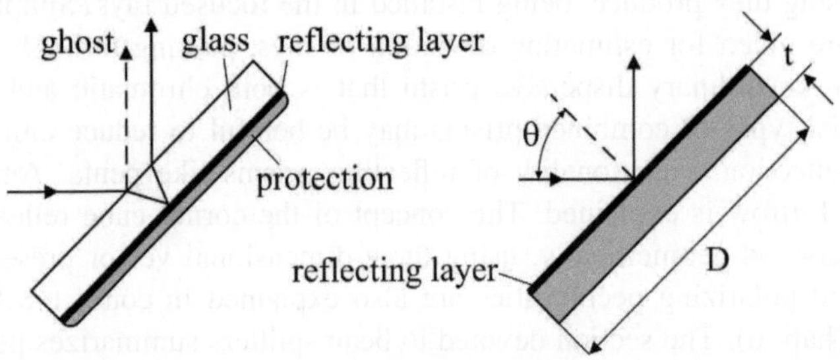

Fig. 1.1 Second-surface (*at left*) and first-surface (*at right*) mirrors

can be critical for applications, requiring diffraction limited performance. It is also worth mentioning that small-scale roughness of a typical mirror surface, measured on distances of a micrometer scale, is about 0.01 nm (standard float glass—0.02 nm), while large-scale roughness measured on a millimeter scale may vary from 0.1 to 1 nm.

Reflecting layer may be either wide-band or narrow-band. The narrow-band or monochromatic mirrors are used mostly with specific laser lines. The reflecting layer then is made as a multilayer stack of many transparent layers designed to reflect at one specific angle of incidence θ and one wavelength λ. Any change of the angle of incidence will lead to smaller reflection coefficient at this particular wavelength. Manufacturers offer mostly $\theta = 45°$ narrow-band mirrors. Monochromatic mirrors are used to suppress background radiation in the systems were only one laser line is used. But their biggest advantage is that highly transparent multilayer stack introduces negligible absorption, which makes maximum reflection coefficient close to 100 %. Therefore, this type of mirrors is used when the beam path is folded many times and the delivered optical power is important. Reflection of monochromatic mirrors is based on interference, therefore, once designed for one particular wavelength such a mirror will also reflect at some other wavelengths, periodically separated from the design wavelength. How strong these ghost reflections are depends on the design, and can never be predicted. Manufacturers never disclose the entire spectral reflectance for multilayer mirrors. Therefore, in laboratory applications, the wide-band metal mirrors with aluminium or silver coatings are the better choice. In order to protect the reflecting surface from damages, a transparent rigid layer is deposited upon it. As a standard, silicon oxide (SiO) is used for protection. Typical spectral reflection for narrow-band and wide-band mirrors is shown in Fig. 1.2. Right mounting options for mirrors are shown in Fig. 1.3. Threaded O-ring should be made of plastic or soft metal like brass or aluminium in order to minimize any damage to glass. Reflecting surface must be turned to frame, otherwise O-ring would damage it during rotation. Examples of wrong mounting are shown in Fig. 1.4.

It may be assumed as a rule that metallic mirrors do not introduce any noticeable depolarization into reflected beam. However, it would be a mistake to assume the same for a system of mirrors. Consider Fig. 1.5. In the laboratory system of coordinates, inputting beam is vertically polarized, but the outgoing beam is horizontally polarized. Since people always navigate in vertical and horizontal directions, the result is also always a surprise.

1.2 Simple Lenses

A single lens is a basic optical element to focus optical beams. For high quality focusing or imaging better to use specially designed and assembled television (TV) lenses, photographic lenses, or microscope objectives, available on the market in

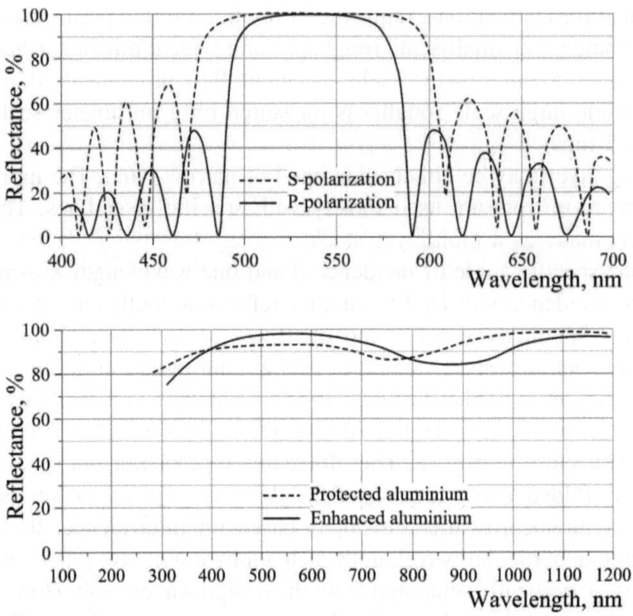

Fig. 1.2 Spectral reflectance curves: 45° narrow-band (*above*) and wide-band (*below*) mirrors

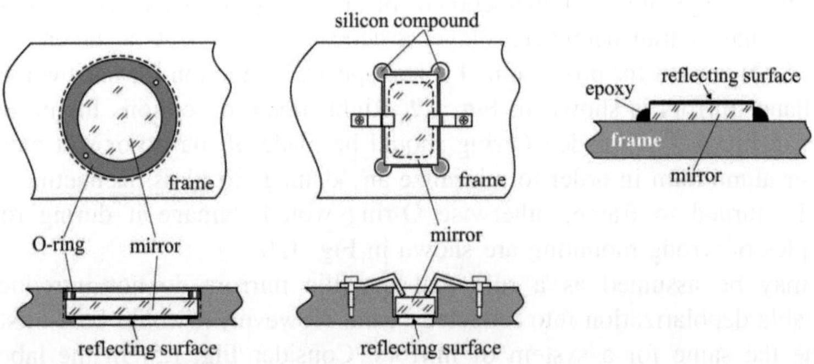

Fig. 1.3 Right mounting. Silicon compound prevents shifting

full variety. All these types of optical elements will be discussed in detail below in Sects. 1.5–1.7. This section is devoted to single lenses—the simplest and cheapest focusing element. Every optical manufacturer offers numerous types of lenses, and it is important to understand what to choose for a particular application. The first thing to understand is the terminology (Fig. 1.6). The effective focal length (EFL) and principal planes P_1, P_2 are more of a theoretical interest than of practical considerations. Usually, the design parameters are back focal length (BFL), lens diameter D, and thickness t. Sometimes, the so-called f-number is used:

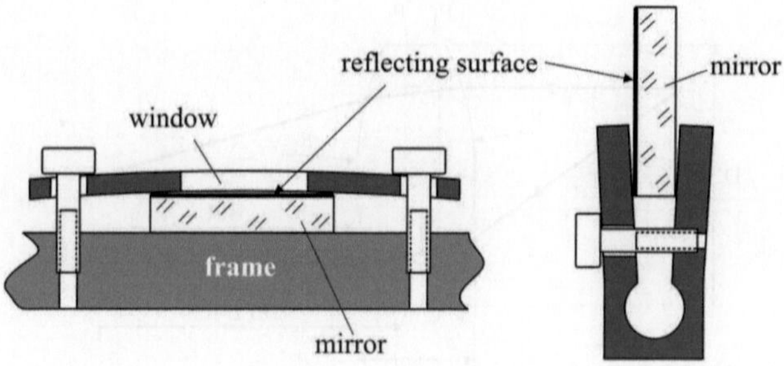

Fig. 1.4 Wrong mounting. Edges will be cleaved and reflecting surface damaged

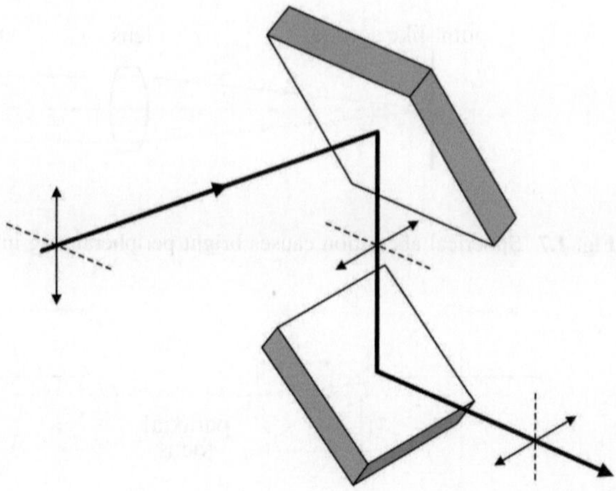

Fig. 1.5 System of mirrors transforms polarization from vertical to horizontal

$$f-\text{number}=\frac{f}{D}.$$

The lens-maker equation determines position of the focus in air:

$$\frac{1}{f}=(n-1)\left[\frac{1}{R_1}-\frac{1}{R_2}+\frac{t(n-1)}{R_1 R_2 n}\right],$$

where n is the refractive index of the lens material. Obviously, various values of R_1 and R_2 make the same focal length. In order to discriminate between the lenses of different shapes but the same focal length, the Coddington shape factor q is useful:

$$q=\frac{R_2+R_1}{R_2-R_1}.$$

What to choose? Choose the lens with the smallest aberrations. Spherical aberration is caused by the fact that ideal spherical shape of a lens surface does not produce ideal spherical wavefront. Practically, it means that parallel laser beam will not be focused in a small point in the focus. Spherical aberration can be easily

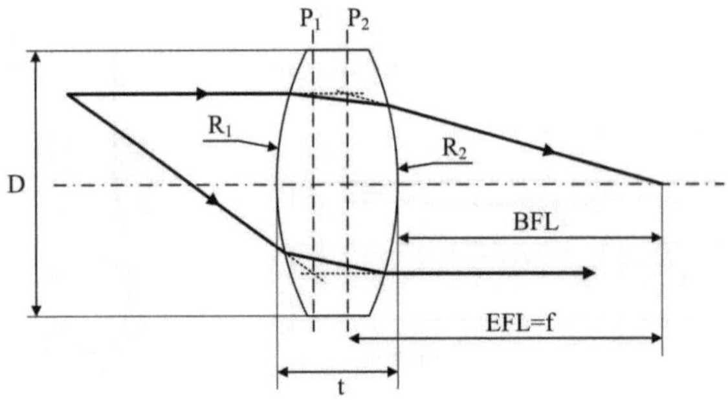

Fig. 1.6 A double convex lens. For the rays coming from left $R_1 > 0$, $R_2 < 0$

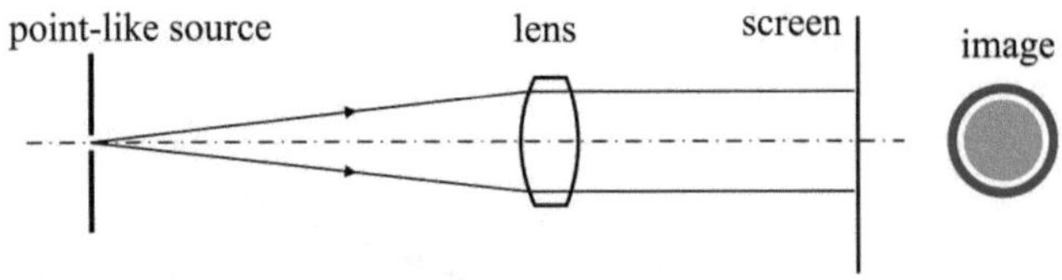

Fig. 1.7 Spherical aberration causes bright peripheral ring in a quasi-parallel beam

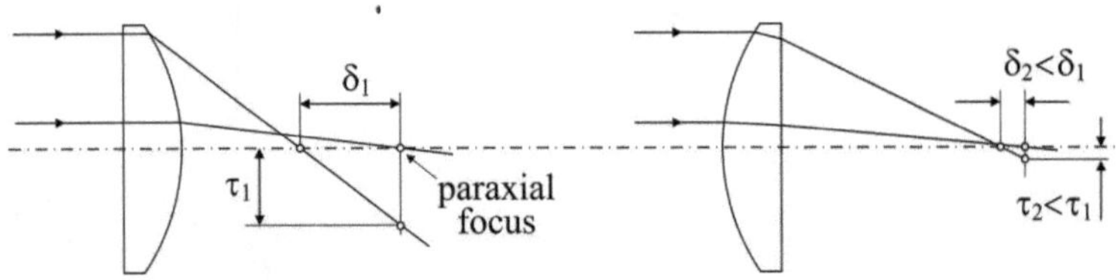

Fig. 1.8 Illumination from the flat surface causes only one refraction. Illumination from the curved surface causes two refractions, which partially compensate spherical aberration

diagnosed as shown in Fig. 1.7. Usually, this type of aberrations is dominant, and the proper choice of lens shape may minimize it. Figure 1.8 shows the difference between two plano-convex lenses of the same shape illuminated from opposite sides. The distance δ along the optical axis between the intercept of the rays that are nearly on the optical axis (paraxial rays) and the rays that go through the edge of the lens (marginal rays) is called longitudinal spherical aberration (LSA). The height τ at which these rays intercept the paraxial focal plane is called transverse spherical aberration (TSA). Figure 1.9 shows how spherical aberration varies with the shape of the lens made of BK7 glass with 100 mm focal length and f-number 10 (another notation f/10).

Plano-convex lens has another important practical advantage: mounting simplicity. One flat surface makes it possible to fix a lens by means of not only a standard threaded O-ring, but also gluing with ultra-violet (UV) bonder (Fig. 1.10).

UV bonder is a transparent compound that hardens when illuminated by UV radiation (UV curing). UV bonder has three important advantages over epoxy:

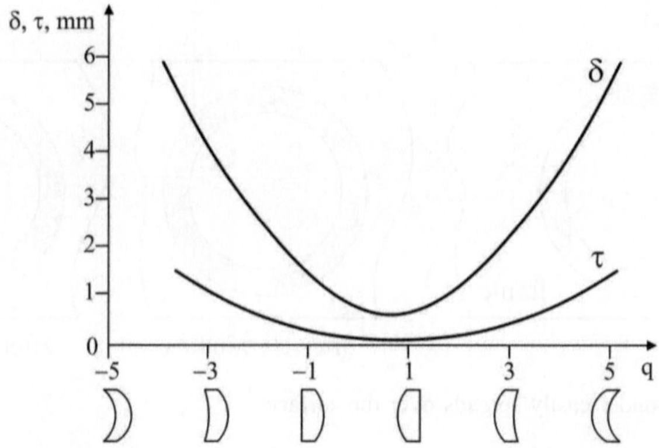

Fig. 1.9 Plano-convex shape ($q = 1$) is nearly the best. EFL is kept constant for all the shapes

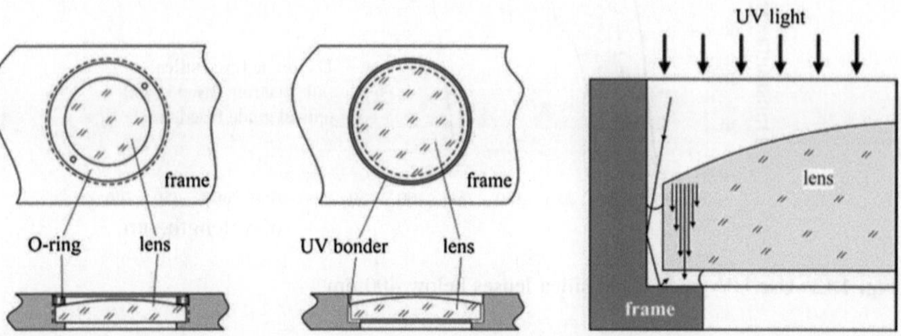

Fig. 1.10 Plano-convex lens mounting options: with threaded O-ring (*at left*), with UV bonder (*in the middle*), and mechanism of curing (*at right*). The lens housing must be deep enough to keep the lens entirely inside, preventing it from possible damage or contamination from outside

it does not hardens without UV curing, thus leaving the opportunity of disassembling the module in case when something goes awry, has low viscosity, allowing the lens to take its position under its own weight, and the ability to spread evenly over the surface due to high wetting. The technique of UV bonding requires certain skills. First, apply three-four droplets of bonder onto the rim of the lens holder (Fig. 1.11). This can be done with the help of a long whisker, like metal wire. Then wait for 10–20 min until the bonder fills the entire rim surface. After that, the lens may be put into its place, checked for its proper position, and UV cured. Standard lens materials used in visible domain, like the BK7 glass, are much less transparent in ultra-violet (UV) wing of the spectrum, below 400 nm (Fig. 1.12). However, strong UV radiation from a powerful lamp will always reach the bonder either by means of multiple reflections or through the glass bulk, as it is shown in Fig. 1.11, and harden the bonder after 1–2 min of exposure.

Manufacturers of optical tables and other mechanical equipment often suggest some standard holders for lenses, among which there are both right and wrong

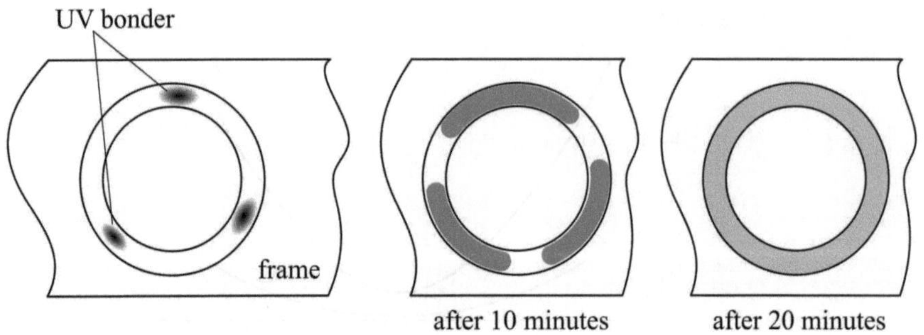

Fig. 1.11 UV bonder easily spreads over the surface

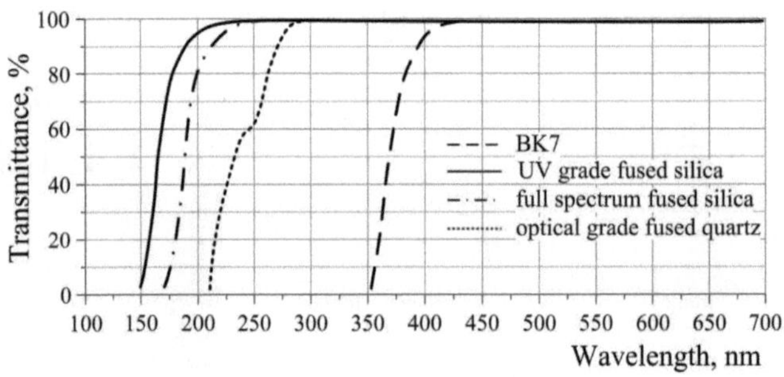

Fig. 1.12 Use UV grade fused silica lenses below 400 nm

options (Fig. 1.13). The wrong option was clearly introduced with good intentions—to hold lenses of diameters different from the standard ½ inch. However, the result is completely unsatisfactory: a lens is always off-axis and never sits tightly in such a tripod.

For plano-concave lenses, the situation is similar (Fig. 1.14): the curved surface should be directed to the source. The only difference is that, when used with high-power lasers, reflection from the curved side may focus reflected beam back into the laser and disrupt its operation. Therefore, with high-power lasers it is recommended to place the flat surface first.

Single-element lenses with spherical surfaces still have too strong aberrations to produce fine focal spots. A possible solution is aspheric lenses, in which the shape of the surface is made to minimize aberrations for parallel input beam. If the input beam is strictly parallel, then it is possible to find exact shape of the surface, producing zero spherical aberration. For manufacturing purposes, this exact shape is usually approximated with the so-called conic formula (Fig. 1.15):

$$z(y) = \frac{y^2}{R\left(1 + \sqrt{1 - (1+k)\frac{y^2}{R^2}}\right)} + A_4 y^4 + A_6 y^6 + A_8 y^8 + \cdots.$$

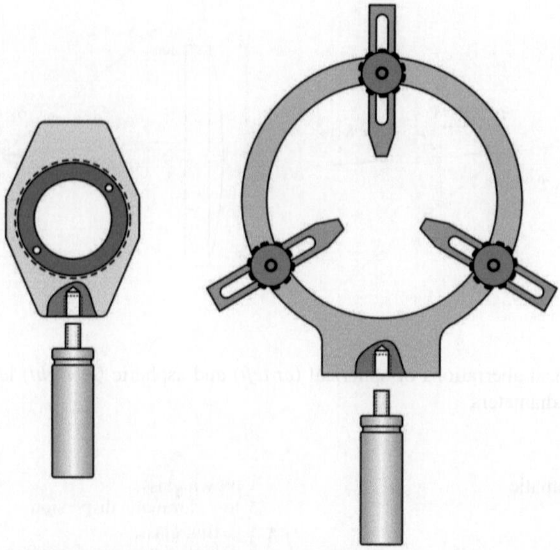

Fig. 1.13 Good (*at left*) and wrong (*at right*) accessories

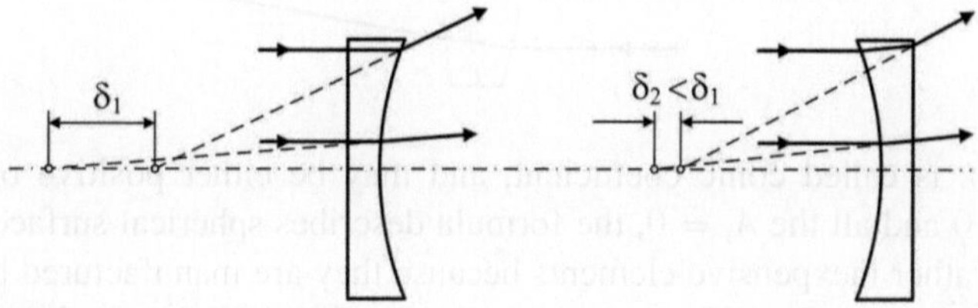

Fig. 1.14 Illumination from the flat surface causes only one refraction. Illumination from the curved surface causes two refractions, which partially compensate spherical aberration

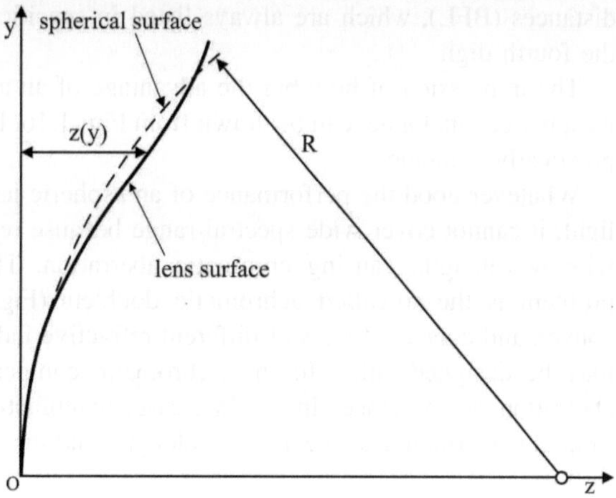

Fig. 1.15 Optimal lens surface $z(y)$ is approximated by a conic formula

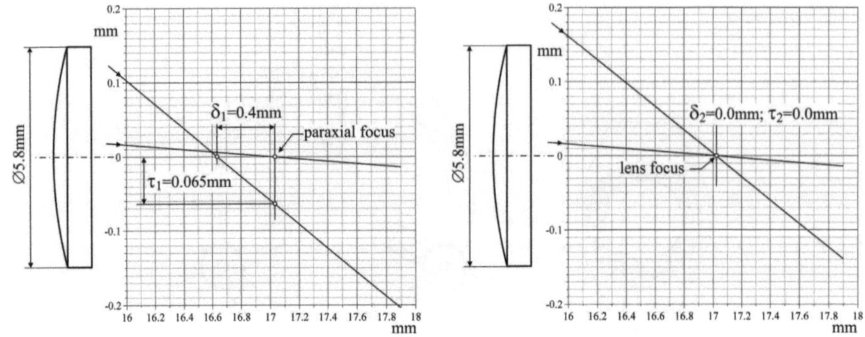

Fig. 1.16 Spherical aberrations of spherical (*at left*) and aspheric (*at right*) lenses of the same focal length and diameters

Fig. 1.17 Achromatic doublet

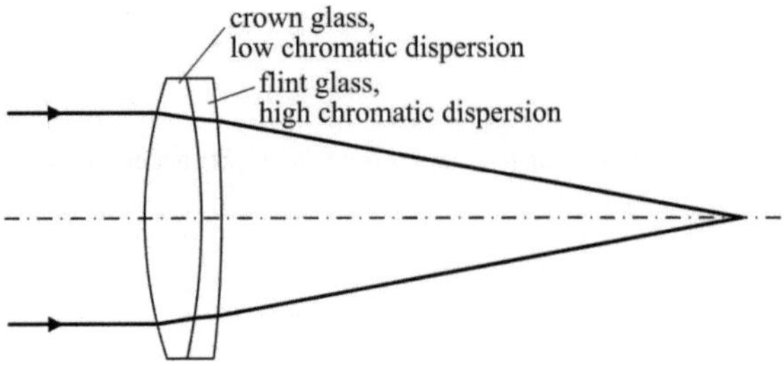

Parameter k is called conic coefficient, and may be either positive or negative. When $k = 0$ and all the $A_i = 0$, the formula describes spherical surface. Aspheric lenses are rather inexpensive elements because they are manufactured by molding technology. Commonly, different manufacturers offer the same product under their own names, bringing confusion to customers. However, comparing conic coefficients, it is easy to realize whether the lenses are the same or not. Another way to discover identity is to compare lens diameters, thickness, and back focal distances (BFL), which are always listed in specifications very accurately, up to the fourth digit.

The impression of how big the advantage of an aspheric lens is, comparing to its spherical analogue, can be drawn from Fig. 1.16. Paraxial foci of the two lenses practically coincide.

Whatever good the performance of an aspheric lens may be in monochromatic light, it cannot cover wide spectral range because refractive index of glass varies with wavelength, causing chromatic aberration. The common solution to this problem is the so-called achromatic doublets (Fig. 1.17)—a pair of cemented convex and concave lenses of different refractive indices. The achromatic doublet may be designed either for best chromatic compensation or for best spherical aberration performance. In the last case, manufacturer often claims diffraction-limited performance at specific wavelength, and the first surface, i.e. closest to the

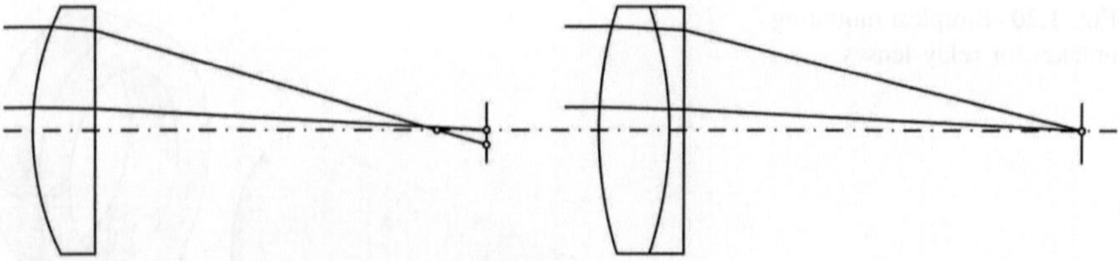

Fig. 1.18 Plano-convex lens compared with achromatic doublet

Fig. 1.19 Relay lens concept. Single lens produces noticeable spherical aberration (*above*). Corrected triplet is a better solution (*in the middle*). With two triplets the length of the system almost halves (*below*)

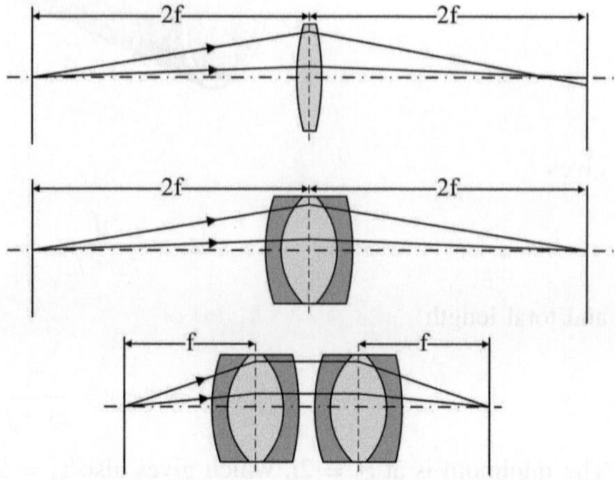

source, is always marked. If the first surface is not marked on the lens, then it is explained in detailed drawing available from the vendor or in its catalog. Lens manufacturers also offer the intermediate design, reasonably good in both chromatic and aberration compensation. Achromatic doublet has much smaller spherical aberration than the plano-convex lens (Fig. 1.18), and therefore should be the choice whenever possible.

A typical problem that may occur in optical laboratory is how to transfer an image from one plane to another. A good example is coupling galvano-scanner to front pupil of an objective. For that purpose, a variety of so-called relay lenses is available on the market. Theoretically, this can be done with only one symmetrical convex lens (Fig. 1.19). The *2f–2f* scheme not only provides unity magnification from entrance to exit but also minimal total length from image to image. Indeed, let z_1 and z_2 be the front image and exit image distances from the lens. Then lens formula

$$\frac{1}{z_1}+\frac{1}{z_2}=\frac{1}{f}$$

Fig. 1.20 Simplest mounting bracket for relay lenses

gives

$$z_1 = \frac{z_2 f}{z_2 - f}$$

and total length

$$L(z_2) = z_1 + z_2 = \frac{z_2^2}{z_2 - f}.$$

The minimum is at $z_2 = 2f$, which gives also $z_1 = 2f$.

However, spherical aberration is strong in such a system. A better solution would be to use symmetrical triplet, for example the so-called Steinheil or Hastings achromats (Fig. 1.19) with very low spherical and color aberrations. However, total length $4f$ of such a system may be too big for a particular application. The length may be roughly halved to $2f$, using the second lens (Fig. 1.19), which represents a typical scheme of a relay lens. Even more complicated optical combinations may be inside relay lenses. From practical point of view, it should be noted that manufacturers and distributors supply the product without any mounting accessories, just like a metal cylinder with optical windows at both sides. In order to install it into a system, a mounting bracket is needed (Fig. 1.20).

1.3 Plates and Prisms

A plane parallel glass plate is the most common element that can be found as optical windows of photodetectors and laser diodes, various cuvettes, beamsplitters, cover glasses, etc. There are two basic things that must be remembered: in general, a plate introduces spherical and chromatic aberration when installed in focused beam, and tilted plate displaces the beam (Fig. 1.21). For practice, it is very important to understand why a glass plate, inserted in a converging beam,

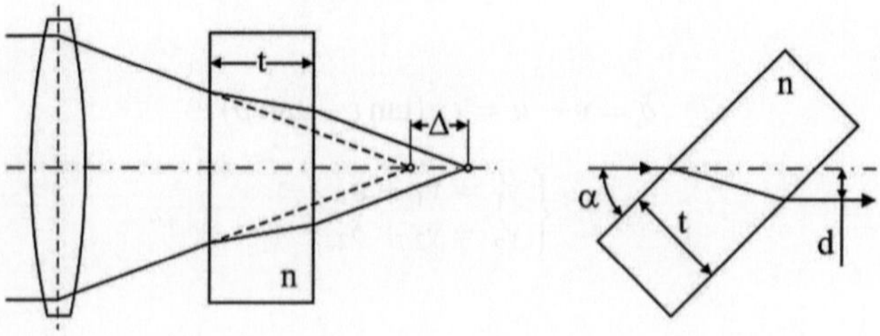

Fig. 1.21 A plate of refractive index *n* causes longitudinal (*at left*) and lateral (*at right*) displacements of rays

Fig. 1.22 Glass plate disrupts conical convergence

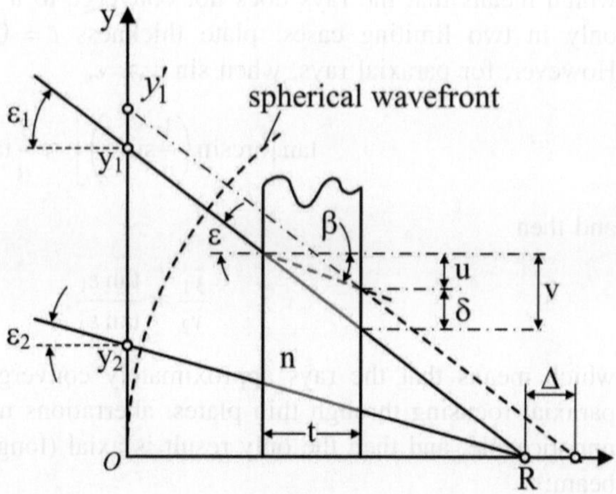

introduces aberration. Aberration means that a converging beam does not converge into one point—the focus. Consider an ideal conical beam, all rays of which come to one point located at a distance R from the vertical axis (Fig. 1.22). If two rays converge to one focus then

$$\begin{cases} \tan\varepsilon_1 = \frac{y_1}{R} \\ \tan\varepsilon_2 = \frac{y_2}{R} \end{cases}$$

or

$$\frac{y_1}{y_2} = \frac{\tan\varepsilon_1}{\tan\varepsilon_2}.$$

When the plate is inserted, the points y_1 and y_2 shift to new locations y'_1 and y'_2. After refraction

$$\sin\varepsilon = n\sin\beta,$$

so that

$$\delta = v - u = t \cdot (\tan \varepsilon - \tan \beta)$$

$$\begin{cases} y_1' = y_1 + \delta_1 \\ y_2' = y_2 + \delta_2 \end{cases}.$$

The ratio

$$\frac{y_1'}{y_2'} = \frac{R \tan \varepsilon_1 + t\{\tan \varepsilon_1 - \tan[\arcsin(\frac{1}{n}\sin \varepsilon_1)]\}}{R \tan \varepsilon_2 + t\{\tan \varepsilon_2 - \tan[\arcsin(\frac{1}{n}\sin \varepsilon_2)]\}} \neq \frac{\tan \varepsilon_1}{\tan \varepsilon_2},$$

which means that the rays does not converge to a focus. Exact focusing occurs only in two limiting cases: plate thickness $t = 0$ or refractive index $n = 1$. However, for paraxial rays, when $\sin \varepsilon \approx \varepsilon$,

$$\tan\left[\arcsin\left(\frac{1}{n}\sin \varepsilon\right)\right] \approx \frac{1}{n}\tan \varepsilon,$$

and then

$$\frac{y_1}{y_2} \approx \frac{\tan \varepsilon_1}{\tan \varepsilon_2},$$

which means that the rays approximately converge to a focus. Therefore, for paraxial focusing through thin plates, aberrations may be quit tolerable or even unnoticeable, and then the only result is axial (longitudinal) displacement of the beam:

$$\Delta = \left(1 - \frac{1}{n}\right) t.$$

Lateral displacement

$$d = t\left[1 - \sqrt{\frac{1 - \sin^2 \alpha}{n^2 - \sin^2 \alpha}}\right] \sin \alpha.$$

It is easy to remember that $d \approx t/3$ for the most common case $\alpha = 45°$ and $n = 1.5$.

An ordinary prism (Fig. 1.23) can be used either for deflection or for spectral dispersion of beams. Spectral dispersion, i.e. separation of white light into components of different wavelengths (different colors), occurs due to dependence of the refractive indexn on wavelength λ. In its simplest form, called the Cauchy equation, this dependence is presented as a series

$$n(\lambda) = B + \frac{C}{\lambda^2} + \frac{D}{\lambda^4} + \cdots,$$

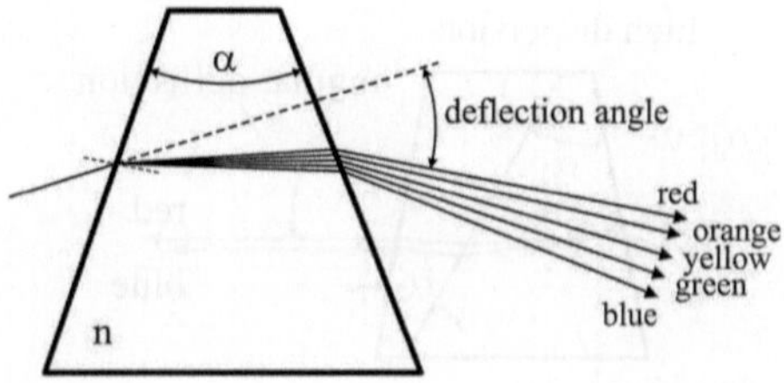

Fig. 1.23 An ordinary prism deflects and disperses light

Table 1.1 Optical properties of common glass types

Material	B	C (μ^2)
Fused silica	1.4580	0.00354
Borosilicate glass BK7	1.5046	0.00420
Hard crown glass K5	1.5220	0.00459
Barium crown glass BaK4	1.5690	0.00531
Barium flint glass BaF10	1.6700	0.00743
Dense flint glass SF10	1.7280	0.01342

in which only the two first terms are important. Coefficients for common optical materials are summarized in the Table 1.1. If used as a deflector, chromatic dispersion is a negative factor, and if used as spectral selector—deflection is undesirable. For deflection of monochromatic laser beams, an ordinary prism will do well. However, for deflection of white-light beams, certain modification should be made in order to minimize spectral divergence (Fig. 1.24). In the direct-vision prism, on the contrary, angular deflection of rays is minimal.

The right-angle prisms are used for bending a beam through 90° or 180° (Fig. 1.25). In each configuration, rays undergo lossless total internal reflection that occurs when the angle of incidence becomes larger than arcsin $(1/n)$. For $n = 1.5$ this value is 41.8°. If the entrance and exit faces are anti-reflection-coated, the total loss may be less than few percent in visible domain. The 90° rotation prism has two main disadvantages comparing to mirrors. Firstly, ghost beams reflected from the faces, even relatively weak, often produce confusing results. Secondly, when used to fold converging beams, prism acts as a glass plate, introducing aberrations. However, being used as 180° rotator, right-angle prism offers significant advantage relative to mirrors: it acts as constant-deviation reflector, redirecting incident rays exactly backwards, regardless of the prism tilt (Fig. 1.26).

In general, the constant-deviation angle is always twice the angle between two reflecting surfaces (in two-dimensional arrangements). For example, if the two reflecting surfaces make 45° then the reflected ray will always be perpendicular to the entering ray irrespective of rotations. This is the property of a penta-prism (Fig. 1.27). Unless refractive index of the penta-prism material is bigger than 2.5,

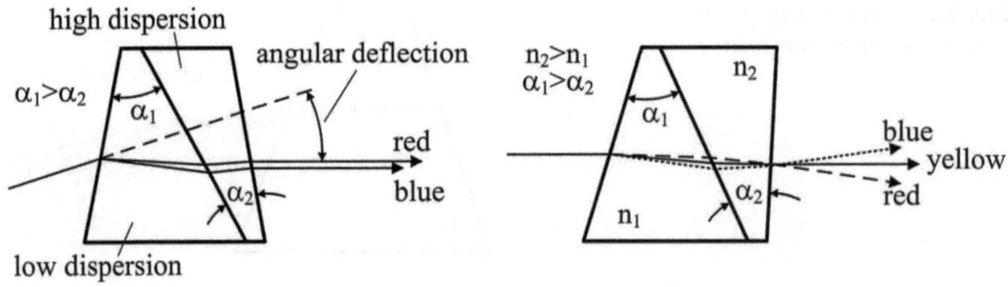

Fig. 1.24 *At Left* achromatic prism. The red and blue rays emerge parallel to each other, deflected from the entrance ray. *At Right* direct-vision dispersive prism

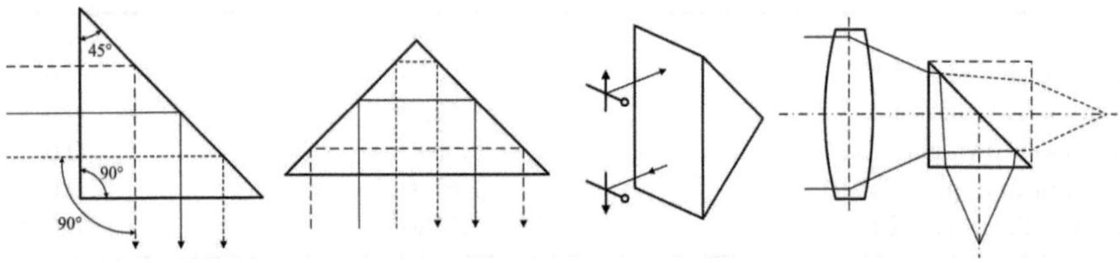

Fig. 1.25 Right-angle prism

Fig. 1.26 In the prism, the entering and exiting rays are parallel, regardless of the initial angle of incidence. Front surface refraction cancels out since the light emerges from the hypotenuse exactly at the same angle as it enters it

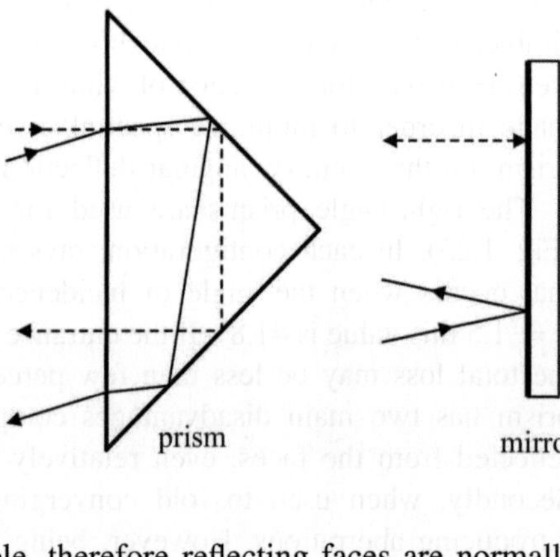

the total internal reflection is impossible, therefore reflecting faces are normally metallized.

Penta-prisms are often used in triangulation range-finders and also in long-travel scanners, like surface profilers, to compensate for inevitable angular rocking of a mechanical scanning stage. The image, emerging from an ordinary right-angle prism, is only half-inverted, i.e., inverted around only one axis. In order to obtain full inversion around both axes, the Amici or roof-prism may be used (Fig. 1.28). It can be obtained from an ordinary right-angle prism by replacing the hypotenuse flat face with two surfaces at 90° whose intersection lies in the hypotenuse. The ray

Fig. 1.27 Penta-prism

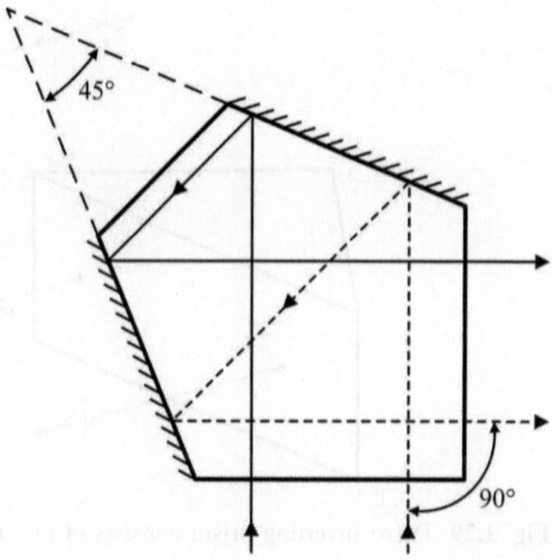

Fig. 1.28 Amici prism

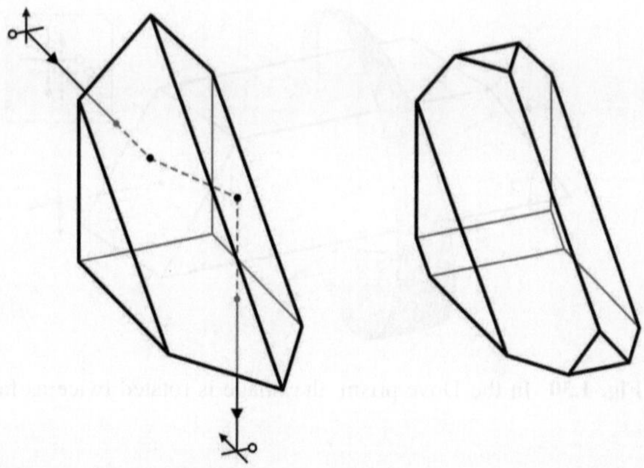

in the figure makes approximately 60° with the roof surface instead of 45° it would be in the right-angle prism. Therefore, total internal reflection takes place for all the rays that would be reflected by an ordinary right-angle prism.

Amici prism changes the beam direction by 90°. When the beam direction must be preserved, the Porro prism may be used, which is actually a combination of two right-angle prisms (Fig. 1.29).

The Dove prism (Fig. 1.30) is an image rotating prism, and it works in total internal reflection. During recent years, this type of a prism became famous in physical laboratories owing to experiments on angular momentum of a photon. It is a common mistake to consider the Dove prism as a right-angle prism, i.e., assuming $\alpha = 45°$. Although $\alpha = 45°$ is a popular geometry, it is not optimal. The shortest design with minimal ratio

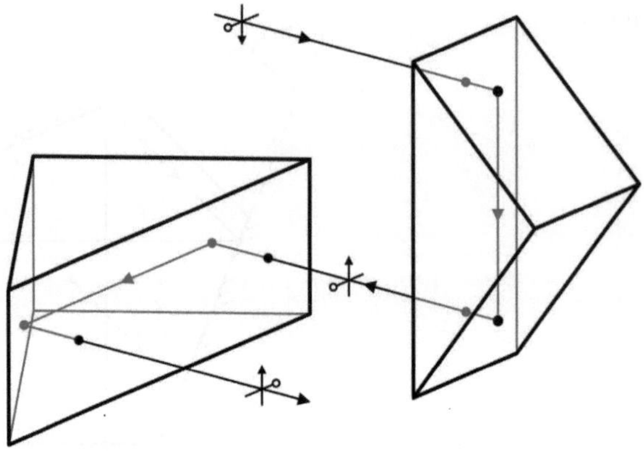

Fig. 1.29 Porro inverting prism consists of two right-angle prisms

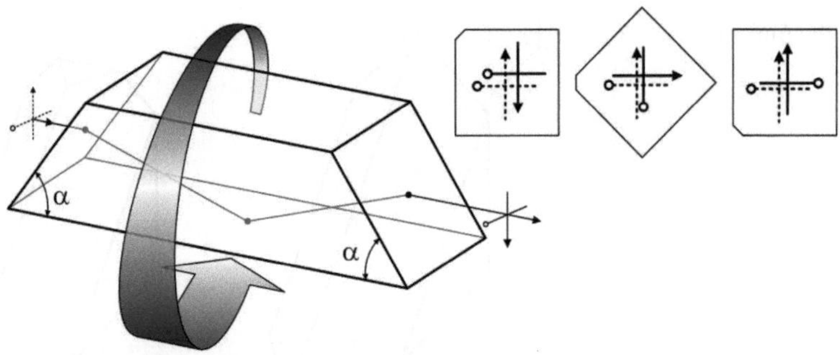

Fig. 1.30 In the Dove prism, the image is rotated twice as fast as the prism

$$\frac{\text{length}}{\text{aperture}} = \frac{1}{\sin(2\alpha)}\left[1 + \frac{\sqrt{n^2 - \cos^2\alpha} + \sin\alpha}{\sqrt{n^2 - \cos^2\alpha} - \sin\alpha}\right]$$

corresponds to $\alpha = 32^\circ$ for $n = 1.5$.

Among the prisms with internal reflections, the Littrow dispersion prism is frequently used for intra-cavity spectral selection in lasers (Fig. 1.31). The prism works only with metallized back face.

1.4 Retroreflectors

An optical arrangement that reflects parallel beam exactly backwards is called a retroreflector. There are two most commonly used types of retroreflectors: a corner cube reflector or a lens with a mirror in its focus. A corner cube reflector may be either a constant 180°-deviation prism that uses total internal reflection, or a

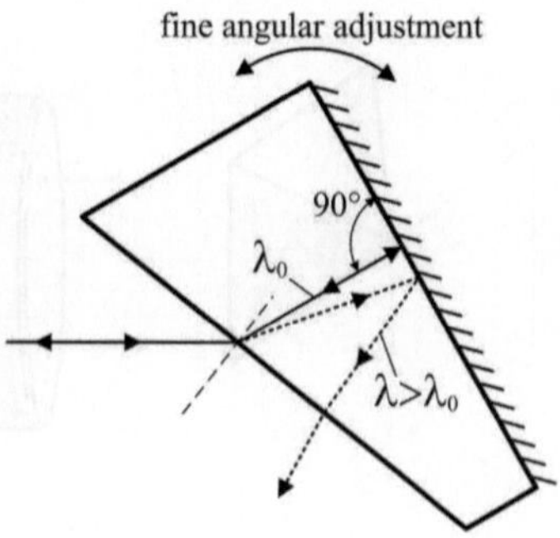

Fig. 1.31 Littrow prism. Exact back-reflection is achieved only for one specific wavelength λ_0, which can be changed by fine rotation of the prism

combination of three mirrors fixed together at 90° to one another (Fig. 1.32). The latter is much more expensive. Since the three-mirror assembly forms a triangular hollow, it is often called a hollow retroreflector.

The theory of a corner cube reflector is simple. In order to understand this mathematically, we have to consider reflection from a single mirror. Let $\vec{k}_0$ be the unity vector of the incoming beam, $\vec{k}_1$—unity vector of the reflected beam, and $\vec{n}$—unity vector of the normal to the mirror. Then

$$\vec{k}_1 = \vec{k}_0 - 2\,(\vec{k}_0, \vec{n})\,\vec{n},$$

where $(\vec{k}_0, \vec{n})$ is a scalar product equal to $\left|\vec{k}_0\right|\,|\vec{n}|\,\cos(\vec{k}_0\,\vec{n}) = \cos(\vec{k}_0\,\vec{n})$. If there are three mirrors arranged with their normals directed along $\vec{n}_1$, $\vec{n}_2$, $\vec{n}_3$ then

$$\begin{cases} \vec{k}_1 = \vec{k}_0 - 2\,(\vec{k}_0, \vec{n}_1)\,\vec{n}_1 \\ \vec{k}_2 = \vec{k}_1 - 2\,(\vec{k}_0, \vec{n}_2)\,\vec{n}_2 \\ \vec{k}_3 = \vec{k}_2 - 2\,(\vec{k}_0, \vec{n}_3)\,\vec{n}_3 \end{cases},$$

and all the three mirrors are orthogonal. Substituting $\vec{k}_1$ into the formula for $\vec{k}_2$, and $\vec{k}_2$—into $\vec{k}_3$, one gets:

$$\vec{k}_3 = \vec{k}_0 - 2\,(\vec{k}_0, \vec{n}_1)\,\vec{n}_1 - 2\,(\vec{k}_0, \vec{n}_2)\,\vec{n}_2 - 2\,(\vec{k}_0, \vec{n}_3)\,\vec{n}_3.$$

This vector is directed oppositely to $\vec{k}_0$. Indeed, multiply $\vec{k}_3$ by $\vec{k}_0$:

$$(\vec{k}_0, \vec{k}_3) = k_0^2 - 2\left[(\vec{k}_0, \vec{n}_1)^2 + (\vec{k}_0, \vec{n}_2)^2 + (\vec{k}_0, \vec{n}_3)^2\right].$$

Since $\vec{n}_1$, $\vec{n}_2$, $\vec{n}_3$ are orthogonal, the three terms in square brackets represent the sum of squared projections of the vector $\vec{k}_0$ onto orthogonal system of coordinate. According to Pithagoras theorem, this sum is equal to k_0^2, and thus

$$(\vec{k}_0, \vec{k}_3) = -k_0^2 = -1,$$

which means that the angle between $\vec{k}_1$ and $\vec{k}_3$ equals to 180°.

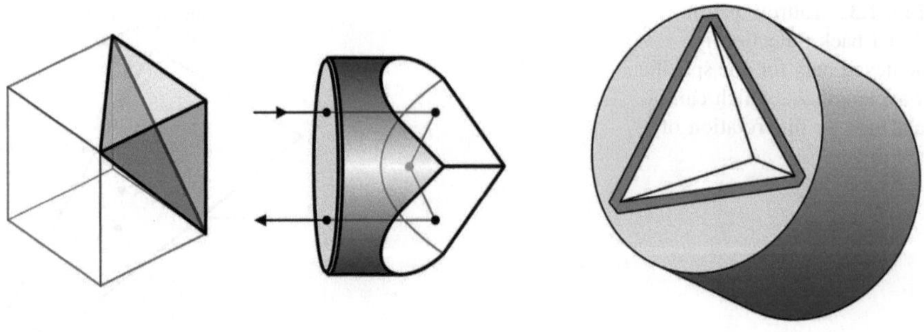

Fig. 1.32 Corner cube reflectors in the forms of a prism (*at left*) and three orthogonal mirrors (*at right*)

Fig. 1.33 For paraxial rays, optical path for all the rays is approximately $2hn$ (*at left*). The reflected beam is inversed over the apex. Edges do not separate the image but are visible as three symmetrical thin *dark lines*—the result of neat chamfering (*in the middle*). Prism-type retroreflectors transform input polarization in an intricate manner, generally making elliptical polarization from the linear. Each of the six visible sectors produces its own ellipticity (*at right*)

The hollow reflectors usually work better, primarily because of better angular tolerances (2 arc seconds against 20–50 of the prismatic ones). They transfer polarization of the input beam to the output without changes. Prismatic retroreflectors suffer from multiple reflections, interference fringes, and also exhibit noticeable polarization artifacts. On the other hand, they are considerably easier to clean and mount. Being illuminated by a plane wave, prismatic retroreflector also returns plane wavefront in the outgoing wave, and optical paths for all the rays are equal to that of the ray, coming to and returning from the apex (Fig. 1.33). However, polarization state of the entire cross section of the beam will not be equal to that of the illuminating wave because total internal reflection changes phases of s- and p-polarizations (p-polarization: electrical vector parallel to the plane of incidence), leading to ellipticity of the reflected beam (see Chap. 5, the Fresnel rhomb theory). This is the reason why for interferometric applications it is better to use either mirror-based or lens-based retroreflectors. Chapter 6 analyzes this topic in detail.

A lens-based retroreflector is shown in Fig. 1.34. According to geometrical optics, all the rays of a parallel beam come to a single point in the focus plane.

Fig. 1.34 In a retroreflector, the mirror is placed in the focal plane of a lens

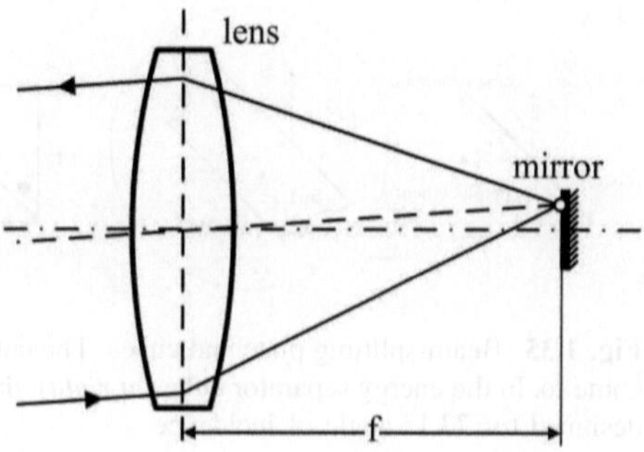

Propagation laws are invariant to time inversion, therefore the same holds true for all the rays outgoing from this point. Therefore, a mirror placed in the focal plane of an ideal lens acts as perfect retroreflector. For a real lens with aberrations, this is true for monochromatic paraxial beams. This property of a lens-mirror combination becomes very disturbing in such applications as scanning laser microscopy, or well known compact disk writers (CD writers), because the reflected beam returns exactly to the laser and causes instability. In many cases, polarization isolation solves this problem.

1.5 Beamsplitters

The two most commonly used types of beamsplitters are the beam-splitting plates and cubes (Fig. 1.35). Almost without exception, they are designed for 45° angle of incidence and transmission ratios 50/50, 70/30, or 90/10 %. The beam-splitting cubes may be either polarizing or non-polarizing. In this section, only the non-polarizing type is considered, leaving the polarizing type for the Chap. 5. Among the beam-splitting cubes, a rarer element is the so-called energy separator cube (ESC) that splits the input beam into two parallel beams. This geometry is sometimes useful in interferomery (Chap. 6).

The beam-splitting plate has only three advantages over the cube: lower price, less aberration when installed in a converging beam, and possibility to completely eliminate the ghost beam when the plate has a wedge. Aberration is smaller simply because the plate is much thinner than the cube, and the role of a wedge is explained in Fig. 1.36.

In all the other components the cube is better: better spectral uniformity of the reflection coefficient, smaller difference between transmission coefficients for s- and p-polarization, less ghosting, no displacement, easier to mount, negligible deformation under mechanical stress. In a beam-splitting plate, the beam reflects from the interface between the air and glass—two materials with very different refractive indices (1.0 and 1.5). Consequently, the reflective multilayer structure cannot be optimized equally well for the two orthogonal polarizations. In the cube,

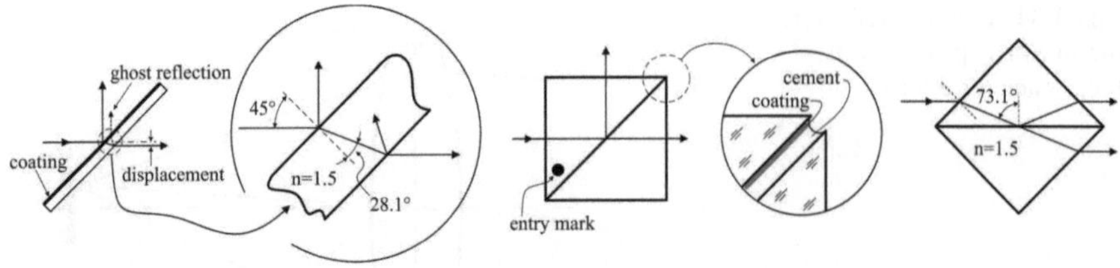

Fig. 1.35 Beam-splitting plate and cubes. The entry mark shows the face where the beam should come to. In the energy separator cube (*at right*), the multilayer coating at the diagonal interface is designed for 73.1° angle of incidence

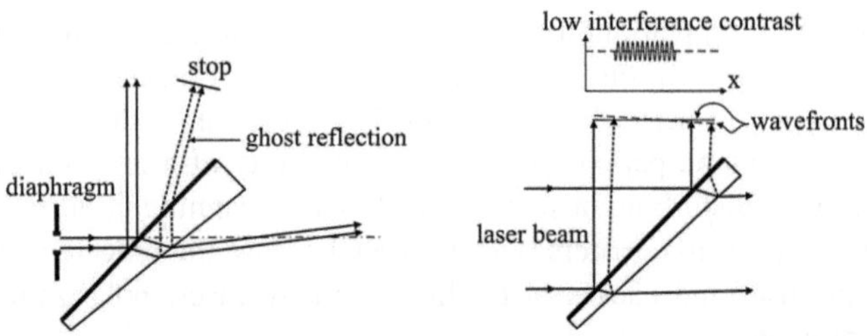

Fig. 1.36 A small wedge spatially separates the ghost beam at the expense of the deflected transmitted beam (*at left*). Much smaller wedge, about 0.5°, does not spatially separate the ghost beam, but is often used for another purpose: to destroy interference in laser beams (*at right*)

however, the reflecting interface is formed by glass and the cement—two materials with matching refractive indices, thus providing better conditions for optimization. Also, the anti-reflection coating on the second face of a 45° plate cannot decrease reflection as well as it can be done at normal incidence in the cube. It becomes clear from Fig. 1.37, showing reflection coefficients for s- and p-polarized waves at different angles of incidence in glass with refraction index $n = 1.5$. Whereas at normal incidence (zero angle) reflectivity is the same for the two polarizations (about 4 %), at 45° oriented plate (28.1° inside the glass, see Fig. 1.35) it is quite different.

Additional problem with ghost reflection is that it may create although subtle but still quite noticeable interference pattern in the form of straight fringes, especially irritating in imaging applications. This phenomenon does not happen in white light because coherence length of white light (on a micrometer scale) is much smaller than double thickness of a cube or a plate. However, with laser beams, it happens all the time, and the one way to mitigate this effect in plates is to make the plate with a tiny wedge (Fig. 1.36, right). Then the ghost wave will be tilted against the primary one by some angle θ, so that the interference between the two waves with the wavelength λ will be seen as sinusoidal modulation

$$\cos\left(\frac{2\pi}{\lambda}\theta x\right)$$

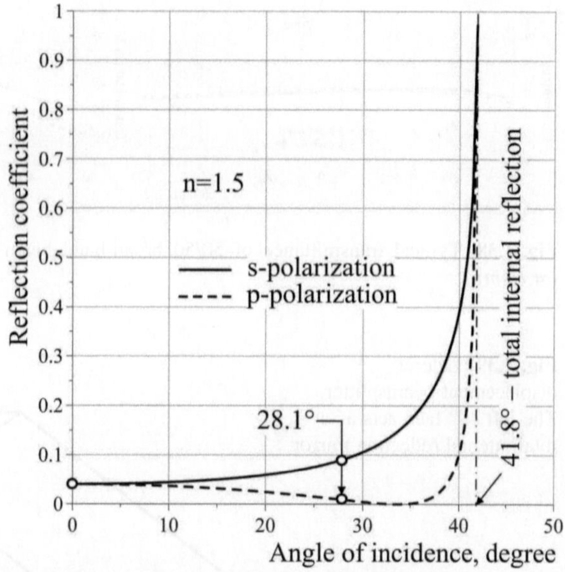

Fig. 1.37 Reflection coefficients for different polarizations of the ghost wave at the interface glass-air

with the period λ/θ. For example, with $\lambda = 0.6\,\mu$ and $\theta = 0.01$ rad (0.6°) the period of the interference pattern would be about 60 μ, and the contrast of such a fine structure is always greatly decreased by finite divergence of the beam and spatial resolution of an optical system.

Manufacturers offer two basic types of beamsplitters: the broad-band and the laser-line ones. While the broad-band type is supposed to provide uniform reflection in the entire visible domain from 400 to 700 nm, the laser-line beamsplitters are designed for a particular laser line. Obviously, the latter ones are much simpler to design and manufacture, and as such, they have practically ideal polarization performance and transmission ratios. As to the broad-band beamsplitters, they may have very different performance, depending on the vendor. The most important factor is the equality of reflection coefficients for s- and p-polarizations. It was already explained why beam-splitting cubes provide better performance relative to beam-splitting plates. Typical curves are shown in Fig. 1.38.

Sometimes, it is preferable to have both split beams going in one direction parallel to each other. This is done by a lateral displacement beam-splitting prism (Fig. 1.39).

Although beam-splitting cubes are easy to mount on any flat surface by means of glue or epoxy, it is better to leave a chance to conveniently replace it if necessary. The first option is to establish a guiding line or plane, to which the cube can be pressed during replacement (Fig. 1.40).

Finally, it is worth mentioning that the third type of beamsplitters also exists: the so called pellicle beamsplitter—a very thin, about 5 μ, polymer membrane

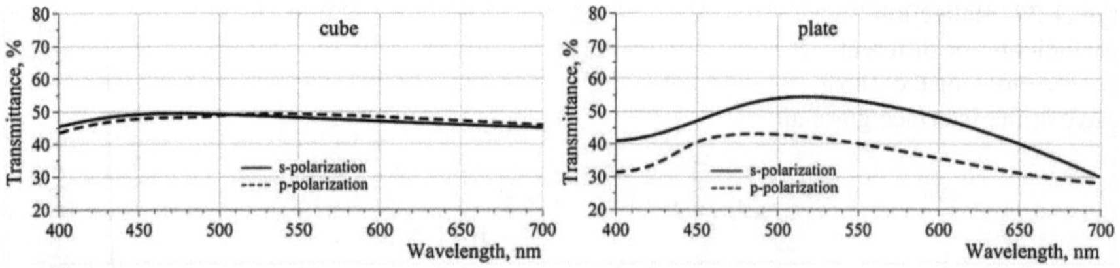

Fig. 1.38 Typical transmittance of 50/50 broad-band beam-splitting cube (*at left*) and plate (*at right*)

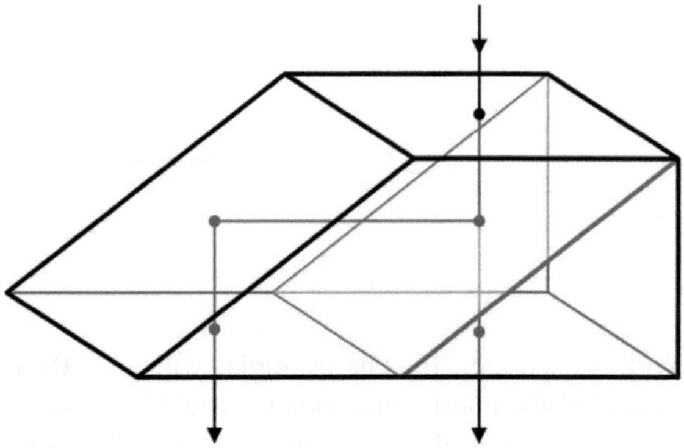

Fig. 1.39 Lateral displacement beamsplitter. The left 45° face acts as a total internal reflection mirror

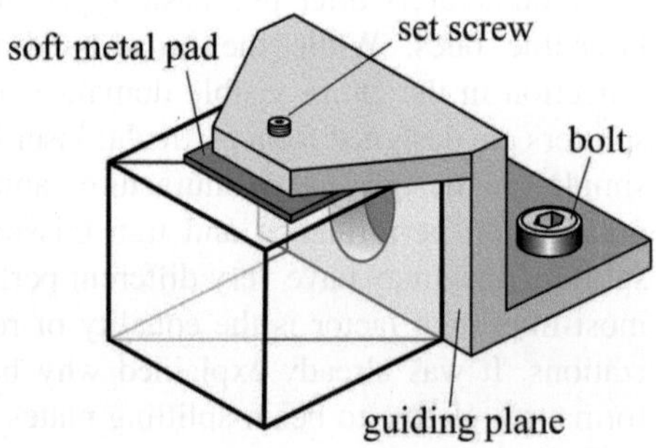

Fig. 1.40 Vertical plane of a clamping bracket accurately guides the cube to its place. Soft metal pad (aluminium, copper, or brass) protects glass from damaging

stretched over the round frame. The foil may be either coated or uncoated. The only advantage of a pellicle beamsplitter is that it practically does not introduce aberrations in focused beams, owing to its extreme thinness. This type of a beamsplitter should be avoided as much as possible: the membrane is extremely vulnerable and hygroscopic, it resonates to vibrations and acoustic noise, sensitive to airflow, produces interference fringes even in white light, and acts as a low-finesse interferometer, causing sinusoidal spectral modulation both in transmitted and reflected beams.

Fig. 1.41 Typical TV lens. The focusing and iris rings adjust image focus and optical flux. Fixing screws clamp the rings after adjustment. The C-mount flange may be removed by releasing the flange screws (usually, three screws)

1.6 Imaging Lenses

High quality white-light imaging with micrometer-scale spatial resolution over the entire field of view can only be achieved with specially designed multi-element optical systems like photographic lenses or microscopy objectives. The simplest and, therefore, cheapest representatives of this class of optical products are the television (TV) lenses, also called the CCTV lenses (charge-coupled device TV lenses). They are chromatically corrected for use with white light, designed to minimize primary aberrations like spherical and astigmatism, with manual focusing of images of distant objects in the range of 100 mm to infinity, and variable iris diaphragm. Compact, easy to mount, light-weighted, with standardized and widely accepted so-called «C-mount» flange, these lenses may be the perfect choice for many applications (Fig. 1.41). The C-mount standard includes flange-to-focus distance 17.52 mm (0.69″), the thread 1″-32TPI-2A (TPI stands for Threads Per Inch), and the threaded part length 3.8 mm (Fig. 1.42). TV lenses are available in a wide variety of fixed focal lengths, ranging from 3.5 to 75 mm. Commonly, no one is surprised that the focal length can be shorter than the flange-to-focus distance: just the ending lens is positioned closer to the image plane (Fig. 1.43, left). However, it is always a surprise that the focal length can be longer than the lens itself. The answer is that the focal length is determined as the distance from the focal point on optical axis to the intercept point with the parallel beam at the input of the lens (Fig. 1.43, right).

TV lenses are designed for various formats of charge-coupled device (CCD) matrices: 1/3″, 1/2″, 2/3″, and even 1″. CCD dimensions for these standards and corresponding image circles in the focal plane are listed in the Table 1.2, which makes it is easy to estimate diameter D of the illuminated area in the image plane, if only CCD format (Chap. 11) is known for the lens. If additionally the focal length f is known, then it is possible to determine angular field of view:

$$\omega = 2\arctan\frac{D}{2f} \approx \frac{D}{f}\,\text{radian}.$$

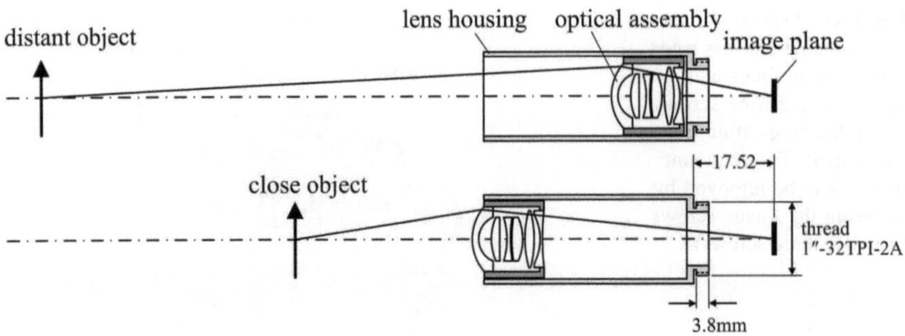

Fig. 1.42 With fixed focal length of optical assembly and constant flange-to-focus distance, the lens still can refocus to various distances due to internal shifts

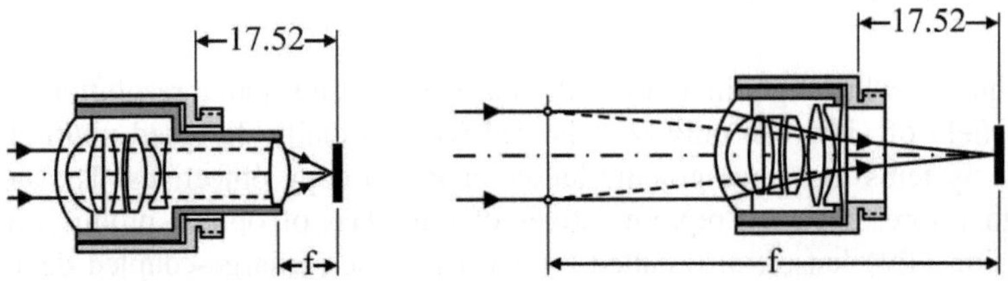

Fig. 1.43 Focal length can be either shorter (*at left*) or longer (*at right*) than the lens itself

Table 1.2 CCD formats

Type	Diagonal (mm)	Width (mm)	Height (mm)	Area (mm^2)	Image circle diameter (mm)
1″	16.0	12.8	9.6	12.3	16
1/2″	8.0	6.4	4.8	30.7	8
1/3″	6.0	4.8	3.6	17.3	6
2/3″	11.0	8.8	6.6	58.1	11
35 mm	43.3	36	24	864	43

For example, the lens is designed for 1/3″ CCD format, and $f = 75$ mm. Then the field of view $\omega \approx 6/75$ radian $\approx 4°$. Or $f = 12$ mm and 1/3″ CCD format. Then $\omega \approx 6/12$ radian $\approx 26°$.

The inner surface of the front rim of a TV lens is often threaded to adopt a filter. Unlike the C-mount thread, the filter thread cannot be standardized for all the models because input optical apertures differ significantly. However, the manufacturer always specifies filter threads for every model. It is also important to know that the C-mount flange itself can be dismantled, giving access to a dovetail flange (Fig. 1.44). In order to do that, use a fine screwdriver to release clamping screws on the C-mount flange (Fig. 1.41).

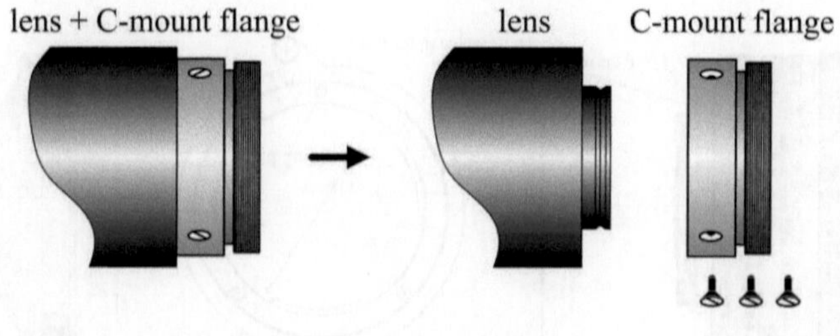

Fig. 1.44 Remove the C-mount flange to use the dovetail flange

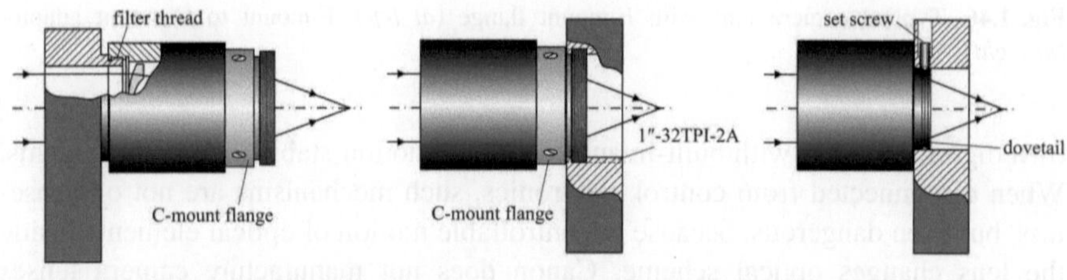

Fig. 1.45 Mounting on filter thread (*at left*), on C-mount (*in the middle*), and on dovetail (*at right*)

TV lenses are very convenient for reliable mounting. There are three options, explained in Fig. 1.45. The dovetail mounting is useful when either the thread cannot be manufactured (poorly equipped mechanical shop), or it is necessary to maintain certain orientation of a lens around its axis in order to gain access to fixing screws on focus and iris rings.

Ordinary TV lenses are not transparent in UV region below 400 nm (Fig. 1.12), and very few manufacturers offer lenses transparent down to 200 nm. But transparency alone is not enough: geometrical and chromatic aberrations must be compensated. Typically, manufacturer claims operation down to the lowest transmittable wavelength, and makes a comment that narrow band-pass optical filter is needed. This means that chromatic aberration is not compensated in the entire spectral region of glass transparency. It is normal, because the list of materials transparent in UV domain is too short to make full compensation of chromatic aberrations.

There are applications, requiring wider image circle than TV lenses can provide. For example, in spectroscopy, linear detectors 30 mm long are common. To project high-resolution images onto such big areas the camera lenses are the perfect choice (Fig. 1.46). Camera lenses were originally designed for 35 mm films (Table 1.2), therefore they can form high-quality images within wide image circle of about 40 mm in diameter. Many companies like Nikon, Canon, Zeiss manufacture camera lenses in large variety, with focal lengths ranging from 15 to 135 mm, and it is necessary to understand what to choose. Here are some guidelines. First of all, Nikon and Canon manufacture their lenses mostly for their

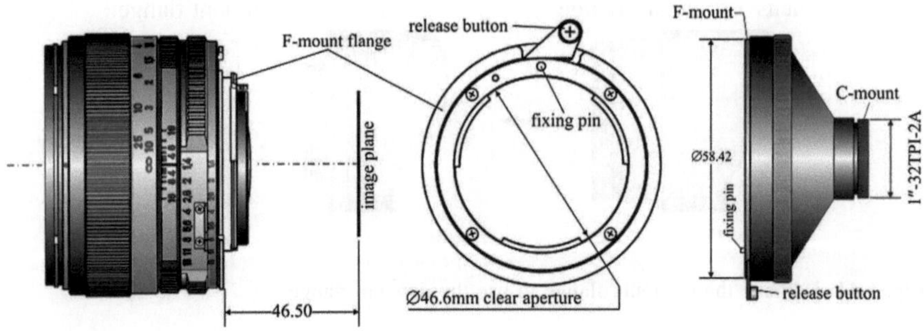

Fig. 1.46 Typical camera lens with F-mount flange (*at left*). F-mount to C-mount adaptor (*at right*)

own digital cameras, with built-in autofocus and motion-stabilization mechanisms. When disconnected from control electronics, such mechanisms are not only useless, but even dangerous, because uncontrollable motion of optical elements inside the lens changes optical scheme. Canon does not manufacture camera lenses without autofocus. Among Nikon camera lenses, it is possible to find the ones without autofocus and stabilization. According to Nikon notation, the lenses are marked as AF (autofocus), AF + S (autofocus plus stabilizer), and M (manual, i.e. no mechanics inside). So, find a lens marked «M». Zeiss, on the contrary, manufactures only lenses without autofocus and stabilization. Secondly, it is necessary to find a lens with appropriate flange. The most common type of flanges for camera lenses is the so-called F-mount, standard for Nikon and also accepted by Zeiss.

F-mount flange is rather precise piece of spring-loaded mechanics, and cannot be easily manufactured in the laboratory. For this reason, it is readily available from distributors of camera lenses or general optical equipment. Moreover, the F- to C-mount adaptors are also on the market.

A very peculiar type of a complex lens, originally developed for machine vision and providing useful properties for physical applications, is the so-called telecentric lens. Consider first the basic optical principle that determines fundamental properties of this type of a lens. Any ray, coming to an ideal lens with the focal length f parallel to its optical axis, intercepts the optical axis in the focal plane (Fig. 1.47). A paraxial parallel beam, coming to the lens at an angle α, focuses in the focal plane in a point r, making the same angle with the lens center:

$$\alpha = \arctan\frac{r}{f} \approx \frac{r}{f}.$$

If a diaphragm of radius r is placed in the focal plane, then the bundle of rays, originating at any point in front of the lens and passing through the diaphragm, form a cone with the generatrix angle α. This may be rephrased in terms of spatial frequencies filtering: since the focal plane of a lens is the Fourier-transform plane, only waves with spatial frequencies within small part $\pm r/f$ of the wave number can pass through. It is actually the basic concept of a telecentric lens.

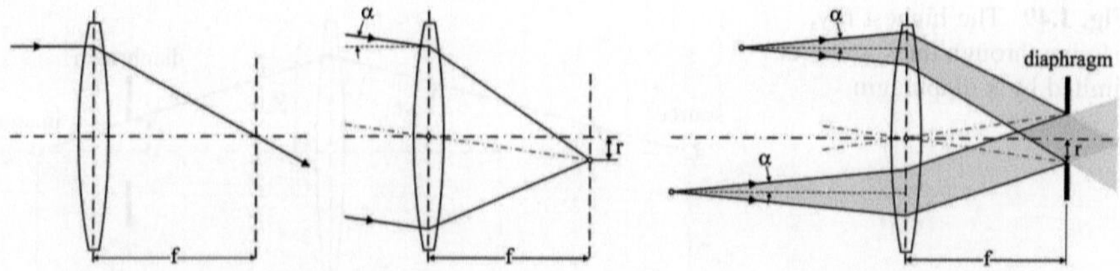

Fig. 1.47 Telecentric configuration relays only those rays that are inside a narrow fixed-angle cone around horizontal axis

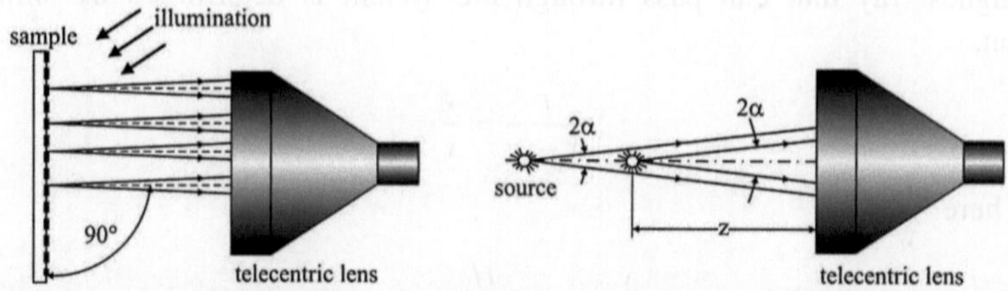

Fig. 1.48 Two basic telecentric concepts: ray cone geometry does not depend on radial position (*at left*) and cone angle is independent of the source axial position (*at right*)

For physical applications, like spectro-photometry, scatterometry, and spectral interferometry, telecentric configuration guarantees two most important concepts, which are portrayed in Fig. 1.48: the axis of the receiving cone of rays is always parallel to optical axis, and the cone angle 2α is constant independently of axial position z of the source. The second one is a very important photometric property, and should be proved with formulas.

Consider a source positioned at z in front of the lens (Fig. 1.49). We are going to analyze how the maximum angle α of rays, passing through the diaphragm, changes with z, and what is the condition for the case when α does not depend on z. Since

$$\alpha = \arctan\frac{y}{z},$$

it is reasonable to analyze the ratio y/z. The image of the source is formed at x behind the lens, and the lens formula

$$\frac{1}{z} + \frac{1}{x} = \frac{1}{f}$$

gives

$$x = \frac{fz}{z - f}.$$

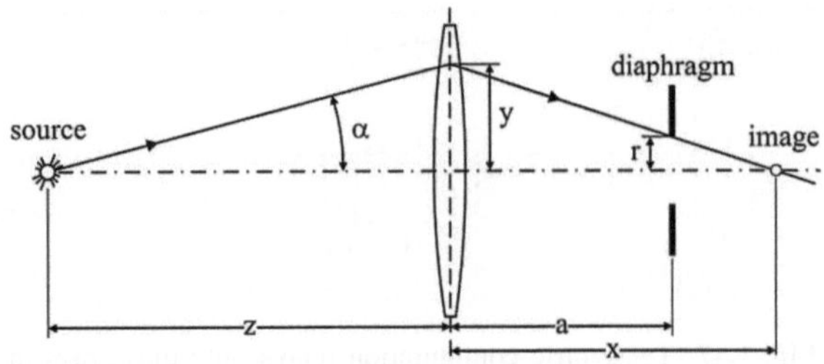

Fig. 1.49 The highest ray, passing through the system, is limited by a diaphragm

The highest ray that can pass through the system is determined by similarity relation:

$$\frac{r}{x-a}=\frac{y}{x}.$$

From here

$$\frac{y}{z}=\frac{rf}{fz-a(z-f)},$$

showing that, in general, y/z depends on z, except for one case $a=f$, i.e. when the diaphragm is placed in the focal plane. Then the solid angle, at which the system receives radiation from the source, is independent of the position of the source, which means constant optical flux through the system. If there is a photodetector behind the diaphragm, then its signal does not depend on the position of the source. This is the photometric property of a telecentric lens.

In order to obtain an image of the source, an imaging lens must be added to the output of the basic telecentric scheme in Fig. 1.47. Thus, conceptually, a telecentric lens may be composed of only two lenses. Real optical schemes of telecentric lenses are much more complicated, and the idea of their design should be explained to avoid confusion. First, consider how the spatial filtering concept is implemented in real design, tracing only parallel rays, coming from infinity (Fig. 1.50). Step number one: a Galilean telescope. Step number two: an short-focus objective lens, focusing the infinity image into its focal plane. Step number three: a diaphragm in this plane, which is better to call a field stop, since it is not a narrow hole any more. This is the end of spatial filtering system. Next, add the final imaging lens that forms an image of the source. The telecentric lens is ready.

There are two types of telecentric lenses basically used in practice: the object-side telecentric and the double-telecentric lenses. The difference is clear from Fig. 1.51. Telecentric lenses always end with the C-mount.

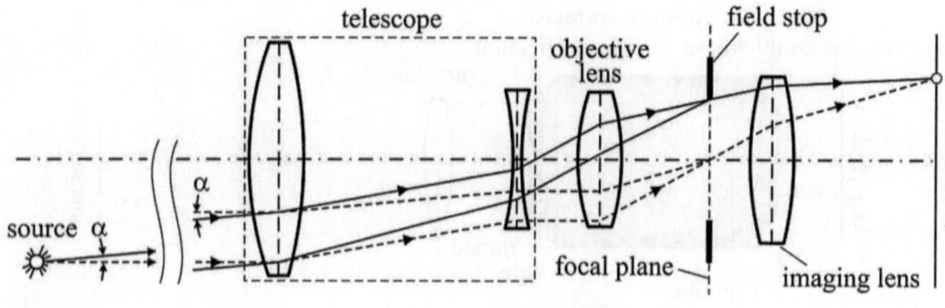

Fig. 1.50 Four steps of telecentric lens design: Galilean telescope, objective lens, field stop, and imaging lens

Fig. 1.51 The object-side (*above*) and double (*below*) telecentric lenses

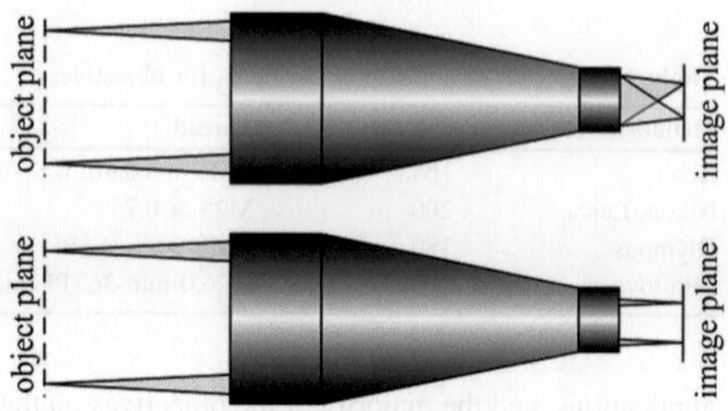

1.7 Microscope Objectives

During centuries since invention of the first optical microscope, microscope objectives have become a very complicated and nearly perfect device. Originally, a microscope objective was supposed to form the image as a traditional lens, i.e. in its image plane conjugated to the object plane. This concept used to produce numerous confusions, because various manufacturers designed their objectives for different conjugation lengths. Nowadays, another concept, known as infinity conjugation, is widely used (Fig. 1.52). In it, the objective is aberration-corrected to produce parallel beam of rays at its output pupil. This means, that no image can be obtained by the objective itself until the so-called tube lens is added. This lens inherited its name from its function: to replace the long tube from the objective to the object image. It is not necessary to specify conjugation distance for every objective any more. Instead, the tube lens focal distance F determines magnification:

$$M = \frac{F}{f},$$

where f is the objective focal length. It is commonly recognized today, that infinity-corrected objectives offer more flexibility to the user in developing specific

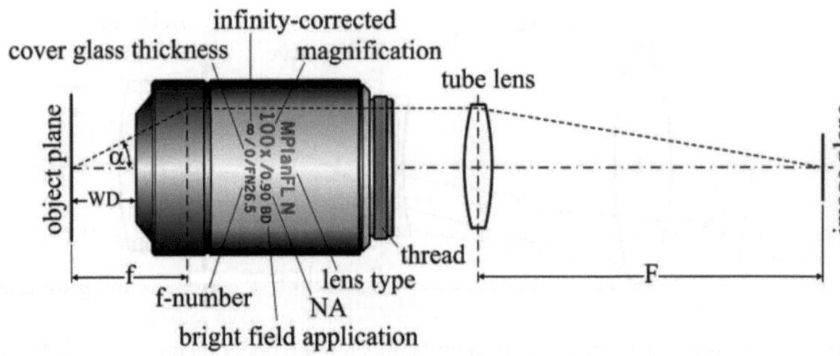

Fig. 1.52 Infinity-corrected objective requires the so-called tube lens to form an image. «WD» is the common abbreviation for working distance, «NA»—numerical aperture

Table 1.3 Specific manufacturer standards for objectives

Manufacturer	F (mm)	Thread
Zeiss	164.5	RMS standard: 0.8″-36TPI (metric 20.23 × 0.706 mm)
Nikon, Leica	200	M25 × 0.75
Olympus	180	RMS 0.8″-36TPI
Mitutoyo	200	∅26.0 mm-36TPI (metric 26 × 0.706)

applications, and the majority of the objectives on the market are of that type. Still, there are very irritating differences between manufacturers about types of the thread on their objectives and values of F used to calculate magnification M. This is explained in Table 1.3. TPI stands for Threads Per Inch, and RMS is the abbreviation for Royal Microscopy Society. Be careful: Mitutoyo defies all the standards—its thread has metric diameter but inch pitch. However, the thread length is short, therefore thread pitch may be set to metric 0.75 without any problem. Considering the aforementioned, there are two main parameters of any objective that define particular application irrelative to the manufacturer: numerical aperture (NA) and working distance (WD). Numerical aperture $\mathrm{NA} = n \cdot \sin \alpha$ is the product of the refractive index n of the medium, interfacing with the object, and sine of the maximum ray angle α, intercepted by an objective. It is essential because, according to famous Abbe formula for an ideal lens, minimum resolving distance d depends on the wavelength λ and numerical aperture NA:

$$d = \frac{\lambda}{2 \cdot \mathrm{NA}}.$$

Working in air, $n = 1$, therefore $\mathrm{NA} < 1$. With immersion (oil between the objective front lens and the object), numerical aperture may reach the value 1.4.

For an ordinary objective, the working distance WD is always shorter than its focal length f (Fig. 1.53). It means that the higher the magnification is, the shorter is the working distance. For example, for $M = 100$, WD ~ 0.3 mm. Many applications are very sensitive to short working distance, therefore manufacturers

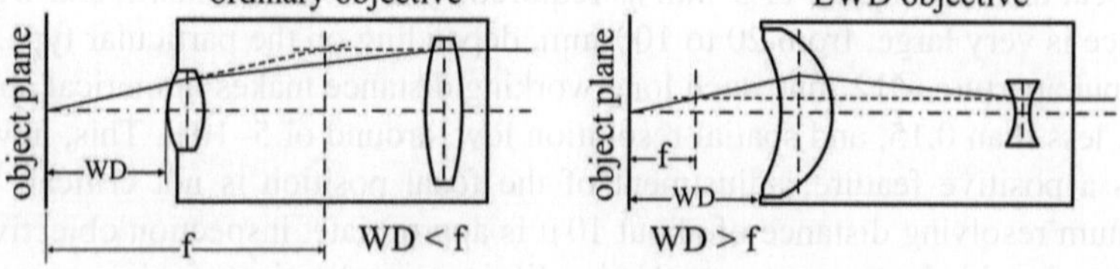

Fig. 1.53 The LWD objective always has a bigger front opening, which makes it easy to distinguish from an ordinary objective. Also, there is no concentration of rays inside the LWD objective, making it safer for laser ablation applications

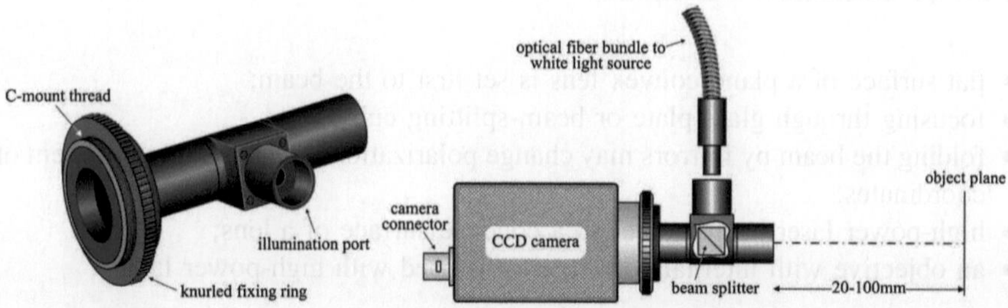

Fig. 1.54 Inspection objective is designed to be combined with a CCD camera through C-mount flange. Knurled ring on C-mount thread allows to fix illumination port at any rotation angle relative to the camera, thus choosing the best position for illumination fiber bundle

offer spring-loaded retraction stopper—a sliding front end—in some objectives with high magnification for safety reasons, in order to protect a rather expensive piece of optics from crashing into the sample during adjustment. However, industrial applications often require long working distances even with high magnification. Such objectives are called the Long Working Distance (LWD) or even Extra Long Working Distance (ELWD) objectives, and may have WD $\sim$ 10 mm with $M = 100$ and $f = 2$ mm. The principle of how to do that is clearly explained in a simplified form in Fig. 1.53, and closely relates to what has been told in Fig. 1.43.

An important conclusion follows for high-power applications: do not use standard objective, because convergence of the beam inside of it may lead to damage on optical surface. Better use the LWD objectives for that purpose.

It is worth mentioning that biological applications require cover glass over a sample, and high numerical apertures of almost all the objectives would make aberrations intolerable unless the objectives were corrected for that. This feature is always marked by a special code, like in Fig. 1.52.

Industrial applications, in particular inspection technology, spawned very peculiar type of microscope objective that may be called the inspection objective (Fig. 1.54). Inspection objective is a special type of a lens designed for machine-vision applications in combination with CCD cameras (Chap. 11). Therefore, its rear focal distance equals 17.5 mm as required by C-mount standard, and working distance is very large: from 20 to 100 mm, depending on the particular type. With the input aperture ∅12 mm, such long working distance makes numerical aperture small, less than 0.15, and spatial resolution low, around of 5–10 μ. This, however, brings a positive feature: adjustment of the focal position is not critical. When minimum resolving distance of about 10 μ is appropriate, inspection objective is a very comfortable instrument, greatly simplifying visualization of micrometer-scale details: all that you need is the CCD camera and a white-light source.

List of Common Mistakes

- flat surface of a plano-convex lens is set first to the beam;
- focusing through glass plate or beam-splitting cube;
- folding the beam by mirrors may change polarization in the laboratory system of coordinates;
- high-power laser beam comes to a concave surface of a lens;
- an objective with internal convergence is used with high-power laser.

Further Reading

W. J. Smith, Modern Optical Engineering, McGraw-Hill Professional; 3rd edition (2000).
M. Born, E. Wolf, Principles of Optics, Pergamon Press; 4th edition (1968).
A. McLeod, Thin Film Optical Filters, McGraw Hill/Adam Hilger; 2nd edition (1989).
E. O'Neil, Introduction to Statistical Optics, Dover Publications; 4th edition (2004).

Chapter 2
Light Sources

What you will face in the market: lamps, lasers, LEDs, and gas-discharge sources. Your cost-effective choice.

Abstract In this chapter, nine sections unfold in front of the reader the physics and details of the main types of light sources that can be found in optical laboratory. The first section summarizes specifications of the tungsten halogen lamp—the most common but not very simple device. Standard dimensions, clamping requirements, marking abbreviations, filament geometry, spectral radiometry, electrical characteristics, installation requirements and other most practical parameters help the user to quickly choose what he needs. Light emitting diodes (LEDs) are the sources of incoherent radiation in wide spectral interval determined by the energy gap of semiconductor material they use. Simple theoretical estimates explain the shape of the spectrum, while practical packaging considerations determine angular distribution. Physics of white-light and resonant-cavity LEDs is explained. Understanding of current–voltage and flux-current characteristics is necessary for efficient use of these devices. Practical rules and electrical schemes of connecting LEDs are essential to avoid typical mistakes. Laser diodes (LDs)—coherently emitting devices—are explained next. Divergence, spectral and polarization properties of LDs are compared to those of LEDs. Coherence of LD radiation results in speckle pattern that may be a serious hindrance in applications. To reduce it, the superluminescent diodes (SLD) may be used. However, both LDs and SLDs show significant astigmatism that can be corrected by means of miniature aspheric lens packages or anamorphic prisms. Electrical drivers for LDs are more complicated than those for LEDs and usually require feedback signals from embedded photodiodes. When utmost beam quality and coherence are needed, helium–neon laser must be the choice, particularly stabilized versions. Principles and practical schemes of stabilization are explained. A special type of stabilized He–Ne lasers widely used in heterodyning and interferometry is the Zeeman laser—two frequency highly stabilized source of coherent radiation. Much more complicated physics, laying in the background of this type of lasers, requires deeper physical explanation, including the Zeeman splitting and Kramers–Kronig relation. Tunable argon and helium-cadmium lasers are explained next. Among the pulsed sources, the most important is the neodymium laser. Narrow and powerful pulses of 1.06 μ radiation are bound to triggering pulses, so that understanding of

V. Protopopov, *Practical Opto-Electronics*, Springer Series in Optical Sciences 184,
DOI: 10.1007/978-3-319-04513-9_2, © Springer International Publishing Switzerland 2014

the pulse sequence is important to organize measurements. Whereas lasers are the sources of coherent radiation, compact and rugged xenon flash lamps are inexpensive sources of broadband incoherent radiation of microsecond duration.

2.1 Tungsten Halogen Lamps

Tungsten halogen lamps are the most common sources of wide-band radiation in visible and near-infrared domains. Like conventional incandescent lamps, they are thermal radiators, with two basic consequences: their spectrum is wide and smooth, without sharp peaks, and they cannot be efficiently modulated. Depending on application, tungsten halogen lamps may take various forms and dimensions, however, in laboratory practice, the compact type shown in Fig. 2.1 is the most popular.

Despite its seeming simplicity, complicated physics and decades of technological development form the basis of this small device. Inside the lamp, the combination of the halogen gas and the tungsten filament produces the so-called halogen cycle chemical reaction, which returns evaporated tungsten from the walls back onto the filament, thus increasing the lamp life and preserving optical clarity of the envelope. The gas fill is complicated and contains either pure xenon, krypton, argon, or their mixture together with pure iodine or bromine, or more complicated chemicals like HBr, CH_3Br, or CH_2Br_2. Halogen cycle together with temperature-resistant quartz glass make the operating temperature higher than in a conventional gas-filled lamp of similar power, increasing the luminous efficacy and color temperature. The internal pressure is also higher, requiring complicated cryogenic technology for filling. The filament is not just raw tungsten wire but the special material underwent complex doping, treatment, and purification processes to ensure ductility and temperature stability. Filament geometry is largely responsible for the photometric properties of tungsten-halogen lamps. When

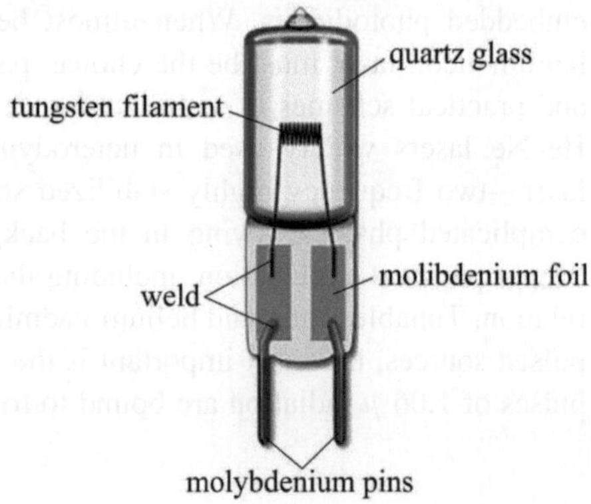

Fig. 2.1 Compact tungsten halogen lamp

highly uniform radiation field is required, like that in microscopy, then a flat-core filament is used, in which the tungsten wire is first wound over the rectangular rod and then flattened across the long axis.

The critical aspect of reliable lamp sealing is the difference between the thermal expansion coefficients of quartz and tungsten. Tungsten has much higher expansion coefficient, which makes it impossible to lead a tungsten pin through the quartz. This problem is solved by intermediate molibdenum foil sections welded between the filament and the contact molibdeium pins. These ribbons, 2–4 mm wide (depending on the lamp current) and 10–20 mm thick, are sealed in the quartz tail. Their longitudinal edges are sharply etched like razor blades, acting in such a way that substantial thermal expansion, occurring only along the wide sides of the bands, is absorbed by quartz without shattering.

The lamp design shown in Fig. 2.1 is often called bi-pin or single-ended. The lamp diameter, distance between the pins, pins diameter, number of turns in the filament, voltage, and other parameters are standardized. For example, the most popular lamp geometries are marked as T3 and T4, where «T» stands for «tubular», and the following integer N determines the diameter in the eighths of an inch: N/8″. Thus, T3 means tubular geometry with the diameter 3/8″ = 9.5 mm, and T4—a bigger tubular lamp with the diameter 4/8″ = 1/2″ = 12.7 mm. For such compact lamps, the standard distance between the pins is either 4 or 6.35 mm, and is marked as G4 or G6.35 respectively. The letter «G» stands for «glass» . The letter «Y» may appear after «G», specifying thicker pins. For both G4 and GY4 lamps, the pin diameters are always 0.7 mm, however, for G6.35 it is 1 mm but 1.3 mm for GY6.35. The combination C6 denotes six turns in the filament («C» means «Coil»). Typical low voltages are 6 or 12 V, and these lamps do not require fuses to ignite.

Designing a light source with a tungsten halogen lamp, it is useful to remember that cooling considerations dictate certain clearance around the lamp (Fig. 2.2, Table 2.1).

A single-ended lamp may be permanently integrated into a reflector, thus presenting the pre-designed optical system, supposed either to focus, or to

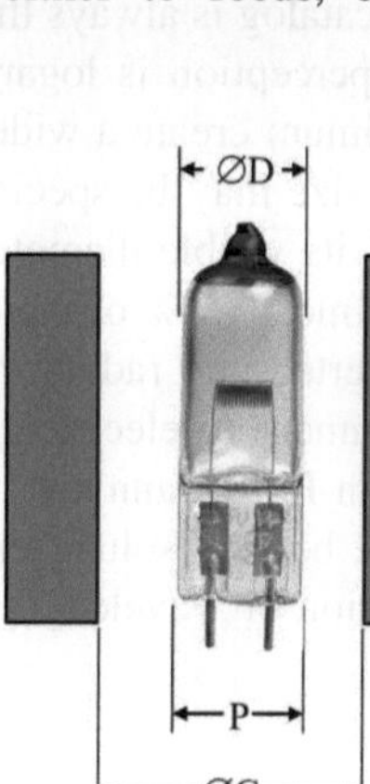

Fig. 2.2 Bulb diameter *D* and pinch width *P* dictate clearance diameter *C* as in Table 2.1

Table 2.1 Cooling clearance for tungsten halogen lamps

∅D_{max} (mm)	P_{max} (mm)	∅C_{min} (mm)
9	11	12
10.5	12.5	13.5
11.5	14	15
13	15	16
13.5	16	16
14	16	17
18	20	21
18.5	20	21

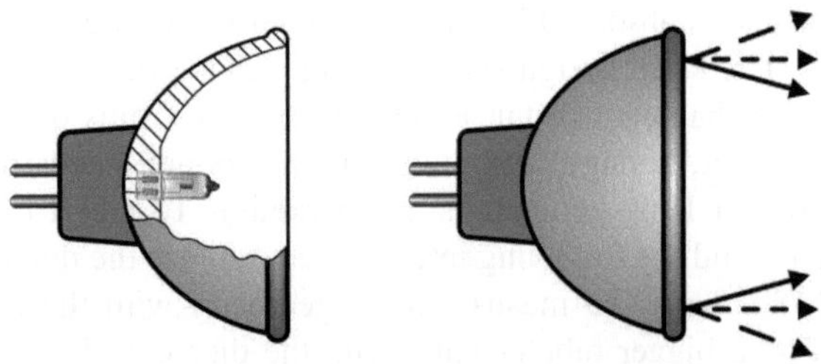

Fig. 2.3 Integrated reflectors may focus, collimate, or diverge the beam

collimate, or to make slightly divergent beam (Fig. 2.3). From the point of view of mechanical design, the diameter of the reflector is the principal dimension, establishing two widely used standards: MR11 and MR16. Originally, the letters «MR» stood for «metal reflector». However, nowadays the halogen lamp reflectors are made of glass or ceramic, so that the initial meaning of this abbreviation is completely forgotten. As to the numbers, it is again the diameter in the eighths of an inch: MR11—11/8″ ≈ 35 mm; MR16—16/8″ = 2″ ≈ 50 mm. In case of the focusing option, the approximate spot size and its distance from the reflector rim are always specified. Manufacturer may claim as small spot size as 3 mm, but in reality, visible spot size is always bigger, because the value specified in a catalog is always the width at half-maximum of the peak intensity, and human eye perception is logarithmic, so that weaker tails of the spot beyond the half-maximum create a wider image. For example, with 50 mm reflector, the smallest spot size may be specified as 10 mm at roughly 30 mm from the reflector, however, its visible diameter will be roughly 20 mm (Fig. 2.4).

Almost 75 % of the electrical power, consumed by a tungsten halogen lamp, is converted into radiation. Its luminous efficacy, defined as the ratio of optical flux (in lumen) to electrical power (in Watt), is about 20–30 lm/W. According to the Stefan-Boltzmann law, the total emitted power is proportional to T^4, where T is the black body absolute temperature (in Kelvin), and its spectral distribution B as a function of wavelength λ may be estimated by the Planck formula

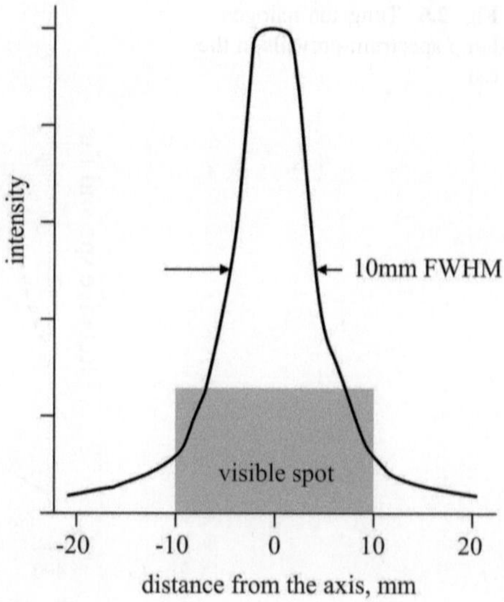

Fig. 2.4 Visible spot is always bigger than the specified one

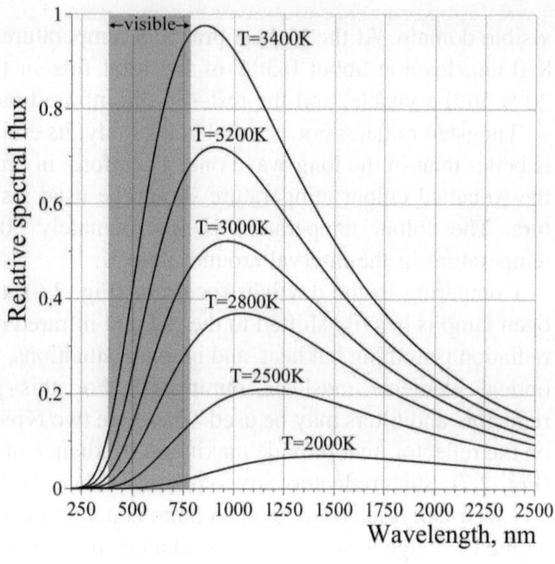

Fig. 2.5 Relative black body spectral flux at different temperatures. Visible region is 380–780 nm

$$B(\lambda) = \frac{2hc^2}{\lambda^5} \cdot \frac{1}{e^{\frac{hc}{\lambda kT}} - 1},$$

where $h = 6.62 \times 10^{-34}$ [J·s] is the Planck constant, $k = 1.38 \times 10^{-23}$ [J/K] is the Boltzmann constant, and c is the speed of light (Fig. 2.5). The maximum of the curve shifts according to the Wien displacement law as $\lambda_{\max} \cdot T = const$, but melting point of tungsten 3383 °C does not permit this maximum to be shifted into

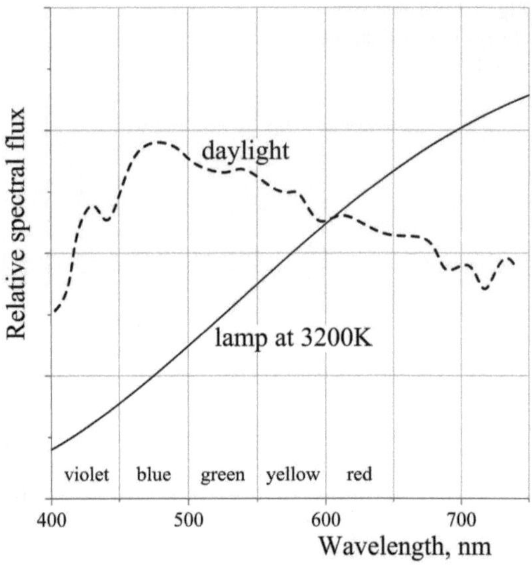

Fig. 2.6 Tungsten halogen lamp spectrum prevails in the red

visible domain. At the highest practical temperatures, the spectral maximum is at 850 nm, leaving about 0.3 % of the total flux in the ultra-violet (UV) domain, 20 % in the visible, and the rest—in the infrared part of the spectrum.

Tungsten radiates not as a real black body: its emission in the short-wave region is better than in the long-wave one. Therefore, in order to use the Planck formula, the so-called colour temperature should be used instead of the physical temperature. The colour temperature is approximately 60 K higher than the physical temperature in the interval around 3000 K.

Comparing to the daylight spectrum (Fig. 2.6), the balance of a tungsten halogen lamp is heavily shifted to the red and infrared parts of the spectrum. Infrared radiation is nothing but heat, and in many situations, the heat load upon subsequent optical elements must be minimized. For this purpose, spectrally selective reflectors and filters may be used. There are two typical multilayer optical coatings on the reflector, that provide maximum irradiance in blue or entire visible domains (Fig. 2.7). Such reflectors are commonly called the blue- and cold-light reflectors.

Electrically, tungsten halogen lamp is not a linear element: a 5 % change of the applied voltage results in 15 % change in optical flux, 8 % change in power consumption, 3 % change in current, 2 % change in colour temperature. The principal source of non-linearity is the temperature dependence of the filament electrical resistance R: the higher the temperature is, the higher is the resistance. For example, if R were constant, then 5 % increase of voltage would result in 10 % increase of power consumption:

$$\frac{V^2(1+0.05)^2}{R} = V^2\frac{1+2\times 0.05+0.05^2}{R} \approx \frac{V^2}{R}(1+0.1).$$

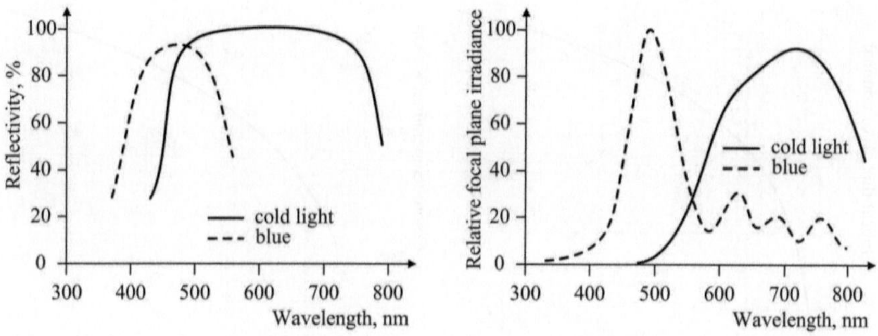

Fig. 2.7 Spectrally selective reflectors block infrared radiation from coming to focal plane. Focal plane irradiance is even more spectrally selective than the reflective coating, because angle of incidence varies significantly over the reflector surface

Thermal increase of R lowers this value to 8 %. Resistance of a cold filament is 20 % smaller than at the operating temperature, making the start-up current bigger than the nominal one. Although the initial sharp current burst lasts less than 0.5 s (only filament heating), final stabilization occurs much later, taking time to warm up the glass bulb and leads.

In laboratory applications, tungsten halogen lamp never works at its nominal voltage because the optical flux is too intense. In order to decrease it, two options may be used: neutral optical filters (attenuators) or lower voltage. The first option is rather cumbersome, and should be applied only in those rare cases when the colour temperature must be constant. Voltage attenuation is a much simpler solution, however, the flux response is non-linear and the light becomes reddish (Fig. 2.8). With variable power supply, the voltage control is not a topic for discussion. The problem may occur when only a stabilized power source is available. First of all, its limiting current must be sufficient to power the lamp. For example, with 6 V 50 W lamp, the current will be up to 10 A at the moment of switching on. The standard power module of a personal computer will do well, providing enough current both at 5 and +12 V outputs. The primitive solution is to use a power variable resistor in series with the lamp (Fig. 2.9, left). The power P dissipated on the resistor

$$P = I^2 r = \frac{V^2 r}{(R + r)^2}$$

reaches its maximum when $R = r$. At this moment, the lamp consumes one quarter of its nominal power—the same as the resistor, so that the resistor maximum power rating must be 25 % of that of the lamp. If electronic control is needed, then the right scheme in Fig. 2.9 will work fine.

Tungsten halogen bulbs operate at high temperatures up to 600 °C, which is far above the melting point of a typical solder ($\sim$350 °C). Therefore, wires soldered to the pins (with application of soldering wax, it is quite possible) will fall off after

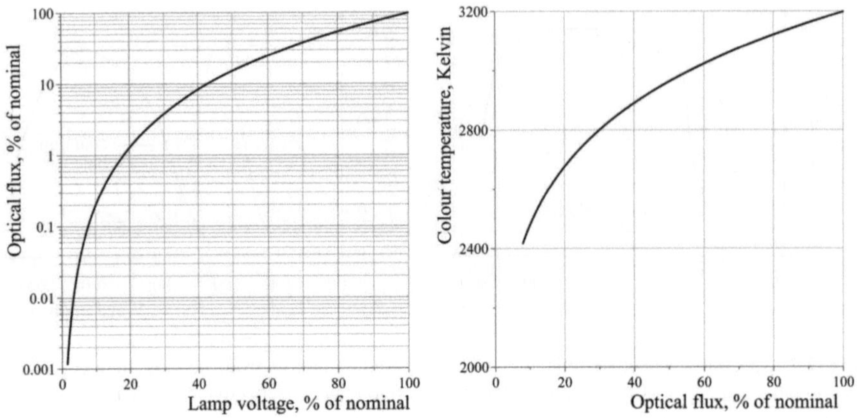

Fig. 2.8 Lamp optical flux responses non-linearly to voltage (*at left*) and colour temperature decreases with flux (*at right*)

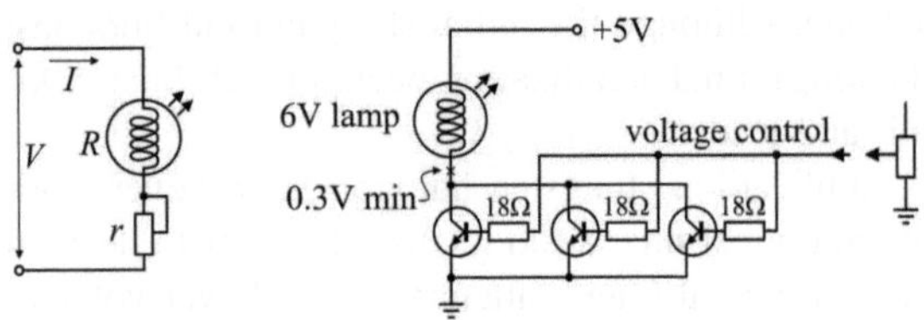

Fig. 2.9 The lamp voltage may be controlled either by a power variable resistor (*at left*) or by power transistors (*at right*). Multiple transistors are only needed to decrease the minimum voltage drop on them (down to 0.3 V) when they are all open and the supply voltage is close to or less than the nominal voltage of the lamp, like in the figure

heating up. Avoid leaving fingerprints on the glass: they will burn in and cause irreversible glass recrystallization, making it opaque and milky. This area then absorbs more heat, and local overheating may even cause the bulb breaking.

2.2 Light-Emitting Diodes

Light-emitting diode (LED) is the semiconductor device that emits incoherent non-polarized and non-collimated light, in contrast with the laser diodes (LDs), which will be discussed in Sect. 2.3. Narrow-band LEDs operate in blue, green, red, and infrared parts of spectrum, and the so-called white-light LEDs cover the entire visible diapason. Manufacturers offer several standard packages for various applications (Fig. 2.10). Although both LEDs and LDs are basically the semi-conductor diodes, the principle of light emission is completely different in these two types of devices. Lasers generate light due to stimulated emission, and this is

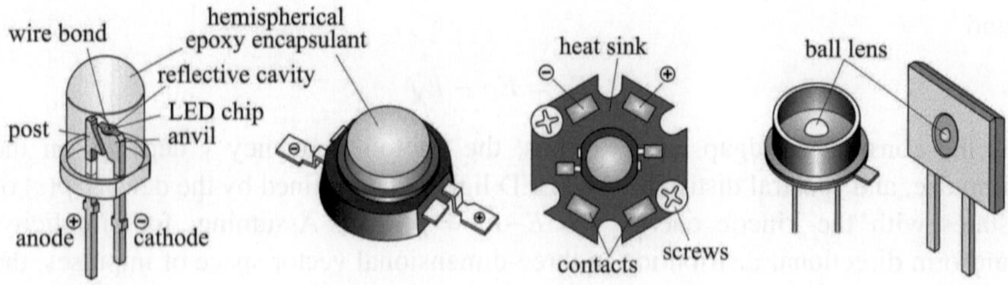

Fig. 2.10 General purpose LED (*left first*); power LED (*second*) may be pre-soldered to an aluminium or copper heat sink plate (*third*); fiber-optic communication LEDs are equipped with ball lenses for better coupling efficiency (*at right*)

the operating principle of LDs. In LEDs, electron–hole pairs recombine spontaneously to produce incoherent emission of photons. Spontaneous recombination has certain characteristics that determine optical properties of LEDs, which may be understood with rather simple but not very rigorous theoretical estimations. It is well known, that electrons in the conduction band and holes in the valence band are the two types of electrical charge carriers. Their energy is composed of the potential energies E_C and E_V in the conduction and valence bands, and the kinetic energies that can be expressed in terms of the impulses p_e and p_h:

$$E_e = E_C + \frac{p_e^2}{2m_e}$$

for electrons, and

$$E_h = E_V - \frac{p_h^2}{2m_h}$$

for the holes. Here m_e and m_h are the electron and hole effective masses. In recombination, both the electron and hole disappear, making the total impulse of the pair zero. Impulse of a photon is negligible relative to that of a particle, therefore impulse conservation law dictates that the impulses of recombining electron and hole are almost equal:

$$p_e = p_h = p.$$

If spontaneous recombination results in photon emission, then the photon energy E and its frequency v are determined by the energy conservation and Planck laws:

$$E = hv = E_e - E_h = E_g + \frac{p^2}{2m},$$

where

$$\frac{1}{m} = \frac{1}{m_e} + \frac{1}{m_h},$$

and

$$E_g = E_C - E_V$$

is the constant bandgap energy. Thus, the photon frequency ν depends on the impulse, and spectral distribution of LED light is determined by the density $\rho(\varepsilon)$ of states with the kinetic energy $\varepsilon = E - E_g = p^2/2\,m$. Assuming, for simplicity, uniform directional distribution in three-dimensional vector space of impulses, the number of states is proportional to the volume of a spherical layer of the thickness *dp*:

$$\rho(\varepsilon)\, d\varepsilon \propto 4\pi p^2\, dp \propto \varepsilon\, dp,$$

so that

$$\rho(\varepsilon) \propto \varepsilon \frac{dp}{d\varepsilon} \propto \sqrt{\varepsilon} = \sqrt{E - E_g}.$$

Probability of the total energy of electron–hole recombination E is determined by Boltzmann distribution

$$e^{-E/kT},$$

where T is the absolute temperature in Kelvin, and $k = 1.38 \times 10^{-23}$ [J/K] is the Boltzmann constant. Thus, the rate of spontaneous emission for the photons with energy $E = h\nu$ is proportional to the function

$$\sqrt{E - E_g} \cdot e^{-E/kT},$$

which is plotted in Fig. 2.11. Its maximum gives the wavelength of LED peak intensity:

$$\lambda_{\max} = \frac{hc}{E_g + kT/2} \approx \frac{hc}{E_g},$$

showing that the peak wavelength is inversely proportional to the bandgap energy E_g. Here $h = 6.62 \times 10^{-34}$ [J·s] is the Planck constant, c is the speed of light. Table 2.2 summarizes bandgap energy of some materials used for LEDs, driving voltage V, and emission spectrum.

The width of the curve

$$\Delta E = 1.8\, kT$$

gives the estimate for spectral width:

$$\Delta\lambda = \frac{1.8\, kT\lambda^2}{hc}, \text{ or } \frac{\Delta\lambda}{\lambda} = \frac{1.8\, kT\lambda}{hc}.$$

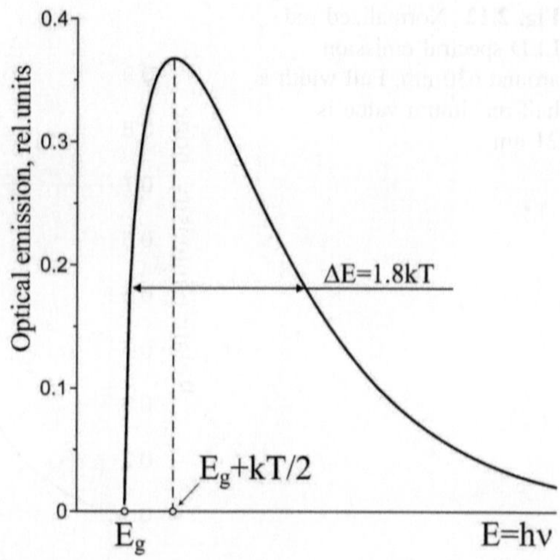

Fig. 2.11 Theoretical estimation of LED spectral intensity. The maximum is located at $E_g + kT/2$, and full width at half-maximum $\Delta E = 1.795\,kT$

Table 2.2 LED materials and emission bands

Material	E_g, eV	Volt (V)	Emission	Wavelength (nm)
InGaAs	0.9	0.8	Infrared	1300
GaAs	1.5	1.4	Infrared	870
AlGaAs	1.8	1.7	Red	700
AlGaInP	1.8–2.2	1.7–2.2	Amber–red	580–700
GaP	2.2	2.4	Green	570
AlGaInN	2.1–2.6	3.2–3.5	Blue–amber	480–600
AlGaN	3.5	3.8	Ultraviolet	350

For example, for red LEDs with $\lambda = 630$ nm at room temperature $T = 300$ K, this formula gives $\Delta\lambda = 15$ nm, which is reasonably close to the real value of 21 nm, as shown in Fig. 2.12.

Coherent properties of LEDs lie in between the tungsten lamps and lasers. Wide relative bandwidth $\Delta\lambda/\lambda$ of LEDs, typically 3–10 %, means that coherence length

$$s = \lambda \cdot \frac{\lambda}{\Delta\lambda}$$

is small, of the order of tens of microns. This makes LED a good choice when image speckles and etalon fringes must be avoided.

Transparent plastic encapsulants that frequently form the entire package of an LED (Fig. 2.10 left), play significant role in extracting and directing the light emission. Without it, only those rays would emerge from the semiconductor die that come to the surface in a narrow cone θ, the other portion being reflected back

Fig. 2.12 Normalized red LED spectral emission around 630 nm. Full width at half-maximum value is 21 nm

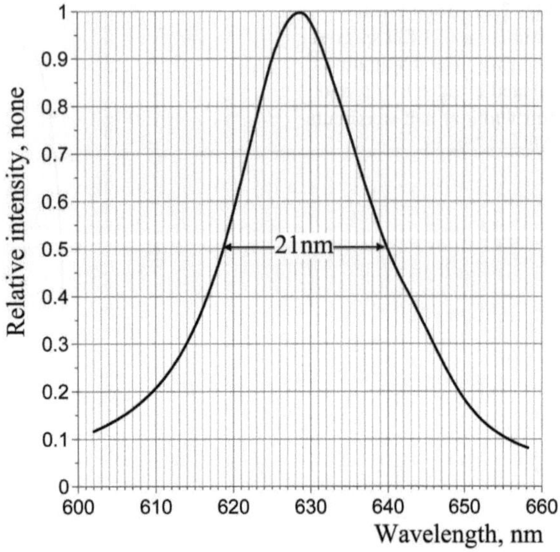

due to total internal reflection, and lost (Fig. 2.13). With the material refractive index $n \approx 3.3$, the critical angle θ on the interface semiconductor-air

$$\theta = \arcsin\frac{1}{n} \approx 18^\circ,$$

which is only 0.3 s radian, or about 2.3 % of a sphere. Encapsulation of the semiconductor die into an epoxy with the refractive index $n' \approx 1.5$ makes the critical angle

$$\theta = \arcsin\frac{n'}{n} \approx 27^\circ,$$

which more than doubles the efficiency. Moreover, making the dome shape vertically elongated, a type of parabolic surface, it is possible to narrow the directional diagram. For collimation, some designs use not only refraction but total internal reflection as well (Fig. 2.14).

Beam divergence is crucial in fiber-optic communication technology, because optical coupling of LED to the fiber core determines overall efficiency of the system. For this particular application, a special LED structure was developed, called resonant-cavity LED. In it, a light-emitting region is placed inside an optical cavity, formed by two flat reflecting layers (Fig. 2.15). These reflectors may be either single metal layers or multilayer stacks, designed to reflect only at working wavelength. The optical thickness of the cavity, taking into account refractive index of the material, is half of the LED wavelength. Thus, interference pattern within the cavity has maximum at the emitting layer, increasing the intensity of light, its spectral purity, and collimation. In visible domain, spectral width of

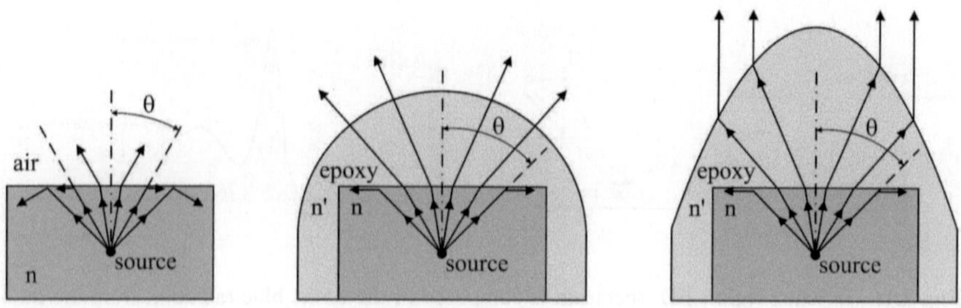

Fig. 2.13 Encapsulation of the semiconductor die (*at left*) into an epoxy dome improves light-extracting efficiency (*center*), and may also improve collimation (*at right*)

Fig. 2.14 Some power LEDs use additional plastic lenses, helping improve collimation

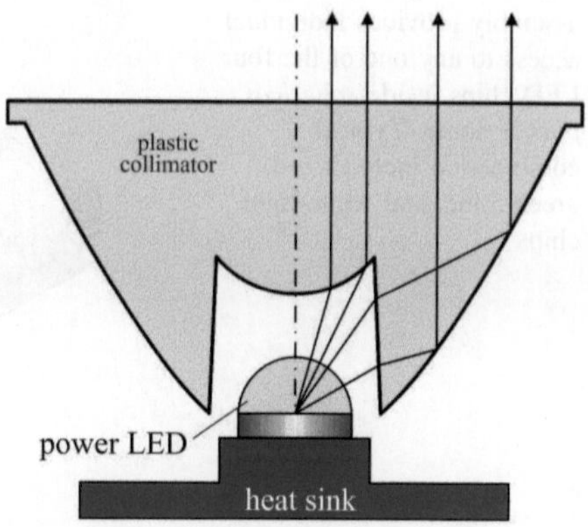

Fig. 2.15 Resonant-cavity LED emits across the thin wide structure. The thickness of the active layer is insufficient to make the optical gain more than unity, as required for lasers. Therefore, diodes of this type emit spontaneous radiation

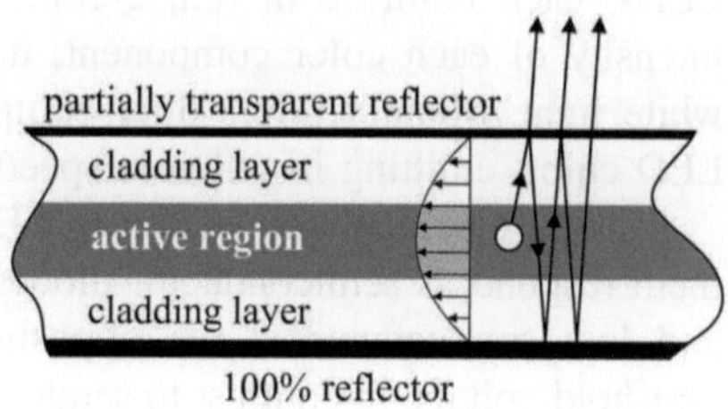

resonant-cavity LEDs is smaller than that of ordinary LEDs by a factor of two, i.e. about 10 at 650 nm. In infrared, this factor is even bigger. Resonant-cavity LEDs are easily recognized by tiny ball-lenses at front panels (Fig. 2.10).

It is clear from the physics of operation that light-emitting diodes are the sources of narrow-band light. A broad-band white-light source may be obtained by a variety of techniques, of which the most common one is immersing a blue LED chip into a pool of fluorescence phosphor that converts narrow blue luminescence of LED into wide red fluorescence (Fig. 2.16). Usually, the cerium-doped ittrium-aluminium garnet (YAG:Ce) is used as a phosphor.

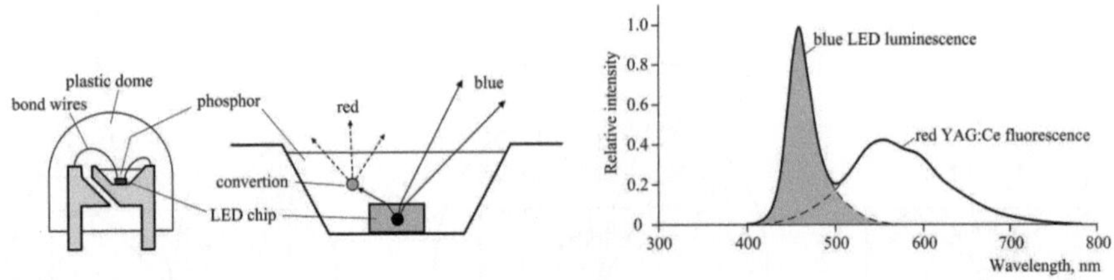

Fig. 2.16 In white-light LED, spectrum is composed of the direct blue radiation from the diode chip and red fluorescence of the YAG:Ce phosphor

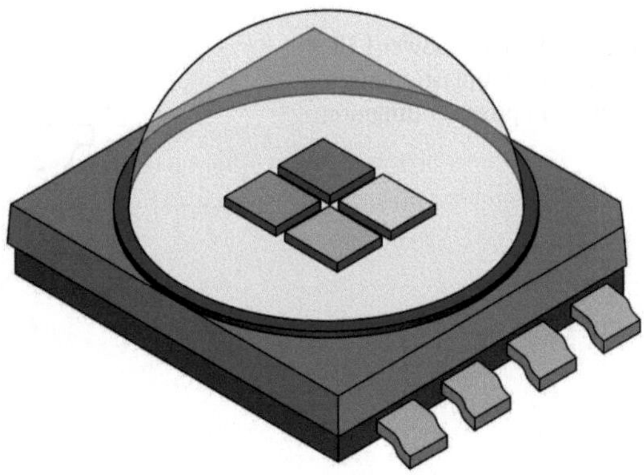

Fig. 2.17 Eight-pin assembly provides individual access to any one of the four LED chips inside *spherical plastic* dome. Typical combination includes red, green, blue, and white-light chips

Another possibility of producing white-light illumination is to combine three LEDs, each emitting in red, green, and blue regions. Even more, controlling intensity of each color component, it is possible to create various rendering of white light. Manufacturers offer compact assemblies, containing four individual LED chips, emitting in different spectral regions (Fig. 2.17).

Making electrical connection to LEDs, it is necessary to remember that all of them respond as semiconductor diodes: high conductance at one voltage polarity and low conductance at the opposite polarity. The only difference is higher threshold voltage in contrast to semiconductor rectifying diodes that have forward voltage drop of only 0.3–0.5 V (Fig. 2.18). Physically, LED optical flux is roughly proportional to the rate of carriers injection, i.e., to the forward current. As such, the electrical connection of a LED should be made so that to precisely control its current, while the forward voltage drop is of no concern for the user. There are two basic schemes for connecting LEDs (Fig. 2.19). The idea is not to allow uncontrollable giant variation of the current that may damage the LED. Figure 2.18 clearly shows that the entire dynamic range of optical flux fits into only $\approx$0.5 V of

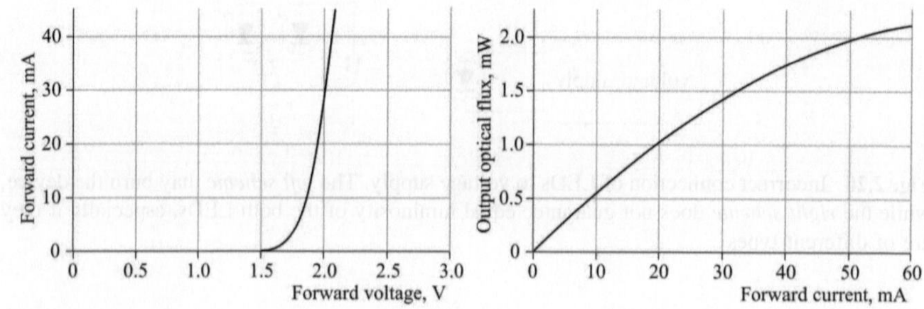

Fig. 2.18 Typical 660 nm LED current-voltage (*at left*) and flux-current (*at right*) characteristics

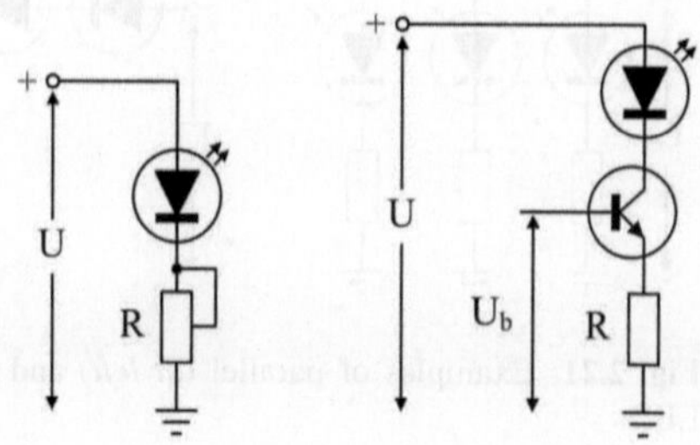

Fig. 2.19 Correct connection of a LED to voltage supply. In both cases, the current is limited and determined either by a variable resistor R (*at left*) or by the base bias voltage U_b (*at right*)

direct voltage drop; but what is even more important, this voltage drop differs significantly for various LEDs and also depends on the temperature. Therefore, the two wrong schemes in Fig. 2.20 should never be applied. In the left scheme of Fig. 2.19, the forward voltage drop on the LED may be assumed to be roughly 1.5 V. Then the resistor value is determined as

$$R = \frac{U - 1.5\,[V]}{I},$$

where I is the LED nominal current. In the right scheme, the transistor accurately sets the current through the LED. The emitter–base voltage drop is roughly 0.5 V for all the silicon transistors. The base current is negligible with respect to that in the collector, therefore the emitter current is approximately the same as the collector current I that drives the LED. Thus, the driving equation for this circuit is

$$U_b = I \cdot R + 0.5\,[V].$$

LEDs may be connected either in series or in parallel (Fig. 2.21). When the series connection is used, the simple scheme in the middle of the figure requires adjustment of the resistor each time when a new LED is added. With the transistor scheme, this is not necessary because the collector current of a transistor does not depend (practically) on the collector voltage.

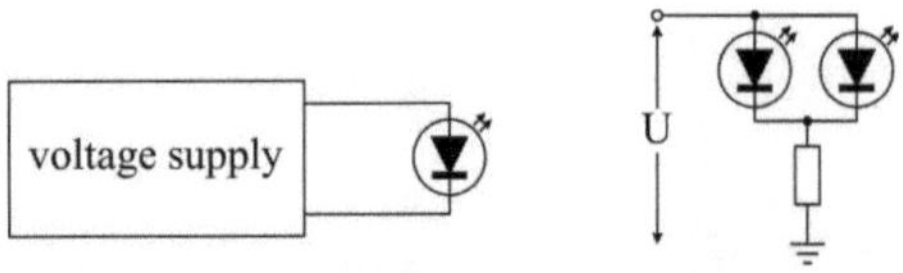

Fig. 2.20 Incorrect connection of LEDs to voltage supply. The *left scheme* may burn the device, while the *right scheme* does not guarantee equal luminosity of the both LEDs, especially if they are of different types

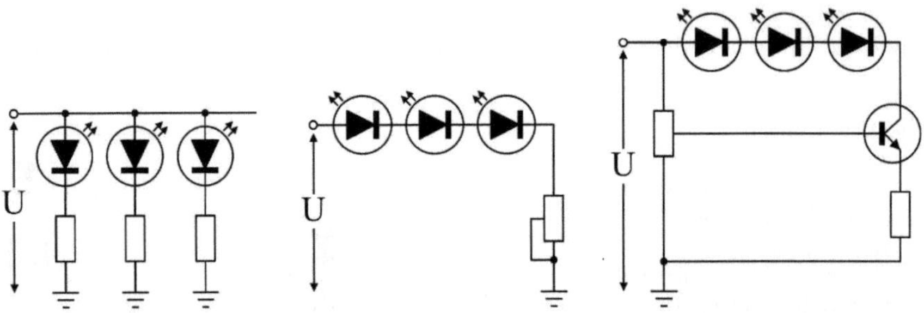

Fig. 2.21 Examples of parallel (*at left*) and series (*in the middle* and *at right*) connections of LEDs

2.3 Laser Diodes

Laser diode (LD) is a small and efficient source of spectrally narrow and spatially coherent radiation in red and near-infrared domains (Fig. 2.22). LD is essentially the light-emitting diode, placed into an optical cavity to establish positive optical feedback. The difference between the LD and the resonant-cavity LED (Fig. 2.15) is the optical length of the cavity and the active area: in LD, it is of a macroscopic size, about 1 mm, which makes the optical gain bigger than unity—the principle condition for stimulated emission. Whereas the resonant-cavity LED is a surface-emitting device, the LD is an edge-emitting source. In it, the electrical current flows across the active layer, $\sim$0.1 μm thick and $\sim$3 μm wide. Microscopic cross-section of the source makes LD a good choice when highly collimated optical beams are needed. The LD cavity is formed by two cleaved walls of a semiconductor crystal (Fig. 2.23). Divergence of the output beam is determined by diffraction on the active layer output aperture: the vertical divergence is bigger than the horizontal one.

The only thing that probably needs explanation is why polarization is horizontal. The reason is that for stimulated emission, optical gain must prevail over losses, which means that only the waves with minimal losses will radiate. For a wave, propagating between two parallel surfaces (Fig. 2.23 right), reflection from the interface with the adjacent medium should be a maximum, and it depends on polarization. According to Fig. 1.37, the s-polarized wave (perpendicular to the

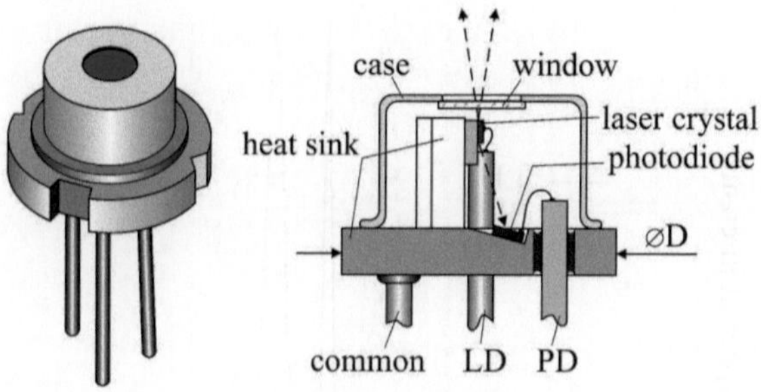

Fig. 2.22 Laser diode package usually has three leads: for the laser diode itself (LD), for monitoring photodiode (PD), and their common. D = 5.6 mm for laser diodes and 9 mm for superluminescent diodes

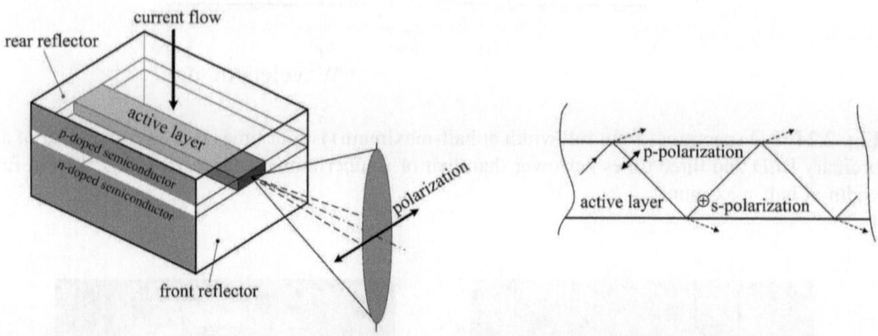

Fig. 2.23 The output beam is spatially coherent, elliptical, and polarized along the short axis of the beam ellipse, i.e. in the direction of the longer side of the active layer cross section (horizontally)

plane of incidence and parallel to the interface) always has bigger reflection. There are two systems of parallel surfaces in Fig. 2.23: the one with vertical walls and small area, and the one with horizontal walls and larger area. Obviously, the wave with polarization parallel to the larger surface will have smaller loss. Therefore, horizontal polarization dominates.

Boundary conditions, imposed by optical cavity, shrink spectral width of the output radiation to the value much smaller than that of a LED (Fig. 2.24).

High spectral purity and, consequently, high spatial and temporal coherence, good for interferometry and spectroscopy, create problems for other applications in the form of speckles and etalon fringes. Effects of LD beam coherence can be directly seen in a microscope focused on the output window (Fig. 2.25). Large number of important applications where speckles are intolerable, like laser autofocusing or laser scanning microscopy, stimulated development of the so-called superluminescent diodes (SLD), with the spectral width of about that of the LEDs. In this type of laser diodes, optical cavity feedback is switched off by means of

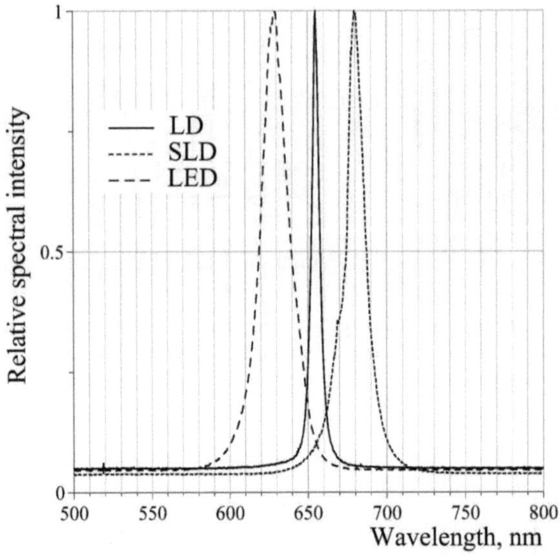

Fig. 2.24 LD spectrum (5 nm full width at half-maximum) is four times narrower than that of an ordinary LED and three times narrower than that of a superluminescent diode (SLD, 13 nm full width at half maximum)

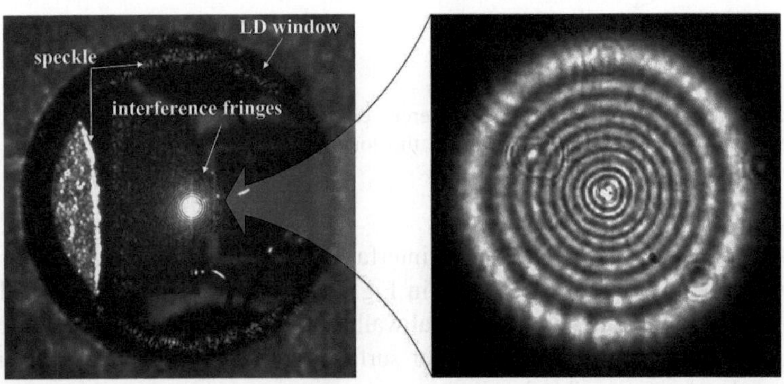

Fig. 2.25 Interference fringes, caused by reflections from the opposite side of the output window, and speckles are the result of high coherence of LD beam

geometrical design and antireflection coatings, thus allowing all the waves propagate with high gain, independently of their wavelength. Physically, the SLD may be described as high-gain traveling-wave optical amplifier. Typical spectrum of an SLD is shown in Fig. 2.24. The SLD package scales the LD package shown in Fig. 2.22 to the diameter 9 mm, and the inner design is essentially the same. The light pattern shows much less ellipticity and may be better described as «banana-like» pattern (Fig. 2.26). Polarization of SLDs strongly depends on particular

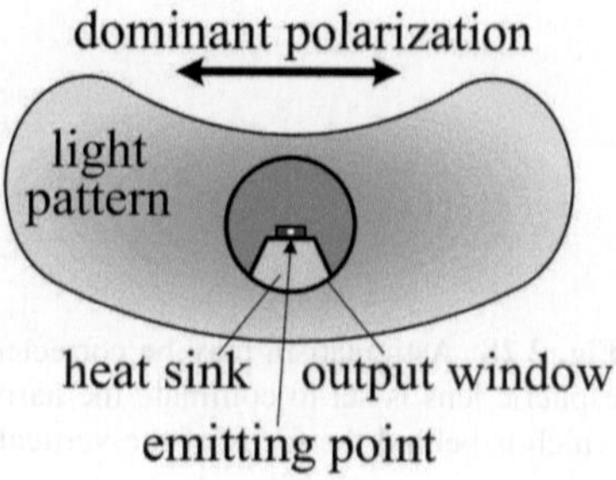

Fig. 2.26 SLD's light pattern

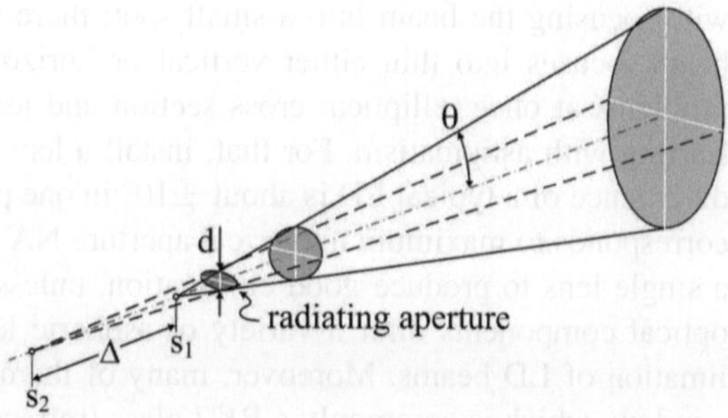

Fig. 2.27 Radiating aperture has different dimensions d along vertical and horizontal axes, making different divergence θ in these planes. In some plane near the aperture, the beam cross section is circular. Unfortunately, this plane is inside the LD case (below the window)

structure, and there are no common rules about the polarization ratio: it may vary from 1:2 to 1:50 with the dominant direction shown in Fig. 2.26. High optical gain in the active region of SLDs makes them extremely sensitive to returning optical power and to temperature variations. The returned wave will be amplified to the saturation level, thus decreasing the output power of the directly traveling wave. It is commonly agreed that less than 1 % reflected power is the limit for LDS normal operation. Temperature stabilization is also important: overheating from +25 to +50 °C decreases optical power by an order of magnitude.

Strong ellipticity of the LD beam causes astigmatism in imaging applications. It can be better understood from the geometrical optics considerations (Fig. 2.27). When a single-mode elliptical beam of the wavelength λ propagates, it diverges diffractionally with different angles θ in horizontal and vertical planes:

$$\theta \approx \frac{\lambda}{d}.$$

The vertical and horizontal asymptotes intercept the optical axis at different points s_1 and s_2, and the difference defines astigmatism:

$$\Delta = s_2 - s_1,$$

which is of the order of 5–20 μm for low-power LDs. If someone wants to create a parallel beam by inserting a lens with the focus at s_1, then vertical rays will be collimated but horizontal rays will be converging. With the focus at s_2, horizontal rays will be collimated but vertical will be diverging. The same situation occurs

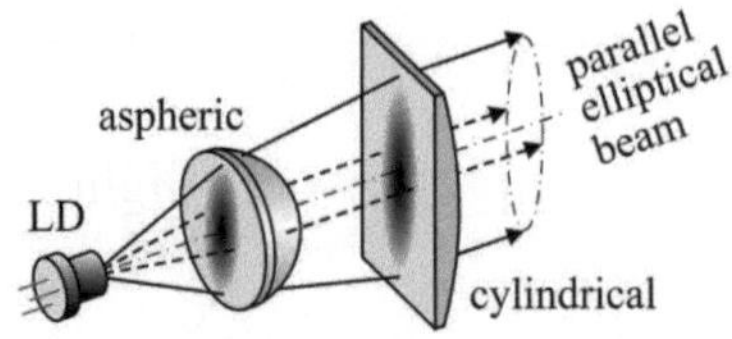

Fig. 2.28 Astigmatism may be corrected by a second cylindrical lens. In this particular figure, aspheric lens is set to collimate the narrow horizontal cone of rays (*dashed lines*), the focus of which is behind the focus of the vertical rays (see Fig. 2.27)

with focusing the beam into a small spot: there will be two planes, in which the beam focuses into thin either vertical or horizontal thread. Thus, there are two problems at once: elliptical cross section and astigmatism. They may be solved, starting with astigmatism. For that, install a lens with the focus at s_1 or s_2. Beam divergence of a typical LD is about $\pm 10°$ in one plane and $\pm 30°$ in another, which corresponds to maximum numerical aperture NA $\approx$ 0.5. This value is too high for a single lens to produce good collimation, unless it is aspheric. Manufacturers of optical components offer a variety of aspheric lenses specially designed for collimation of LD beams. Moreover, many of them are corrected for the LD output window, which is commonly a BK7 glass (refractive index 1.517) 0.25 mm thick. Glass refractive index and thickness are always specified by both the lens and LD manufacturers. Then rays, coming from one focus, will be perfectly collimated, whereas those from the second one–not. To correct this natural misalignment, install a weak cylindrical lens after the aspheric one (Fig. 2.28). Now, the beam is perfectly collimated in two planes. However, it is elliptical in cross section. The simplest way to correct ellipticity is to install a circular diaphragm, cutting approximately uniform portion from the beam cross section. Regretfully, this results in significant loss of optical power. If optical power is crucial, use anamorphic pair of prisms as shown in Fig. 2.29. Both mounted modules and un-mounted pairs are available on the market. The mounted option is convenient because it does not require adjustment and the input and output beams are parallel. However, only certain magnifications, like 2×, 3×, 4×, and 6×, are available. On the contrary, un-mounted pairs provide any desirable magnification (within geometrical limits) by means of angular adjustment of the two prisms, but require additional mechanical components, and parallelicity of the beams requires adjustment.

LDs require special electrical connection and drivers. The current–voltage characteristic of an LD is the same as that of an LED (Fig. 2.18 left). With the current increasing from zero, until the optical gain in the laser cavity reaches unity, the output optical flux increases almost linearly as in LEDs (Fig. 2.30). The threshold is reached when amplification exceeds losses, i.e. the optical gain exceeds unity. After that, optical flux increases sharply with the current, and tiny variations of the current produce strong variations of optical flux. Temperature dependence introduces positive feedback, making operation unstable. For the

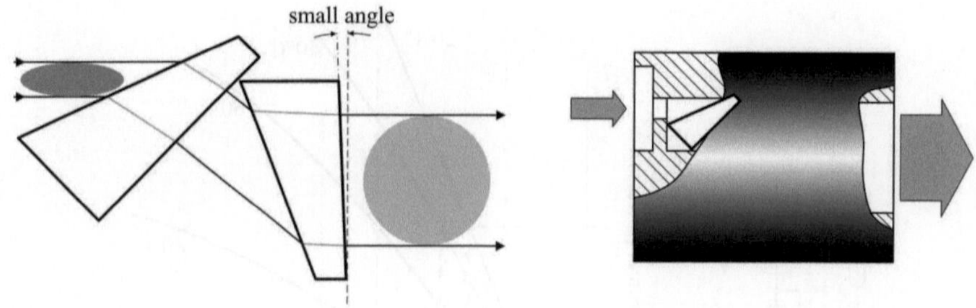

Fig. 2.29 Note that rear faces of the prisms must not be perpendicular to the beams in order to avoid reflection back into the laser cavity

Fig. 2.30 In the after-threshold region, optical characteristic of an LD is much steeper than that of an LED

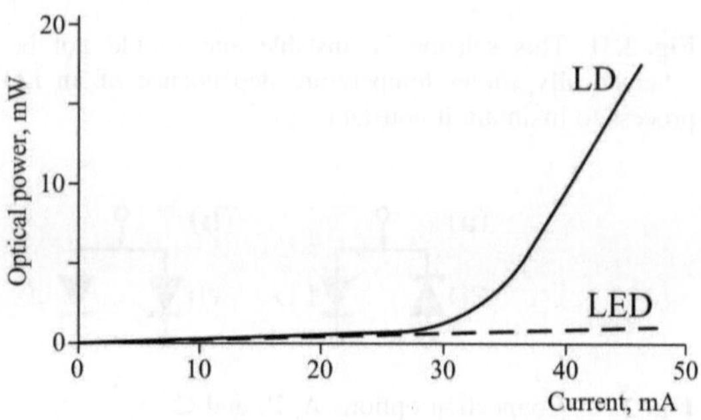

purpose of explanation only, consider the scheme in Fig. 2.31 that must never be used in practice. The LD current, whatever small it is, develops some heat on the diode, increasing its temperature. It causes drop of the optical power. After a minute or two after switching the circuit on, the user realizes that his LD is not bright enough (the first arrow down in Fig. 2.31), and trims the resistor to rise the optical power to the initial level (the second arrow up). The larger current now develops even bigger amount of heat. However, due to bigger temperature gradient relative to the ambient air or the heat sink, the heating rate is smaller than in the very beginning. Nonetheless, after some time, the trimming process needs to be repeated. Finally, the work day is over, and the user switches the mains off. The next day, he switches the LD on, and a terrible burst of light (long vertical pointing arrow to 20 °C) shows that something is wrong in the circuit. Even if the LD has not been fatally damaged, the trimming process needs to be repeated again. An opposite situation may occur when the LD current–voltage characteristic is more sensitive to temperature than the optical power: then the LD forward voltage quickly decreases with temperature, increasing the current and light output, and the trimming is needed to bring it down to the initial level. If this is the case, then in the morning the LD does not step over the threshold. Either way, the optical control loop is necessary to ensure stable operation. This is the reason why a tiny photodiode (PD) is always included in the LD case (Fig. 2.22). Its front surface is

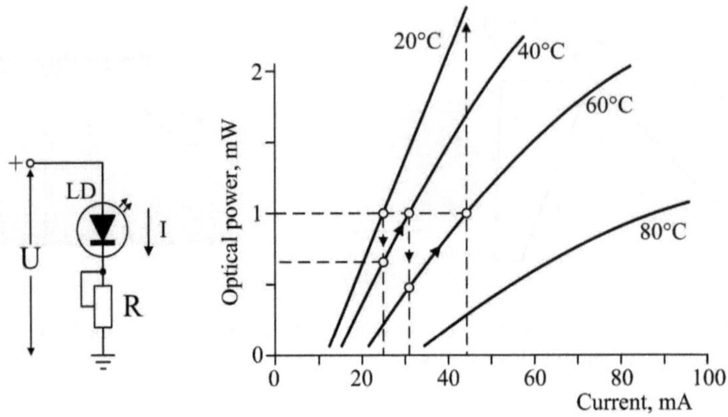

Fig. 2.31 This scheme is unstable and should not be used in practice. The plot *at right* schematically shows temperature dependence of an LD optical flux and manual adjustment process to maintain it constant

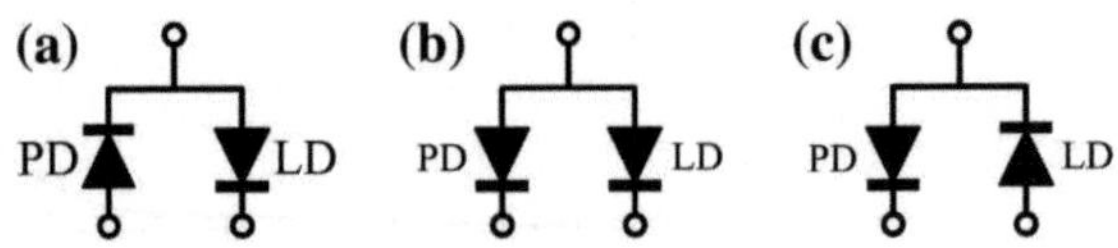

Fig. 2.32 Connection options A, B, and C

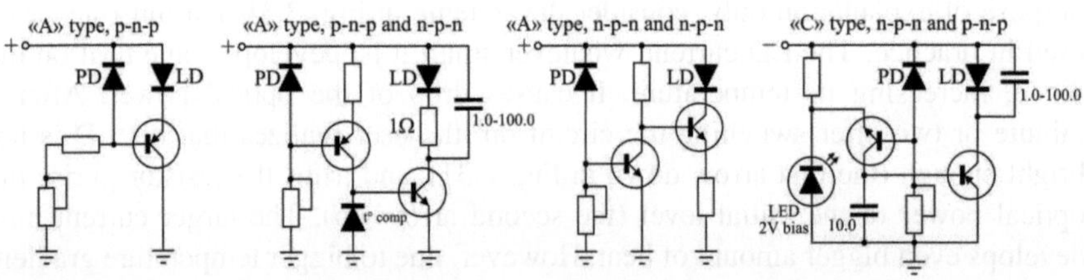

Fig. 2.33 PDs are optically coupled to LDs inside the case, and lock on the negative feedback, fixing the LDs output optical flux. Variable resistors tune the flux. Capacitors absorb first burst during switching on, and may be applied to all the circuits. Not all the circuits work equally well

tilted relative to optical axis, preventing possible coupling to the laser cavity. In order to minimize the number of leads from four to three, the laser diode chip and the photodiode are always connected, and there are three main options of this connection shown in Fig. 2.32. A series of simplest (but not the best) control circuits (drivers) is presented in Fig. 2.33. A better performance can be achieved with operational amplifiers, in more complicated schemes like the one shown in Fig. 2.34. The capacitor in the negative feedback of the operational amplifier sets the response time to about 100 ms, averaging all the possible noise, coming from

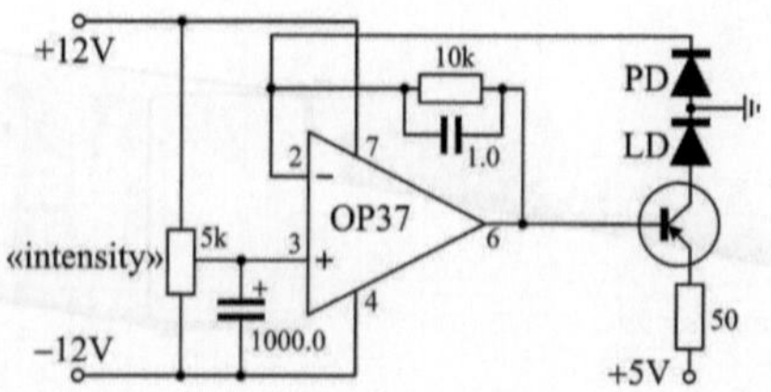

Fig. 2.34 As a design feature, the common point of the PD and LD is always electrically connected to metal case, which is the only part that can be placed in contact with the heat sink or the cooler. For safety reasons, it is always better to have bulky heat sink grounded. This is why the PD-LD common point is grounded in this scheme («C» type connecton)

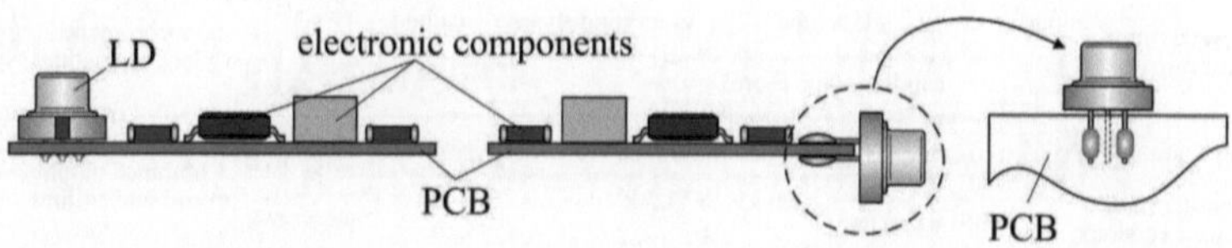

Fig. 2.35 Vertical (*at left*) and horizontal (*in the middle* and *at right*) options of mounting LDs on a printed circuit board (PCB)

the power supplies. The capacitor at the non-inverting input of the operational amplifier blocks sharp voltage bursts at the first moment of switching on.

Manufacturers of electronic components offer a variety of LD drivers in integrated circuit packages, suitable for each one of the three connection types A, B, or C. Even finalized printed circuit boards, with all the necessary electronic components installed, are available, ready for soldering the LD. The user has the options to connect the LD either with intermediate wires or directly to the printed circuit board (PCB) (Fig. 2.35).

2.4 Helium–Neon (He–Ne) Lasers

In visible domain, helium-neon (He–Ne) lasers operate at five wavelengths: 543, 594, 604, 612, and 633 nm, of which the line 632.8 nm is the strongest one. They are the sources of choice when highest spectral and polarization purity, coherence, and perfect beam geometry are needed. He–Ne lasers are commonly used in spectroscopy as reference and calibrations sources, in interferometry and holography, as well as in phase- and frequency-resolved measurements. Manufacturers offer two basic geometries: the round and the table-top ones (Fig. 2.36). In a He–Ne laser, the light-emitting component is neon, whereas helium serves to create population inversion at the neon atoms. The proportion of helium to neon is typically from 5:1 to 10:1, and the working pressure is between 2 and 4 Torr.

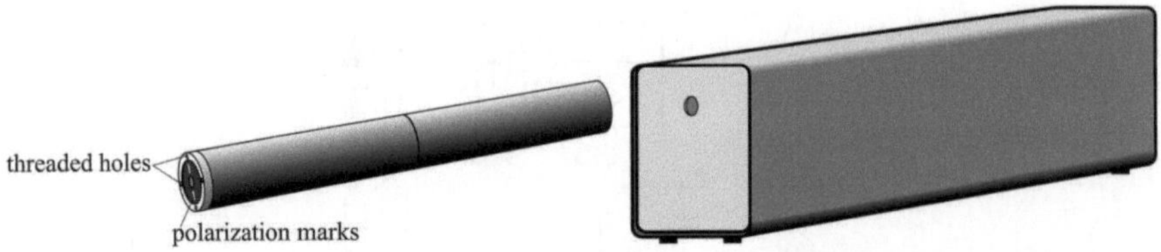

Fig. 2.36 If the beam of a table-top laser is polarized, then the polarization is always vertical. For the round-shaped lasers, interface flanges for beam expanders are in assortment. Some of the table-top options are also equipped with flanges for beam expanders, mainly of the C-mount type (see Chap. 1 for dimensions)

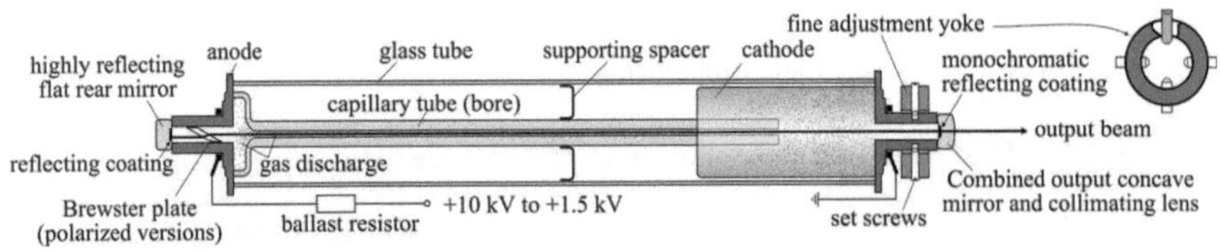

Fig. 2.37 Narrow capillary tube stabilizes the discharge and serves as the transversal mode selector for fundamental Gaussian TEM_{00} mode. Resonant multilayer dielectric cavity mirrors select the wavelength. Rear nearly 100 % reflecting mirror is always flat in order to reserve the opportunity of installing a Brewster plate for polarized versions of lasers. The output mirror (coupler) is always concave, formed by depositing narrow-band (monochromatic) reflection coating onto the concave surface of a weak concave-convex positive lens. Thus, the output window serves as both the cavity coupler and the collimating lens for the output beam. The output metal tubing may also have a narrow circular trench for fine adjustment of the coupler. For that, a circular yoke with four or three screws is put onto it, and the screws produce tiny deformations, bending the end to a desirable angle so as to maximize the output power

Special isotopes are used for utmost performance, although natural gases also produce stimulated emission but on a lower scale. A standard He–Ne laser without special stabilization electronics (stabilized lasers will be discussed in the next section) consists of the glass laser tube and a ballast resistor with wires (Fig. 2.37). Typical ignition voltage on the tube is about 10 kV, which should be downsized after ignition to about 1.5 kV. It is generally believed that the He–Ne laser produces linearly polarized output beam. However, it is not a rule: without the Brewster plate, the laser generates randomly polarized beam. The output power does not depend on whether the beam is polarized or not. It is a common mistake to consider the Brewster plate as a polarizer for the output beam because the beam transmitted through the Brewster plate is not totally polarized. Reflection curves at the interface air-glass for s- and p-polarized waves are shown in Fig. 2.38 as a function of the angle of incidence. At the Brewster angle, polarization parallel to the plane of incidence (p-polarization) does not reflect at the first interface. It is easy to compute that this wave, refracted into the glass, does not reflect from the second interface either (Fig. 1.37, angle of incidence 33.7°). Therefore, there is no p-component in the reflected beam, which means that the reflected beam is 100 %

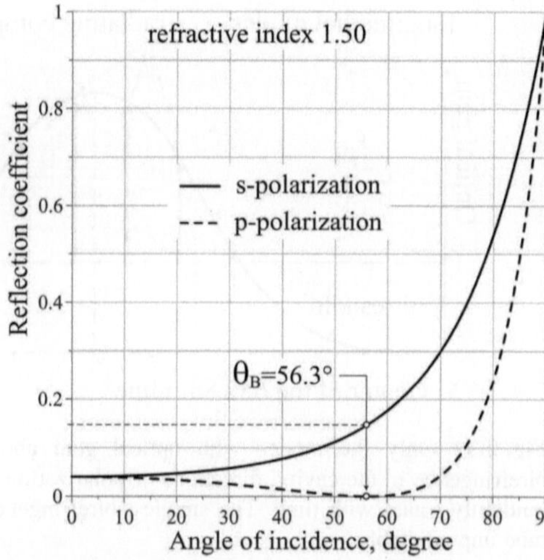

Fig. 2.38 Only ≈15 % of the p-polarized wave is reflected at the Brewster angle θ_B

s-polarized. However, this is not the case for the transmitted beam: it is composed of 100 % of p- and ≈(100–2 × 15) % = 70 % of s-polarized waves, which means that the Brewster plate is far not the polarizer for the transmitted beam. Nonetheless, this ≈30 % difference is enough to almost totally suppress stimulated emission for the s-polarized wave, making the laser beam well polarized, with the polarization ratio about 500:1. The most convenient place to install the Brewster plate is near the flat mirror, where radial displacement of the beam due to finite plate thickness does not influence the mode structure.

With the Brewster plate installed, all the stimulated emission power is radiated into a linearly polarized mode. Without the Brewster plate, the same power will be radiated randomly polarized, which means that polarization vector is randomly distributed between the horizontal and vertical directions. If someone wants to make linear polarization from randomly polarized laser beam, it will result in two negative consequences: the power will be two times lower, and additional amplitude noise will appear as the result of random projections of the polarization vector onto the polarizer axis.

A well-designed He–Ne laser always works at fundamental Gaussian mode TEM_{00}, and has the beam diameter 0.5–1 mm and divergence 1.5–2 mrad. As to the output power, it depends on the number of longitudinal modes. Each longitudinal mode satisfies the standing-wave condition and, therefore, is separated from its neighbors by frequency spacing

$$\Delta f \approx \frac{c}{2L},$$

where c is the speed of light and L— the cavity length (refractive index of the gas mixture is assumed to be approximately unity). The neon gain curve has full width at half-maximum (FWHM) of the order of 1500 MHz at room temperature and work pressure ∼5 mTorr (Fig. 2.39). Thus, for a 500 mm long (high gain) tube

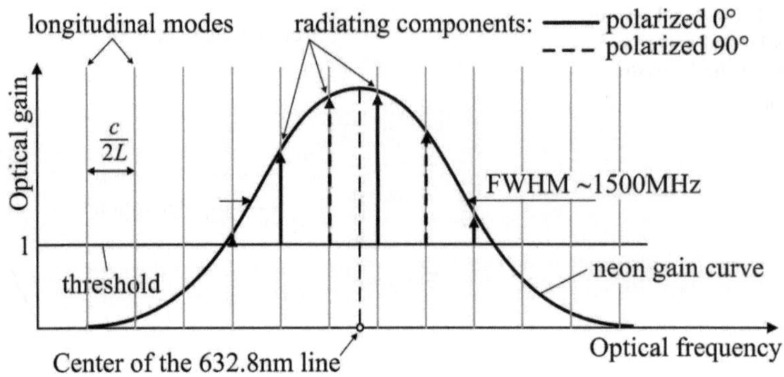

Fig. 2.39 Only the modes with optical gain above the threshold may radiate. With no birefringence in the cavity, directions of polarization are not fixed relative to the tube, and may randomly rotate with time. The smallest birefringence fixes polarization directions against the tube unpredictably

with its mode spacing of about 300 MHz, as many as 5–6 lines may radiate simultaneously. The most curious and unpredictable fact is that adjacent longitudinal modes have orthogonal polarization. For that, there are both theoretical backgrounds, like saturation-induced anisotropy of the medium, and practical explanations like birefringence of the multilayer mirrors. However, it is not a rule, and other experiments show equal polarization states for adjacent modes. Nevertheless, in those lasers where such polarization orthogonality does exist (including Zeeman lasers), frequency stabilization may be applied, as described in the next section. Obviously, the longer the laser tube is, the bigger are the gain and the number of longitudinal modes, contributing to output power. Figure 2.40 summarizes this result in a plot. Single-mode He–Ne lasers can hardly deliver more than 0.5–0.8 mW of optical power, while the multi-mode ones are capable of producing 15–20 mW at 633 nm. At 543 nm, even the multi-mode He–Ne laser cannot produce more than 0.5–1.5 mW.

Mode composition is directly related to laser coherence. The coherence length l_c is determined by the time interval τ, during which the phase of the wave does not change more than by π:

$$l_c \approx c \cdot \tau,$$

where c is the speed of light. In a single-mode operation, coherence length of an He–Ne laser may extend to a kilometer. However, for standard lasers with $m > 1$ longitudinal modes separated by Δf from each other, the entire spectral width $m \cdot \Delta f$ may be close to the width of the neon gain curve, i.e. about 1500 MHz. In this case,

$$l_c \sim \frac{c}{m \cdot \Delta f} \sim 20\,\text{cm}.$$

Fig. 2.40 Output power as a function of mode spacing. *Dots* represent available data from various He–Ne lasers; the *curve* shows the trend

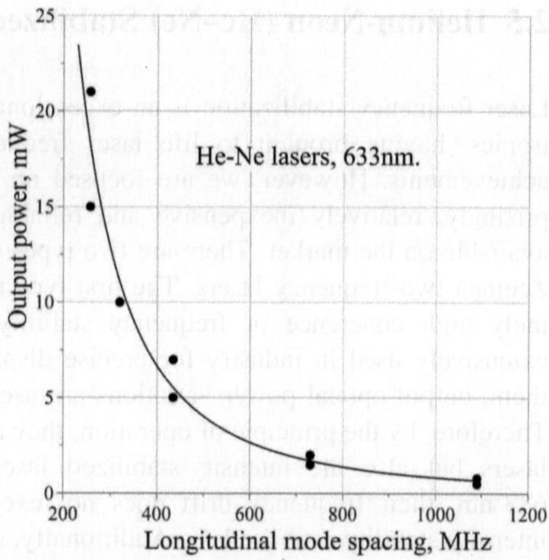

Fig. 2.41 The He–Ne and laser diode (LD) spectral curves as recorded by a spectrometer with 0.5 nm spectral resolution

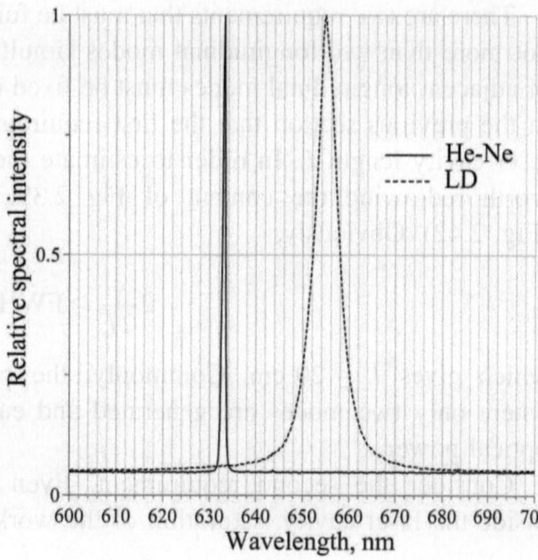

But even the entire neon gain spectral width ~0.002 nm is below spectral resolution of ordinary spectrometers. Therefore, the He–Ne lasers are a convenient tool for spectral calibration. Figure 2.41 compares spectral curves of a laser diode and a He–Ne laser.

2.5 Helium-Neon (He–Ne) Stabilized Lasers

Laser frequency stabilization is an exceptionally elaborated branch of opto-electronics, having brought to life laser frequency standards and other superior achievements. However, we are focused on routine laboratory equipment. Surprisingly, relatively inexpensive and reliable He–Ne stabilized lasers are also available on the market. There are two types of such lasers: single-frequency and Zeeman two-frequency lasers. The first type is good in applications where extremely high coherence or frequency stability are needed. The second type is extensively used in industry for precise displacement measurements. In both of them, output optical power variations are used to stabilize the optical frequency. Therefore, by the principle of operation, they are not only the frequency-stabilized lasers but also the intensity-stabilized lasers. With about 0.5 mW power at 633 nm, their frequency drift does not exceed 3 MHz during hours, and the intensity stability is about 1 %. Additionally, some single-frequency He–Ne lasers provide polarization ratio about 4000:1, i.e. an order of magnitude better than the ordinary ones. Consider this type first.

There are two requirements that must be fulfilled: the cavity length must sustain not more than two longitudinal modes simultaneously, and polarization direction of adjacent longitudinal modes must be fixed relative to the tube. It was explained in the previous section that the first requirement imposes the upper limit on the laser cavity length L. In order to estimate the maximum length of the tube, it is worth redrawing the content of Fig. 2.39 in the form relevant to our case (Fig. 2.42). Obviously,

$$2\frac{c}{2L} \geq \text{FWHM},$$

which gives $L \leq 20$ cm. Commonly, the range 17 cm $\leq L \leq$ 23 cm is used, where only two modes are generated and each of them delivers over 1 mW of optical power.

Consider the second requirement. Even without any optical imperfections inside the laser cavity, saturation of the working gas during stimulated emission

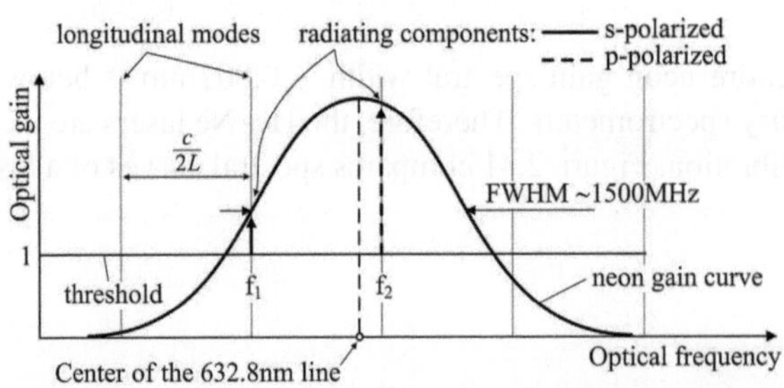

Fig. 2.42 In a stabilized He–Ne laser, not more than two longitudinal modes with frequencies f_1 and f_2 must exist within the neon gain curve

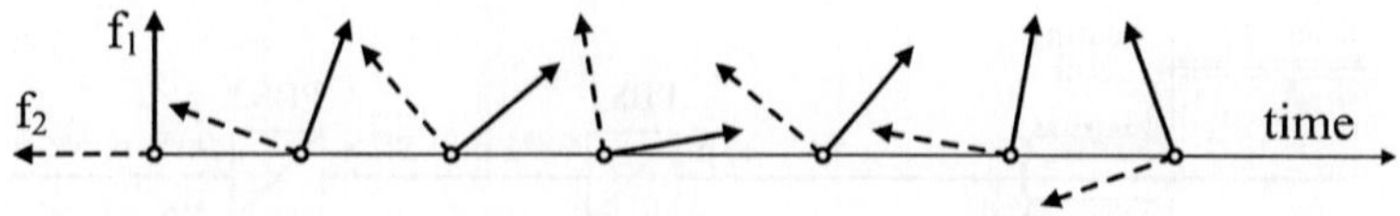

Fig. 2.43 Random evolution of orthogonally polarized mode pairs

creates infinitesimal birefringence, establishing polarization priorities to longitudinal modes. Therefore, the mode with f_1 will have one polarization direction, and the modes with f_2—the orthogonal one. However, there is no preferable direction for them in the laboratory system of coordinates, so that the orthogonally polarized mode pairs would arbitrarily evolve with time, like it is shown in Fig. 2.43, and it would be impossible to select them, using a fixed polarizer. On the contrary, a strong polarization selector inside the cavity, like the Brewster plate for example, will only override weak polarization priorities for different frequencies, and all the longitudinal modes with all the frequencies will be polarized in one direction, making it impossible to select any one of them. Luckily, cavity mirrors always display some weak after-manufacturing birefringence that is sufficient to establish predominant direction. This predominant direction can be identified after the laser is ignited and properly fixed relative to laser housing by rotating the laser tube.

With all the aforementioned, it is easy to understand the principle of single frequency stabilization as it is presented in Fig. 2.44. The control algorithm maintains equal intensities of the two longitudinal modes with the frequencies f_1 and f_2, which is possible only when they are localized symmetrically around the center of the neon gain curve. Clearly, f_1 and f_2 do not depend on the absolute value of the gain, and therefore do not depend on possible deviations from its nominal value, like ageing, for example. Thus, the laser beam is frequency stabilized. Of course, optical intensities, coming to the photodetectors, are different because they pass through quite different polarizing beam splitters. Therefore, respective electrical signals are scaled to equality by trimming a variable resistor in the non-inverting input of the operational amplifier. Proper positioning of longitudinal modes within the neon gain curve is accomplished by heating the metal tubing at one end of the laser tube, using thermal expansion to change the cavity length. Stabilization occurs only when the feedback is negative, which, in terms of Fig. 2.42, means that the heating coil must cool when intensity of the p-polarized mode is bigger than that of the s-polarized one. Then the cavity contracts, moving the two modes rightward and equalizing their intensities. If polarity of the photo-diodes is opposite to that, then the coil will be heated, increasing L, and the mode structure will contract. This is the positive feedback, and instead of equalization, intensities of the s- and p-polarized modes will diverge. However, this is not fatal because, after some time, the low-frequency s-mode will be pushed outside the neon gain dome and the new s-mode will be sucked from the right. When this happens, the s- and p-modes are in the right order, and the feedback is

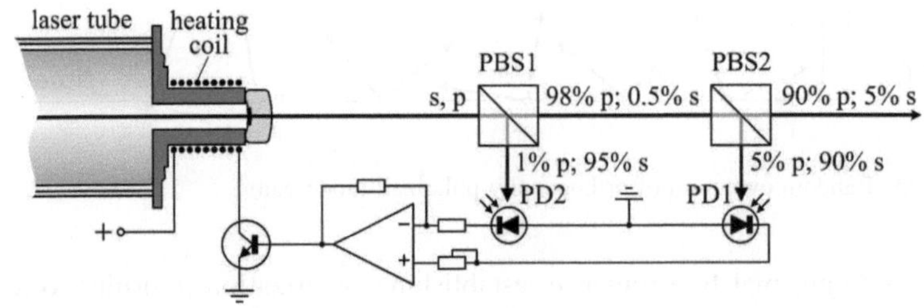

Fig. 2.44 Generalized schematic of frequency/power stabilization loop. Polarizing beam splitters PBS1 and PBS2 separate orthogonal polarizations and direct them towards the photodiodes PD1 and PD2 that convert optical power into electrical signals to control the cavity length via heating

negative again. Such flips over the instability are common for stabilized He–Ne lasers, and may occur several times during the warm-up process. The stabilized temperature of the heating coil may be around 80 °C.

Since applications of the stabilized lasers require high coherence and spectral purity, one of the two modes must be eradicated. For that, the first beam-splitter is designed to almost totally reflect the s-mode (horizontal polarization in Fig. 2.44) and the second one does almost the same but with the priority to reflect certain amount of p-mode as well. With the values of the Fig. 2.44, the output relative intensity of the s-polarized mode is about $5 \times 10^{-3} \times 5 \times 10^{-2} = 2.5 \times 10^{-4}$ and $0.98 \times 0.9 = 0.9$ for the p-mode, making polarization ratio ~4000:1–almost an order of magnitude higher than in a conventional He–Ne laser. The price for that is much higher cost.

In another type of single-frequency He–Ne lasers, even shorter tubes are used, in the range 9 cm $\leq L \leq$ 12 cm, keeping only one single longitudinal mode within the neon gain curve. The shorter tube leads to smaller output power less than 0.5 mW. Since there is no need to select any one mode, polarization may be random, and an ordinary non-polarizing beam-splitter samples the output beam (Fig. 2.45). Random polarization is not the only option, and with the Brewster plate inside the tube, the laser will produce linear polarization without sacrificing the output power.

As in the previous scheme, stabilization algorithm maintains the output power constant at the level set by the reference voltage U_{ref}. However, the same level of amplitude stabilization does not produce the same level of frequency stabilization as before (Fig. 2.46) because the optical gain may vary with time, influenced by ageing of the gas mixture or by instabilities of the discharge current. Nevertheless, such incomplete stabilization is better than no stabilization at all and, considering slowness of the drift, may be excused in those applications where only coherence and amplitude stability are important, like interferometry.

Zeeman lasers are low-power He–Ne lasers with axially applied magnetic field and a feedback for frequency stabilization. The laser cavity is chosen as short as to sustain only one longitudinal mode, therefore output optical power does not

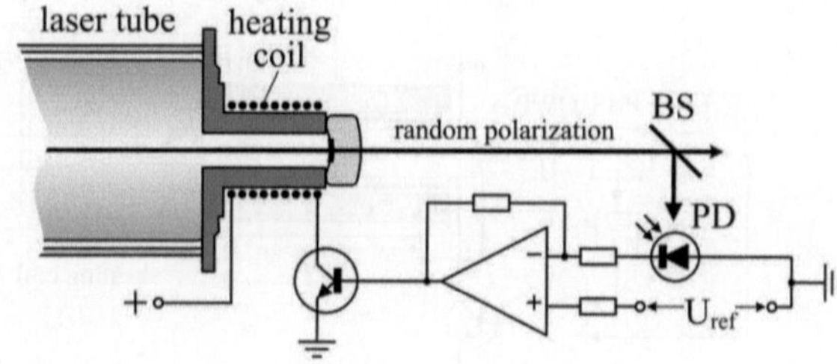

Fig. 2.45 In a single-mode cavity, only one photodiode (PD) may be used, coupled to a non-polarizing beam-splitter (BS). Reference voltage U_{ref} sets the working point for stabilization

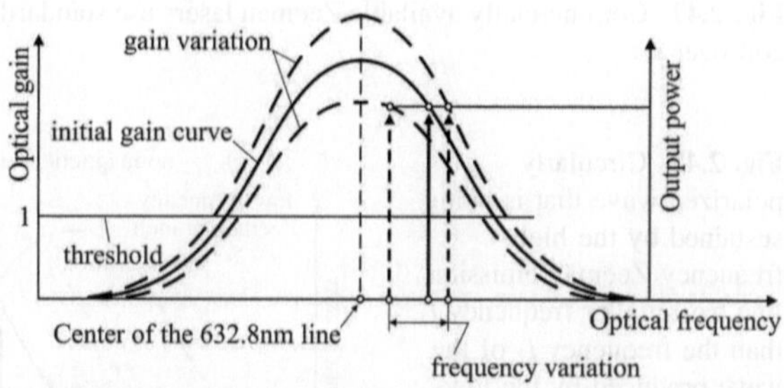

Fig. 2.46 In a single-mode scheme, constant output power does not guarantee constant frequency when the gain varies

exceed 0.5 mW. When axial magnetic field is applied, the originally single neon emission line splits in two separate lines with the frequency difference about several hundred megahertz and opposite circular polarizations. This is the Zeeman effect, which also creates infinitesimal differences in refractive indices n_1 and n_2 for the oppositely polarized waves. Then the standing-wave condition for the only one longitudinal mode

$$m \cdot \lambda = 2L,$$

where λ is the wavelength, L—length of the laser cavity, m—an integer number, gives the same λ for the two oppositely polarized standing waves but slightly different frequencies f_1 and f_2 for them:

$$f_{1,2} = \frac{c}{\lambda \cdot n_{1,2}},$$

where c is the speed of light. For commercially available Zeeman lasers, this mode split $|f_1 - f_2|$ lies in the range 1–3 MHz. It is of a primary importance for interferometry and heterodyning that the two independent waves are of the same mode structure and travel same paths, experiencing same optical heterogeneities, and therefore, are supposed to have identical wavefronts.

Combined frequency and amplitude stabilization is performed by comparing intensities of the two waves, traveling inside the laser cavity (Fig. 2.47). These waves, originally circularly polarized inside the cavity, are transformed into linearly polarized waves with the help of a quarter wave plate (QWP) (Chap. 5) and split into two at the polarizing beam splitter (PBS). Photo-detectors PD_1 and PD_2 measure intensities of the two waves, and the differential signal controls the

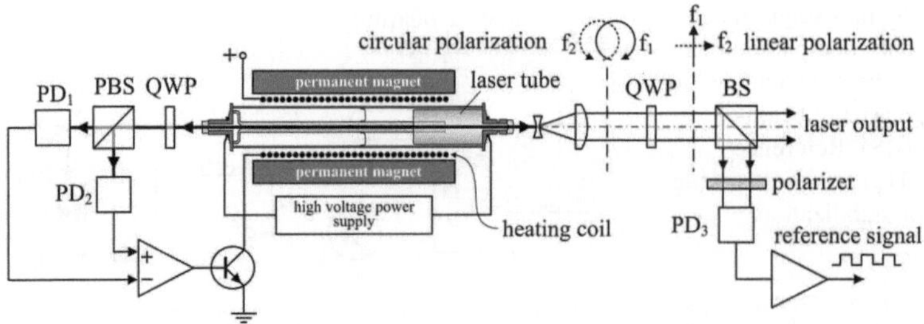

Fig. 2.47 Commercially available Zeeman lasers use standard He–Ne laser tube with the heating coil over it

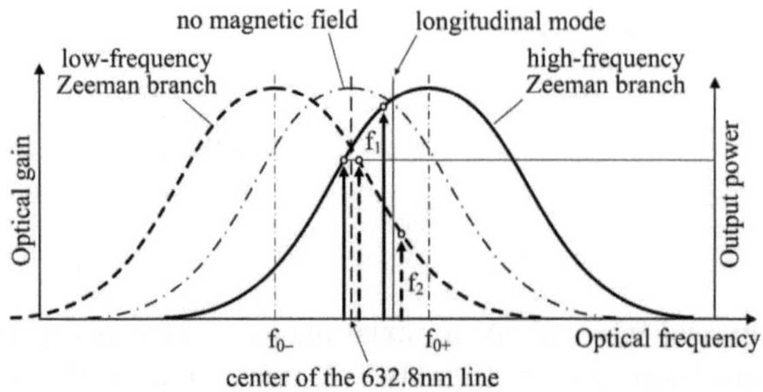

Fig. 2.48 Circularly polarized wave that is being sustained by the high-frequency Zeeman emission line has smaller frequency f_1 than the frequency f_2 of the wave produced by the low-frequency Zeeman emission line

current through the heating coil, maintaining the length of the laser cavity so as to equalize intensities of the two components (Fig. 2.48). Without magnetic field, the gain curve of the width ~1500 MHz is centered at 632.8 nm. A single longitudinal mode is supposed to be somewhere within it, not necessarily in the middle. Axial magnetic field splits the neon emission line in two: the low-frequency line centered at f_{0-} and the high-frequency one centered at f_{0+}. The Zeeman split $f_{0+} - f_{0-}$ is of the same scale as the gain curve itself. For the He–Ne lasers, the splitting constant is equal to 1.82 MHz/Gs, which gives $f_{0+} - f_{0-} \sim 500$ MHz for the typical field of about 300 Gs. Thus, the Zeeman split is two orders of magnitude bigger than the mode split:

$$|f_{0+} - f_{0-}| \gg |f_1 - f_2|.$$

Each of the two modes with frequencies f_1 and f_2 is amplified either by the low- or by the high-frequency Zeeman gain branches. Suppose that f_1 mode is produced by the high-frequency Zeeman branch, and f_2—by another one. Now it is time to ask what is bigger: f_1 or f_2? This is far not the trivial question, which requires the Kramers–Kronig relation, establishing a very general form for the dependence of the refractive index n on frequency f:

$$n(f) - 1 \approx -(f - f_0) \cdot \alpha$$

where f_0 is the center of the spectral line and α—a factor, depending on absorption. Strictly speaking, α also depends on frequency, but on much lower scale. Now, the frequencies f_1 and f_2 can be estimated with respect to each other:

$$f_1 = \frac{mc}{2L \cdot [1 - \alpha(f_1 - f_{0+})]}, \quad f_2 = \frac{mc}{2L \cdot [1 - \alpha(f_2 - f_{0-})]}.$$

Introducing positive quantities

$$\Delta_+ = |f_{0+} - f_1| \text{ and } \Delta_- = |f_2 - f_{0-}|,$$

and using smallness of $n-1$, one obtains:

$$f_1 \approx \frac{mc}{2L}(1 - \alpha\Delta_+), \quad f_2 \approx \frac{mc}{2L}(1 + \alpha\Delta_-).$$

Thus,

$$f_1 < f_2 .$$

This configuration is shown in Fig. 2.48. Thermal expansion of the cavity shifts both f_1 and f_2 to the stabilization point in the middle of the superposition of the low- and high-frequency Zeeman spectral lines. On its way to this point, the amplitude of the lower frequency component f_1 decreases, while that of the higher frequency f_2 increases. Equal intensities correspond to constant frequencies f_1 and f_2 if only magnetic field is constant. Stability of the magnetic field is very important for frequency stability. For example, any massive magnetic parts on the optical table, positioned close to the laser, may noticeably change its frequency split $f_2 - f_1$.

To provide a reference for phase measurements, a non-polarizing beamsplitter (BS) (Chap. 1) directs a small portion of the output linearly polarized beam to the third photo-detector PD_3. Being orthogonally polarized, these two waves do not interfere on the photo-detector until they are coupled by a polarizer.

Zeeman lasers present several advantages, especially when used in optical heterodyne interferometry. The first is that the two interfering beams have same optical axes and coinciding wavefronts. This feature is crucial for heterodyne technology since any wavefront misalignment leads to unreliable phase measurements. The phase difference map measured in the cross-section of the two interfering waves is presented in Fig. 2.49. It shows that the maximum phase shift is less than 6° across the beam of 6 mm in diameter, being on the average even smaller—~2°. The second advantage is that the frequency split of several megahertz is small enough to be easily processed by conventional electronics. The third advantage is that stability of the output intensity and frequency is much better than that of a non-stabilized laser.

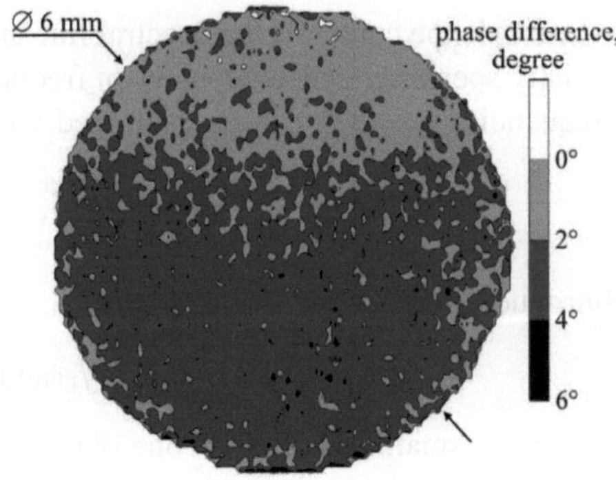

Fig. 2.49 Cross-section map of f_1 and f_2 phase difference in a Zeeman laser

2.6 Helium-Cadmium (He-Cd) Lasers

When a compact table-top source of moderately coherent (~10 cm) blue or UV radiation is needed, then He-Cd laser, emitting at either 441 or 325 nm, is the best choice for laboratory applications. Its geometry is basically the same as that of a table-top He–Ne laser (Fig. 2.36 right), however power requirements are much higher, which is necessary to maintain evaporation of metallic cadmium. Output optical power of about 40 mW at 441 nm and 15 mW at 325 nm is typical for both the random and linear polarization options.

2.7 Tunable Lasers

Regretfully, there are no laboratory-size lasers whose wavelength can be continuously tuned in visible domain. However, spectral tuning from one discrete spectral line to another, covering relatively wide range of wavelengths in visible, is quite possible with argon lasers (Fig. 2.50). In them, stimulated emission takes place on electron transitions in argon ions, basically in singly ionized argon. In order to extend tuning range to the red wing of the spectrum, krypton-argon mixture may be used. An adjustable intra-cavity prism, commonly the Littrow prism (Chap. 1, Fig. 1.31), serves as a spectral selector. A micrometer screw is used to manually adjust the wavelength. With the prism angle about 30°, the incidence angle is close to the Brewster angle, which creates ~15 % preference for the p-polarized wave (polarization in the plane of incidence), sufficient to suppress the orthogonal polarization and to ensure linear polarization of the output beam. The gain curve of argon ion transitions is around 2500 MHz, so that the number of longitudinal modes is bigger than in a He–Ne laser, making coherence poorer: of the order of several centimeters. Some manufacturers claim coherence of argon lasers of the order of 10 cm.

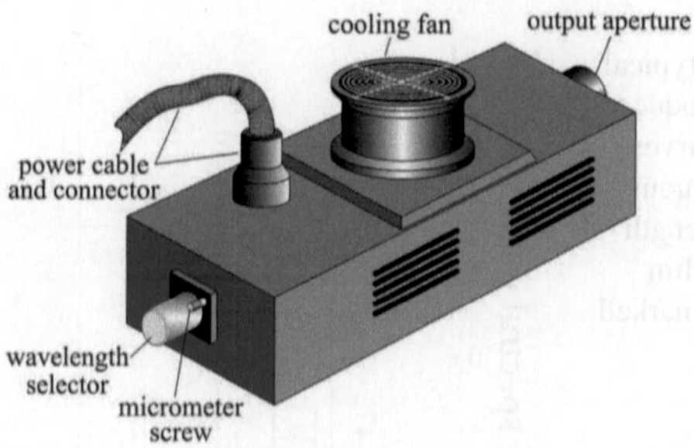

Fig. 2.50 Even the smallest air-cooled argon lasers are much bigger than He–Ne lasers. Thick power cable connects the laser to the power supply. Cooling fan creates a lot of acoustical noise and vibrations

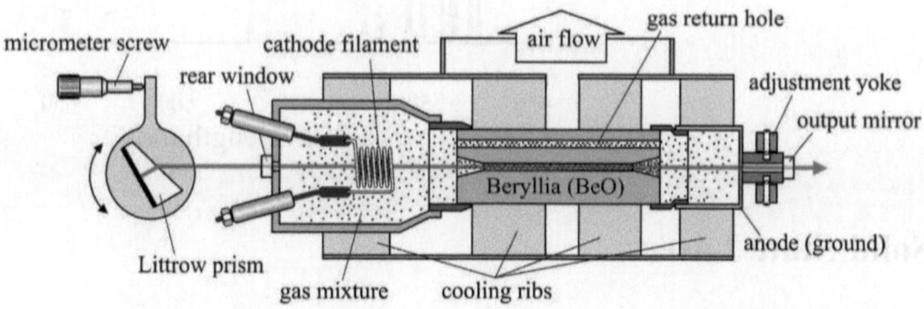

Fig. 2.51 Generalized structure of an argon ion laser. Sometimes, additional magnetic coil is used to create strong axial magnetic field during start, contracting plasma to the center of the capillary and assisting to initial ignition

Argon ionization and pumping population inversion requires much more current than for He–Ne laser: $\sim$10 A in contrast to maximum 10 mA in a He–Ne laser. Only a heated cathode filament can produce such a big current. With around 100 V voltage drop on the tube, this develops about 1 kW of heat, which must be evacuated from relatively small volume inside the laser housing. For this reason, even small argon lasers are always equipped with powerful and noisy fans that also transfer vibrations to optical tables they stand upon. Therefore, argon lasers are more suitable for illumination purposes rather than for precise optical work.

Domination of the fundamental transversal mode TEM_{00} is supported by a 0.5–0.8 mm capillary bore in a beryllia (BeO, beryllium oxide) ceramic tube—one of the very few materials that can withstand argon plasma temperature and effectively conduct the heat to the outer walls (Fig. 2.51). Beryllium oxide conducts heat five times better than any metal.

Adjusting the micrometer screw, it is possible to select one spectral line at a time. Figure 2.52 gives a good impression of how many lines can be selected from a typical argon-krypton laser. With such a wide spectral coverage, maximum power from one line does not exceed 20 mW, although other narrow-band model may deliver up to 200 mW at the strongest lines, like 514 nm.

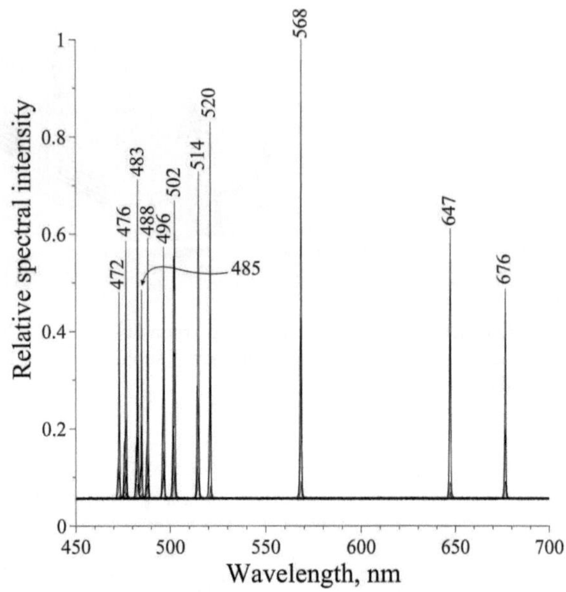

Fig. 2.52 Continuous scan over all the lines of a typical argon-krypton laser produced this picture. Several curves recorded during continuous adjustment of a wavelength selector are visible within each line. Each line is marked by its wavelength in nanometers

2.8 Solid-State Lasers

Solid-state lasers, including the most widely used neodymium-doped yttrium aluminium garnet (Nd:YAG) lasers, are more the industrial tools rather than the light sources of an opto-electronic laboratory. However, some fine analytical techniques, like laser-induced breakdown spectroscopy, depend on this type of lasers, which is the reason to dwell upon them. First of all, the Nd:YAG lasers are water-cooled devices, i.e., hoses, coolant canisters and a pump module must be somewhere around it. Some manufacturers offer integrated power-supply, control, and cooling units, which greatly simplifies deployment of the entire laser system. Commercially available Nd:YAG lasers generate $\sim$50 mJ at the wavelength 1064 nm in approximately 10 ns pulse train with repetition rate up to 20 Hz. Such a beam, being focused on any material, produces evaporated plasma plume, emitting characteristic spectrum that may be used to identify chemical composition of the material. This is essentially the principle of a laser-induced breakdown spectroscopy. To produce such short and energetic pulse, the resonator losses are modulated in phase with optical pumping—a technique commonly referred to as resonator quality switching or Q-switching (Fig. 2.53). Front end of the external or internal triggering pulse fires the lamp that pumps population inversion at neodymium energy levels (Fig. 2.54). The population inversion increases to its maximum during some short time $\sim$150 μs after the lamp has fired. During this interval, there is no additional retardation (phase difference for the two orthogonal linear polarizations) in the lithium-niobate crystal that is the heart of the Q-switch. Linearly polarized wave passes through it, preserving polarization, passes through

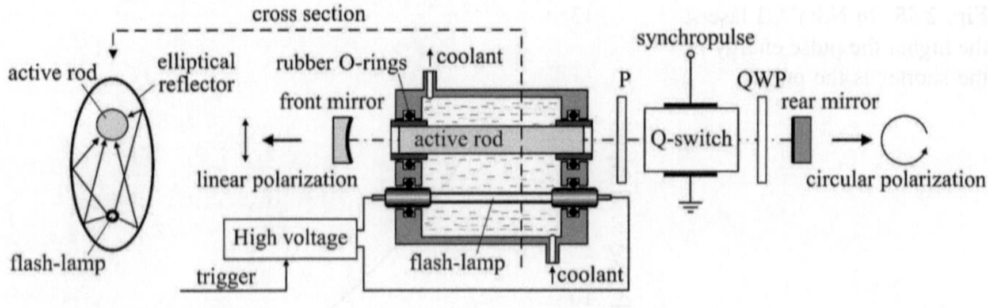

Fig. 2.53 For better efficiency, internal chamber is elliptical in cross section, having both the active rod and the lamp in its opposite foci. The ends of the active rod are permanently sealed into precisely machined metal tubings, making replacement easy. Tightly fit rubber O-rings prevent coolant leakage. The lamp is also replaceable. «P» stands for «polarizer» , «QWP» –for «quarter wave plate» (Chap. 5). The laser beams, emerging from the front and rear mirrors, are polarized differently: linear at the front and circularly at the rear mirrors. Specification always identifies polarization of the output beam

Fig. 2.54 Typical Nd:YAG laser cyclogram

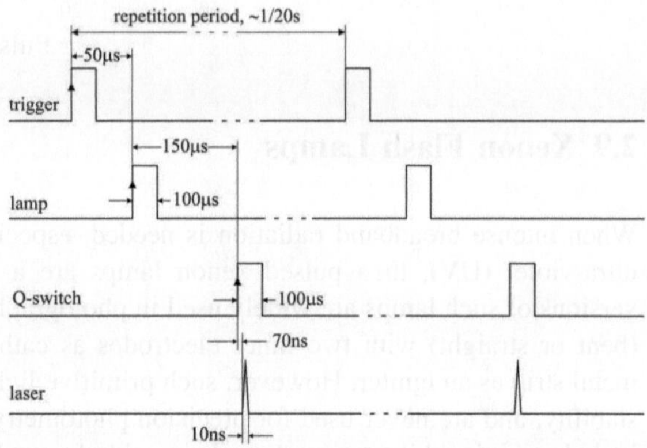

the quarter-wave plate (QWP), changing polarization to the circular one, and reflects back from the cavity mirror. At the way back, the wave transforms to linear polarization orthogonal to the initial one, and the polarizer stops it. The cavity is blocked, preventing premature stimulated emission. At the moment when population inversion reaches its maximum, the Q-switching synchropulse activates additional retardation in the lithium-niobate crystal, so that the reflected wave transforms to initial polarization direction and passes through the polarizer. The cavity is unlocked, and short strong optical pulse is generated. Carefully chosen delay time between the lamp and Q-switch pulses is essential for maximum output power. Therefore, some manufacturers offer the function of its manual adjustment. It is also worth mentioning that the width of the laser pulse, ~ 10 ns, is a complicated function of intra-cavity kinetics, and is not a constant value. In particular, it depends on the pulse energy (Fig. 2.55).

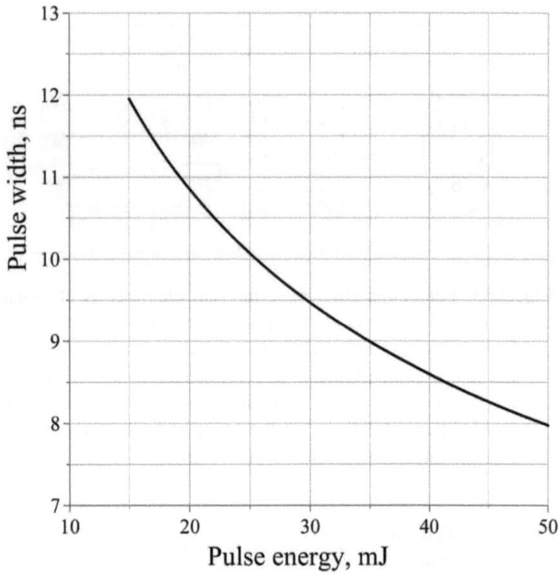

Fig. 2.55 In Nd:YAG lasers, the higher the pulse energy is, the shorter is the pulse

2.9 Xenon Flash Lamps

When intense broadband radiation is needed, especially with strong emphasis in ultra-violet (UV), then pulsed xenon lamps are a proper choice. The simplest versions of such lamps are widely used in photography and industry: a quartz tube (bent or straight) with two inner electrodes as cathode and anode and an outer metal strip as an igniter. However, such primitive light sources suffer from low arc stability, and are never used for precision photometry. High-stability xenon lamps include additional igniters and sparkers, with the total number of electrodes four or even five. Among all the variety of form-factors available on the market, the compact front-emitting lamps are very convenient for laboratory applications (Fig. 2.56). Despite their smallness, xenon flash lamps are physically complicated devices, requiring sophisticated high-voltage electronics for ignition. In practice, rarely someone braves to make his own high-voltage transformer to ignite the xenon lamp. Understanding that, manufacturers offer assembled flash-lamp modules with standard connectors for low-voltage power and triggering (Fig. 2.57). The only that the customer needs to do is to choose the options, and the following information helps to do it rightly.

The four principal electrodes of a high-stability flash lamp are the cathode, anode, igniter and sparker (Fig. 2.58). The idea of the controlled ignition is to create pre-determined ionized canal between the cathode and anode. There are two separate branches of the electrical circuit: the trigger circuit with high-voltage transformer for ignition and the main discharge circuit for creating powerful light pulse. The trigger circuit is powered by positive voltage $U_{tr} \approx$ 100–300 V, and

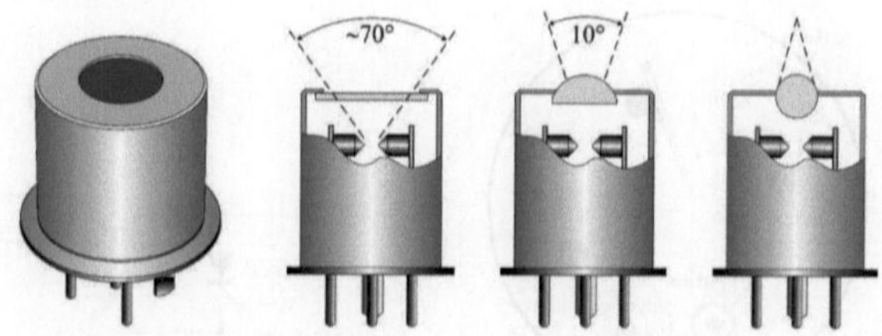

Fig. 2.56 Xenon flash lamp with three types of windows

Fig. 2.57 Common packages of xenon flash lamps

the main discharge circuit is connected to a higher voltage source $U_d \approx 0.4$–1 kV. In the flash-lamp module with a single power line, these two voltages are created by two different switching generators.

In static position, some initial positive voltage is already spread between the anode and the igniter through the resistor divider and the diode. This voltage drop is insufficient to ignite the arc. When the triggering pulse comes to the thyristor, it discharges the capacitor through the transformer, creating 5–7 kV pulse in the secondary coil. This voltage is then applied simultaneously to the sparker, igniter, and anode. In the sparker, due to small separation between the central electrode and the metal cylinder filled with dielectric ceramic, peak electric field is higher than between the igniter and the cathode. Therefore, the sparker ignites first, ionizing the gas and creating an UV radiation. The UV radiation releases photoelectrons from the cathode, igniter, and anode, thus preparing the ionization path from the cathode to the igniter and next to the anode. Now the igniter can fire, creating well-ionized canal to the cathode, and concentrating the U_d voltage drop only on the shorter part of the gap between the anode and the cathode. The electric field here rises to the breakdown value, and the main discharge fires between the anode and cathode, applying all the electrical power stored in the discharge capacitor C_d to the arc. In high-power xenon lamps, the second, third, and even the fifth igniter may be used to create well-determined discharge canal (Table 2.3). After the light pulse has been ignited, it continues until the discharge capacitor supplies current. Therefore, light pulse duration is approximately proportional to C_d capacitance (Fig. 2.59). Maximum repetition rate is inversely proportional to C_d, varying from 10 Hz to 1 kHz. Xenon flash lamp spectrum has two distinct maxima: in UV around 300 nm, and in visible around 500 nm (Fig. 2.60). The UV portion increases with U_d. Maximum electrical input to the lamp in a single pulse may be up to 1 J, but the average power dissipation must not exceed its maximum value, which is of the order of 50 W. For small xenon lamps, typical values are 40 mJ of electrical pulse and 100 Hz repetition rate.

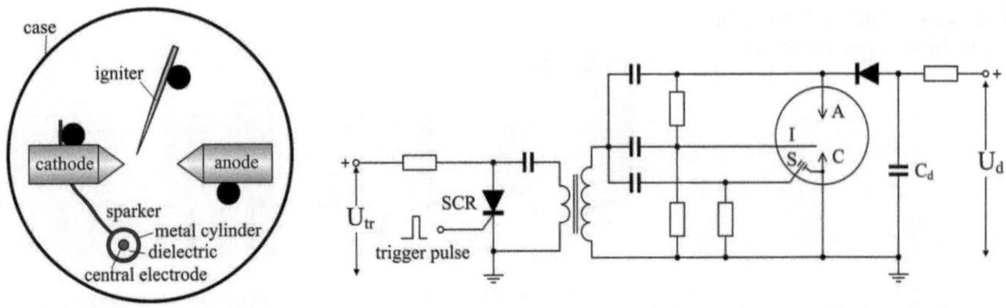

Fig. 2.58 Xenon flash lamp design and electrical circuit. The «A», «C», «I», and «S» stand for the «anode», «cathode», «igniter», and «sparker». The SCR is a standard abbreviation for the thyristor: silicon-controlled rectifier

Table 2.3 Typical number of igniters used in xenon flash lamps

Arc size (mm)	Number of igniters
1.5	1
3	2
8	5

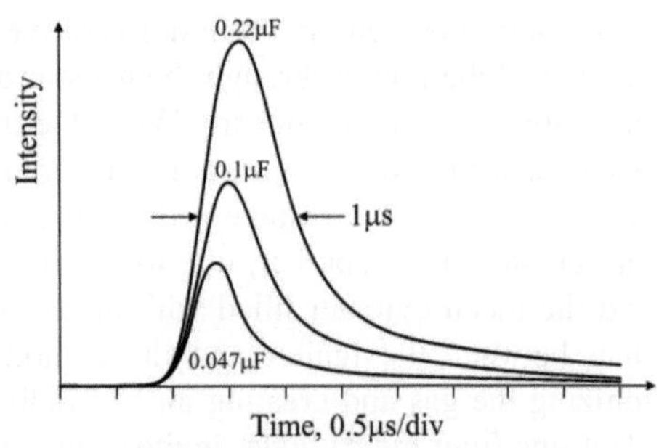

Fig. 2.59 Light pulses of a xenon lamp with different C_d. $U_d = 1$ kV

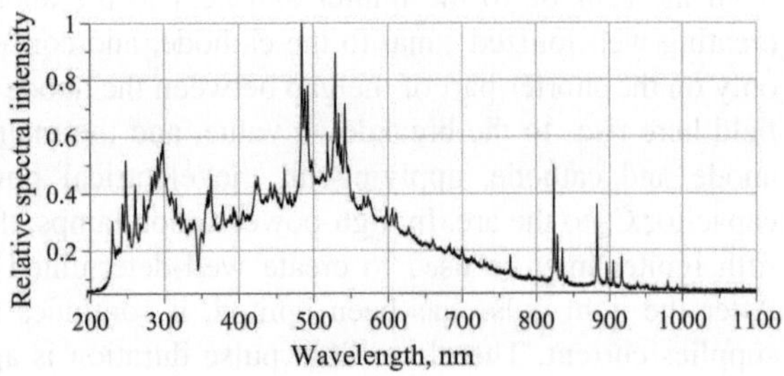

Fig. 2.60 Typical xenon lamp spectrum

List of Common Mistakes

- soldering wires directly to pins of a tungsten halogen lamp;
- leaving fingerprints on tungsten halogen lamp bulb;
- different light-emitting diodes connected in parallel;

- connecting laser diodes to a voltage source without a driver;
- using a polarizer to produce polarized beam from randomly polarized He–Ne laser.

Further Reading

E.F. Schubert, Light-Emitting Diodes, 2nd ed., Cambridge University Press, 2006.
A. Yariv, Quantum Electronics, 3rd ed., John Wiley and Sons, 1989.
M.A. Heald, J.B. Marion, Classical Electromagnetic Radiation, 3rd ed., Dover Publications, 2012.

Chapter 3
Photoreceivers

If you need a photodetector not only as a safety switch, then inevitably it becomes a key component of your new system together with an amplifier and data-acquisition electronics. Careful choice is crucial.

Abstract Physics, functionality, design concept and electronic circuits of photoreceivers are explained in eight sections. The chapter opens with photodiodes—the most reliable, compact, and inexpensive sensing component of photoreceivers. Physics of a *pn*-junction and carrier separation mechanism of light detection are carefully explained. The basic knowledge of electrodynamics is necessary to understand this subject. Electrical field distribution inside the *pn*-junction, open and short circuit characteristics of a photodiode are explained in detail. Advantages and disadvantages of photovoltaic and reverse bias connections are clearly identified. Current–voltage curves determine linearity and sensitivity of response. Functionality and field distribution inside the PIN structure—widely used type of photodiodes—is explained. Antireflection coating increases quantum efficiency up to 90 %. Simple theoretical formulas help to determine quantum efficiency from experimentally measured spectral sensitivity curves. Phototransistors, although being highly sensitive elements, are highly non-linear as well, therefore they are mostly suitable for switching applications. Avalanche photodiodes possess inner amplification but suffer from instability and noise. Multi-element sensors are typically used in beam stabilization circuits, like the one described in Chap. 10. Electronic read-out circuit is a very important component of a photoreceiver, and several practical recommendations and detailed schemes are presented in the fifth section of the chapter. Moving from the simplest switching solutions to highly linear circuits with operational amplifiers, this section introduces the concepts of negative feedback, reverse and zero bias schemes, cut-off frequency, gain-bandwidth product, and transient response. Some practical tricks are suggested to improve frequency characteristic of the photodiode receiver. Practical schemes for connecting multi-element sensors are also examined. Operational amplifiers, although being very efficient instrument for building photodiode receivers, are not the only solution, and several high-speed circuits on discrete transistors are presented. Helping to avoid bitter mistakes, this section also presents typical examples of wrong circuits that do not work. Photomultipliers (PMTs)—the most sensitive photodetectors based on inner amplification—are explained in detail in the sixth section. Some mathematics and knowledge of statistics are needed to explain why inner amplification increases signal-to-noise ratio. Special features, like the total

V. Protopopov, *Practical Opto-Electronics*, Springer Series in Optical Sciences 184,
DOI: 10.1007/978-3-319-04513-9_3, © Springer International Publishing Switzerland 2014

internal reflection prism, non-linearity on saturation, and inner design of compact PMT modules, are presented. Microchannel plates (MCPs) are the subject of the seventh section. This information helps to understand performance of the gated intensified spectrometers in Chap. 9. The chapter ends with PMT receivers—complete devices, incorporating both the PMT sensor and subsequent electronics. This section recommends the simplest and most reliable transimpedance solution based on operational amplifiers. Very essential rules must be observed in order to avoid common mistakes that make the entire system not working.

3.1 Photodiodes

Photodiodes (PDs) are small, least expensive, and most reliable opto-electronic elements. Available in very few standardized cases (Fig. 3.1), they may contain single-, two-, four-, or multi-element configurations.

Although all the PDs are essentially the semiconductor *pn*-junction diodes, their design differs from the design of traditional rectifying diodes in a way to maximize light absorption efficiency. Figure 3.2 shows how it works. Metallurgical contact between the *p*- and *n*-doped silicon layers creates electrical field that separates positive (holes) and negative (electrons) carriers, depleting the nearby region of initially existed free electrical carriers. Thermal generation may be considered to be negligible, as it will be shown below. With no photons, coming to the structure, there is no current in the outer circuit. Photons absorbed within the depletion region generate new carriers, being quickly separated by the electrical field, thus contributing to the current in the outer circuit. This is the so-called photovoltaic regime of the PD, when no external voltage is applied, and the PD itself acts as a source of current. Photons absorbed outside the depletion region (*p*-layer or *n*-bulk silicon) generate holes and electrons that are not separated from each other quickly, tend to remain close to each other due to Coulomb force, and eventually

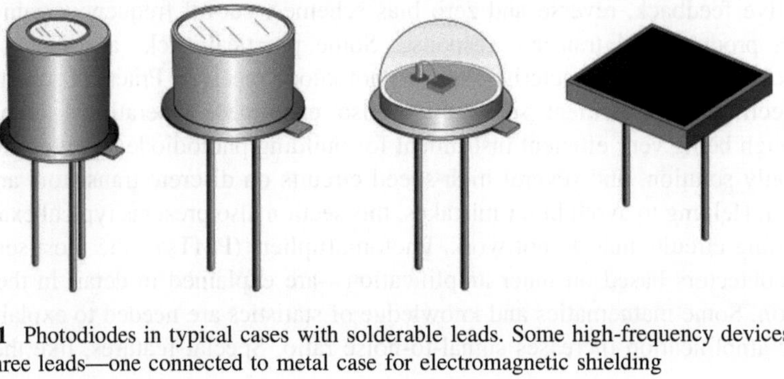

Fig. 3.1 Photodiodes in typical cases with solderable leads. Some high-frequency devices may have three leads—one connected to metal case for electromagnetic shielding

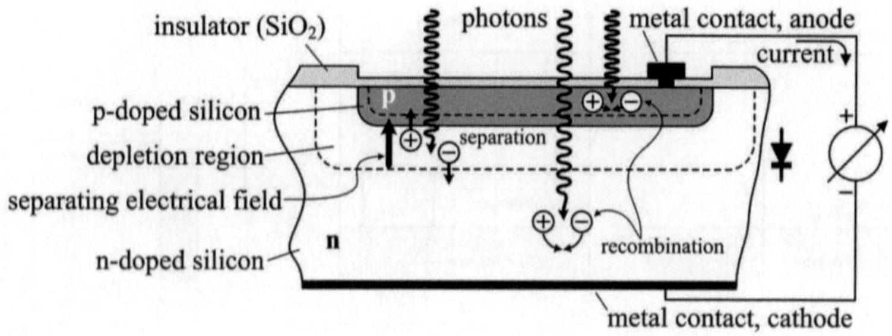

Fig. 3.2 Basic features of a PD structure. Diffusion-created *p*-layer is relatively thin, about few micrometers only, producing least possible absorption of photons. Separating electrical field inside the *pn*-junction directs positively charged carriers (*holes*) to anode, and negatively charged (*electrons*)—to cathode

recombine, making no contribution to the outer current. Thus, the depletion region should be thick and the front *p*-layer thin for better efficiency of a PD. For clear understanding of this subject, a deeper insight into the physics of a *pn*-junction is needed.

In the *p*-type silicon, boron (B) atoms are embedded into silicon crystalline structure by means of diffusion. Boron has three electrons at the energy level with quantum number 2, and therefore can accept five more electrons at this level to fill the electron structure to the maximum electron number of 8. Silicon atoms, surrounding the boron atom, display the valence four, making four bonds with the neighbors and with the boron atom as if it is the silicon atom. Thus, the boron atom can accept one more electron, and therefore it is called acceptor. The energy of this fifth electron is slightly above the energies of the four other electron bonds. This energy difference is around 0.02–0.05 eV $\approx$ 3–8 $\times$ 10^{-21} J, which is about kT $\approx$ 4.1 $\times$ 10^{-21} at room temperature. Therefore, almost all the acceptor energy levels are populated at room temperature (Fig. 3.3). When any electron from the valence band is excited to the acceptor level, it leaves a positively charged vacancy called hole. Electrical neutrality requires that the number of holes equals the number of acceptors. However, while acceptors are fixed in the crystal lattice and cannot move, the holes can move like electrons, but with lower mobility.

In silicon, the energy band, corresponding to freely moving conduction electrons, lies 1.1 eV = 1.76 $\times$ 10^{-19} J above the valence band, which is three orders of magnitude above the kT, thus leaving conduction band unpopulated. The *n*-type silicon is doped with phosphor (P) or arsenic (As), which are the five-valence elements. Therefore, for them, one electron bond is excessive in silicon crystal lattice. The energy difference between this one excessive electron and the conduction band is on the same order of 0.02–0.05 eV, therefore it may be easily donated to the conduction band at room temperature, leaving one positively charged hole after him (Fig. 3.3). Again, this hole is fixed while the excited electron can freely move in the above-lying conduction band.

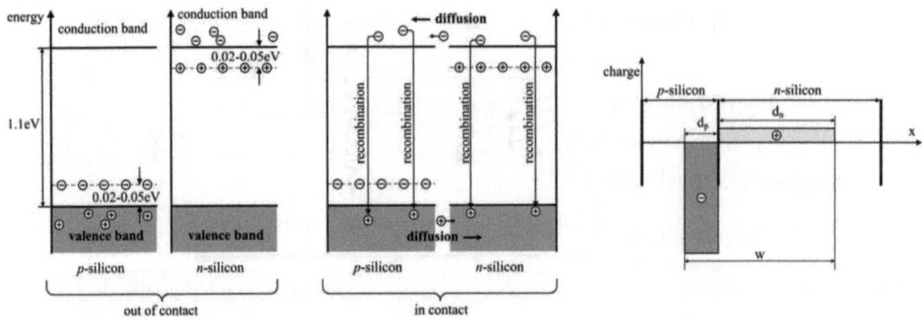

Fig. 3.3 In a disconnected (unbiased) and not illuminated *pn*-photodiode, negative charge inside the *p*-layer is created solely by the acceptors, while positive charge in the *n*-layer—solely by donors. The charged region is devoid of free carriers due to electron–hole recombination. Positively and negatively charged volumes are designed to be of different thickness in order to make maximum total thickness w and minimum thickness of the front *p*-layer

After the *p*- and *n*-doped materials are connected, diffusion of free carriers begins: electrons from high concentration in the *n*-layer to zero concentration in *p*-layer, and holes from *p*-layer to *n*-layer. Donors and acceptors remain in place completely charged. Free electrons recombine with holes. When equilibrium reaches, there are no free carriers in some areas in the *p*- and *n*-doped layers, and these areas become oppositely charged: positively by donors in *n*- and negatively by acceptors in the *p*-layer (Fig. 3.3). These charges are determined by concentration of acceptors N_p and donors N_n. Thus, in the volume of the thickness $w = d_p + d_n$, there are no free electrical carriers, and this area is called the depletion area. Electrical neutrality requires that

$$N_n d_n = N_p d_p.$$

Commonly $N_p > N_n$ in order to ensure as big w as possible with the thinnest possible front *p*-layer:

$$w = d_p \left(1 + \frac{N_p}{N_n} \right).$$

Charge density ρ and electrical field $\vec{E}$ obey the equation

$$\operatorname{div} \vec{E} = 4\pi \rho.$$

Therefore, the electrical field E inside the PD is localized within the depletion region, and is directed from cathode to anode (Fig. 3.4). The photons absorbed inside the depletion region generate electron–hole pairs that are immediately separated by the electric field that directs holes to the anode and electrons to the cathode. Therefore, illumination generates positive voltage at the anode. Suppose now that illuminated PD is loaded by a resistor R (Fig. 3.5 left). This is the photovoltaic regime. The current I flows in the outer circuit, imposing positive

Fig. 3.4 In a disconnected PD, light generates positive voltage at the anode

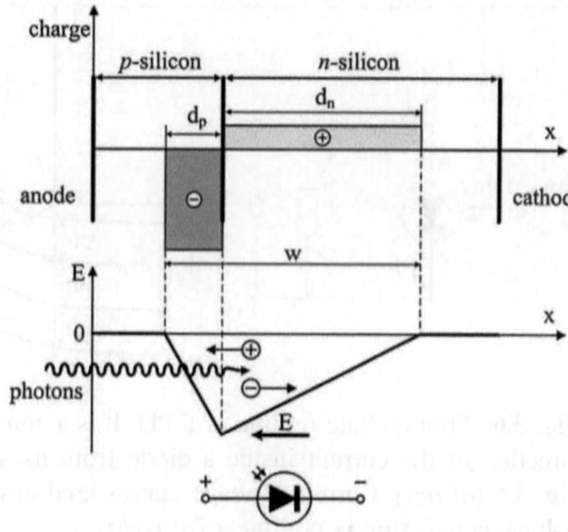

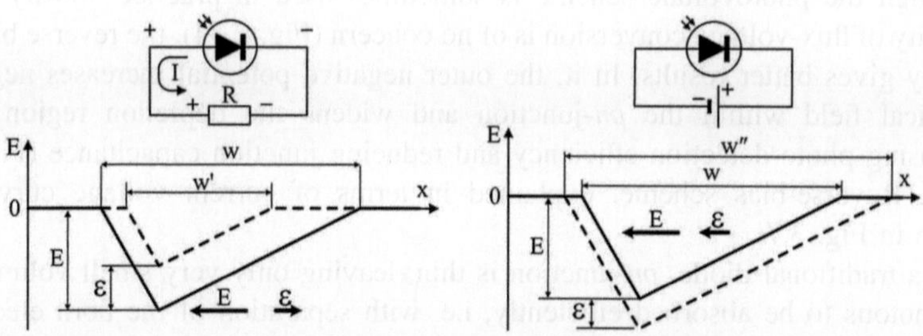

Fig. 3.5 Connecting a PD to a load *R* decreases separating electrical field *E*, the width of the depletion region *w*, and the efficiency (*at left*). The current *I* is small and natural conductivity of both the *p*- and *n*-silicon is high, so that the voltage drop on non-depleted areas may be considered zero as well as the electrical field there. Reverse bias increases separating electrical field *E*, the width of the depletion region *w*, and the efficiency (*at right*)

voltage drop $I \times R$ between the anode and cathode and introducing an oppositely directed field

$$\varepsilon \approx \frac{I \cdot R}{w}.$$

It lowers initial field intensity *E* by the value ε and narrows the depletion region to a value $w' < w$, thus decreasing photo-detection efficiency. Lower efficiency displays itself as smaller response of the output current to variations of the optical flux Φ. Figure 3.6 explains this effect in a more clear way, using the current–voltage characteristics of a PD. The positive voltage drop on the PD is commonly referred to as the direct bias, while the opposite is called the reverse bias. Thus, in the photovoltaic regime, the PD is directly biased. The first disadvantage of direct biasing of the *pn*-photodiode is its non-linearity and lower photo-detection efficiency. Another negative feature is the increase of the capacitance due to

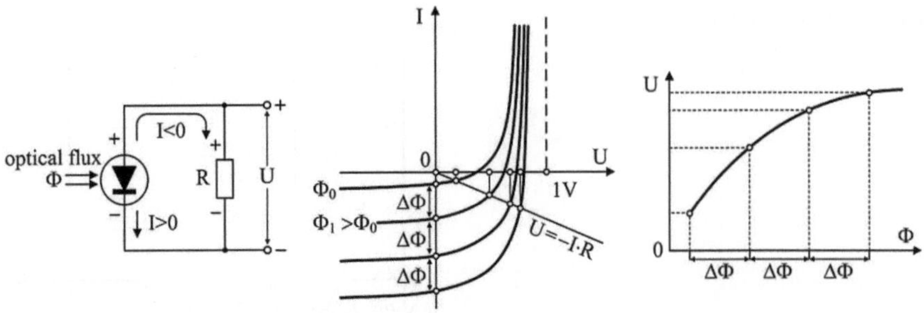

Fig. 3.6 Photovoltaic regime of a PD. It is a convention in radio engineering to choose positive direction of the current inside a diode from its anode to cathode, which is opposite to that in Fig. 3.5 (*at left*). Current–voltage curves tend down as optical flux increases (*center*). The flux-voltage conversion is nonlinear (*at right*)

narrowing of the depletion region, which reveals itself at high frequencies. Although the photovoltaic scheme is sometimes used in practice, mostly when linearity of flux-voltage conversion is of no concern (Fig. 2.34), the reverse biasing usually gives better results. In it, the outer negative potential increases negative electrical field within the *pn*-junction and widens the depletion region, thus increasing photo-detection efficiency and reducing junction capacitance (Fig. 3.5 right). Reverse-bias scheme, explained in terms of current–voltage curves, is shown in Fig. 3.7.

In a traditional diode, *pn*-junction is thin, leaving only very small volume for the photons to be absorbed efficiently, i.e. with separation of the born electrical carriers of the opposite polarity (holes and electrons). If the oppositely charged carriers are not separated quickly, they recombine again, making no contribution to photo-current. The carriers can be separated by electrical field, which exists between the *p*- and *n*-doped regions. To increase this volume and thus to improve detection efficiency, the PDs are made with relatively thick silicon layer of low intrinsic conductivity (1–10 kΩ cm), separating the *p*- and *n*-doped layers (Fig. 3.8). For comparison, the *n*-doped substrate has much higher conductivity ~5 mΩ cm. This is the so-called *pin*-diode structure. It gives two basic advantages: better photo-absorption efficiency and lower capacitance for better frequency response. Although conductivity of the intrinsic silicon (*i*-silicon) is very low, it is not a perfect insulator at room temperature, and some number of free electrons in the conduction band and holes in the valence band does exist due to very low but finite concentration of donors $N_{i\mathrm{donor}}$ and acceptors $N_{i\mathrm{accept}}$. As such, depletion regions around *p*-*i* and *i*-*n* interfaces also exist, but unlike the *pn*-junction, they are extremely narrow. Consider for example the *p*-*i* interface. Holes, penetrated from the *p*-layer into the *i*-layer, recombine with the electrons in the conduction band of the *i*-layer. For this process, electrical neutrality requires $N_p d_p = N_{i\mathrm{donor}} d_{i\mathrm{donor}}$, where $d_{i\mathrm{donor}}$ is the depletion depth of the conduction-band electrons in the *i*-layer. Thus,

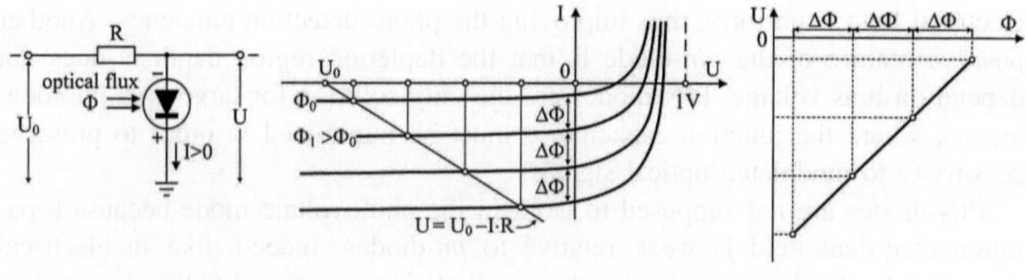

Fig. 3.7 Reverse-bias connection of a PD (*at left*). Current–voltage curves tend down as optical flux increases (*in the middle*). The flux-voltage conversion is linear (*at right*)

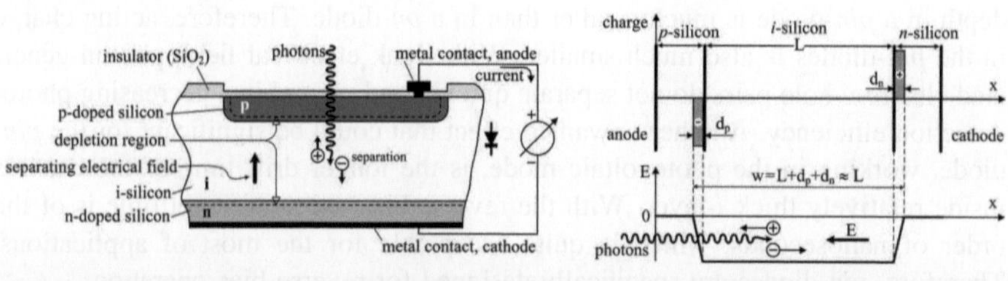

Fig. 3.8 PIN-diode structure. The intrinsic (i) silicon layer receives holes from the *p*-layer and electrons from the *n*-layer; they recombine with the *i*-silicon own electrons and holes, depleting the entire *i*-layer. Charge of the opposite polarity is localized in thin layers adjacent to the *p-i* and *n-i* interfaces, making the electrical field uniform but weaker than in a *pn*-junction

$$d_{i\,\mathrm{donor}} = d_p \frac{N_p}{N_{i\,\mathrm{donor}}}.$$

With the inequality $N_p/N_{i\mathrm{donor}} \gg 1$ in several orders of magnitude, the $d_{i\mathrm{donor}}$ upper limit is simply the depth of the *i*-layer itself: $d_{i\mathrm{donor}} \approx L$. Then in return

$$d_p = L \frac{N_{i\,\mathrm{donor}}}{N_p} \ll L.$$

The same consideration for the *i-n* interface gives $d_n \ll L$.

Now consider the balance of free electrical carriers in the *i*-layer. The conduction-band electrons totally recombine with the holes diffused from the *p*-layer. The valence-band holes totally recombine with the electrons diffused from the *n*-layer. The initial number of donors and acceptors in the *i*-layer is roughly the same: $N_{i\mathrm{donor}} \sim N_{i\mathrm{accept}}$. Therefore, total electrical charge here is approximately zero. Since charge density ρ determines the electrical field E

$$\operatorname{div} \vec{E} = 4\pi \rho,$$

integration of the electrical charge over the depth gives distribution shown in the right picture of the Fig. 3.8. Comparing this result to the *pn*-diode (Fig. 3.4), it is easy to see that, in the *pin*-diode, depletion region is thicker and separating electrical field is uniform, thus improving the photo-detection efficiency. Another positive feature of the *pin*-diode is that the depletion region depth w does not depend on bias voltage. PIN-diodes are the only solution for large-area photodetectors, where the junction capacitance must be minimized in order to preserve sensitivity to modulated optical signals.

PIN-diodes are not supposed to work in the photovoltaic mode because separation electrical field is weak relative to *pn*-diodes. Indeed, like in electrical capacitor, the field intensity inside the *pin*-diode is proportional to the charge at its borders. This charge is proportional to concentration of acceptors N_p in the *p*-layer (donors N_n in the *n*-layer) multiplied by the charge depth d_p (d_n). Concentration of acceptors is roughly the same in both the *pn*- and *pin*-diodes, whereas the charge depth in a *pin*-diode is much smaller than in a *pn*-diode. Therefore, acting charge in the *pin*-diodes is also much smaller. With weak electrical field, photon-generated electron–hole pairs do not separate quickly and recombine, decreasing photodetection efficiency. Another unwanted effect that could be significant for the *pin*-diode, working in the photovoltaic mode, is the longer drift time of the carriers inside relatively thick *i*-layer. With the reverse bias, carriers travel time is of the order of nanoseconds, which is quite acceptable for the most of applications. Therefore, *pin*-diodes are specifically designed for reverse-bias operation.

Performance of a PD greatly depends on electronic circuit it is connected to, and PD receivers will be considered in the next section. However, there are three parameters that characterize the PD alone: size, quantum efficiency, and reponsivity. The size of a sensitive element may range from ∅0.1 mm for high-frequency PDs to 10 × 10 mm when response time is not a factor. Small size does not necessarily mean that in a particular application the photodetector will receive low flux: focusing optics can collect much more optical flux that the photodetector alone. However, the size of a photodetector does determine the angular field of view of the system. Quantum efficiency of a PD is the ratio of the number of electron–hole pairs generated to the number of photons incident. In a certain spectral interval between 400 nm and 1 μm, where photons freely penetrate to the *pn*-junction and have sufficient energy to excite electron–hole pairs over the silicon energy bandgap, quantum efficiency of PDs is high: up to 0.9 in anti-reflection-coated devices and ~0.6 in the non-coated ones. In the ultra-violet (UV) region, major part of photons is absorbed in thin upper *p*-doped layer, where the carriers are not separated and recombine quickly with no avail to the current. Thus, quantum efficiency in the UV domain falls to zero. In the infra-red region beyond 1.1 μm, photon energy is insufficient to excite carriers over the silicon bandgap, abruptly turning quantum efficiency to zero. For this reason silicon is well transparent in infrared.

Reflection of photons at the interface air-semiconductor may seriously impair quantum efficiency. The Fresnel formulas give the reflection coefficient at normal incidence as

$$R = \left| \frac{n-1}{n+1} \right|^2,$$

where $n = n_r + in_i$ is the complex refractive index of a semiconductor. For silicon, for example, $n_r \sim 3.5$ in visible domain, which gives $R \sim 0.3$. Thus, without antireflection coating, one third of the input optical flux would be reflected back. However, even a single optical layer over the silicon may significantly reduce reflection. For a transparent material with refraction index n_1, transmission coefficient equals

$$T = 1 - R = \frac{4n_1}{(n_1+1)^2}.$$

Total transmission through the cover layer into silicon is the product of transmission coefficients through two interfaces: air-layer and layer-silicon:

$$\frac{4n_1}{(n_1+1)^2} \cdot \frac{4n_1 n_r}{(n_1+n_r)^2 + n_i^2}.$$

The minimum of this product is reached when $n_1 \approx \sqrt{n_r}$, which for silicon gives the value 1.87. Two most frequently used materials are silicon oxide SiO_2 ($n = 1.54$) and silicon nitride Si_3N_4 ($n = 1.98$). Such a layer, which also serves as a protector against diffusion of water vapor and other gases into the silicon, reduces reflection by several times.

Quantum efficiency η is the parameter that cannot be measured directly. It is always computed, using other measurable quantities like current, voltage, or optical power. The standard parameter commonly listed in datasheets is the spectral sensitivity S, which measures as the ratio of the PD short-circuit current I to the input optical power P at a given wavelength λ. Since the current is the number of electrons per unit time, and the optical power is the number of photons per unit time, the spectral sensitivity is equal to

$$S \left[\frac{\mathrm{A}}{\mathrm{W}} \right] = \eta \frac{e\,\lambda}{h\,c},$$

where $e = 1.602 \times 10^{-19}$ (Coulomb) is the elementary charge, $h = 6.626 \times 10^{-34}$ [W·s^2]—Planck constant; c—speed of light. Substituting the values, obtain

$$S \left[\frac{\mathrm{A}}{\mathrm{W}} \right] = \eta \cdot 8.06 \cdot 10^{-4} \times \lambda\,[\mathrm{nm}] = \eta \cdot 0.806 \cdot \lambda\,[\mu\mathrm{m}].$$

Measuring S and comparing to this formula, it is possible to determine quantum efficiency. Figure 3.9 presents typical spectral sensitivity of a silicon PD.

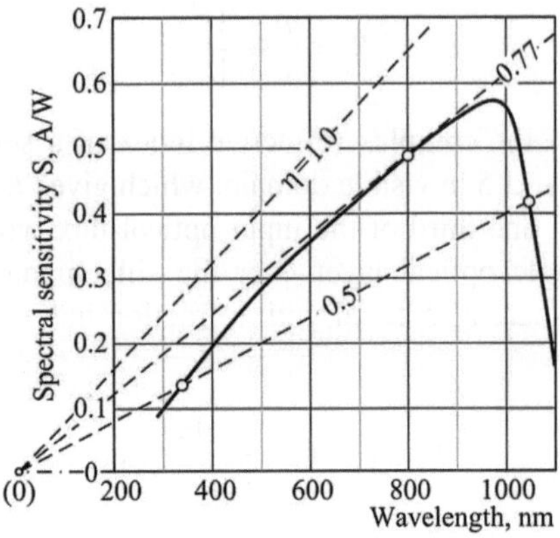

Fig. 3.9 Spectral sensitivity of a typical silicon PD reaches maximum around 980 nm. *Dashed lines* show spectral sensitivity, corresponding to constant quantum efficiency η. From these data it follows that quantum efficiency of this particular PD reaches maximum value 0.77 around 800 nm

Fig. 3.10 Structure of an *npn*-phototransistor (*at left*) and its functional scheme (*at right*). PT can be viewed as a photodiode whose current drives the base of a conventional small-signal transistor

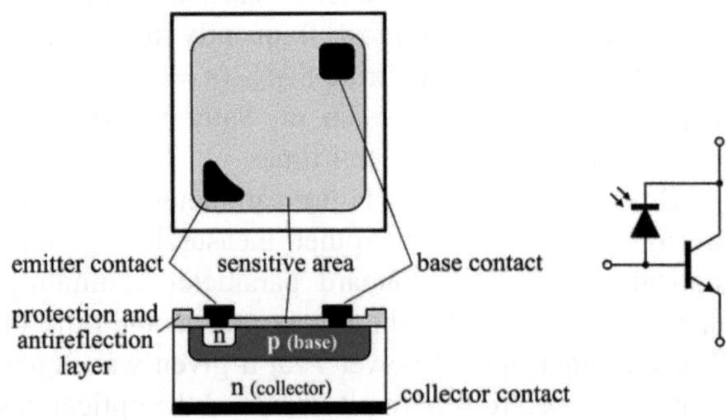

3.2 Phototransistors

Phototransistor (PT) is a transistor with its collector–base *pn*-junction exposed to light. Even an ordinary transistor, sealed in a metal case, may be easily converted to a photodetector if the top wall of its case is removed. The idea of such a combination is to deliver much bigger output current than a photodiode alone can provide. Of course, photodiode current may be amplified to any desired level by means of external circuitry, and the following section explains how to do it in a most efficient way. However, when linearity and speed of response are of no concern, a PT may be the simplest solution. After the detailed explanation made in the previous section, Fig. 3.10 does not require much of discussion. The peak response of the PTs is at a somewhat shorter wavelength than that of a typical silicon photodiode (Fig. 3.11) because the *pn*-junctions in a PT are formed in epitaxial rather than crystal grown silicon as in photodiodes.

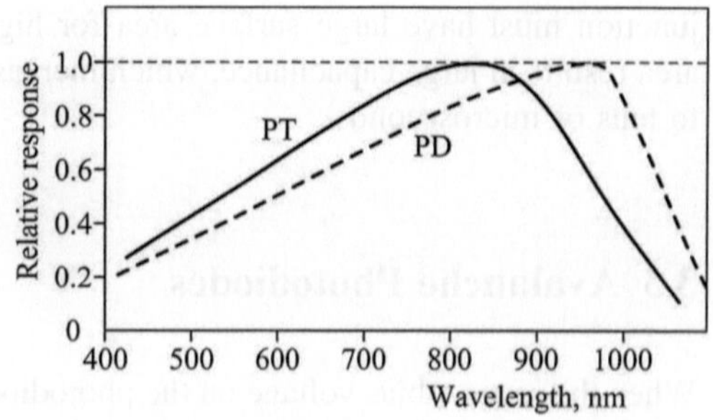

Fig. 3.11 Phototransistors (PT) show maximum spectral response at a shorter wavelength than photodiodes (PD)

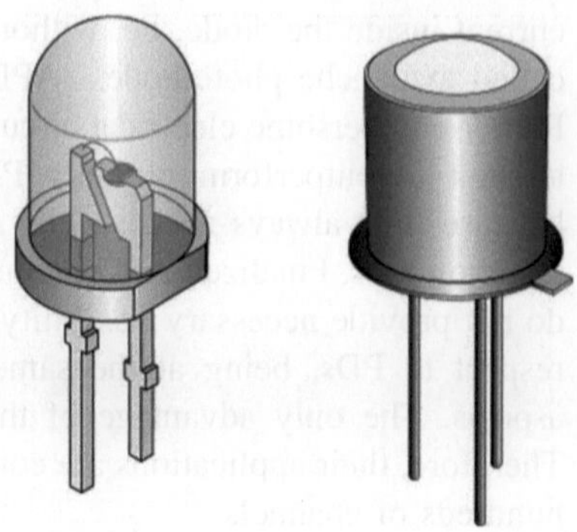

Fig. 3.12 Two-lead PTs look exactly like light-emitting diodes, while the three-lead ones may be mixed up with photodiodes

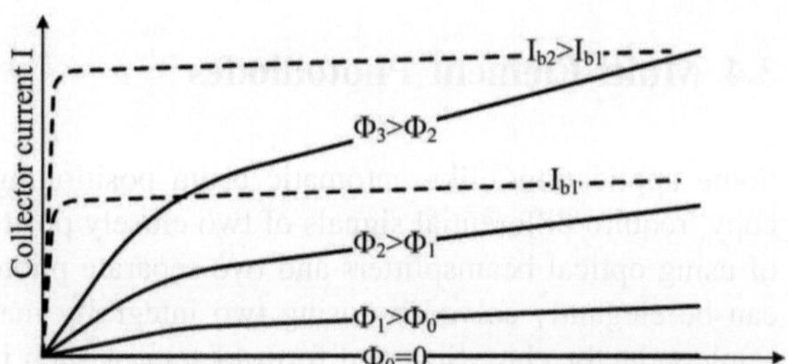

Fig. 3.13 With optical flux Φ increasing, dynamic resistance dU/dI of the PT decreases (*solid curves*), while in small-signal transistors it remains practically constant and big for all the base currents I_b (*dashed curves*)

The external base connection is not actually required for operation. However, some users want to have this option in order to optimize the performance by base biasing. Therefore, manufacturers offer both two- and three-lead packages (Fig. 3.12). Commonly, the PT gain is about 500–1000, which makes the output current on the scale of milliamps rather than microamperes. Nonetheless, some additional amplification is still necessary for convenient handling of the signal.

With the PT design optimized for photo-detection, the current–voltage curves are far from the ideal ones of the small-signal transistors (Fig. 3.13). Another drawback of the PTs is relatively slow response. Whereas in speed-optimized small-signal transistors collector–base capacitance is minimized, in PTs this junction must have large surface area for high photo-detection efficiency. Large area results in large capacitance, which increases the time constant of the response to tens of microseconds.

3.3 Avalanche Photodiodes

When the reverse bias voltage on the photodiode (PD) exceeds some certain value, the breakdown takes place by impact ionization and avalanche multiplication of free carriers. This phenomenon, when controlled, may be used to amplify photocurrent inside the diode, i.e. without external amplifiers. Such photodetectors are called avalanche photodiodes (APDs). However, avalanching is hard to control. Rather cumbersome electrical circuits were invented to stabilize this regime in an attempt to outperform ordinary PDs. APDs are rarely used in the laboratory because they always need special electronic modules that cannot be made by a common user. Finalized APD modules can be purchased for higher price, but they do not provide necessary flexibility and, what is more important, superiority with respect to PDs, being at the same time inferior to photomultipliers in all the aspects. The only advantage of the APDs over photomultipliers is lower cost. Therefore, their applications are confined to micro-channel detecting systems with hundreds of channels.

3.4 Multi-Element Photodiodes

Some applications, like automatic beam positioning or autofocusing in microscopy, require differential signals of two closely positioned photodetectors. Instead of using optical beamsplitters and two separate photodetectors, the tasks like that can be elegantly solved by using two integrally made sensitive elements in one semiconductor chip. Size and form-factors of such photodetectors may be different, but typically one device contains two or four sensitive elements. The last option is commonly organized in a form of four adjacent sectors and is called the quadrant photodetector (Fig. 3.14).

3.5 Photodiode Receivers

A photodiode receiver is a combination of a photodiode, an amplifier, and a demodulator, if necessary. When linearity and response speed are of no concern, then phototransistors (PTs) may be the simplest solution (Fig. 3.15).

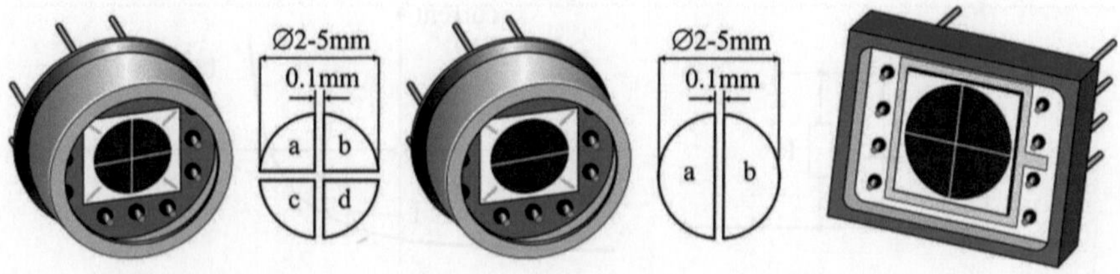

Fig. 3.14 Quadrant photodiodes are the most popular option because they can be easily reconnected to the two-element option. However, the two-element version is also available from the manufacturers

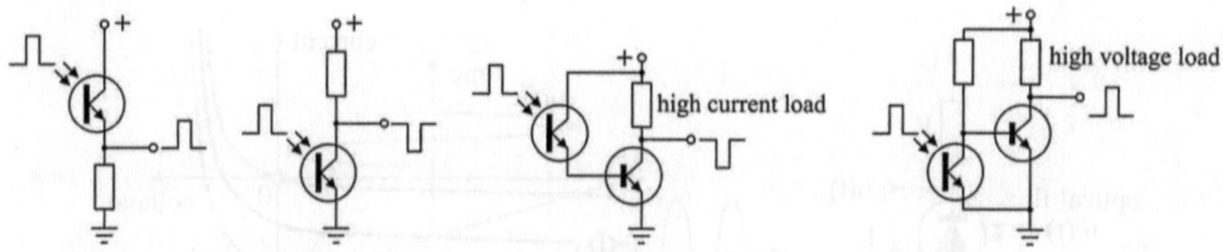

Fig. 3.15 PTs are mostly used as switching solutions

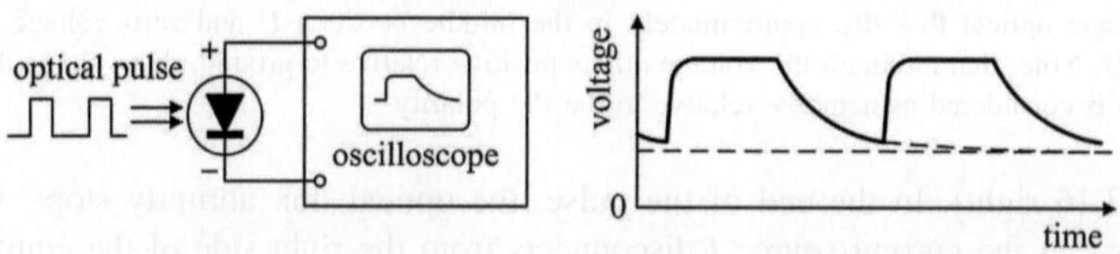

Fig. 3.16 Direct connection of a PD to a recorder. Constant voltage pedestal shown in a straight *dashed line* is the initial PD thermal voltage accompanied by optical background

Photodiodes (PDs) are used in all the other cases. What the user always wants from a photoreceiver, is high output voltage amplitude that could be easily observed at an oscilloscope or digitized in a computer. Then the simplest and the worst solution that requires nothing but a PD is shown in Fig. 3.16. This is the already known photovoltaic regime of a PD, and from Fig. 3.6 it follows that the output voltage can be no more than ~0.5 V. But what is not known is that such a scheme displays very slow response in pulsed operation, especially at the pulse tails (Fig. 3.16 right). This feature follows from the equivalent scheme of a PD shown in Fig. 3.17. When the rectangular optical pulse comes, the current source I quickly charges the capacitance C, and the voltage drop over it increases positively, applying positive bias to the junction (Fig. 3.5). This process is quick because C is relatively small, not more than hundreds of picofarads even for large-area PDs, and the voltage rises almost linearly with the front edge slope I/C. Under positive bias, efficiency of photo-detection falls, decreasing the photo-current I. In return, the rate of the voltage rise also falls to almost zero at the top of the pulse

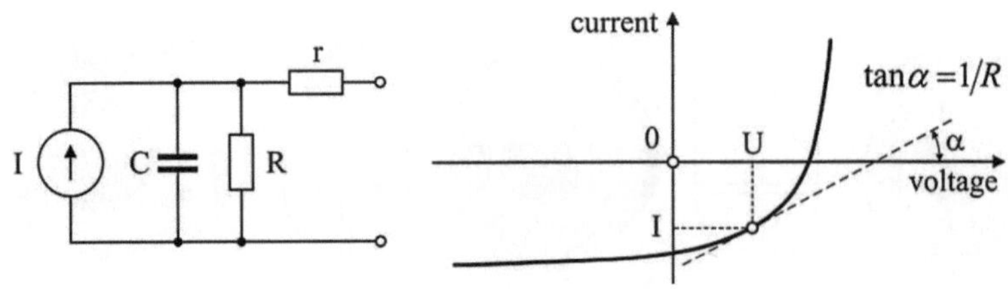

Fig. 3.17 Equivalent circuit of a PD includes the source of a photo-current I, junction capacitance C, junction dynamic resistance R defined as the derivative of the voltage over the current (reciprocal slope of the current–voltage curve), and the series resistance r. All the parameters depend on the bias voltage U

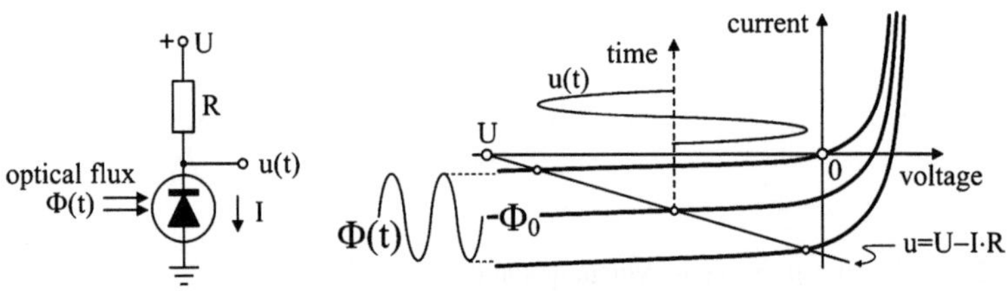

Fig. 3.18 The series resistor R should be chosen so as to place the working point, corresponding to average optical flux Φ_0, approximately in the middle between U and zero voltage, i.e. at $\approx 0.5\ U$. Note, that although the voltage $u(t)$ is positive relative to ground, this voltage drop on the PD is considered as negative relative to the PD polarity

(Fig. 3.16 right). In the end of the pulse, the optical flux abruptly stops, which means that the current source I disconnects from the right side of the equivalent circuit. Then the junction capacitance C starts to discharge through R. The value of R, of the order of kilohms under the positive bias, quickly increases to hundreds of kilohms when the voltage tends to zero (Fig. 3.17 right). Therefore, the tail of the pulse is not exponential, decaying slower then its front.

The reverse-bias scheme in Fig. 3.18 gives much better results. In order to preserve equal voltage swing for both the positive and negative parts of a sinusoid, the load resistor R must be roughly equal to the reverse bias resistance of the PD, which may be of the order of hundreds of kiloohms. For the open circuit, the larger the R is, the larger is the voltage swing $u = I \cdot R$. On the other hand, large R greatly reduces load capacity: a normal load resistor $r \sim 1\ \text{k}\Omega$, connected in parallel to the PD, shunts the voltage drop on the PD to the value $\approx U \cdot r/R$, thus reducing modulation amplitude to one half of that value, i.e. roughly 100 times relative to initial $0.5U$. In practice, this problem is solved by using operational amplifiers (Fig. 3.19). Load capacity of an operational amplifier is very high because its output resistance is low: hundred ohms output series resistance that goes down to milliohms with closed feedback. Any operational amplifier can easily drive a standard 50 Ohm load. The only that we need now is to derive a relation between the photocurrent and the output voltage u.

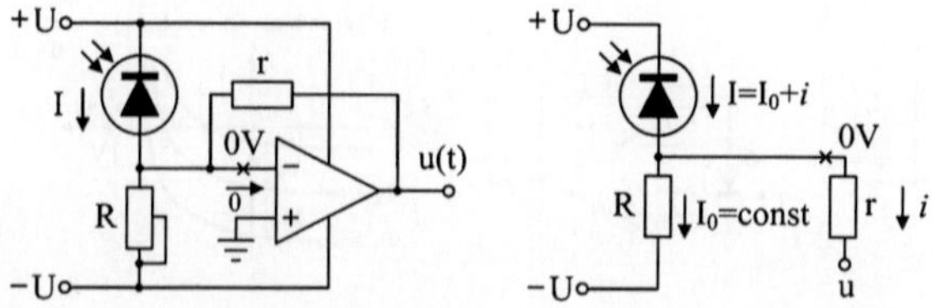

Fig. 3.19 Negative feedback, established by the resistor *r*, maintains the voltages at the inverting and non-inverting inputs equal to each other, i.e. equal to zero. Current does not flow into the inverting input: because of the negative feedback for transimpedance (current-to-voltage) operational amplifiers and because of very high input resistance for ordinary (voltage-to-voltage) operational amplifiers. In both cases, the equivalent circuit at right holds true

Fig. 3.20 Simplified reverse-bias scheme, suitable for low dark current PDs

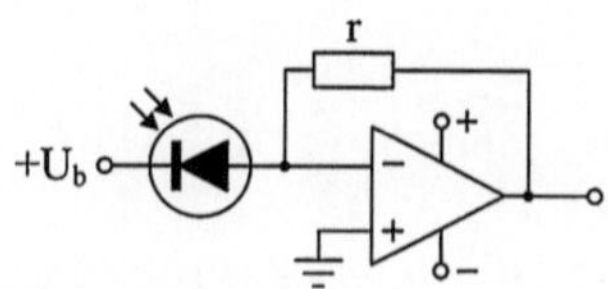

Let the working point be characterized by the PD current I_0. Conventionally, the output voltage *u* at the working point is adjusted to zero by trimming the variable resistor *R* so that

$$I_0 R = U.$$

Now let the input optical flux change so that $I = I_0 + i$. Since the voltage at the inverting input is maintained at zero level, the current through *R* remains $I_0 = const$. Therefore, the entire variation of the photocurrent *i* flows through the feedback resistor *r*, making the output voltage equal to $u = i \cdot r$. Since the positive current *i* flows from zero voltage to *u*, it means that *u* is negative. Thus, the scheme in Fig. 3.19 acts as an inverting current-to-voltage converter. In the literature, it is sometimes called the transimpedance amplifier, which is, in fact, misleading as it may be mixed up with a completely different thing called transimpedance operational amplifier. The scheme in Fig. 3.19 works pretty well with both types of operational amplifiers, may they be the ordinary voltage-to-voltage or the transimpedance current-to-voltage ones. The only thing that should be checked, choosing an operational amplifier, is its input capacitance: the lower the better. As the reverse biasing of a PD decreases its capacitance, thus improving the high-frequency performance, it would be unwise to loose this advantage by shunting the PD with a bigger capacitance.

Some applications, like optical power meters, require the working point be set at zero photocurrent, which corresponds to zero optical flux. In that case, the reverse biasing scheme may be simplified to what is shown in Fig. 3.20, providing that the PD has low dark current.

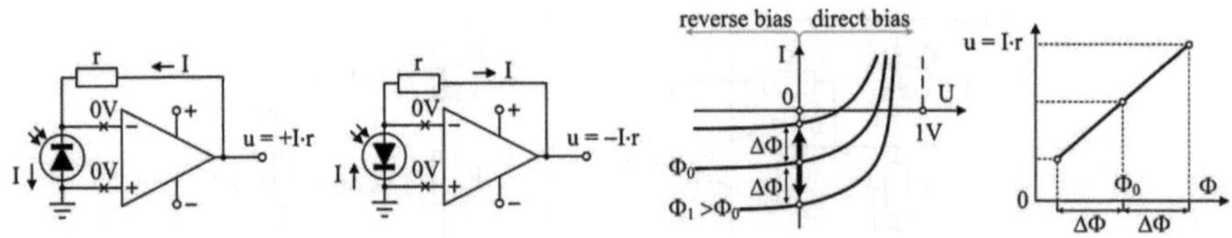

Fig. 3.21 Zero-bias scheme is also linear to optical flux

Fig. 3.22 When $R_1 = R_2$, this scheme compensates for bias currents in the inputs of a voltage-to-voltage operational amplifier

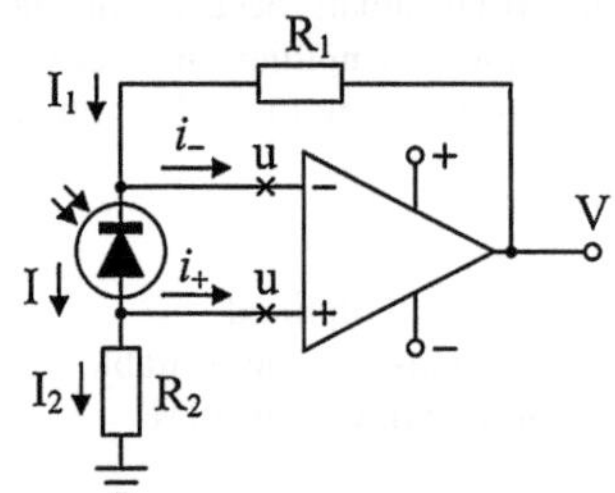

The scheme may look strange to those experienced in operational amplifiers but new to photodiodes, because it is well known that positive bias applied to inverting input of an operational amplifier results in negative voltage at its output, thus driving the working point far below zero. The answer is that the dark resistance of a good PD may be about megaohms, whereas the feedback resistor $r \sim$ 10–100 kOhm. Thus, with the bias voltage $U_b = +5$ V, the working point at the output of the operational amplifier will be shifted by less than 0.5 V, which may be considered as tolerable with the entire voltage swing $\sim\pm10$ V.

Analysis of the photovoltaic regime, as it is shown in Fig. 3.6, must not lead us to a conclusion that the non-biased scheme is always nonlinear, and as such, inferior to the reverse-biased scheme. The truth is that the photovoltaic regime always results in direct biasing of a PD. But what happens if the PD is not biased at all, i.e. at zero bias? This is exactly what we have in Fig. 3.21. The negative feedback established by the resistor r maintains the voltage difference between the inverting and non-inverting inputs practically zero. Therefore, the voltage drop on the PD is also zero, so that variations of optical flux make the working point moving up and down the current axis without entering the highly nonlinear direct-bias region (Fig. 3.21 right). Consequently, linearity remains good. The output voltage u and the PD current I are connected as before: $u = I \cdot r$, with the only difference that the voltage is proportional to the entire PD current I and not to its variation i.

Modification of zero-bias scheme shown in Fig. 3.22 presents some theoretical and historical interest, but does not introduce any real advantages with respect to the simpler scheme in Fig. 3.21.

Consider the output voltage V. Since potentials at both the inverting and non-inverting inputs are approximately equal for an operational amplifier with large gain, the following equations hold true:

$$\frac{V-u}{R_1} = I + i_-; \quad (I - i_+)R_2 = u.$$

Excluding u, one obtains

$$V = I(R_1 + R_2) + i_- R_1 - i_+ R_2.$$

Commonly, this scheme is considered with $R_1 = R_2 = R$:

$$V = 2IR + (i_- - i_+)R.$$

Historically, first operational amplifiers were of voltage-to-voltage type with input resistance of about hundreds of kilohms in both inputs. With roughly the same value of the dark resistance of the PD, input currents i_- and i_+ were considered as significant contribution to the output signal, requiring compensation. The scheme in Fig. 3.22 does exactly that: with $i_- \approx i_+$, they cancel each other almost completely:

$$V \approx 2I \cdot R.$$

However, years of developments of more and more advanced versions of operational amplifiers, such as those with field-effect transistors (FET) in the inputs, with input resistance of the order of hundreds of megohms, made this feature insignificant. Another peculiarity is the two times bigger output voltage relative to the formula for the scheme in Fig. 3.21. This factor of two acts almost magically on inexperienced users, making them to believe that it is a real advantage. But it is not. If someone wants to increase the overall gain by two times, he would increase one resistor R from, say, 100–200 kOhm, rather than attach a second large-value resistor to non-inverting input, henceforth introducing additional capacity and collecting electromagnetic interference on the voltage-sensitive point. We say voltage-sensitive because the non-inverting input of any operational amplifier has large input resistance, in contrast to the inverting input, which may be as small as tens of Ohms for the trans-impedance operational amplifiers. As the development moved on, it was found that trans-impedance operational amplifiers, also called current-to-voltage amplifiers, provide much better frequency performance. They respond to current in the inverting input and to voltage in the non-inverting one, making i_- and i_+ completely different. Nowadays, the majority of high-frequency operational amplifiers, with the gain-bandwidth products of hundreds of megahertz, are of this type, thus making the scheme in Fig. 3.22 inappropriate.

To summarize photodiode receivers with operational amplifiers, it is useful to consider connection of a PD to the non-inverting input (Fig. 3.23). The non-inverting input of any operational amplifier, whatever its type is, always has very big input resistance of about 1–10 MOhm. Therefore, even weak PD photocurrent will develop sufficient voltage on the amplifier input. As such, the operational

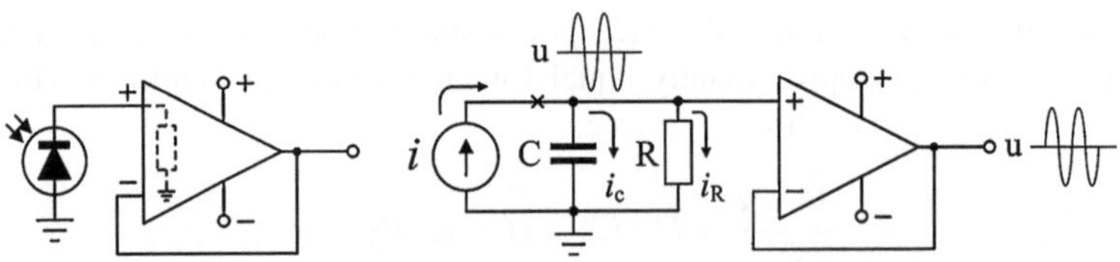

Fig. 3.23 Input resistance of the operational amplifier serves as the load to the PD (*at left*). Frequency response may be better analyzed with the equivalent circuit of a PD (*at right*). Here R is a combined resistance of the PD and the operational amplifier input

amplifier itself serves only to increase load capacity of the PD, and its gain may be made equal to unity. This explains why the inverting input and the output of the amplifier are just shunted. Obviously, the PD works again in the photovoltaic regime, whose drawbacks were already discussed. However, we are now interested in more detailed analysis of the effect that the PD capacitance has on frequency response of the circuit. Since the gain of the amplifier is unity, the output voltage u is the same as the voltage at the input:

$$u = i_R R = i_C \frac{1}{j\omega C},$$

where traditionally $j = \sqrt{-1}$ and ω is the angular frequency of modulation. Another equation holds true for the sum of the currents:

$$i = i_C + i_R.$$

Solving this system of equations for u, one obtains

$$|u| = \frac{iR}{\sqrt{1 + (RC\omega)^2}}.$$

Thus, the output voltage drops with frequency, and the so-called cutoff frequency $\omega = (RC)^{-1}$ indicates when the voltage drops to 0.7 of its maximum value, which emphasizes the importance of smallness of the PD capacitance for high-frequency operation. All this happens only because a portion of the PD photocurrent flows into the virtual capacitor, formed by the *pn*-junction. The current i_C through a capacitor is the derivative of the capacitor charge $Q = Cu$ over time t:

$$i_C = \frac{dQ}{dt} \approx C\frac{du}{dt}$$

The last equality is written as approximation because the *pn*-junction capacitance C also depends on u. Anyway, it means that if the voltage on the PD is maintained constant $u = const$ then no current flows into the capacitor, as if the capacitor itself is excluded from the circuit. This is exactly the case of the circuits in Figs. 3.19, 3.20 and 3.21, where the PD is connected to the inverting input. Thus,

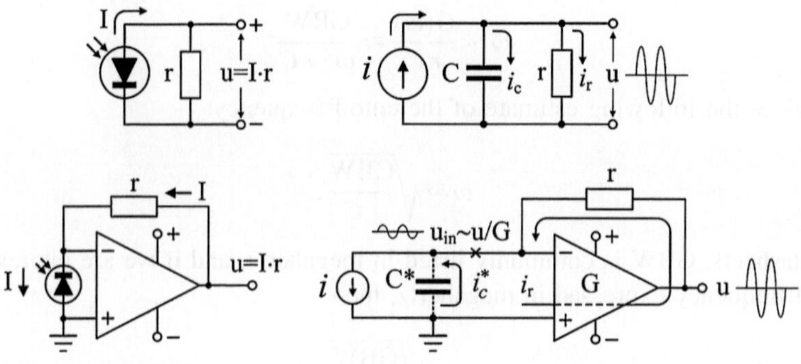

Fig. 3.24 The photovoltaic (*above*) and operational amplifier (*below*) schemes with their equivalent circuits

connecting the PD to the inverting input of an operational amplifier is advantageous because it improves high-frequency performance. This advantage, however, may be easily lost if the wiring itself has considerable capacitance, which is the case when the connecting leads are long or the PD is connected through a coaxial cable. Therefore, place all the components on the board as close together as possible.

It is instructive to compare frequency responses of the photovoltaic scheme and the scheme with the inverting operational amplifier (Fig. 3.24). Although the output voltages $u = I \cdot r$ are the same, cutoff frequencies are dramatically different. The cutoff frequency for the photovoltaic scheme is $\omega = (rC)^{-1}$, where C is the PD junction capacitance. With the operational amplifier, situation changes. In the last paragraph we assumed that the voltage on the inverting input is maintained constant: $u = const$. However, this is only an approximation: with the finite open-loop gain $G < \infty$, the input voltage is of the order of $u_{\text{in}} \sim u/G$. Therefore, the PD capacitance is not entirely excluded from the circuit, but swapped by a much smaller equivalent capacitance $C^* \sim C/G$. Since the current is now divided between r and C^*, the cutoff frequency is equal to

$$\omega = \frac{1}{r\,C^*} \sim \frac{G}{r\,C}.$$

But here we have some complication: the open-loop gain G depends on frequency ω. For operational amplifiers, it is believed that the gain-bandwidth product (GBW) is roughly constant at high frequencies. GBW is one of the most important parameters of an operational amplifier, always being included in a datasheet. Using the GBW, we then can write for high frequencies:

$$G(\omega) \approx \frac{\text{GBW}}{\omega},$$

where GBW measures in radians per second. This leads to an equation

$$\omega \sim \frac{G(\omega)}{rC} \approx \frac{\text{GBW}}{\omega \cdot rC}$$

that gives the following estimate of the cutoff frequency:

$$\omega \sim \sqrt{\frac{\text{GBW}}{rC}}.$$

In datasheets, GBW is commonly listed in megahertz, and if we are interested in cutoff frequency expressed in megahertz, then

$$f \sim \sqrt{\frac{\text{GBW}}{2\pi rC}},$$

where the $1/rC$ constant must also be converted to megahertz. For example, if the PD with the capacitance $C = 100$ pF is connected to $r = 100$ kOhm load resistor in a photovoltaic scheme, then $rC = 10$ μs, and the cutoff frequency $f = 1/rC = 0.1$ MHz. On the other hand, if the same PD is connected to an inverting input of an operational amplifier with, say, 62.8 MHz GBW (a multiple of 2π for easy computations) and the same feedback resistor, then the cutoff frequency increases to 1 MHz.

Not only the capacitance of a PD itself but also parasitic capacitance of the feedback resistor C_f may significantly impair transient response of the receiver. For an operational amplifier with the feedback resistor r shunted by a parasitic capacitance C_f, the cutoff frequency $\omega = (rC_f)^{-1}$. A standard 0.25 W axial metal-film resistor is believed to have parasitic capacitance around 0.2 pF. With $r = 100$ kOhm, it gives $rC_f \sim 20$ ns, which may be considered insignificant. But someone may desire to have much larger output signal, and install the 10 MOhm or even 100 MOhm feedback resistor. Then the time constant may increase to 2 or even 20 microseconds—a quite noticeable rise- and fall-time at rectangular pulses. It is even more important because the most advanced current-to-voltage operational amplifiers do not work with that large feedback resistors. Therefore, choose the feedback resistor less than 1 MOhm, and the higher output voltage can be easily gained by adding a second amplifier.

All the aforementioned is exemplified by experimental results presented below. Two schemes with roughly the same current-to-voltage gain were tested according to Fig. 3.25, with two different types of operational amplifiers. The first set of data shows transient responses with AD845—a voltage-to-voltage JFET operational amplifier with 10^5 MOhm input resistance in both inputs, 4 pF input capacitance, and 16 MHz GBW. The second set is for AD844—a current-to-voltage operational amplifier with input resistances 50 Ohm in the inverting and 10 MOhm in the non-inverting inputs, 2 pF input capacitance, and 60 MHz GBW. In general, transimpedance operational amplifier AD844 is superior to AD845 in response speed. However, it requires bigger current in its inverting input than 10 MOhm feedback resistor can provide, leaving the PD capacitance uncompensated. It increases rise and fall times almost to half of the pulse width. The JFET AD845 is not as critical

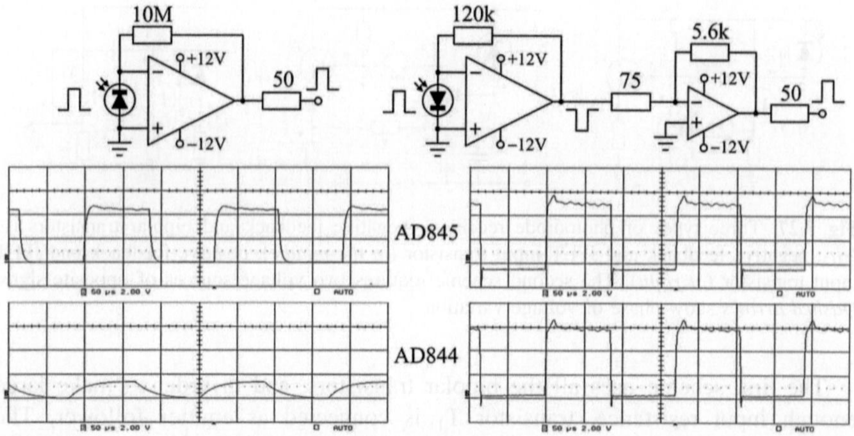

Fig. 3.25 Comparison of two schemes. Fast laser diode generates the pulse train. Scale in the left lower corner of each oscilloscope trace reads: 50 μs; 2.00 V. *Horizontal dotted lines* show the trigger level

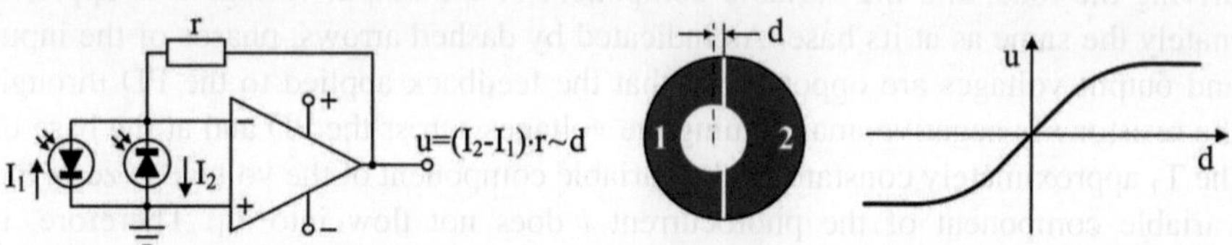

Fig. 3.26 Two-element PD connected to operational amplifier works as a displacement meter, with the output voltage *u* proportional to the displacement *d* of an optical beam across the PD border

to feedback resistor as the AD844, however, the fronts and tails of pulses are clearly softened with 10 MOhm, whereas 120 kOhm resistor keeps them sharp. This example shows again that smaller feedback resistor is better, not only improving response time, but also making the scheme insensitive to types of operational amplifiers.

Operational amplifiers are also useful for multi-element PDs, providing simple and efficient solutions for differential circuits (Fig. 3.26).

Although operational amplifiers are very simple and reliable instrument to construct photodiode receivers at frequencies up to tens of megahertz, higher frequencies require special circuits built on discrete transistors. For instance, photoreceivers for fiberoptic communications systems do not require high linearity but only high slew rate, which may be achieved by simple circuits with high-frequency transistors (Fig. 3.27). In all of them, the PD is reversely biased and the feedback is applied to maintain constant voltage across it. As we already know, these are two necessary tricks to minimize the effect of the PD capacitance. However, the type of the feedback and the type of transistors may be different.

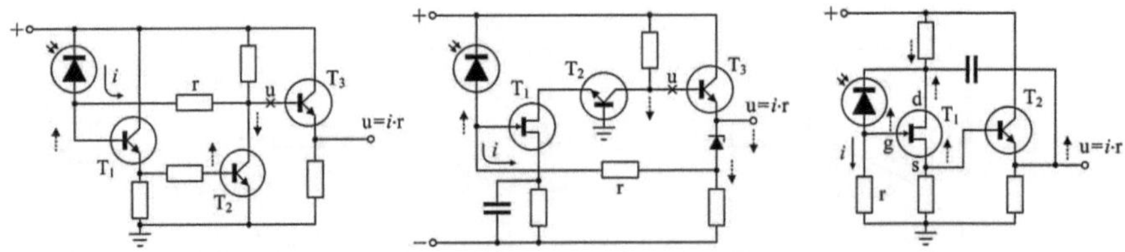

Fig. 3.27 Three types of photodiode receivers: negative feedback and bipolar transistors (*at left*); negative feedback and JFET input transistor (*in the middle*), positive feedback and JFET input transistor (*at right*). The second scheme requires two voltage sources of opposite signs. *Dashed arrows* show phase of voltage variation

The first scheme uses all the bipolar transistors, and in order to make large enough input resistance, transistor T_1 is connected as emitter follower. The emitter–base junction of T_2 must not shunt the follower output resistor, therefore it is connected through a resistor bigger than that of the follower. The voltage gain of the emitter follower is roughly unity, therefore the total voltage gain is acquired by T_2 at its collector resistor. The last transistor T_3 also forms emitter follower for driving the load, and the variable component of the output voltage u is approximately the same as at its base. As indicated by dashed arrows, phases of the input and output voltages are opposite, so that the feedback applied to the PD through the resistor r is negative, maintaining the voltages across the PD and at the base of the T_1 approximately constant. If the variable component of the voltage is zero, the variable component of the photocurrent i does not flow into T_1. Therefore, it entirely flows through the feedback resistor r, making $u = i \cdot r$.

The second scheme differs from the first one by the type of the input transistor. Although JFET transistors provide lower gain relative to bipolar transistors, they show lower noise. Therefore, when noise is an essential factor, JFET transistors should be given a priority. Dashed arrows show the phase of the variable component of the voltage, making it clear that the feedback established by the resistor r is negative. Again, the last transistor T_3 works as an emitter follower with the Zener diode subtracting excessive portion of the constant voltage in order to connect the variable component u to the input with lower voltage level. Zener diode must not necessarily be of a high-frequency type: even with high capacitance, the high-frequency current will filter down through it. The only capacitor in the scheme shunts the source resistor for higher gain. The main difference between this scheme and the first one is that the entire gain is delivered by the second bipolar transistor connected with common base. This provides better frequency separation between the input and the output parts of the circuit. Since the voltage at the gate of the T_1 is kept constant, the variable component of the PD current flows entirely through the feedback resistor r, making the output signal $u = i \cdot r$.

Unlike the first two, the third scheme uses positive feedback through a capacitor, which would inevitably make the scheme to oscillate if it were applied to the input. However, it is applied not to the input but to the opposite contact of the PD, thus preserving overall stability and, again, maintaining constant the

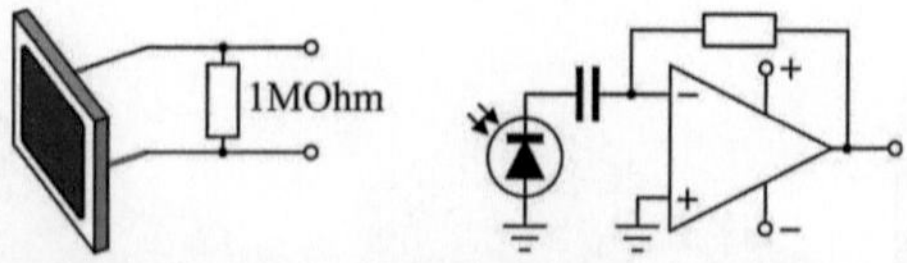

Fig. 3.28 Common mistakes with PDs mostly relate to detection of modulated signals. Large-area PD usually has big capacitance, and therefore should not be connected unbiased to a big resistance of megohm scale (*at left*). Unshunted capacitance will eventually be charged by unipolar PD current and lock the PD directly biased (*at right*)

voltage drop over the PD. When the voltage at the T_1 gate goes up, the voltage at its drain attempts to go down, as indicated by the upper dashed arrow. Nonetheless, the voltage swing that comes from the output through the capacitor, drives this point up (lower dashed arrow). Since the emitter follower T_2 applies much bigger current than T_1, the voltage on this contact of the PD goes up, thus preserving the voltage drop over the PD constant. Input resistance of the JFET is about hundreds of megohms, so that the photocurrent goes mostly through r. As the emitter followers T_1 and T_2 make the gain about unity, the output signal $u \approx i \cdot r$.

There are two most frequent mistakes with PDs (Fig. 3.28). First, connecting large-area unbiased PD ($\sim$1 cm^2) to a large-value resistor ($\sim$1 MOhm, like oscilloscope input) in the intention to observe modulated optical signals. The motivation is straightforward: the larger the area—the bigger the signal, and the larger the load resistor—the bigger the voltage swing on it. Yes, it is correct for non-modulated signals, but modulation requires small time constant, which in this case may be expected to be of the order 10^4 pF $\times$ 1 MOhm $=$ 0.01 s—a value sufficient to detect only frequencies below 60 Hz.

Another attempt may be made to install a capacitor after the PD in order to block the direct-current signal from entering operational amplifier, thus keeping the output at zero level. But it will not work: the capacitor will immediately be charged and will stop the PD current by the potential of opposite polarity.

3.6 Photomultiplier Tubes

A photomultiplier tube (PMT) is the most sensitive high-speed photodetector available. In the laboratory, its time comes when optical flux becomes too low for photodiodes or phototransistors. Amazingly, at low signals, PMTs with quantum efficiency of only 15 % completely outperform photodiodes with their quantum efficiency up to 90 %. The reason is low-noise inner amplification by secondary electron emission. Therefore, PMT requires high voltage of a kilovolt scale, and voltage-dividing system of resistors connected in series. All this sounds like too much for easy use in the laboratory, and it is true. Understanding that, manufacturers offer completely finalized ready-to-use devices called PMT modules,

Fig. 3.29 PMT modules come in a variety of form-factors. They require low-voltage power, usually 5 or 12 V direct current, and can be easily interfaced with simple electronic components to adjust the PMT gain

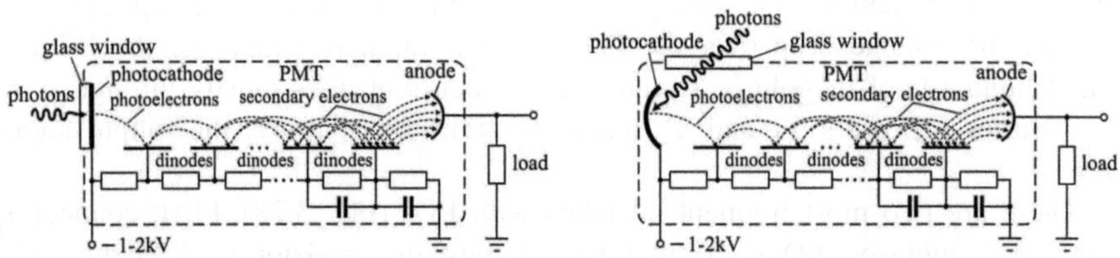

Fig. 3.30 Schematic diagram of a PMT: transmission (*at left*) and reflection (*at right*) types. They are also called front- and side-illumination types. The dynodes form an electron multiplier with typical gain about millions. Capacitors maintain constant voltage on the last dynodes during pulsed operation, when current increases in avalanche from photocathode to anode. Anode is not connected inside the vacuum tube

looking like small, the size of a matchbox, metal boxes with only few wires/ connectors for low-voltage power, gain adjustment, and the output (Fig. 3.29). All other components, like low-to-high voltage converter, resistive divider, and the PMT itself, are hidden inside. The only thing the user has to do is to interface the PMT module with an amplifier. But to do that correctly, it is necessary to understand some basic rules and limitations, discussed in this section.

PMT is a vacuum tube, containing the photocathode followed by focusing electrodes, electron multiplier with many accelerating electrodes (dynodes), and an electron collector (anode). Specific design of the electrodes may vary from one type to another, but the basic concept is the same (Fig. 3.30). The only classifying difference is the type of the photocathode: the reflection or the transmission type. The transmission type requires less space for the electrode system, being therefore the choice for ultra-miniature designs used in PMT modules.

The key parameter of a PMT is its quantum efficiency that includes quantum efficiency of the photocathode itself and optical efficiency of collecting the photons. The most advanced photocathode material, commonly referred to as S20 or multialkali photocathode, is a Na–K-Cs-Sb thin-film compound, manufactured by depositing consecutive layers of sodium, potassium, and cesium onto the thick

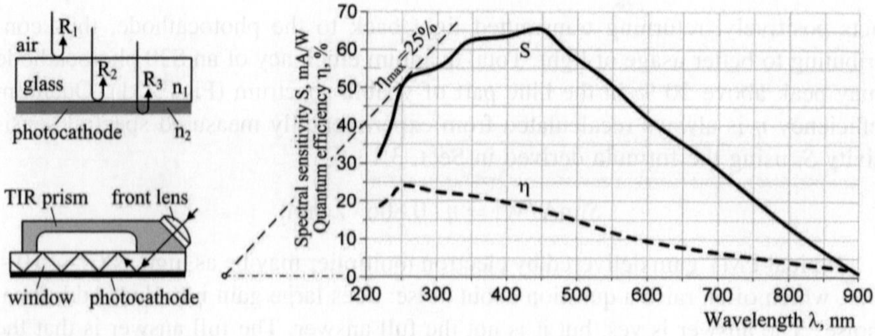

Fig. 3.31 Reflection budget of a PMT window and typical sensitivity curves. Some additional improvement can be made by a total internal reflection (TIR) prism glued to the window. The aim of the prism is to direct the rays at approximately 45° to the window surface without considerable reflection at the front interface. The second leg of the prism serves only for stronger hold on the window. Manufacturers of PMTs do not offer this option

antimony layer in high vacuum. Neither the exact chemical composition or structure of this sandwich, nor true functions of individual components are known. The manufacturing process is controlled by measuring light transmission coefficient and photoemission current during deposition. Refractive index of the final film varies, depending on the manufacturing process, and is known to range from 2.5 to 4.5.

There are three reflecting interfaces in the optical path: air-glass, glass-photocathode, and photocathode-vacuum (Fig. 3.31). The first two contribute negatively, attenuating the optical flux that reaches the photocathode. Refractive index of glass or fused silica (for UV applications) is typically $n_1 = 1.5$. The Fresnel formula then gives reflection coefficient at the first interface

$$R_1 = \left|\frac{n_1 - 1}{n_1 + 1}\right|^2 \approx 4\,\%,$$

which can be reduced to less then 1 % by additional antireflection coating onto the glass. Relatively high refractive index of the photocathode film $n_2 \sim 3$–4 adds much stronger reflection from the second interface

$$R_2 = \left|\frac{n_2 - n_1}{n_2 + n_1}\right|^2 \sim 10\text{–}20\,\%,$$

which cannot be compensated for by any coating. Finally, reflection from the bottom interface with the reflection coefficient

$$R_3 = \left|\frac{n_2 - 1}{n_2 + 1}\right|^2 \sim 20\text{–}30\,\%$$

acts positively, returning transmitted light back to the photocathode, thus contributing to better usage of light. Total quantum efficiency of an S20 photocathode may peak above 20 % in the blue part of visible spectrum (Fig. 3.31). Quantum efficiency η is always recalculated from experimentally measured spectral sensitivity S, using the formula derived in Sect. 3.1:

$$S[\mathrm{mA/W}] = \eta \cdot 0.806 \cdot \lambda\,[\mathrm{nm}].$$

Typical PMT gain delivered by electron multiplier may be as high as $G \sim 10^6$–10^7, which often raise a question about noise: does large gain introduce additional noise? The answer is yes, but it is not the full answer. The full answer is that the electron multiplier is the least noisy amplifier that can be made, and the electrical signal-to-noise ratio at the output of the PMT remains practically the same as the optical one at its input. This conclusion requires some mathematics. With optical intensity F, exposure time T, and photocathode quantum efficiency η, the average number of photoelectrons is

$$v = F \cdot T \cdot \eta.$$

The actual number of photoelectrons m is a random variable with Poisson probability distribution

$$p(m) = \frac{v^m e^{-v}}{m!}.$$

Each photoelectron creates on the average G secondary electrons at the output of the multiplier, so that the average number of electrons after multiplication is $G \cdot v$. Probability of creating M secondary electrons from exactly m photoelectrons may again be assumed Poissonian with the average value $G \cdot m$:

$$p(M|m) = \frac{(m\,G)^M}{M!} e^{-mG}.$$

Probability of generating M secondary electrons at the PMT output with the input optical flux, creating v photoelectrons on the average, is the sum of partial probability functions:

$$P(M) = \sum_{m=0}^{\infty} p(M|m) \cdot p(m) = \sum_{m=0}^{\infty} \frac{(m\,G)^M e^{-mG}}{M!} \cdot \frac{v^m e^{-v}}{m!},$$

understanding that $0! = 1$. The average number of secondary electrons $\overline{M}$ must be $G \cdot v$, and it is indeed so, which can be proved by rearrangement of the sum

$$\overline{M} = \sum_{M=0}^{\infty} \sum_{m=0}^{\infty} M \frac{(m\,G)^M e^{-mG}}{M!} \cdot \frac{v^m e^{-v}}{m!} = \sum_{m=0}^{\infty} \frac{v^m e^{-v} e^{-mG}}{m!} \left[\sum_{M=0}^{\infty} M \frac{(m\,G)^M}{M!} \cdot \right]$$
$$= G \cdot v.$$

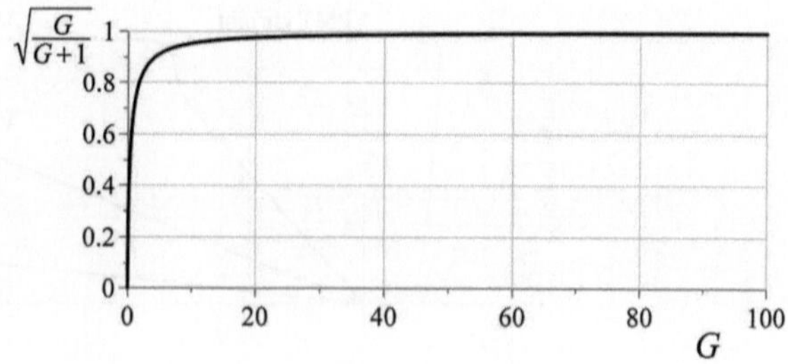

Fig. 3.32 Function $\sqrt{G/(G+1)}$ quickly reaches its maximum value

We call the noise the real-mean-square variation σ of photo- or secondary-electrons:

$$\sigma = \sqrt{\overline{m^2} - \overline{m}^2},$$

and the signal-to-noise ratio SNR

$$\mathrm{SNR} = \frac{\overline{m}}{\sigma}.$$

For Poisson distribution of photoelectrons, $\overline{m} = \nu$, $\sigma = \sqrt{\overline{m}} = \sqrt{\nu}$, and

$$\mathrm{SNR}_{\mathrm{input}} = \sqrt{\nu}.$$

For the secondary electrons, some mathematical manipulation gives their average square number

$$\overline{M^2} = \sum_{M=0}^{\infty} M^2 P(M) = G^2(\nu^2 + \nu) + G\nu$$

that leads to $\sigma = \sqrt{G\nu(G+1)}$ and

$$\mathrm{SNR}_{\mathrm{output}} = \frac{G\nu}{\sqrt{G\nu(G+1)}} = \sqrt{\frac{G}{G+1}} \cdot \mathrm{SNR}_{\mathrm{input}}.$$

Theoretically, since always $\sqrt{G/(G+1)} < 1$, the signal-to-noise ratio at the PMT output is always smaller than at its input (Fig. 3.32). However, practically, they are almost the same already at as small gains as 10^2.

PMTs are designed for low-light applications, therefore they saturate quickly as the optical flux increases, especially the PMT modules that use miniature photomultipliers with small photocathode area (Fig. 3.33). The smaller the photocathode and dynodes area is, the smaller is the maximum current delivered to the anode. With the input optical flux increasing, the photocurrent at the last sections of the dynode system increases to values comparable to those in the resistive divider circuit. This causes voltage redistribution, resulting in visible increase of

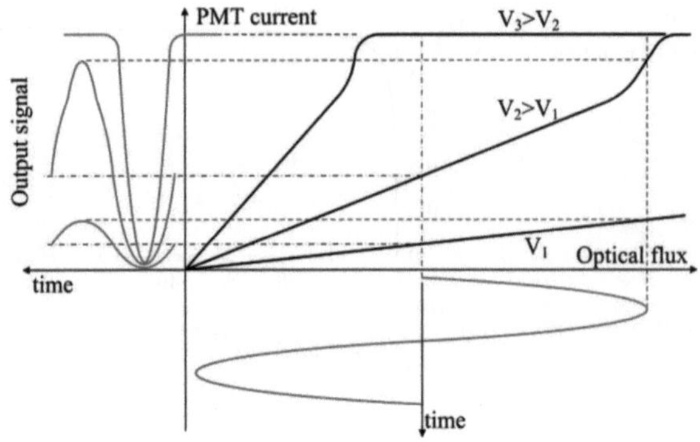

Fig. 3.33 With moderate optical fluxes, PMT shows exceptional linearity. Saturation is preceded by sharp increase of gain. Higher cathode voltage *V* makes higher gain and leads to earlier saturation. Saturation current may be roughly considered as a constant for a particular PMT

the PMT gain. As it happens, further increase of optical flux quickly leads to saturation, when output current cannot follow variations of the input optical flux. The gain of the PMT, i.e. the slope of the current-flux curve, can be tuned by changing cathode voltage V. Higher V results in earlier saturation, producing non-linear response. Sometimes, it may be difficult to realize what has happened, when you clearly see the signal at the oscilloscope, but the amplitude does not react to light variations. The solution is simple: set lower PMT gain.

As PMTs are mostly used in low-light applications, the dark current is a very important factor. Dark current originates from thermal electron (thermionic) emission from the photocathode. As such, it can be reduced by cooling the photocathode or the entire PMT. There are two processes, associated with the cooling: reduction of the dark current and change of spectral sensitivity (Fig. 3.34). Luckily, the second process acts positively in UV–visible domain, making cooling an efficient technique, when utmost sensitivity is required. Cooling is not a simple procedure, especially on large-area PMTs. In order to avoid condensation on the input window, the PMT should be placed inside a sealed housing filled with dried gas. Thermoelectrically cooled PMT modules are available on the market. The most efficient way to create cooled PMT modules is to use miniature metal-package photomultipliers: small size eases requirements to thermoelectric cooler, and metal case significantly improves thermal conductivity. Not only this type of PMTs is most suitable for cooling, but it is used in the majority of compact PMT modules as well (Fig. 3.35).

PMT is a very fast photodetector, with time resolution about several nanosecond. Metal-package PMTs and PMT modules are among the best on this scale because small size of the electrodes makes transient time minimal. However, microchannel plates (MCPs) offer even better time response. These devices are described in the next section.

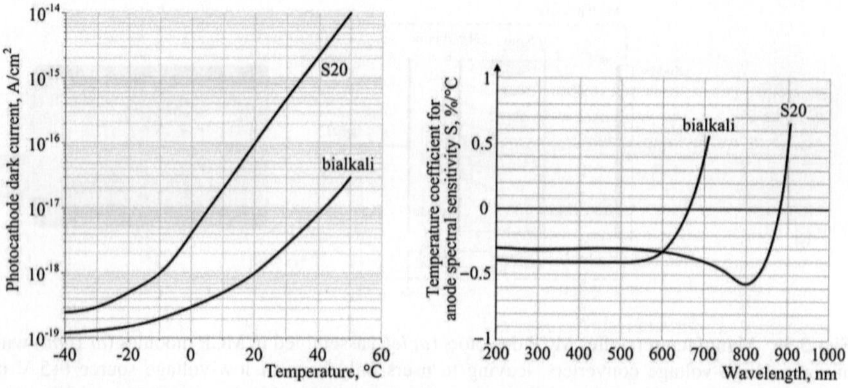

Fig. 3.34 Cooling reduces dark current by orders of magnitude. Spectral sensitivity increases with cooling in UV–visible domain (temperature coefficient is negative), but decreases in near infrared around 900 nm. Note that the *left figure* shows the dark current of the photocathode alone, whereas the *right figure* shows anode sensitivity, which includes temperature changes of the electron multiplier

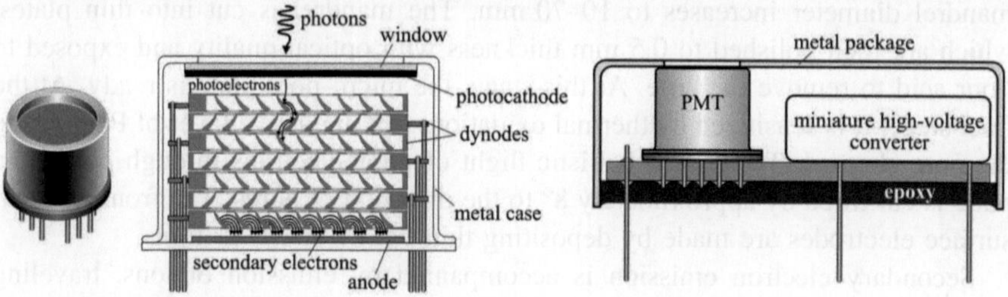

Fig. 3.35 Metal-package PMTs (*at left*) are used in compact PMT modules (*at right*). Dynode system is made of thin finely perforated metal foil (*in the middle*)

3.7 Microchannel Plates

In metal-package photomultipliers, dynode system is composed of perforated foil with windows of macroscopic size ~0.5 mm. Microchannel plates (MCPs) work in exactly the same way but differ by the size of windows, which are better characterized by the word "pores", with the diameter of about ten micrometers (Fig. 3.36). The technology of manufacturing microchannel plates is an example of lucky physical coincidence that brings the desired result. It is based on the fact that the flint glass, a sort of glass with high refractive index, contains much of lead oxide (PbO)—up to 24 %. Therefore, it is resistant to flour acid—a common etching agent for glass. And the lucky coincidence is that, at the same time, PbO is one of the best electron-emissive materials. First, a hollow billet of lead oxide cladding glass is filled with a sort of glass core that can be easily removed by

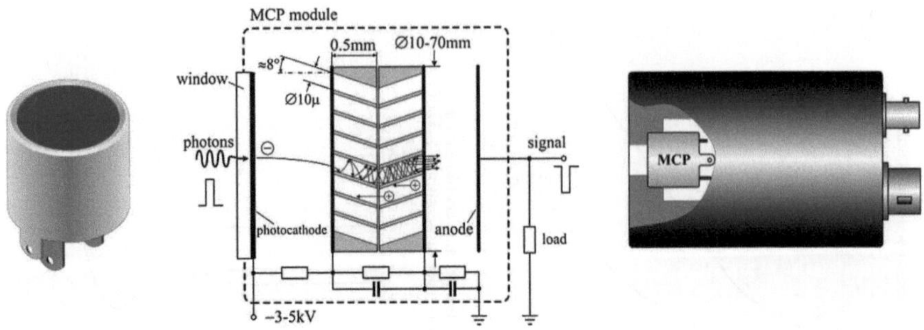

Fig. 3.36 Manufacturers offer MCP detectors (*at left*) assembled in MCP modules (*at right*) with miniature high-voltage converters, leaving to users only connect low-voltage source (+5 V or +12 V) and signal cable. Anode is not connected to ground inside the module

chemical etching. Such a piping has initial diameter of about 1 mm. It is then drawn through a furnace for the first time to implode, cut into many sections of about 10 cm long, and packed into a tight bundle. The bundle is drawn through the furnace again for softening, stretched, and cut into sections once more. This procedure is repeated many times until the core diameter shrinks to 10 μ and the mandrel diameter increases to 10–70 mm. The mandrel is cut into thin plates, which are then polished to 0.5 mm thickness with optical quality and exposed to flour acid to remove the core. At this stage, the micro-pore plate is ready. At the final stage, it is sensitized by thermal oxidation, making active layer of PbO inside the pore channel. To prevent ballistic flight of photoelectrons through pores, the plate is cut tilted by approximately 8° to the direction of pores. The front and rear surface electrodes are made by depositing thin Fe–Cr or Ni–Cr films.

Secondary electron emission is accompanied by emission of ions, traveling back to photocathode. Ions can poison the photocathode if they reach it. To prevent this from happening, two microchannel plates are combined mirror-like, which is commonly known as the chevron design (Fig. 3.36).

Small size of the pores and thinness of plates result in extremely high temporal resolution of less than 1 ns, which is almost ten times better than that of PMTs. MCPs produce the same gain as PMTs, i.e. about one million in a single-channel devices, may have large front area with micrometer-scale spatial resolution, and therefore are widely used as image intensifiers (Chap. 9). Other advantages of MCPs over PMTs are compactness and insensitivity to magnetic field. The disadvantages are high price, noticeable aging, and long dead time. Aging displays as lower gain at the same cathode voltage, which depends on the total charge transferred through the plate. Some measurements show that the charge density of 0.1 C/cm^2 decreases the MCP gain by 3–10 times. Aging is also known for PMTs, but on a much lower scale: gain loss factor of 2 following withdrawal of 200 C/cm^2. The dead time is a parameter, showing how long the MCP does not respond to input light, after having reacted to strong short optical pulse. When optical pulse comes, it wipes electrons away from the pores, leaving them positively charged. This positive charge blocks amplification of the next pulses, until the electro-neutrality is restored by volume

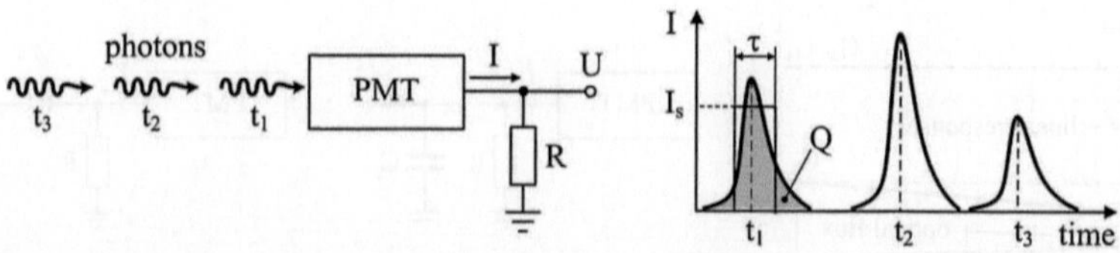

Fig. 3.37 The width τ of a single-photon pulse, of the order of nanoseconds, may be considered as a constant parameter for a given PMT. The area under a single-photon pulse of current is the charge delivered to the output by a single photon: $Q = \int I(t)dt$. Then the single-photon amplitude of the PMT pulse may be estimated as $I_s = Q/\tau$

currents inside the plate. This phenomenon is especially important for photon-counting regimes. Manufacturers never list this parameter in datasheets, but some measurements claim the dead times in the range from microseconds to milliseconds.

3.8 Photomultiplier Receivers

If the gain of photomultipliers is of the order of millions, do we really need an additional amplifier? The answer is yes to all the applications, although in some cases it is not very obvious. Consider the least obvious case of photon counting, when the input optical flux is so weak that single-photon electrical pulses at the anode of a PMT are resolved in time (Fig. 3.37).

The charge Q delivered to the output by a single photon is the product of the PMT gain G by elementary charge $e = 1.6 \times 10^{-19}$ [A·s] $= 1.6 \times 10^{-19}$ (Coulomb):

$$Q = G \cdot e.$$

The output single-photon voltage amplitude U_s is the product of the single-photon current amplitude I_s (Fig. 3.36) by the load resistance R:

$$U_s = \frac{Ge}{\tau} R.$$

With $G = 10^7$, $R = 50$ Ohm, and $\tau = 5$ ns, it gives $U_s = 16$ mV—a value that is quite detectable by a low-noise recorder. These calculations are often considered as the main argument against the amplifier: if the pulse is detectable without an amplifier, why should we add any electronic circuit that brings in additional noise? However, two other effects, accompanying this result, are never addressed: the life time and saturation. With the detecting photon rates $\sim 10^8$ s^{-1}, the average output current is 160 μA. As was told in the end of the previous section, total charge of 200 C transferred through the PMT halves its gain, which in our case will occur after 350 h of operation, i.e. after a month of everyday work. Another effect is

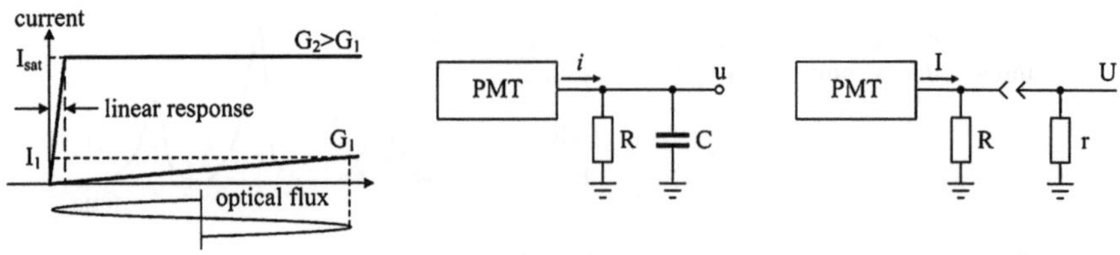

Fig. 3.38 Three reasons, why PMT requires additional amplifier. With high PMT gain, the interval of linear response, also called dynamic range, may be too narrow for the input optical swing (*at left*). High output voltage obtained with large load resistor R may be compromised by low cutoff frequency due to finite load capacitance C (*in the middle*). High load resistor R may be shunted by input resistance r of the recording device, which is typically 50 Ohm, thus lowering the voltage to the noise level (*at right*)

saturation: under the gain $G = 10^7$, PMT works in saturation, leaving no chance to analyze signal amplitude of continuous signals.

Applications, dealing with continuous signals, definitely require additional amplifier after the PMT in order to increase dynamic range, improve frequency response and load capacity. These three reasons are summarized graphically in Fig. 3.38 and briefly commented below.

Gain G_2 gives large current swing up to I_{sat} but narrow range of linear response. Gain $G_1 < G_2$ may give necessary linear response range but smaller current swing I_1. The solution is to set PMT gain at G_1 and use additional amplifier with the gain I_{sat}/I_1. Thus, optimal performance requires adjusting of two parameters: the PMT gain and the gain of the amplifier. Understanding the necessity of additional amplification, some manufacturers offer PMT modules with built-in preamplifier. Be advised, however, that built-in preamplifier with fixed factory-set gain does not make big difference. Even if such a combination is designed correctly (we may only hope), then probably the gain of the preamplifier is set to match its upper voltage limit to saturation current of the PMT. This would ensure no truncation on the upper limit of the PMT signal. However, there may be a problem on the lower limit. For example, if the user wants wider linear optical response, he sets smaller PMT gain, which can possibly make output electrical signal below useful threshold, like 0.5 V for a diode rectifier or 3.6 V for TTL logic circuit.

At high frequency, variable component of the output voltage

$$|u| = \frac{iR}{\sqrt{1 + (RC\omega)^2}}$$

is proportional to the load resistance R, but cutoff frequency $\omega = (RC)^{-1}$ is inversely proportional to it. The voltage amplitude can be raised by additional amplifier, but the cutoff frequency cannot. There are two solutions to this problem, both using amplifiers. In the simplest one, the amplifier does not provide negative feedback to PMT. Therefore, first priority is to guarantee cutoff frequency by lowering R, and then compensate for the amplitude by setting proper amplifier gain. In the more efficient scheme discussed in Sect. 3.5, the amplifier sets

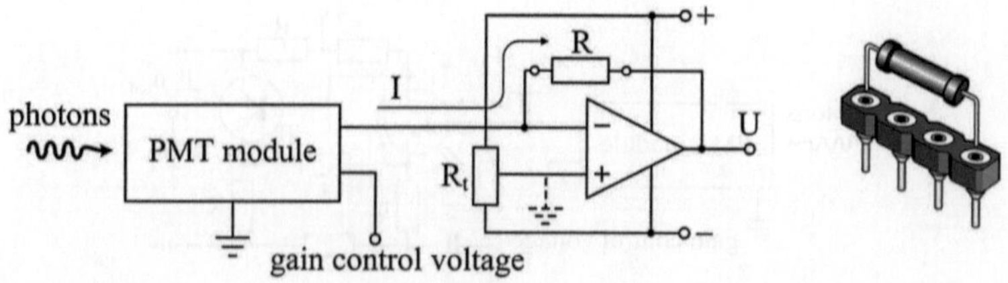

Fig. 3.39 Trimming resistor $R_t \sim 10$ kOhm sets zero output voltage. If small output bias is not important, the non-inverting input may be grounded (*dashed line*). Exchangeable feedback resistor $R < 1$ MOhm sets the amplifier gain. A four-pin section of standard single-in-line package (SIP) socket works well for 0.25 W axial resistor. Avoid potentiometer as the feedback resistor as it introduces additional capacitance and flicker

negative feedback to PMT, not only amplifying the signal but also improving cutoff frequency.

Without the load, PMT output voltage $U = I \cdot R$ may be high, if R is chosen large. Connection of PMT to outer circuits with lower input resistance $r < R$ lowers the signal to $U = I \cdot Rr/(R + r)$. Therefore, connect PMT to an operational amplifier, which always has high load capacity, and forget about low loads. The basic circuit that may be recommended for all the applications below 10 MHz is shown in Fig. 3.39. Due to high open-loop gain of the operational amplifier, only negligible current flows into the inverting input. Therefore, almost whole PMT current I flows through the feedback resistor R, making output voltage $U = I \cdot R$ regardless of the load. All the conclusions about cutoff frequency, drawn in the Sect. 3.5 for photodiodes, remain true for PMTs. Therefore, choose operational amplifiers with gain-bandwidth product more than 100 MHz for high-speed applications. Second amplification stage may be applied as in Fig. 3.25 when the highest bandwidth is needed.

Operational amplifiers are designed with the priority for high open-loop gain $\sim 10^6$, which inevitably sacrifices speed. This high gain plays its role with photodiodes, when it is necessary to shunt relatively large junction capacitance about hundreds of picofarad. Output capacitance of PMTs is much smaller, only few picofarad. Therefore, large negative gain is not needed, and even faster response can be obtained with discrete transistors (Fig. 3.40).

Common mistakes that are known with PMTs are mostly associated with the use of capacitors (Fig. 3.41). When strong background is present together with modulated component, it is always better to filter it out somehow. The straightforward solution is to insert a capacitor somewhere, and the simplest mistake that could be made is to place it right after the PMT anode (Fig. 3.41 at left). Since the current flows in one direction, the capacitor gets charged immediately to the voltage that blocks electrons from reaching the anode. The current stops. Having realized that mistake, a user may move the capacitor to the right (Fig. 3.41 in the middle). This works for a while, but soon the user faces another challenge: inexplicably long tails on short pulses (Fig. 3.41 at right). The

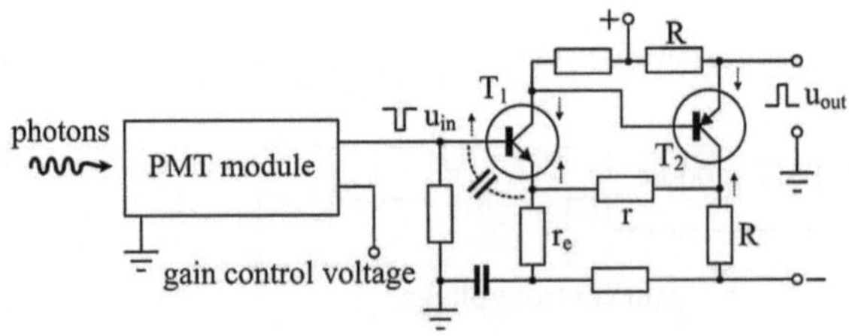

Fig. 3.40 Opposite polarity transistors T_1 and T_2 form an efficient amplifier with an open-loop gain proportional to the product of h_{21} of the transistors. The resistor r establishes an in-phase feedback, which is actually a negative feedback because it minimizes voltage drop on the emitter–base junction of T_1. This feedback virtually disconnects the emitter–base capacitance from the circuit, improving frequency response. Voltage gain is set by the product r/r_e, which may be ~10–30. Rise time of the order of 1 ns is achievable. *Dashed arrows* show phase of voltage variation

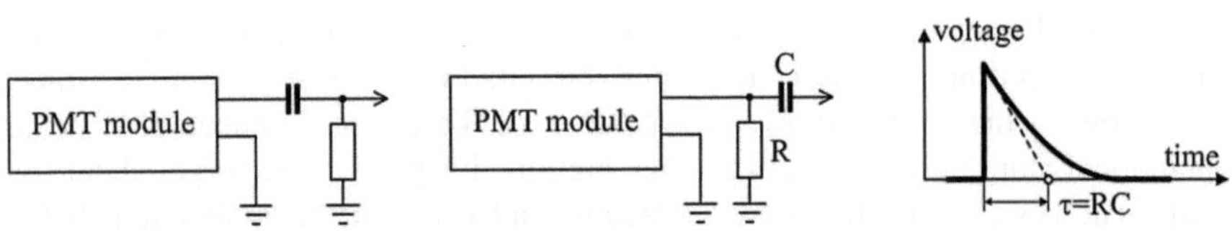

Fig. 3.41 Wrong use of capacitors in the PMT circuits

reason is simple: when optical pulse stops, the anode becomes disconnected from any source of electrons, and the capacitor can discharge only through the resistor R. The time constant is at best RC (if the impedance of the right part of the circuit is much less than R). On sinusoidal signals, however, the circuit will work without noticeable errors, if not to mention phase delay.

List of Common Mistakes

- connecting large-area photodiode to high-resistive load without bias;
- connecting photodiode to a load through a capacitor;
- connecting photomultiplier to a load through a capacitor.

Further Reading

M. Johnson, Photo-detection and measurement: maximizing performance in optical systems, McGraw-Hill, 2003.

S. Donati, Photodetectors. Devices, Circuits, and Applications, Prentice Hall PTR, 2000.

P. Horowitz, W. Hill, The Art of Electronics, Cambridge University Press, 2nd ed., 2001.

Chapter 4
Modulation-Demodulation Techniques

Not everything is possible in optics, and optical modulation has many limitations. This chapter advises on how to choose the proper technique for the specific application.

Abstract This chapter, divided in six sections, begins with the naturally modulated sources: lasers that are modulated due to principle of operation—pulsed modulated solid-state lasers and sinusoidally modulated Zeeman two-frequency lasers. The Zeeman laser presents an exceptional possibility of creating monochromatic sinusoidally modulated light source. With some straightforward mathematics, the conditions of most efficient modulation are derived and experimental oscilloscope traces are presented. This forms the basis for the technology of measuring the cutoff frequency of photoreceivers. Mechanical modulators with perforated rotating wheel are simple devices that are suitable for modulation of any optical beams. But even they can be further optimized by some simple practical tricks. Electro-optical modulators (EOMs) described in the third section are compact, reliable, and surprisingly versatile devices being used for amplitude, polarization, phase and frequency modulation of laser beams. However, their performance can only be fully understood with detailed mathematics, explaining interaction of laser beams with electrically active birefringent crystals. Various types of artifacts may compromise modulation, and this section shows how to avoid it. Numerical computations help to realize these phenomena. Another widely used and also physically complicated device is the acousto-optical modulators (AOMs). The nature of interaction between light and acoustic wave can only be understood with differential equations and special functions, describing periodical solutions for the diffracted beam. Nonetheless, even with this theoretical complexity, it is possible to explain the principle of acousto-optical diffraction in simple terms of reflection from a periodical structure. In this section, deep theory is followed by very practical explanation of the design features of a typical AOM. The next section describes simple and efficient practical solutions for modulating LEDs and LDs, supporting the conclusions by real oscilloscope traces. It is not widely known that even white-light LEDs can be modulated at high frequencies, and temporal evolution of their spectrum is presented, obtained with the help of the gated spectrometer outlined in Chap. 9. The last section of this chapter guides the reader through basics of demodulation techniques and filtering: LC filters, crystal filters, diode rectifiers, active rectifiers, synchronous demodulation (lock-in amplifiers). Along with necessary theoretical explanation, practical circuits and schemes are presented, ready for implementation.

V. Protopopov, *Practical Opto-Electronics*, Springer Series in Optical Sciences 184, DOI: 10.1007/978-3-319-04513-9_4,

4.1 Naturally Modulated Sources

Naturally modulated optical sources, coherent and incoherent, were considered in Chap. 2. All of them provide device-specific type of modulation that cannot be changed (Table 4.1).

Although the fundamental wavelength of Nd:YAG laser is 1064 nm, manufacturers offer very efficient conversion modules with second harmonic in visible domain at 532 nm. Being the pulsed sources, Nd:YAG lasers and Xe flash lamps must be synchronized by trigger pulses, as the generalized scheme in Fig. 4.1 shows.

Delay time is an important parameter, always specified by manufacturers. For Xe flash lamps the delay is typically several microseconds, whereas for the Nd:YAG lasers it is longer: 150–200 μs (see Chap. 2).

An exception from the pulse family are only Zeeman lasers that may be used as nearly perfect source of sinusoidal modulation. It was explained in Chap. 2 that the output of the Zeeman laser is composed of two orthogonal linearly polarized waves with optical frequencies shifted by 1–3 MHz, depending on magnetic field intensity.

Consider Fig. 4.2. Two orthogonal linearly polarized waves do not produce any interference pattern. Indeed, let the two vector waves be

$$\vec{E}_1 = \vec{e}_1 A_1 \exp(i\,\omega_1 t), \quad \vec{E}_2 = \vec{e}_2 A_2 \exp(i\,\omega_2 t)$$

Table 4.1 Modulated optical sources

Source	Type	Pulse width	Repetition rate (Hz)	Frequency	Wavelength (nm)
Nd:YAG laser	Pulsed	7–10 ns	0–20	–	1064; 532
Xe flash lamp	Pulsed	1–10 μs	0–500	–	200–1000
He-Ne Zeeman laser	Sinusoidal	–	–	1–3 MHz	633

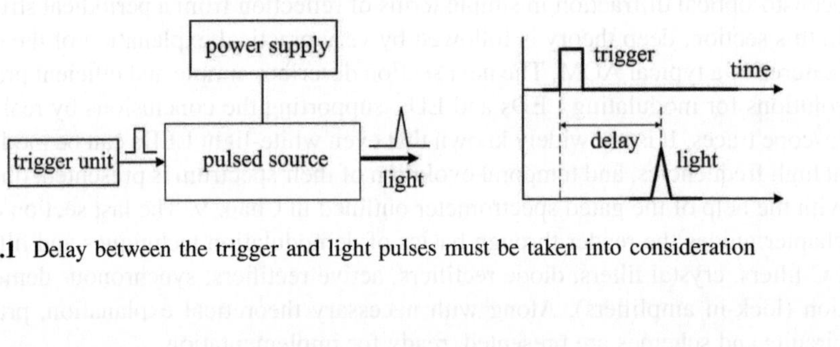

Fig. 4.1 Delay between the trigger and light pulses must be taken into consideration

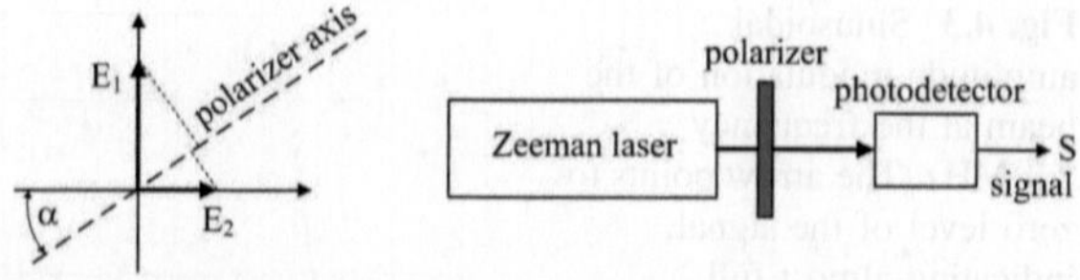

Fig. 4.2 Polarization coupling of the two waves of the Zeeman laser

with unity directional vectors $\vec{e}_1$ and $\vec{e}_2$, real amplitudes A_1 and A_2, and optical angular frequencies ω_1 and ω_2 (cycling frequencies of the order of 10^{15} Hz). As usual, $i = \sqrt{-1}$. Orthogonality means that the scalar product $\vec{e}_1 \cdot \vec{e}_2 = 0$. Photodetector reacts to intensity I, so that the signal S is

$$S = C \cdot I = C \cdot \left|\vec{E}_1 + \vec{E}_2\right|^2,$$

where C is the conversion coefficient.

$$I = \left|\vec{E}_1 + \vec{E}_2\right|^2 = I_1 + I_2 + \vec{e}_1 \cdot \vec{e}_2\, 2A_1A_2 \cdot \cos(\Delta\omega \cdot t),$$

$$I_{1,2} = \left|\vec{E}_{1,2}\right|^2; \Delta\omega = |\omega_1 - \omega_2|.$$

Since $\vec{e}_1 \cdot \vec{e}_2 = 0$, $I = I_1 + I_2$—just the sum of intensities without any interference. In order to make the waves interfere, a polarizer must be inserted between the laser and the photodetector. After the polarizer, directional vectors of the two waves coincide, giving the intensity

$$\begin{aligned} I &= A_1^2 \sin^2\alpha + A_2^2 \cos^2\alpha + \sqrt{A_1A_2}\,\sin 2\alpha \cdot \cos(\Delta\omega \cdot t) \\ &= I_{\text{average}} \cdot \left[1 + \frac{\sqrt{A_1A_2}\,\sin 2\alpha}{A_1^2 \sin^2\alpha + A_2^2 \cos^2\alpha} \cdot \cos(\Delta\omega \cdot t)\right] \end{aligned}$$

with $I_{\text{average}} = A_1^2 \sin^2\alpha + A_2^2 \cos^2\alpha$ standing for the average intensity. The term in square brackets represents typical interference pattern with the amplitude

$$\frac{\sqrt{A_1A_2}\,\sin 2\alpha}{A_1^2 \sin^2\alpha + A_2^2 \cos^2\alpha},$$

oscillating at the intermediate frequency $\Delta\omega/2\pi \sim$ 1–3 MHz. This happens when wavefronts of the two waves perfectly coincide within the photodetector sensitive area, which is called perfect spatial coherence. Any wavefront mismatch leads to smaller amplitude, which may be phenomenologically described by a factor $r \leq 1$:

$$I = I_{\text{average}} \cdot \left[1 + r\frac{\sqrt{A_1A_2}\,\sin 2\alpha}{A_1^2 \sin^2\alpha + A_2^2 \cos^2\alpha} \cdot \cos(\Delta\omega \cdot t)\right].$$

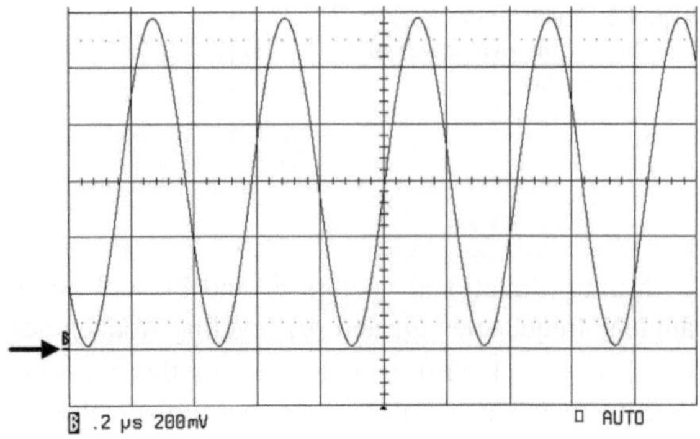

Fig. 4.3 Sinusoidal amplitude modulation of the beam at the frequency 2.3 MHz. The arrow points to zero level of the signal, indicating almost full modulation. Scale in the left lower corner reads: 0.2 μs; 200 mV per division

The conclusion is that the intensity of light oscillates sinusoidally with the intermediate frequency $\Delta\omega$. The second term in square brackets represents modulation, and the modulation depth is

$$r\frac{\sqrt{A_1A_2}\sin 2\alpha}{A_1^2\sin^2\alpha + A_2^2\cos^2\alpha}.$$

Remarkably, Zeeman lasers give the only possibility in optics to produce perfect sinusoidal modulation with the depth close to unity. For that, three independent conditions have to be fulfilled:

$$r = 1,\ A_1 = A_2,\ \text{and}\,\alpha = 45°.$$

The first two relate to the quality of a particular Zeeman laser, and the third one is entirely in hands of the user. Luckily, the Zeeman lasers provide good wavefronts matching and equality of amplitudes. Figure 4.3 shows the typical signal.

Sinusoidal modulation may be useful for characterization of photoreceivers. The idea is explained in Fig. 4.4. With the modulation depth close to unity, light intensity oscillates as

$$I = I_{\text{average}} \cdot (1 + \cos\omega t),$$

where ω stands instead of $\Delta\omega$ for angular modulation frequency ~1–3 MHz.

Let the photoreceiver frequency response be $G(\omega)$. Then the electrical signal

$$S(t) = G(\omega)\cdot I_{\text{average}}\cdot(1+\cos\omega t) = G(0)\cdot I_{\text{average}} + G(\omega)\cdot I_{\text{average}}\cdot\cos\omega t$$
$$= G(0)\cdot I_{\text{average}}\left(1+\frac{G(\omega)}{G(0)}\cos\omega t\right)$$

is sinusoidally modulated with the depth $G(\omega)/G(0)$. Modulation depth can be easily measured as it is shown in Fig. 4.4 (center). We are interested in the cutoff frequency ω_c, which determines high-frequency response $G(\omega)$ (see Chap. 3):

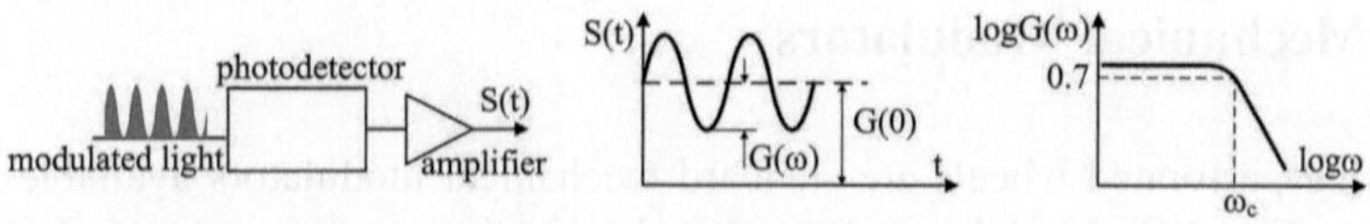

Fig. 4.4 Modulation depth $G(\omega)/G(0)$ of the signal determines cutoff frequency ω_c

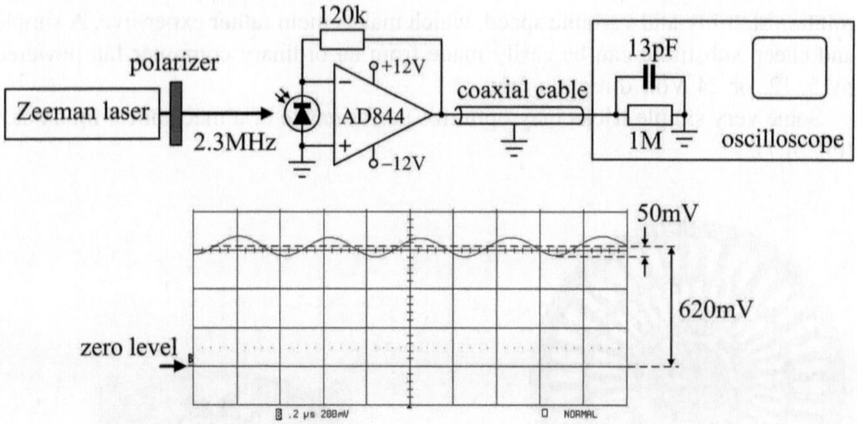

Fig. 4.5 Photoreceiver based on current-to-voltage operational amplifier AD844 with gain-bandwidth product 60 MHz. With shown polarity of the photodiode, the signal is positive. Measured cutoff frequency ≈200 kHz

$$G(\omega) = \frac{G(0)}{\sqrt{1+\left(\frac{\omega}{\omega_c}\right)^2}}.$$

At cutoff frequency $\omega = \omega_c$, signal decreases to $1/\sqrt{2} = 0.7$ of its static value $G(0)$. Inverting this relation, one gets

$$\omega_c = \frac{\omega}{\sqrt{\left(\frac{G(0)}{G(\omega)}\right)^2 - 1}},$$

where ω—known modulation frequency, and modulation depth $G(\omega)/G(0)$ is measured. As an example, consider measurement of the cutoff frequency of the photoreceiver shown in Fig. 4.5.

Modulation depth is ≈0.08, which gives cutoff frequency

$$f_c \approx \frac{2.3\,\text{MHz}}{\sqrt{150}} \approx 200\ \text{kHz}.$$

4.2 Mechanical Modulators

Rotating perforated wheels are standard mechanical modulators available on the market (Fig. 4.6). Modulation frequency lies in the range from several Hertz to several kilohertz. If a mechanical modulator is used for synchronous detection of weak optical signals with a lock-in amplifier, then the reference channel should be added. Carefully designed factory-made modulators are supposed to have high rotation stability and variable speed, which makes them rather expensive. A simple and cheep substitute can be easily made from an ordinary computer fan powered by 5, 12, or 24 Volt direct current.

Some very simple tricks may optimize performance of a mechanical modulator (Fig. 4.7).

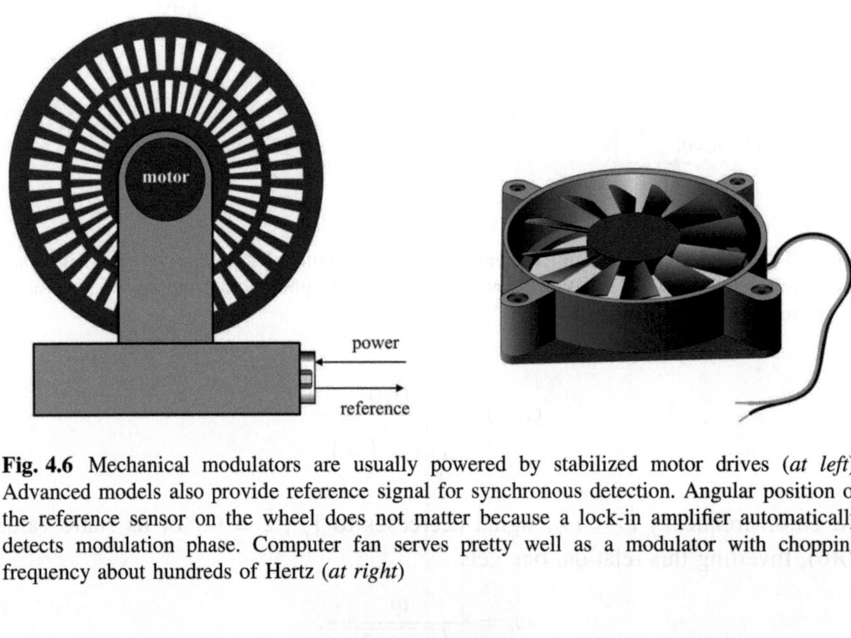

Fig. 4.6 Mechanical modulators are usually powered by stabilized motor drives (*at left*). Advanced models also provide reference signal for synchronous detection. Angular position of the reference sensor on the wheel does not matter because a lock-in amplifier automatically detects modulation phase. Computer fan serves pretty well as a modulator with chopping frequency about hundreds of Hertz (*at right*)

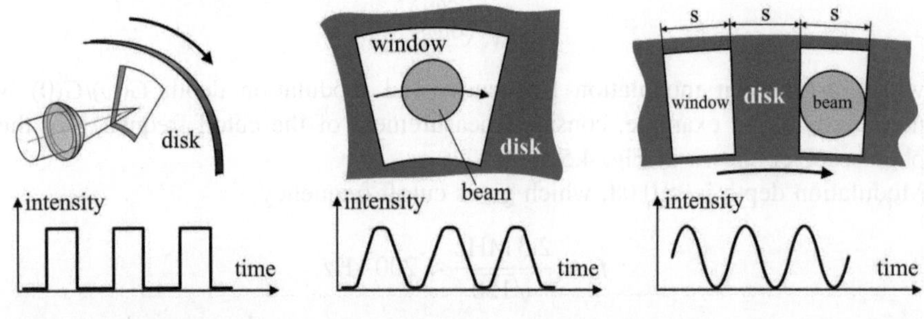

Fig. 4.7 For sharp rectangular signal, the beam must be focused into the plane of the disk (*at left*). Quasi-trapezoidal signal (*in the middle*). Quasi-sinusoidal signal (*at right*)

4.3 Electro-Optical Modulators

Electro-optical modulators (EOMs) are compact, reliable, and surprisingly versatile devices being used for amplitude, polarization, phase and frequency modulation of laser beams (Fig. 4.8). Due to physics of operation, EOMs require linearly polarized narrow (1–3 mm) beams of small angular divergence. For that reason, they are mostly used with lasers as light sources.

Birefringent crystals, like lithium niobate ($LiNiO_3$), lithium tantalate (LiTaO), potassium dihydrogen phosphate (KDP), form the active medium of an EOM. In such a crystal, waves with different polarizations propagate with different speed, i.e. experience different indices of refraction. Specifically, lithium niobate—the most widely used uniaxial crystal on the market—has two definite orthogonal axes of birefringence, called z and x axes (Fig. 4.9). Propagating perpendicular to $z-x$ plane, i.e. in y direction, the wave with the electric field vector along z experiences refractive index n_1, whereas for the orthogonally polarized wave, with its electric filed along x, the refractive index is n_2. These refractive indices change

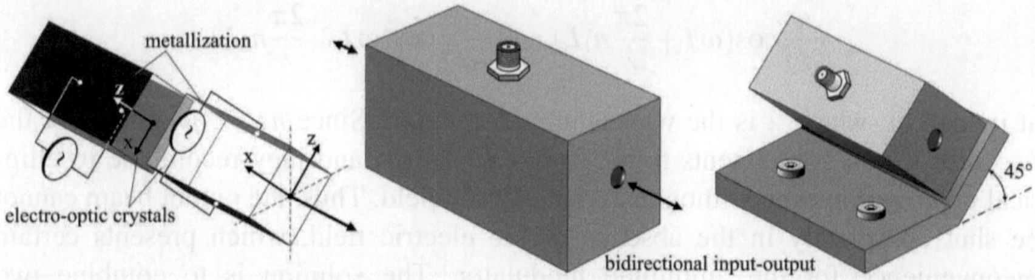

Fig. 4.8 EOMs in vertical orientation are used for amplitude and polarization modulation (*at left*). Bolted to 45° bracket, they may be used for phase and frequency modulation (*at right*). Vertical orientation is the common convention due to predominant vertical polarization of the majority of lasers. Specially made EOMs for only phase/frequency modulation also have vertical orientation like in the left figure, but it is difficult to use them for amplitude modulation. No polarizers are installed inside the EOM. With the refractive index of the crystal about 2, the reflection would be around 30 % from each surface, making antireflection coating necessary

Fig. 4.9 In $LiNbO_3$, $n_1 = 2.20240$, $n_2 = 2.28647$ at the wavelength of HeNe laser 632.8 nm

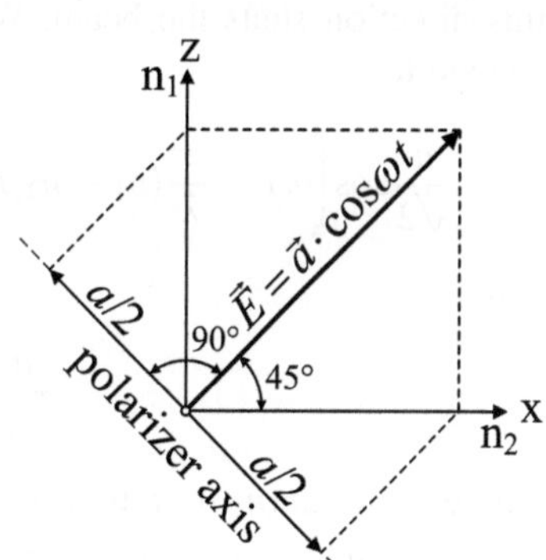

when the external electric field is applied to the crystal. The idea of an amplitude modulator is to change refractive indices so as to transform linear polarization at the input into orthogonal linear polarization at the output. Then the crosswise-oriented polarizer installed in the output beam (Fig. 4.9) transmits the beam when the external electric field is applied, and shuts it when the electric field is zero.

Consider this idea in more detail. When the external electric field is applied to the crystal along z axis, refractive indices change to

$$n_1' = n_1 + \varepsilon_1, \quad n_2' = n_2 + \varepsilon_2,$$

with ε_1 and ε_2 proportional to the electric field. Consider now what happens at the output of a crystal of the length L when the input wave with optical angular frequency $\omega \sim 10^{15}$ Hz is polarized 45°, as in Fig. 4.9. Projections on z and x axes are the same

$$\frac{a}{\sqrt{2}} \cos(\omega t)$$

at the input of the crystal and

$$\frac{a}{\sqrt{2}} \cos(\omega t + \frac{2\pi}{\lambda} n_1 L) \text{ and } \frac{a}{\sqrt{2}} \cos(\omega t + \frac{2\pi}{\lambda} n_2 L)$$

at its output, where λ is the wavelength in vacuum. Since $n_1 \neq n_2$, phases of the two orthogonal components differ from each other, and they recombine to elliptical polarization even without external electric field. Thus, the output beam cannot be shut completely in the absence of the electric field, which presents certain inconvenience for the amplitude modulator. The solution is to combine two identical crystals in series, like in Fig. 4.8, and apply opposite voltage polarity to them. Then with zero voltage, both the output projections of the wave field are the same

$$\frac{a}{\sqrt{2}} \cos\left[\omega t + \frac{2\pi}{\lambda}(n_1 + n_2)L\right],$$

recombining to the initial wave vector. Now, the polarizer installed crosswise to this direction shuts the beam. With the external voltage applied, the projections change to

$$\frac{a}{\sqrt{2}} \cos\left[\omega t + \frac{2\pi}{\lambda}(n_1 + n_2)L + \frac{2\pi}{\lambda}(\varepsilon_1 - \varepsilon_2)L\right] = \frac{a}{\sqrt{2}} \cos(\omega t + \varphi_0 + \varphi)$$

and

$$\frac{a}{\sqrt{2}} \cos(\omega t + \varphi_0 - \varphi),$$

where $\varphi_0 = 2\pi(n_1 + n_2)L/\lambda$ and $\varphi = 2\pi(\varepsilon_1 - \varepsilon_2)L/\lambda$. It must be noted that the signs of ε_1 and ε_2 may be opposite, which increases the effect, or not, depending on

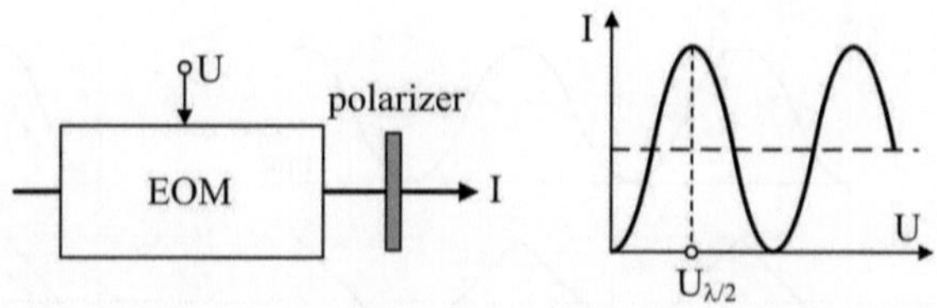

Fig. 4.10 Polarizer installed in the output beam of an EOM converts polarization modulation into amplitude modulation

the particular type of the crystal, but it does not change the final result. Now we have to compute the projection of the output wave field onto the polarizer axis. It is composed of the wave that has come through z axis

$$\frac{a}{2}\cos(\omega t + \varphi_0 + \varphi)$$

and the wave that has come through the x axis

$$-\frac{a}{2}\cos(\omega t + \varphi_0 - \varphi).$$

The opposite sign is ascribed to the latter one because these components act in opposite directions, as it is clearly seen from Fig. 4.9. Applying well-know trigonometric identity for the cosine difference, obtain:

$$-a \cdot \cos(\omega t + \varphi_0) \cdot \sin\varphi.$$

The optical intensity I is then

$$I = \frac{a^2}{2}\sin^2\varphi = I_0 \cdot \sin^2\varphi,$$

where $I_0 = a^2/2$ is the intensity of the input beam. Since the electro-optic effect is linear (the so-called Pockels effect), phase is proportional to voltage U with the proportionality coefficient k:

$$\varphi = k \cdot U.$$

Therefore, the voltage applied to EOM controls its output intensity from zero to its maximum value (Fig. 4.10). The voltage required to fully open the EOM is called the half-wave voltage $U_{\lambda/2}$ because it makes optical path difference Δ between the orthogonal projections equal to half of the wavelength:

$$\varphi - (-\varphi) = 2\varphi = \pi = \frac{2\pi}{\lambda}\Delta, \quad \Delta = \frac{\lambda}{2}.$$

The half-wave voltage $U_{\lambda/2}$ is the main parameter of the EOMs because it directly determines attenuation:

$$\frac{I}{I_0} = \sin^2\left(\frac{U}{U_{\lambda/2}} \cdot \frac{\pi}{2}\right).$$

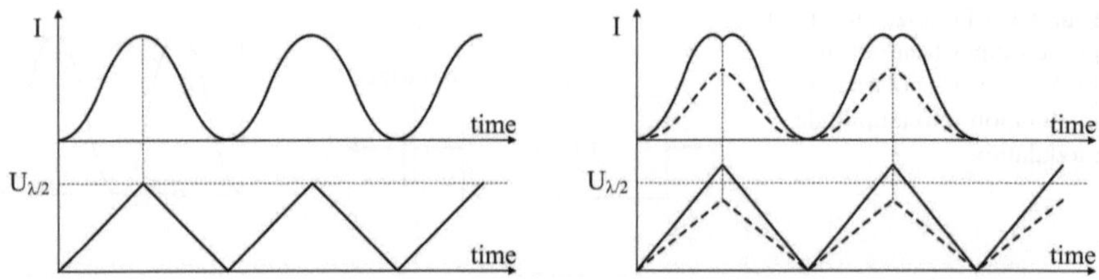

Fig. 4.11 For sinusoidal modulation, voltage amplitude must be kept exactly equal to $U_{\lambda/2}$

The half-wave voltage depends on many factors, such as the crystal electro-optic response, its length L, beam diameter. Actually, the crystal, with a certain voltage U applied to it, reacts to electrical field inside it, which is inversely proportional to cross-sectional size of the crystal. In turn, the crystal cross-section must be as big as not to truncate the laser beam. Therefore, the bigger the laser beam diameter is, the higher is the half-wave voltage. Typically, for 2 mm beams, $\lambda = 633$ nm, and $LiNbO_3$ crystals, $U_{\lambda/2} \sim 200$ V for the amplitude-type EOMs. The beam diameter increases the half-wave voltage proportionally. On the contrary, the crystal length L reduces the half-wave voltage proportionally, since φ is proportional to L.

With EOM, rectangular modulation is not a problem if only the rising/falling edges are not too short. But with pulse edges of nanosecond scale, electronic driver must meet serious requirements. The capacitance of the EOM itself is not too big: about ten picofarad. But 50 Ohm coaxial cable adds ~100 pF per meter of length. Therefore, if the driver is connected to EOM through 1 m cable, then it must pump 200 V through 100 pF in 1 ns. How big current is it? Since the charge is the product of the capacitance by voltage, and on the other hand, it is also the product of current by time, the peak current that the driver must deliver into the EOM is about

$$\frac{10^{-10}[\mathrm{F} = \mathrm{A}\cdot\mathrm{s/V}]\cdot 200\,[\mathrm{V}]}{10^{-9}\,[\mathrm{s}]} = 20\,[\mathrm{A}].$$

Not that it is too much for contemporary technology, but simply not every driver can do that. Therefore, it is always better to make the cable as short as possible.

Theoretically, since $\sin^2 x = (1 - \cos 2x)/2$, EOM can also be used for sinusoidal modulation. For that, the voltage must repeat triangular shape with the amplitude exactly equal to $U_{\lambda/2}$ (Fig. 4.11). Such a signal can be obtained from a digital waveform generator, or by integrating bar-step signal. Additional amplifier will be needed to drive the swing to ~200 V.

Regretfully, upper frequency limit for triangular signals of even the best function generators is about 1 MHz. Nonetheless, there is another possibility to make good high-frequency sinusoidal modulation, if we may sacrifice output intensity. Consider what happens when EOM is driven by a standard sinusoidal signal with the amplitude $u < U_{\lambda/2}$ and angular frequency Ω:

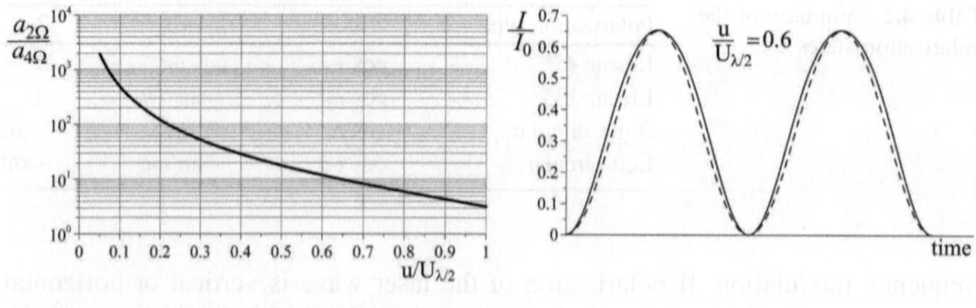

Fig. 4.12 Ratio of harmonics amplitudes $a_{2\Omega}/a_{4\Omega}$ as a function of $u/U_{\lambda/2}$ *(at left)*. The figure at right shows that even with as big ratio $u/U_{\lambda/2}$ as 0.6, the signal I/I_0 is practically sinusoidal (*solid line*), comparing to an ideal sinusoid (*dashed line*)

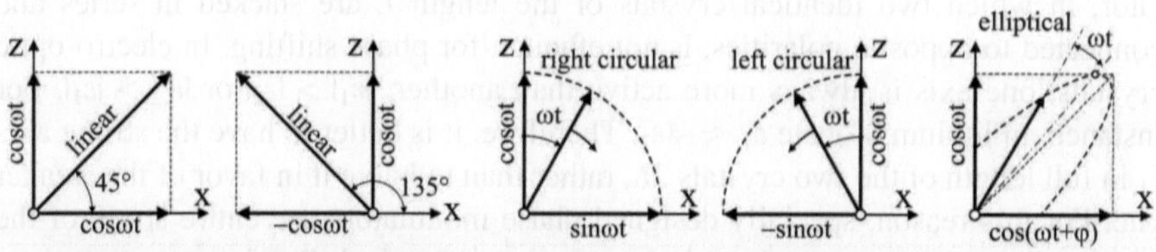

Fig. 4.13 Basic polarization states at the output of the amplitude-type EOM

$$\frac{I}{I_0} = \sin^2\left(\frac{u}{U_{\lambda/2}} \cdot \frac{\pi}{2} \cdot \sin \Omega t\right).$$

Higher harmonics rapidly vanish when $u/U_{\lambda/2} \to 0$, making the signal I/I_0 practically sinusoidal already with $u/U_{\lambda/2} < 0.6$ (Fig. 4.12):

$$\frac{I}{I_0} \approx \left(\frac{u}{U_{\lambda/2}} \cdot \frac{\pi}{2} \cdot \sin \Omega t\right)^2 = \left(\frac{u}{U_{\lambda/2}} \cdot \frac{\pi}{2}\right)^2 \cdot \frac{1}{2}(1 - \cos 2\Omega t).$$

Thus, the smaller is the driving voltage u, the better is the shape of the signal. The only disadvantage is the smaller output intensity. However, with bright laser at the input, this may be not the biggest problem.

If the polarizer in Fig. 4.10 is removed, the EOM acts as the polarization modulator, transforming linear polarization of the input laser beam into elliptical polarization at its output. Both linear and circular polarizations may be considered as the particular cases of elliptical polarization. This variety is represented graphically in Fig. 4.13 and summarized in Table 4.2.

Configuration of the polarization modulator can also be converted to a phase or frequency modulator, if polarization direction of the input wave coincides with z or x axis. In this case, there is no birefringence in the crystal, linear polarization remains intact, and the only electro-optical phenomenon is the change of the optical path due to small variances of the refractive indices ε_1 and ε_2. This leads to variation of optical phase only, which may be used in interferometry or for

Table 4.2 Summary of the polarization states

Polarization type	Z axis	X axis	2φ
Linear 45°	$\cos \omega t$	$\cos \omega t$	0
Linear 135°	$\cos \omega t$	$-\cos \omega t$	π
Right circular	$\cos \omega t$	$\sin \omega t$	$-\pi/2$
Left circular	$\cos \omega t$	$-\sin \omega t$	$+\pi/2$

frequency modulation. If polarization of the laser wave is vertical or horizontal, like in many commercially available lasers, then the EOM should be tilted 45° (Fig. 4.8). With the voltage applied, phase changes by $\varphi = +2\pi(\varepsilon_1 - \varepsilon_2)L/\lambda$ or $\varphi = -2\pi(\varepsilon_1 - \varepsilon_2)L/\lambda$, depending on which one of the two axes z or x coincides with the polarization direction. However, configuration of the amplitude modulator, in which two identical crystals of the length L are stacked in series and connected to opposite polarities, is not efficient for phase shifting. In electro-optic crystals, one axis is always more active than another: $|\varepsilon_1| > |\varepsilon_2|$ or $|\varepsilon_2| > |\varepsilon_1|$. For instance, in lithium niobate $\varepsilon_1 \approx 4\varepsilon_2$. Therefore, it is better to have the strong axis ε_1 in full length of the two crystals $2L$, rather than to halve it in favor of the weaker one. For this reason, specially designed phase modulators use entire space of the housing for only one crystal, with the strong axis vertical. Retaining our previous notations, the phase change is now $\varphi = 4\pi\varepsilon_1 L/\lambda$, i.e. more than two times bigger than before (we assume ε_1 to be the strong axis). This advantage comes from two simple physical reasons: although the total length of the electro-optic medium remains the same $2L$, the entire crystal is now connected to the voltage of one polarity, and the strongest axis is used through the entire length. The term ↔ half-wave voltage ≈means different things for the amplitude and phase modulators, and sometimes may be confusing. For amplitude modulation, the half-wave voltage $U = U_{\lambda/2}$ makes phase difference between two projections equal to π, which means $\varphi = \pi/2$:

$$\varphi = \frac{\pi}{2} \cdot \frac{U}{U_{\lambda/2}}.$$

Therefore, when the amplitude modulator is used for phase modulation, it requires two times higher voltage U to produce phase change $\varphi = \pi$ than its datasheet value $U_{\lambda/2}$. For phase modulation, $U = U_{\lambda/2}$ means $\varphi = \pi$:

$$\varphi = \pi \frac{U}{U_{\lambda/2}}.$$

The conclusion is that specially designed phase modulator is more efficient that the amplitude modulator tilted by 45°. In visible domain, typical half-wave voltage for phase-type EOMs is about one hundred volts.

For applications associated with heterodyning, it is always necessary to produce constant frequency shift about megahertz between two laser beams. Solutions with acousto-optical modulators are not very convenient because they deviate the beam.

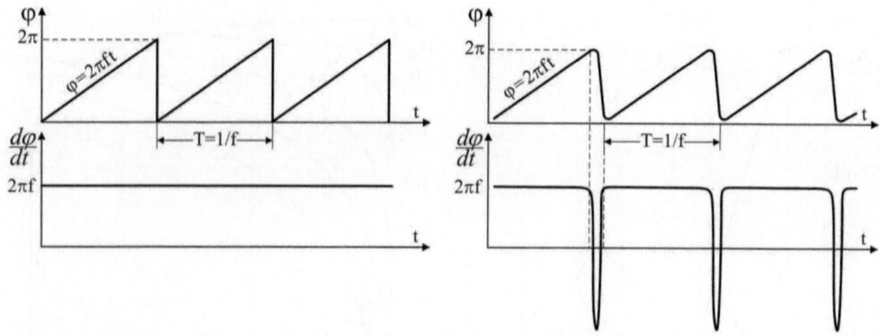

Fig. 4.14 Theoretically, the saw-tooth phase modulation produces constant frequency shift (*at left*). In practice, however, the falling edge cannot be made abrupt and vertices are rounded, which distorts the spectrum (*at right*)

The phase-type EOMs could possibly make this job without angular deviation, but how efficient could it be? Since frequency is the derivative of the phase, theoretically straightforward solution looks like this: make periodical saw-tooth phase modulation $\varphi(t) = 2\pi f \cdot t + \varphi_0$ (Fig. 4.14). Then the frequency shift is exactly f [Hz]. From the practical point of view, however, this approach looks unreal, because the saw-tooth signal has wide spectrum that would be inevitably filtered by lead cables and input capacitance of the EOM, thus feeding the crystal with already a distorted waveform, substantially widening the optical spectrum. Moreover, even creating the saw-tooth signal itself, with the swing about 200 V and the period T of the order of microseconds, is a challenge. Therefore, laboratory limitations dictate the only way: drive the EOM with sinusoidal signal and be aware of how strong the higher harmonics are. Mathematically, the spectrum of the signal with sinusoidally modulated phase can be derived analytically in terms of Bessel functions of the first kind with integer order J_n. Figure 4.15 presents the result graphically. This figure shows amplitudes $A_n(x)$ of the four first spectral components in the following decomposition ($i = \sqrt{-1}$):

$$e^{i\varphi(t)} \equiv e^{i\,[\pi x \sin(\Omega t)]} = \sum_{n=-\infty}^{+\infty} J_n(\pi x)\, e^{in\Omega t},\ J_{-n}(x) = (-1)^n J_n(x),\ A_n = |J_n(\pi x)|.$$

Here the argument $x = u/U_{\lambda/2}$, where u is the modulation voltage. It means that an optical wave E_0 with frequency $\omega \sim 10^{15}$ Hz

$$E_0 = e^{i\omega t},$$

after the EOM will be split in a series of optical waves with slightly different optical frequencies:

$$E = E_0 \cdot e^{i\varphi(t)} = a_0 e^{i\omega t} + a_{\pm 1} e^{i(\omega \pm \Omega)t} + a_{\pm 2} e^{i(\omega \pm 2\Omega)t} + \ldots.$$

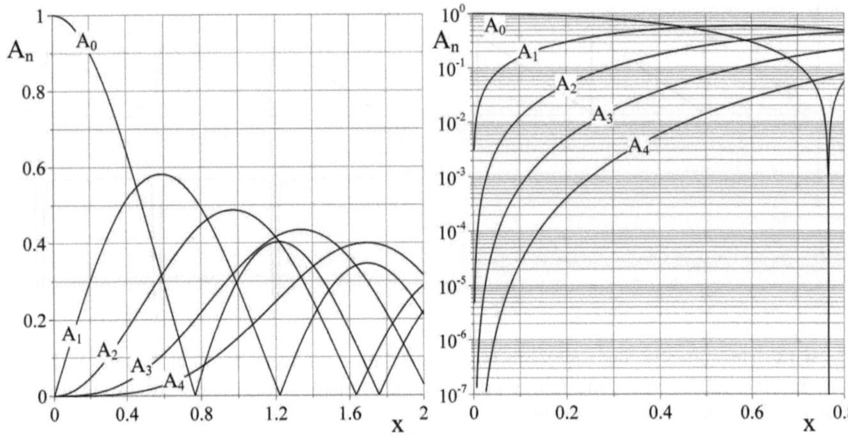

Fig. 4.15 Spectral composition of the signal with sinusoidally modulated phase in linear (*at left*) and logarithmic (*at right*) scales

It is an interesting feature of sinusoidal phase modulation that initial wave with optical frequency ω vanishes completely at $x = 0.765$. This is exactly what is needed for heterodyning: the output beam is shifted in frequency and does not contain initial frequency ω. The multiples of modulation frequency Ω can be easily filtered out by electronic filters. However, acousto-optical modulators (discussed in the next section) do this job better.

EOMs require fine axial adjustment to the input beam.

4.4 Acousto-Optical Modulators

Acousto-optical modulators (AOMs) are based on interaction of an optical wave on artificial phase grating created by an acoustic wave in transparent optical medium like glass. Although they perform only two basic functions—optical switching and frequency shifting—AOMs are widely used in numerous applications. Electro-optical modulators also may be used for switching and frequency shifting, but AOMs do it without special polarization requirements and make frequency shifting naturally, without limitations discussed in the preceding section. In order to understand functionality of AOMs, consider first physical principles of diffraction on acoustic waves in the so-called phase screen approximation (Fig. 4.16). This will require some mathematics, but the result is worth it.

Plane optical wave $A_0 e^{-i\omega t}$ with the wavelength λ comes normally to the acoustic cell excited by angular frequency Ω, with acoustic wave propagating upwards with the acoustic wavelength Λ. Regions of tiny contraction and

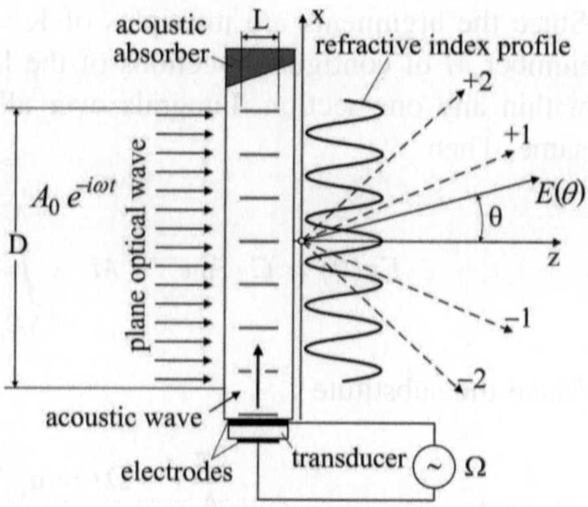

Fig. 4.16 Phase screen approximation assumes no divergence of the optical wave within the acoustic wave region. This approach may also be called the thin layer approximation

expansion, induced by traveling acoustic wave, form sinusoidal modulation of the material refractive index:

$$n(x,t) = n_0 + \delta \cdot \sin(\mathrm{K}x - \Omega t), \quad \mathrm{K} = \frac{2\pi}{\Lambda},$$

so that the optical wave emerges from the cell with the phase

$$\varphi_0 + \varphi \cdot \sin(\mathrm{K}x - \Omega t), \quad \varphi_0 = k n_0 L, \quad \varphi = k\delta L, \quad k = \frac{2\pi}{\lambda}.$$

In the Frounhofer region, i.e. far away from the cell, the optical field is a function of the diffraction angle θ:

$$E(\theta) = C \cdot \int_{-\infty}^{+\infty} E(x,t) \cdot e^{-ik\theta x} dx = C \cdot A_0 e^{-i\omega t} \int_{-\infty}^{+\infty} e^{i\varphi \cdot \sin(\mathrm{K}x - \Omega t)} \cdot e^{-ik\theta x} dx,$$

where unimportant constant C contains all the constant phase terms. Formally, the integral is taken over infinity, understanding that the optical field is zero outside the cell. From general physical phenomenology, we know that optical wave diffracts on periodical structures with the diffraction angle, separating diffraction orders, being

$$\theta_d = \frac{\lambda}{\Lambda}.$$

We are now interested in finding optical intensities along these directions, making the angles $m\theta_d$; $m = 0, \pm 1, \pm 2, \ldots$ with the z axis. For that, evaluate the integral alone:

$$\int_{-\infty}^{+\infty} e^{i\varphi \cdot \sin(\mathrm{K}x - \Omega t)} \cdot e^{-ik\theta x} dx = \int_{-\infty}^{+\infty} e^{i\varphi \cdot \sin\left(\frac{2\pi}{\Lambda}x - \Omega t\right)} \cdot e^{-i\frac{2\pi}{\Lambda} m x} dx.$$

Since the arguments are multiples of $K = 2\pi/\Lambda$, we may divide x axis into big number M of contiguous sections of the length Λ, and evaluate the integral only within any one section. Integrals over all the other sections will be exactly the same. Then

$$E_m(\theta) = C \cdot A_0 e^{-i\omega t} M \cdot \int_{-\Lambda/2}^{+\Lambda/2} e^{i\varphi \cdot \sin\left(\frac{2\pi}{\Lambda}x - \Omega t\right)} \cdot e^{-i\frac{2\pi}{\Lambda}mx} dx.$$

Make the substitute

$$\frac{2\pi}{\Lambda}x - \Omega t = u, \quad dx = \frac{\Lambda}{2\pi}du,$$

to obtain

$$E_m(\theta) = C \cdot A_0 e^{-i(\omega+\Omega)t}(M\Lambda) \cdot \frac{1}{2\pi} \int_{-\pi-\Omega t}^{+\pi-\Omega t} e^{i\varphi \cdot \sin u} \cdot e^{-i\,mu} du.$$

Since the integral of a periodical function over its period does not depend on the shift, specifically Ωt in our case, we may finalize:

$$E_m(\theta) = C \cdot A_0 e^{-i(\omega+m\,\Omega)t} D \cdot J_m(\varphi),$$

with

$$J_m(\varphi) = \frac{1}{2\pi} \int_{-\pi}^{+\pi} e^{i\varphi \cdot \sin u} \cdot e^{-i\,mu} du$$

being the Bessel functions of the first kind. Parameter $D = M \cdot \Lambda$ is equal to vertical dimension of the illuminated area, which have the meaning of the cross-sectional area since the y axis is not involved in computations. This result brings us to two most important conclusions. First, the optical frequencies of diffraction maxima are shifted by $m \cdot \Omega$, which explains why AOMs are so important for heterodyning. Second, amplitudes (intensities) of the diffraction maxima depend on the strength of acoustic wave, which determines maximum variation δ of the refractive index:

$$\varphi = \frac{2\pi}{\lambda} \cdot \delta \cdot L.$$

Finally, the interaction length L is as important as the strength of the acoustic wave. Now, consider intensities of diffraction maxima

$$I_m = |E_m|^2 \sim J_m^2(\varphi)$$

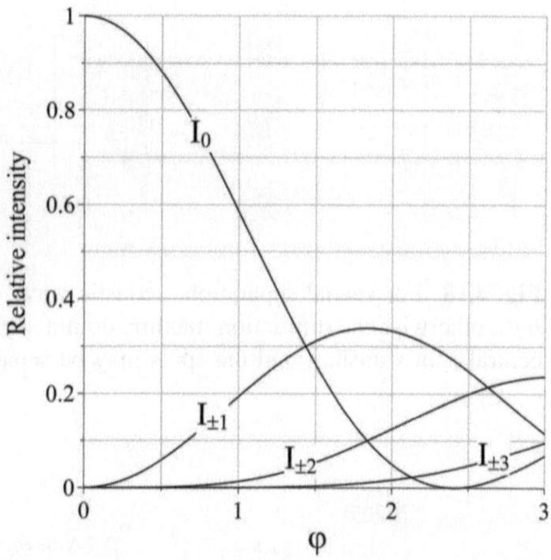

Fig. 4.17 Acoustic wave pumps the intensity out from zero-order optical wave into higher orders

as functions of φ (Fig. 4.17). Mathematical identity that holds true for Bessel functions

$$\sum_{m=-\infty}^{+\infty} J_m^2(\varphi) = 1$$

plays the role of the energy conservation law in our case. It is impossible to create big variations of refractive index δ, as it would lead to cracks, as well as heterogeneities of acoustic wave make it useless to work on long interaction paths L. Therefore, only small φ should be analyzed. As the two parameters δ and L are combined in φ, any increase of one of them leads to redistribution of diffracted intensities from zero-order maximum to higher orders. Phenomenologically, it looks like gradual emerging of bright spots near the main laser beam, while acoustic power increases, until the central (main) spot totally quenches at $\varphi = 2.41$. In order to be able to spatially separate the diffraction maxima, the diffraction angle θ_d should be bigger than angular divergence of the laser beam. Assuming that the input beam is Gaussian with the diameter D, this gives (Fig. 4.18)

$$\frac{\lambda}{D} < \frac{\lambda}{\Lambda}, \text{ or } \Lambda < D.$$

For example, for the dense flint glass as an active medium, the speed of sound is ~3.6 km/s. Being excited by $\Omega/2\pi = 40$ MHz signal, this gives $\Lambda = 0.09$ mm. Thus, a laser beam with, say, $D = 1$ mm will clearly be resolved in all the maxima. However, being focused to a spot of about 0.1 mm, it will not, until the acoustic intensity is increased to the value where $\varphi = 2.41$. Full quench of the zero beam means that the acousto-optical element can work like modulator: without acoustic wave the element is totally transparent, with the acoustic wave of certain amplitude the central beam is fully blocked. However, the first-order

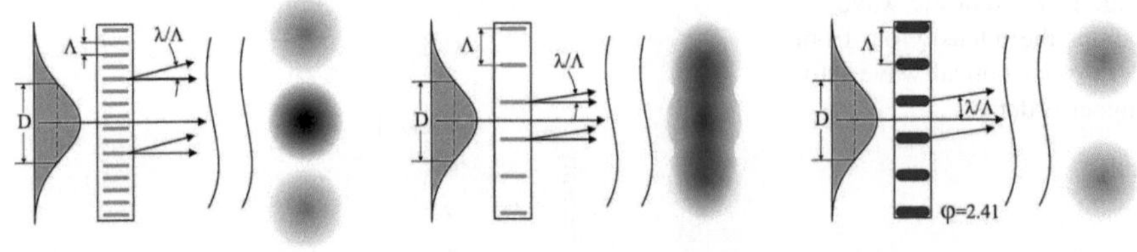

Fig. 4.18 For spatial separation, acoustic wave must be much shorter than the beam width (*at left*), otherwise the diffraction maxima do not separate (*in the middle*). With the phase $\varphi = 2.41$, central spot vanishes, and the spots may be separable again (*at right*)

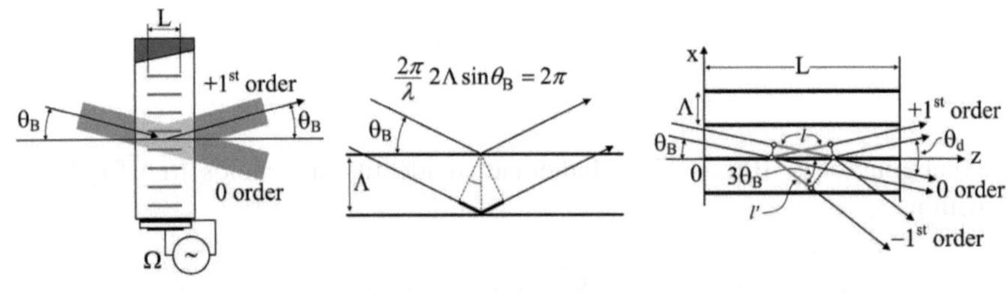

Fig. 4.19 AOM in Bragg geometry. The –1st order wave quenches along the acoustic planes due to destructive interference: the optical path difference $l - l' = l(\cos 3\theta_B / \cos\theta_B - 1)$ is not a multiple of λ. The same for all the higher orders, so that only +1st and zero orders compete at the output. At some value of acoustic power, the zero order totally converts into the +1st order

diffraction peaks $I_{\pm 1}$, with optical frequencies $\omega \pm \Omega$, are always of a primary importance, and Fig. 4.17 shows that they never reach the intensity of the zero-order peak. Another optical modality, called the Bragg scheme or Bragg geometry, solves this problem.

The AOM in Bragg geometry is explained in Fig. 4.19. In this scheme, the input optical beam comes to acoustic planes at the so-called Bragg angle θ_B:

$$\sin\theta_B = \frac{\lambda}{2\Lambda}.$$

Typically, the Bragg angle is small, about 2–3 milliradians, so that it may be written $\theta_B \approx \lambda/2\Lambda$. The smallness of θ_B makes the reflectance at the acoustic planes high even though variations δ of the refractive index are small. Indeed, the Fresnel formulas give the reflection coefficient at the interface with refractive indices n_1 and n_2 as (s-polarization)

$$\left|\frac{n_1 \sin\theta_1 - n_2 \sin\theta_2}{n_1 \sin\theta_1 + n_2 \sin\theta_2}\right|^2 = \left|\frac{1-p}{1+p}\right|^2,$$

where $\theta_{1,2}$ are the inclination angles of the coming and refracted waves, and $p = n_2 \sin\theta_2 / n_1 \sin\theta_1$. Clearly, the bigger p, the bigger the reflectivity. For this case, the Snell law gives $n_1 \cos\theta_1 = n_2 \cos\theta_2$, and assuming $n_2 = n_1 + \delta$ and $\delta \ll 1$, some manipulations give

$$p^2 \approx 1 + \frac{2\delta}{n \sin^2 \theta_1},$$

where $n_2 \approx n_1 = n$. The same result holds true for p-polarization since p may be changed to $1/p$. In glass, variation of the refractive index δ caused by stress F is determined by photoelastic constant β:

$$\delta = \beta \cdot F.$$

Maximum stress that can be applied to glass is limited by its tensile strength, which is about 50 MPa for high-quality medium. The photoelastic constant of an ordinary optic-quality glass $\beta \approx 3 \times 10^{-6}$ MPa^{-1}, which gives $\delta \sim 10^{-4}$. It looks like infinitesimal value relative to glass refractive index $n \approx 1.5$. However, with small θ_1, for example 3 milliradians, $p \approx 4$, and the reflection coefficient even on a single acoustic plane is around 0.5. With M acoustic planes within the beam, total reflection coefficient

$$\left| \frac{1 - p^{2M}}{1 + p^{2M}} \right|^2$$

may reach unity. This is what makes AOM in Bragg geometry efficient.

It is important that the diffraction angle θ_d is twice the Bragg angle: $\theta_d = 2\theta_B$. Under this condition, the input beam and the +1st order diffracted maximum are in the mirror reflection relative to acoustic planes (Fig. 4.19). Therefore, all the partial waves reflected from the acoustic planes are in phase, interfering constructively in the +1st order diffraction maximum along the entire length L of the modulator. On the contrary, the −1st order waves and all the higher orders are not in phase along z axis, and they are eliminated by destructive interference. Hence, only two waves at the output of the Bragg type AOM should be considered: the zero and +1st diffraction orders. We are interested in how they convert into each other. With the notations of Fig. 4.16, the wave at the input plane of the AOM is now

$$E(\vec{r}, t) = A_0 e^{i(\omega t - \vec{k}_0 \vec{r})}$$

where wave vector $\vec{k}_0$ comes at the Bragg angle (Fig. 4.20). Inside the AOM, only two waves propagate with high intensities: the +1st order wave

$$A_1(z) e^{i\,[(\omega + \Omega)t - \vec{k}_1 \vec{r}]}$$

and the zero order wave

$$A_0(z) e^{i\,(\omega t - \vec{k}_0 \vec{r})}.$$

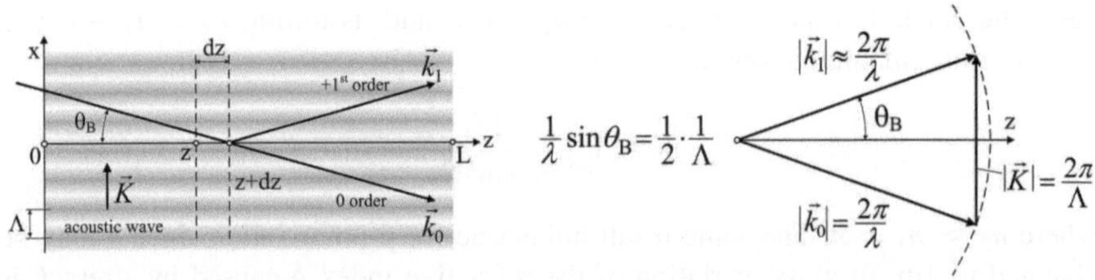

Fig. 4.20 In dynamic theory, the phase screen approximation is used only to describe propagation through thin layer of thickness *dz*. The entire acousto-optic length *L* is big compared to *dz*. Only those waves propagate with high intensities that satisfy the Bragg condition (*at right*)

Since $\Omega \ll \omega$, we may assume $k_1 \approx k_0 = 2\pi/\lambda$. As before, refractive index is modulated:

$$n(x,t) = n_0 + \delta \cdot \sin(\Omega t - \mathrm{K}x), \quad \mathrm{K} = \frac{2\pi}{\Lambda}.$$

Propagating through thin layer *dz*, the combined wave $E(\vec{r},t)$ passes through the phase screen with the phase distribution

$$\varphi(x,t) = k_0 n(x,t)\, dz,$$

and transforms to

$$E(\vec{r}+dz,t) = E(\vec{r},t)\cdot e^{i\phi(x,t)} \approx E(\vec{r},t)e^{i\phi_0}\left[1+\frac{k_0\delta}{2}dz\, e^{i(\Omega t-\vec{K}\vec{r})} - \frac{k_0\delta}{2}dz\, e^{-i(\Omega t-\vec{K}\vec{r})}\right]$$

as both δ and *dz* are small. At the next step, we need to substitute the sum

$$E(\vec{r},t) = A_0(z)e^{i\,[\omega t-\vec{k}_0\vec{r}]} + A_1(z)e^{i\,[(\omega+\Omega)t-\vec{k}_1\vec{r}]}$$

into both left and right parts of the equation above, neglect the terms that do not satisfy the Bragg condition

$$\vec{k}_1 = \vec{k}_0 + \vec{\mathrm{K}},$$

group separately the terms with ω and $\omega+\Omega$, and make use of

$$A(z+dz) - A(z) = dA.$$

The phase term $e^{i\varphi_0}$ does not affect the amplitudes, and must be dropped. The result is the system of differential equations:

$$\begin{cases} \frac{dA_0}{dz} = -\gamma A_1 \\ \frac{dA_1}{dz} = \gamma A_0 \end{cases}, \quad \gamma = \frac{k_0\,\delta}{2} = \frac{\pi\,\delta}{\lambda}.$$

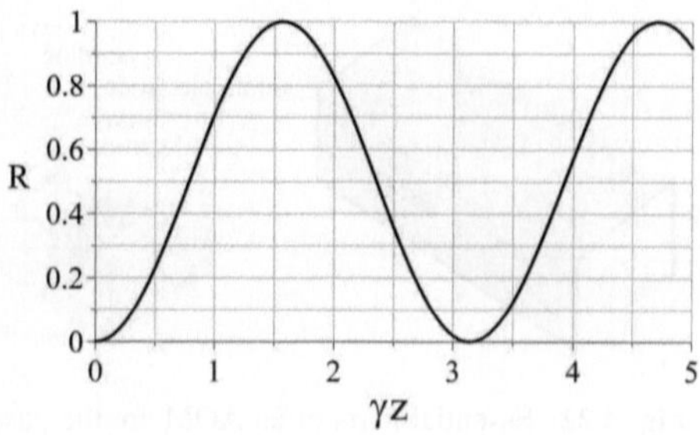

Fig. 4.21 Bragg conversion coefficient reaches maximum when $\gamma z = \pi/2$

The solution is simple:

$$A_0(z) = A_0(0)\ \cos\gamma z,\quad A_1(z) = A_0(0)\ \sin\gamma z,$$

which gives the conversion coefficient for intensities

$$R(z) = \left|\frac{A_1(z)}{A_0(0)}\right|^2 = \sin^2\gamma z.$$

This result is shown graphically in Fig. 4.21. Theoretically, in a Bragg cell of the length L, 100 % conversion occurs when

$$L = \frac{\lambda}{2\delta}.$$

Assuming the crystal being acoustically pumped at the verge of its breakdown, and using the value $\delta \sim 10^{-4}$, we get the lower estimate of the cell length $L \sim 3$ mm at $\lambda = 633$ nm. With an order of magnitude lower power and $\delta \sim 10^{-5}$, a quite reasonable value $L \sim 3$ cm is an estimate for the propagation length needed to achieve maximum conversion in glass. In reality, diffraction efficiency cannot exceed 90 % primarily because of the both optical and acoustic waves divergence.

Typical AOM package incorporates an active crystal with a transducer glued to one side of it, and a radio-frequency resonant excitation circuit (Fig. 4.22). Manufacturers offer a variety of active materials, some listed in the Table 4.3. Typically, AOMs are designed for narrow beams of about 2–3 mm in diameter. The reason is divergence of the sound wave in the crystal, which changes the Bragg condition over the beam diameter, thus reducing efficiency. In order to improve acoustic divergence, transducers are sometimes made with special profile over the crystal length.

In the switching mode, the time constant of an AOM is determined by the time in which acoustic wave crosses the beam. Therefore, in order to achieve fast switching, the beam must be focused into the crystal to a spot of about 50 μ. Then switching time may be about 10 ns. However, focusing makes conversion efficiency poorer, because the Bragg condition cannot be fulfilled equally well for all the spatial components of the focused beam. With 10 ns switching time, efficiency may expected at the level 70–80 %. For the estimates of rise times, use the value 200 ns per one millimeter of beam diameter.

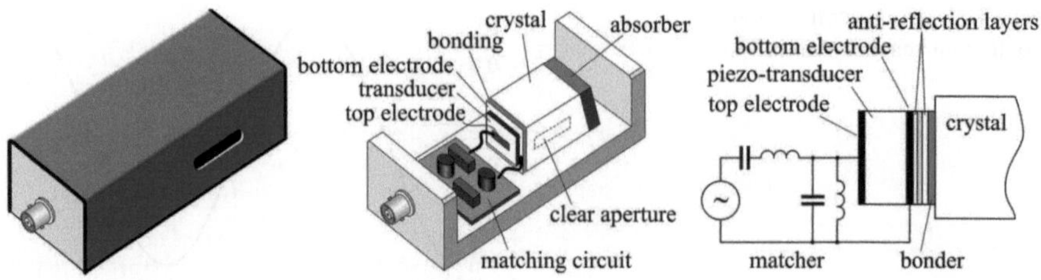

Fig. 4.22 Essential parts of an AOM are the passive radio-frequency matcher and acoustic anti-reflection layers between the transducer and the crystal. The top electrode may also be specially shaped, optimizing divergence of the acoustic wave

Table 4.3 Acousto-optical materials

Material	Type	Refractive index	Density (g/cm^3)	Speed of sound (km/s)
Fused quartz	Glass	1.457	2.20	5.96
Dense flint	Glass	1.6–1.9	3.5–6	3.2–3.6
$LiNbO_3$	Crystal	2.2	4.64	6.5
TeO_2	Crystal	2.26/2.41	6.0	4.2
$PbMoO_4$	Crystal	2.26/2.38	6.95	3.6

AOMs might be considered as ideal instruments for heterodyne applications as they produce diffracted beams shifted in frequency by a definite value equal to the acoustic frequency. However, this frequency, usually above 40 MHz, is always too high for demodulation, presenting significant complications for electronic circuitry. The standard solution to this problem is two AOMs connected in series, as it is explained in Fig. 4.23. All AOMs require fine angular adjustment.

4.5 Electronically Modulated Sources

The easiest way to make electronically modulated light source is to use light-emitting diodes (LEDs) and laser diodes (LDs). Chapter 2 explaines how to connect these devices in continuous-wave mode, and here are some advises on how to make modulated sources. One straightforward solution for rectangular pulse modulation consists of just a transistor and an LED (Fig. 4.24). It works well up to several megahertz modulation rate, if the LED itself responds at this frequency. At zero input voltage, the transistor is closed and no current flows into the LED. When the positive voltage edge comes to the base of the transistor, it opens, but no light is emitted yet because the voltage drop on the LED is below its cutoff voltage, which is commonly 1.5 V for the red- and 3 V for the green-emitting LEDs. This delay happens because some time is needed to charge the capacitance of LED junction from zero voltage to the cutoff voltage. Therefore, for faster

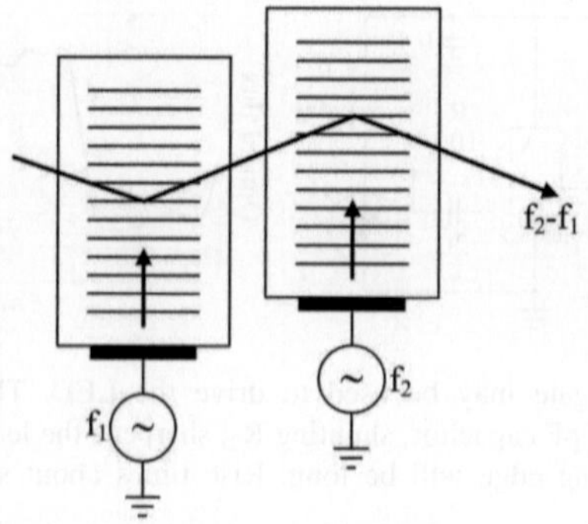

Fig. 4.23 Low frequency shifts. Two AOMs, two function generators, two acoustic drivers, and two adjustable optical tables are needed

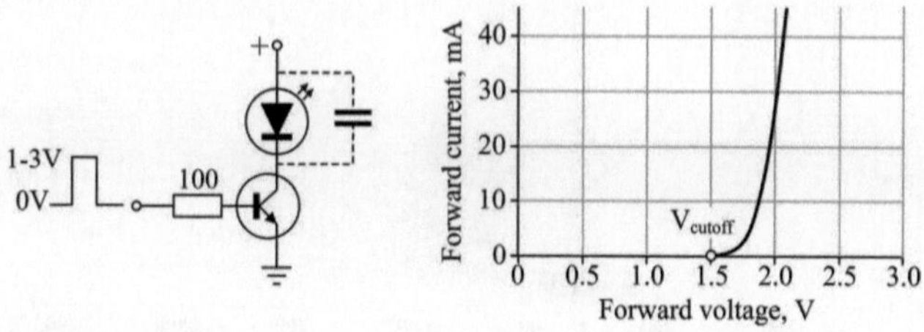

Fig. 4.24 Simplest modulation circuit. The 100 Ohm resistor separates the base from the generator, lowering the base voltage to its normal 0.5 V, which otherwise would be the generator voltage, decreasing the voltage over the LED

modulation, initial voltage on the LED must be preset to the cutoff voltage, which is usually done by the resistive divider R_1–R_2 (Fig. 4.25).

Picking up LEDs from occasional stocks, you never know whether it is fast enough for your application. The speed does not depend on the colour or form-factor: some may work up to megahertz, others may not. The only way to be sure is to order specific fast-response type from a vendor. Surprisingly, white-light LEDs are remarkably fast despite they are based on fluorescence, usually considered as slow mechanism (Chap. 2). Therefore, one may expect that not all the spectral components of a white-light LED are equally fast, and it is true (Fig. 4. 26). Nonetheless, spectrally integrated rise time may be estimated as 30 ns, followed by much longer fall time of about 200 ns.

Solutions for pulse modulation are straightforward, but things complicate when it is necessary to obtain sinusoidal optical flux. It is a common mistake to think that sinusoidal electrical signal applied to a LED will produce sinusoidal optical flux. Deep non-linearity of the current–voltage characteristic of a LED (Fig. 4.24) transforms sinusoidal electrical signal into a series of narrow optical pulses at the output (Fig. 4.27). A closer approximation to sinusoidal optical modulation may

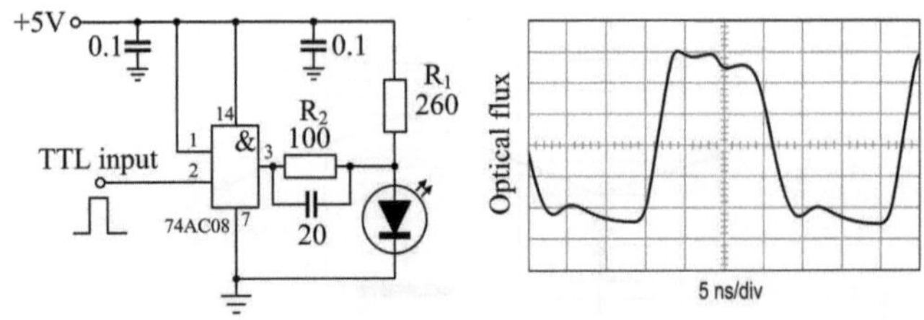

Fig. 4.25 Logical AND gate may be used to drive the LED. The TTL voltage levels are 0.5 «0»–3 «1» V. The 20 pF capacitor, shunting R_2, sharpens the leading edge of the pulse. If it is too large then the falling edge will be long. Rise times about several nanoseconds can be obtained with this circuit

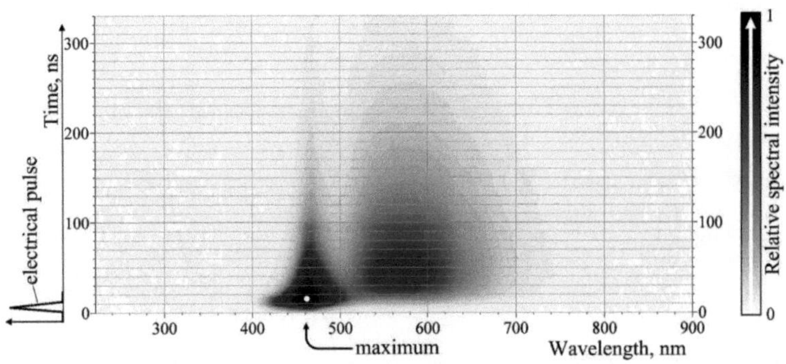

Fig. 4.26 Time-resolved spectrum of a white-light LED excited by a 10 ns electrical pulse. White spot shows position of an absolute maximum at 460 nm. The blue component lasts approximately 30 ns with less than 10 ns rise time, whereas the red fluorescence continues over 100 ns, but not more than 300 ns. The picture was obtained with a gated spectrometer (Chap. 9)

be achieved only with small-amplitude modulation at some pedestal, which should be experimentally adjusted for a particular LED.

Laser diodes (LDs) are far faster sources than LEDs. However, they require more complicated electronic circuits for stabilization (Chap. 2), which should be modified for modulation like it is shown in Fig. 4.28.

For only short-pulse operation, when extremely narrow optical pulses must be generated with large duty ratio, heating and long-term flux stability may be not the key issues, and then a simplified circuitry like that shown in Fig. 4.24 may be used. It is hard to say how short the optical pulse really is because the optical shape is always masked by photodetector response, oscilloscope amplifier, excitation transistor, lead cables, etc. Anyway, with high-speed photomultiplier and 5 ns electrical pulse, as short as 6 ns optical response can be obtained (Fig. 4.29).

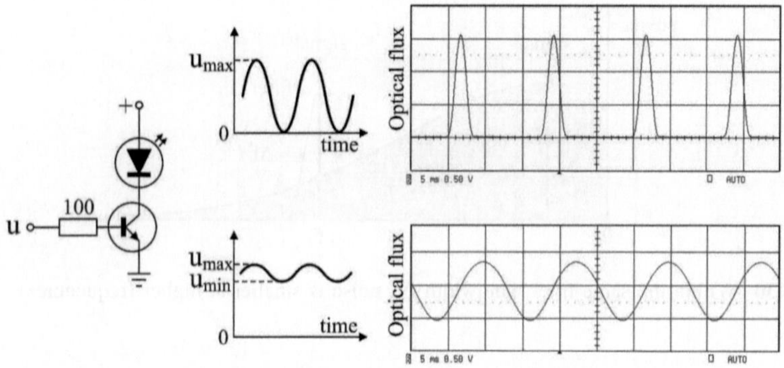

Fig. 4.27 With the input sinusoid oscillating between 0 V and $u_{max} = 1.9$ V, the LED optical emission is just a train of narrow peaks (*upper picture*). Approximately sinusoidal optical wave may be obtained with the pedestal $u_{min} = 1.8$ V (*lower picture*)

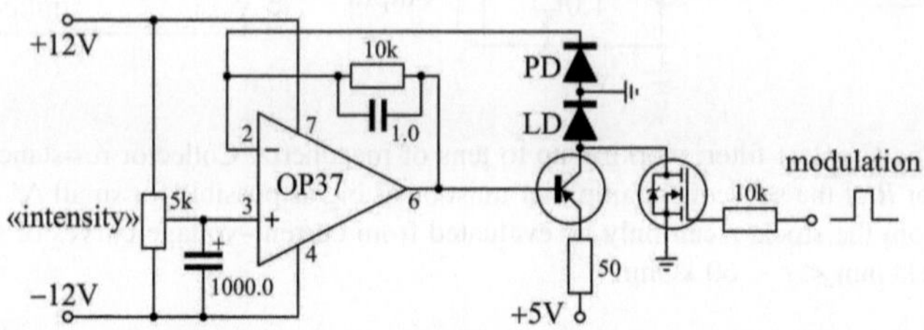

Fig. 4.28 N-channel MOSFET transistor shunts the LD to interrupt optical flux. Interrupt voltage is positive

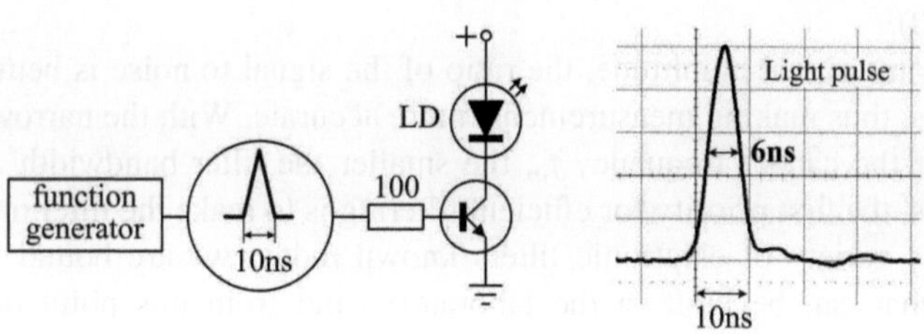

Fig. 4.29 Triangular pulse from the function generator has 10 ns at the base and full width at half-maximum 5 ns. It may be believed that the LD (653 nm) repeats the electrical pulse without significant broadening

4.6 Demodulation

Optical modulation and subsequent demodulation are used basically for the two reasons: either to reduce background and noise, or to extract phase information. Noise reduction can be achieved either by high-frequency filtering or by

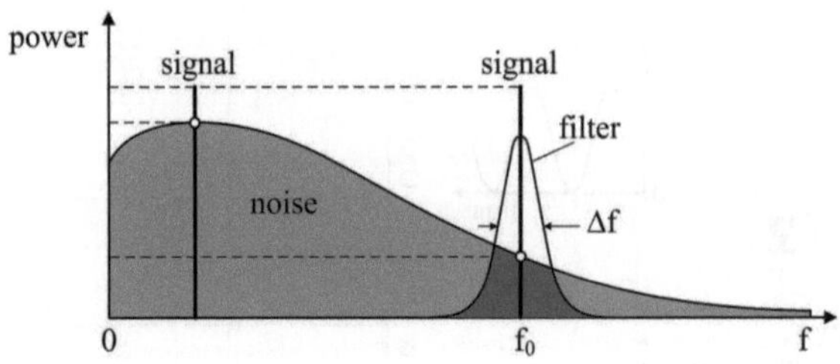

Fig. 4.30 Within the same filter bandwidth Δf, noise is smaller at higher frequencies

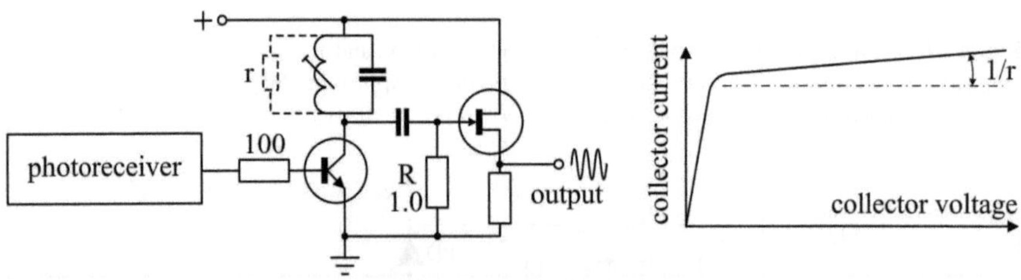

Fig. 4.31 The simplest filter, working up to tens of megahertz. Collector resistance r and input active resistor R of the subsequent amplifier must be as big as possible for small Δf. While R may be chosen from the stock, r can only be evaluated from current–voltage curves of the transistor. Typically, 6 kOhm $< r <$ 60 kOhm

synchronous detection, whereas phase can be extracted only by synchronous detection. Consider filtering first.

Noise in electronic devices decreases with frequency f approximately as $1/f$ (Fig. 4.30).
With the same signal amplitude, the ratio of the signal to noise is better at higher frequencies, thus making measurements more accurate. With the narrow-pass filter centered at the carrier frequency f_0, the smaller the filter bandwidth Δf, the less noise. Thus, the first priority for efficient filtering is to make the filter narrow. With tremendous variety of electronic filters known today, we are bound to practical solutions that can be built in the laboratory, and from this point of view, the simplest is not always the worst (Fig. 4.31). With all the practical limitations, quality factor Q of such a filter may be up to 20 at $f_0 = 20$ MHz, meaning the bandwidth is $\Delta f = f_0/Q \sim 1$ MHz. Filters with the highest possible quality factor $Q \sim 10^6$ can be made using quartz resonators. Factory tuned to a specific frequency, marked on the metal case, they are cheap, reliable, and available for a variety of frequencies in the megahertz range. Quartz resonators, also commonly called quartz crystals, should not be mixed up with crystal oscillators—electronic generators stabilized by a quartz crystal. Resonators can be easily recognized by only two leads, while crystal oscillators always have four. With $Q \sim 10^6$, the bandwidth of a crystal-stabilized filter $\Delta f = f_0/Q$ is of the order of hertz. Therefore, careful stabilization of modulation frequency is crucial to ensure coincidence

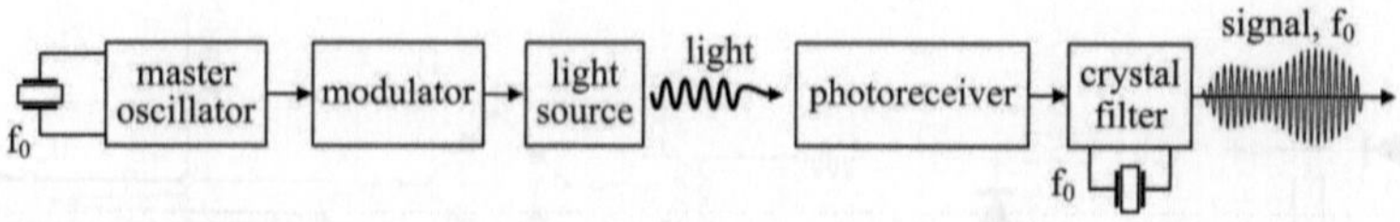

Fig. 4.32 Crystal-based filter provides strong rejection of noise, but requires fine frequency tuning to the modulator frequency

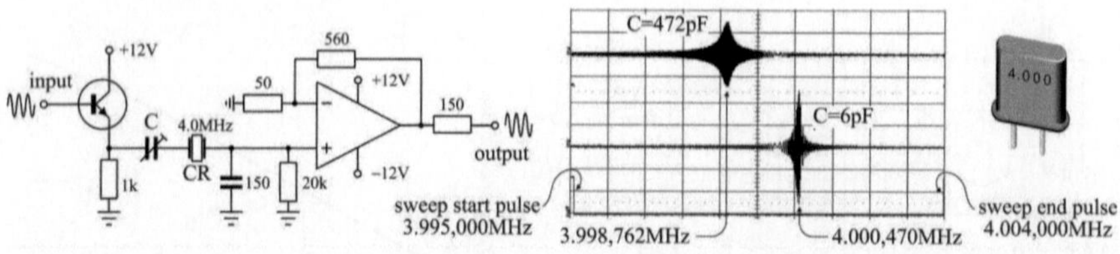

Fig. 4.33 Adjusting the capacitor *C*, the 4 MHz crystal resonator (*CR*) can be tuned by ±850 Hz, i.e. about ±0.02 % around central frequency. *Right panel* shows frequency sweeps from 3.995,000 to 4.004,000 MHz with $C = 6$ pF and 472 pF. Typical quartz resonator is shown in the *upper right* corner

with the central frequency of the filter. Generalized scheme of such a measuring system is shown in Fig. 4.32. The problem is that actual central frequencies f_0 of the two equally labeled crystal resonators may not exactly coincide because of finite manufacturing margins. The solution is to adjust central frequency of the crystal filter to modulation frequency. Although the fundamental resonant frequency of a crystal is strictly defined by its geometry, some tiny variations are still possible, and this margin is sufficient to adjust filter frequency to that of the modulator. A sample circuit is shown in Fig. 4.33.

Now, when sinusoidal signal is filtered out, it is time to demodulate it, i.e. to convert to a direct-current voltage proportional to the amplitude. Simple schemes of half- and full-wave diode rectifiers shown in Fig. 4.34 will not work, because we expect small amplitudes below 0.5 V—the cutoff voltage of a typical silicon diode. The solution is the so-called active rectifier—an operational amplifier with a diode in the feedback loop (Fig. 4.35). Its performance is worth its complexity: with the output voltage swing of 5 V, it provides linear response down to 1 mV amplitude of the high-frequency input signal. Operational amplifiers must be of voltage-to-voltage type, and have maximum possible gain-bandwidth product.

The advantage of a classic filtering is that it does not require information about the phase of the signal. The synchronous detector, which will be considered next, is also a kind of a narrow-pass filter, but it needs special reference signal, oscillating in constant phase relative to the useful signal. Some applications require not only amplitude measurements but also phase of modulation. For this, synchronous detection is the only suitable technique, implemented in the so-called lock-in

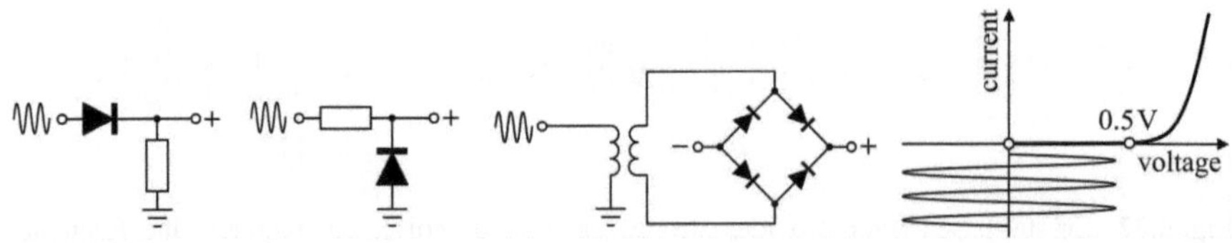

Fig. 4.34 These simple diode rectifiers do not work with signals below 0.5 V

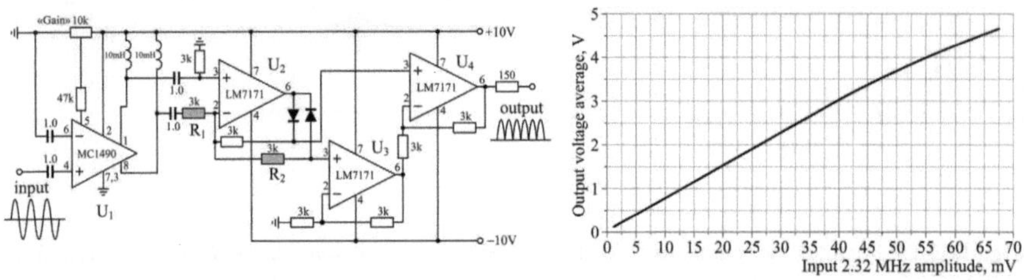

Fig. 4.35 One possible implementation of a full-wave active rectifier for detecting 2.32 MHz signal from HP5517B Zeeman laser. The U_1 is a high-frequency video-amplifier with variable gain and two opposite-phase outputs. The U_2–U_4 are voltage-to-voltage operational amplifiers with gain-bandwidth product 200 MHz. The U_2 works as a full-wave rectifier, while U_3 and U_4 equalize and sum the signals. The U_2 supplies as high voltage as necessary to maintain equal voltages at its inputs. Therefore, the voltage at the divider R_2–R_1 is always proportional to input voltage, virtually excluding the diodes from the circuit. To protect the circuit from spontaneous oscillation, the separating low-pass filters should be included into the +10 V line between the U_1 and U_2–U_4 (not shown for clarity)

amplifiers. However, not each lock-in amplifier can measure phase: this option is available only in more complicated and, consequently, more expensive products. The basic idea of synchronous detection as a narrow-pass filter allows simple practical implementation without such a specialized equipment. In order to proceed further, we need some mathematics, which could be done simpler with the assumption that the input signal is a sinusoid, although this is not necessary in practice. Figure 4.36 explains the principle. We assume for the beginning that both the useful and reference signals have exactly the same frequency ω, i.e. $T_1 = T$, which mathematically means that their relative phase does not change. From now on, the useful signal will be called just signal, and the reference signal—the reference. The signal then is a sinusoid

$$y = a\,\sin(\omega t + \varphi),$$

and the output x is the sum of integrals over the halves of a period:

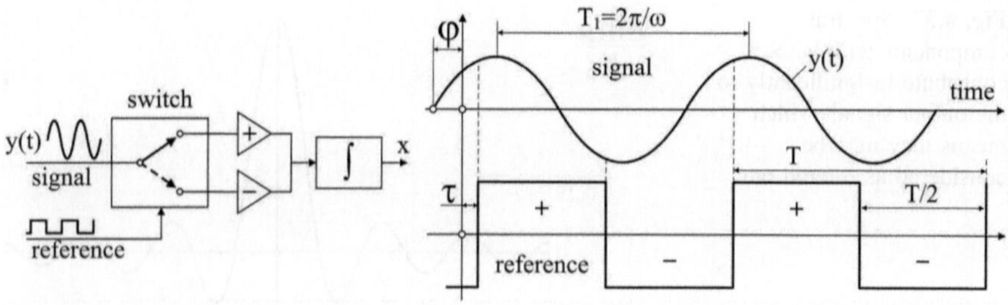

Fig. 4.36 The idea of synchronous detection is to connect input signal to the output integrator periodically: positively during one half of the period and negatively during another one. Figure 4.38 explains how to produce this switching electronically

$$x = \sum_{n=1}^{M} (S_n^+ + S_n^-), \quad S_n^+ = \int_{\tau+nT}^{\tau+nT+T/2} a \sin(\omega t + \varphi)\, dt,$$

$$S_n^- = -\int_{\tau+nT+T/2}^{\tau+(n+1)T} a \sin(\omega t + \varphi)\, dt,$$

where $M = t/T$ is the number of periods in the integration time t. Since $\omega nT = 2\pi n$ and $\omega T/4 = \pi/2$, after some trigonometric manipulations we get

$$x = \frac{2at}{\pi} \cos(\omega\tau + \varphi).$$

Please, be attentive and do not mix up τ with the running time t: the argument of the cosine is the total phase $\psi = \omega\tau + \varphi$, and it does not depend on time. Thus, the first conclusion is that synchronous detector, with sinusoidal signal at its input, produces constant signal proportional to integration time t and the signal amplitude a. The second conclusion is that the output depends on the total phase ψ, and may be positive, negative, or even zero. This uncertainty is important for high-frequency applications, where phases and delays are hard to control, but it may be considered as insignificant for low-frequency modulators such as rotating wheels (Sect. 4.2), where total phase $\psi = \omega\tau + \varphi$ can be adjusted to zero purely mechanically.

The next thing we are going to analyze is the filtering capability of the synchronous detector. For that, assume small difference between frequencies of the signal and reference, i.e. $T_1 = T(1 - \delta)$. Since we assume $\delta \ll 1$,

$$\frac{\omega T}{4} = \frac{\pi}{2} \cdot \frac{T}{T_1} \approx \frac{\pi}{2}.$$

However, for the second term ωnT, the number of periods may be large $n \gg 1$, and the term itself may differ significantly from the multiple of 2π:

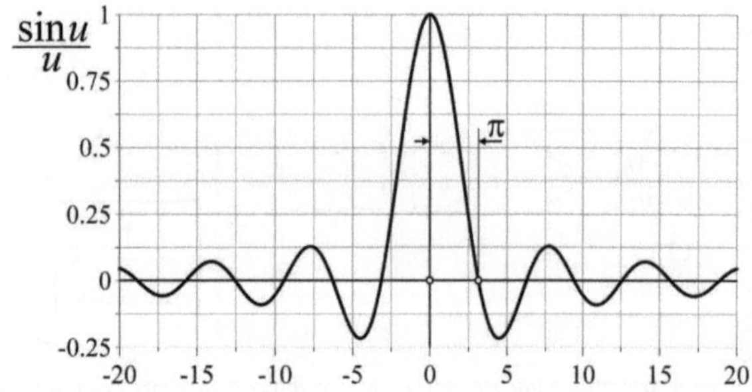

Fig. 4.37 Spectral components with $|u| > \pi$ contribute insignificantly to the output signal, which means they may be considered as filtered out

$$\omega nT = 2\pi n\frac{T}{T_1} = 2\pi n\,\frac{T}{T - \delta\cdot T} = 2\pi n + n\beta,$$

where $\beta = 2\pi\delta$. Repeating the previous series of calculations, assuming $\omega T/4 \approx \pi/2$ and $3\omega T/4 \approx 3\pi/2$, we get:

$$x \approx \frac{4a}{\omega}\sum_{n=1}^{M}\cos(\omega\tau + \varphi + n\beta) = \frac{4a}{\omega}\mathrm{Re}\left[e^{i(\omega\tau+\varphi)}\sum_{n=1}^{M}e^{in\beta}\right].$$

For geometric series use

$$\sum_{n=1}^{M} q^n = q\frac{1-q^M}{1-q}$$

and $\sin(\beta/2) \approx \beta/2$ to obtain the final result:

$$x \approx \frac{4a}{\omega}\mathrm{Re}\left[e^{i(\omega\tau+\varphi)}\sum_{n=1}^{M}e^{in\beta}\right] = \frac{2at}{\pi}\cos\left[\omega\tau + \varphi + (M+1)\frac{\beta}{2}\right]\cdot\left(\frac{\sin\frac{M\beta}{2}}{\frac{M\beta}{2}}\right).$$

It is composed of rapidly varying cosine term and a well-known slowly varying function

$$\left(\frac{\sin u}{u}\right),\ u = \frac{M\beta}{2} = \pi M\delta,$$

shown in Fig. 4.37. It may be considered as filtering function because it determines maximum frequency detuning at which the synchronous detector is still responding to the input signal, regardless the total phase difference:

$$u < \pi,\quad \delta < M^{-1};\quad \frac{\delta\cdot T}{T} = \frac{\Delta f}{f} = \frac{1}{Q} < \frac{1}{M}.$$

Summarizing, the synchronous detector acts simultaneously as a narrow-pass filter with the quality factor $Q = M \equiv t/T$ and a rectifier. With a megahertz modulation frequency and integration time one second, theoretical quality factor is about one million, which is the scale of a quartz filter. Combination of filtering and

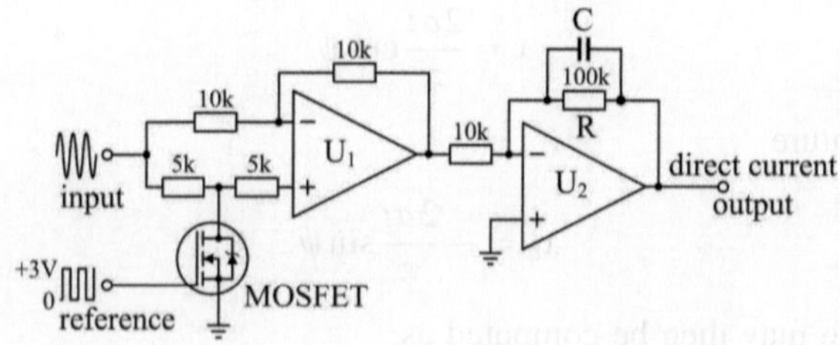

Fig. 4.38 Simple synchronous detector. The N-channel MOSFET switch alters the U_1 gain between +1 (*open switch*) and −1 (*closed switch*). The on-state resistance of MOSFET is very low: about 0.03 Ohm. The U_2 amplifies the signal and integrates it with the time constant *RC*

Fig. 4.39 To determine phase, the quadrature signal is needed

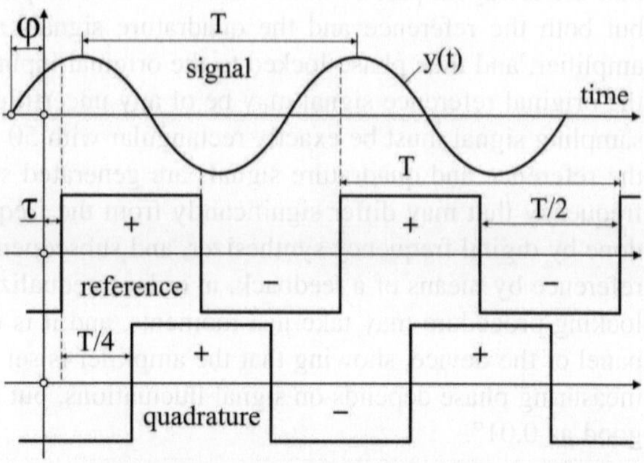

demodulation properties is very attractive, but we pay for it by the requirement to have a reference signal.

Synchronous detector can be readily built with very few standard electronic components as shown in Fig. 4.38. The only comment that should be added to the figure caption relates to the operational amplifier, fed with the same signal in both the inverting and non-inverting inputs. This option may look unusual, but it is easy: since operational amplifier maintains voltage difference at its inputs close to zero, the current does not flow through the resistor in the inverting input, neither it flows through the feedback resistor. As such, the output equalizes the input, and the total gain is +1.

The simple schemes like above are very attractive, but they do not measure phase of the input signal. In order to measure phase, the so-called quadrature signal must be generated, which represents a piece of high technology, available only in advanced versions of lock-in amplifiers. First, briefly consider the idea (Fig. 4.39). The quadrature signal is a duplicate of the reference, shifted exactly by one quarter of the period. It forms the second sampling channel with the hardware exactly like the first one. Thus, two identical channels are synchronized by the quarter-wave shifted references, and they deliver two quarter-wave shifted signals: the primary one

$$x = \frac{2at}{\pi} \cos \psi$$

and the quadrature

$$x_q = -\frac{2at}{\pi} \sin \psi.$$

The total phase may then be computed as

$$\psi = -\arctan \frac{x_q}{x}.$$

Such computations can be accomplished only by onboard digital signal processors that are always a part of an advanced lock-in amplifier. It may sound surprising, but both the reference and the quadrature signals are also generated inside the amplifier, and then phase-locked to the original input reference. The reason is that the original reference signal may be of any uncertified periodical shape, while the sampling signal must be exactly rectangular with 50 % duty ratio. Therefore, both the reference and quadrature signals are generated simultaneously at some initial frequency that may differ significantly from the frequency of modulation. This is done by digital frequency synthesizer, and subsequently phase-locked on the input reference by means of a feedback, in order to equalize the frequencies. The phase-locking procedure may take just moments, and it is usually indicated on the front panel of the device, showing that the amplifier is set for work. Actual precision of measuring phase depends on signal fluctuations, but on stable signals it may be as good as 0.01°.

List of Common Mistakes

- obtaining amplitude modulation from Zeeman laser without a polarizer;
- driving light-emitting diodes with sinusoidal voltage to produce sinusoidally modulated optical flux;
- using diode rectifiers on small signals below 0.5 V.

Further Reading

M. Born, E. Wolf, Principles of Optics, Cambridge University Press, 7th ed., 1999.
V. Protopopov, Laser Heterodyning, Springer, 2009.
B.E.A. Saleh, M.C. Teich, Fundamentals of Photonics, Wiley, 1991.
E.F. Schubert, Light-Emitting Diodes, 2nd ed., Cambridge University Press, 2006.
P. Horowitz, W. Hill, The Art of Electronics, Cambridge University Press, 2nd ed., 2001.

Chapter 5
Polarization Optics

When you need to care about polarization: interferometry, heterodyning, optical isolation, variable attenuation.

Abstract Physics of wide variety of polarization components and techniques available on the market is explained and summarized in six sections with details that cannot be found in textbooks or primers. The chapter opens with classification of polarizers, introducing basic parameters: extinction ratio and transmission. Brief insight into design and appropriate clamping technology of the simplest types of polarizers—the sheet and dichroic polarizers—precedes the discussion of much more sophisticated crystal polarizers. Significant portion of mathematics needed for understanding theory of beam propagation in birefringent media is dispatched to Chap. 12, leaving only phenomenological explanation based on the Huygens principle. Table 5.2 in this section presents the fullest summary of crystal polarizers—Nicol, Foucault, Glan-Taylor (Glan-laser), Glan-Thompson, Glan-Taylor (Brewster angle)—and separating prisms—Wollaston, Nomarski, Rochon. Birefringence is explained using the concept of ellipsoid of refractive indices, and practical geometrical technique of finding propagation direction for the extraordinary ray is presented. Numerous practical aspects that determine efficiency of a particular polarizer are discussed, and real comparative measurements of the three basic types of polarizers are summarized in Table 5.3. Section 5.2 analyzes polarization separators: beam-splitting cubes of various types and Nomarski prism. The Nomarski prism is a very tricky element, often being explained erroneously, therefore this section may serve as a good introduction into the subject, supported also by detailed mathematics and Fortran codes in Chap. 12. The subject of Sect. 5.3—phase elements—requires mathematics of a complex variable, especially the Fresnel rhomb and its modifications. Some simple but detailed analytical computations are also needed to realize spectral limitations of phase elements. However, this portion of analysis arms the reader with valuable practical ability to discriminate between right and wrong use of waveplates and understanding of some practical tricks like their tilting. Compensators—the Babinet-Soleil and quarter-wave plate—is the topic of the forth section. The figures in this section are mostly self-explanatory. Polarization isolators are the essential part of many delicate optical experiments, and therefore Sect. 5.5 sheds some additional light on this subject that cannot be learned from textbooks because many ideas were developed and brought to market only recently,

V. Protopopov, *Practical Opto-Electronics*, Springer Series in Optical Sciences 184,
DOI: 10.1007/978-3-319-04513-9_5,

Fig. 5.1 Sheet, dichroic, and crystal polarizers. *White marks* show polarization direction

Table 5.1 Polarizers types

Type	Material	Spectral range (nm)	Transmission (%)	Extinction ratio	
				Datasheet	Measured
Sheet	Plastic	400–700	90	5×10^3–10^4	4×10^4
Dichroic	Plastic/glass	400–700	30	10^4	3×10^5
Prisms	Crystals	350–2000	90	10^5	9×10^5

like broadband Faraday isolators or polarization-independent isolators. Section 5.6 of this chapter presents a useful practical example of using polarization components to create simple variable attenuator.

5.1 Polarizers

There are three basic types of polarizers most frequently used in practice: sheet polarizers, dichroic polarizers, and various types of prisms with birefringent crystals (Fig. 5.1). Table 5.1 summarizes their principle features, and detailed description.

Sheet polarizers are thin (0.5–0.7 mm typically) pieces of dichroic plastic, being sold in a variety of sizes. The least expensive polarizing elements can be conveniently cut out of such sheets in any desirable shape. Ordinary scissors is a proper tool for that. The plastic is flexible but still rigid enough to keep its shape after cutting. Residual curviness may be a problem for keeping the cut pieces flat, therefore mechanical clamping over the perimeter is always a plus (Fig. 5.2). Although transparent bent plastic foils always exhibit large birefringence, flexures or strain in the polarizing sheets do not produce any noticeable deterioration to performance because of absorbing nature of their polarizing properties. Sheet polarizers are manufactured from plastic compound like polyvinyl-alcohol, treating it with iodine-containing chemicals and stretching uniformly in one direction. As a result, iodine atoms spread within the polymer and align among its macromolecules in thread-like conducting chains, forming absorbing canals for the light whose electrical field vector is parallel to them. Since there is no electron mobility in the

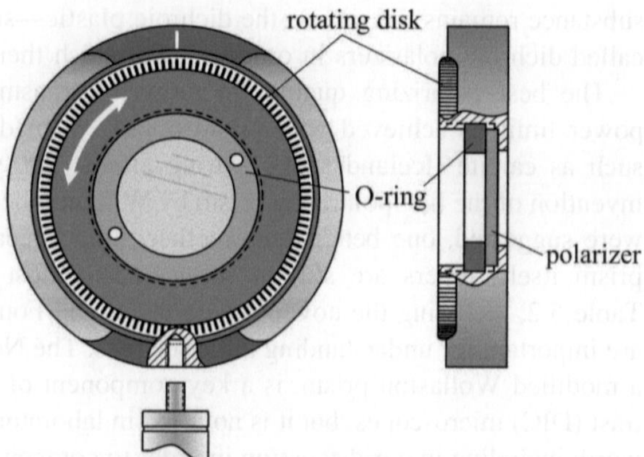

Fig. 5.2 The best option for mounting sheet polarizers is to use threaded holders and O-rings, available in a variety from manufacturers of opto-mechanical kits

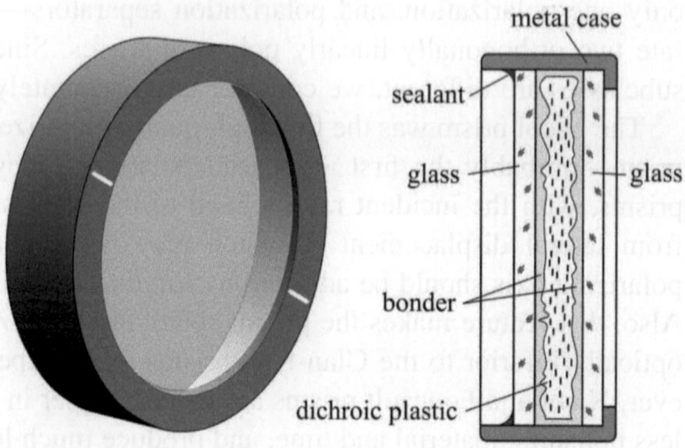

Fig. 5.3 Dichroic polarizer. Refractive index of the transparent bonder matches that of the polarizing plastic, making roughness invisible. The bonder may be hardened by UV curing, for example. Roughness of the plastic layer is exaggerated

perpendicular direction, orthogonally polarized light is transmitted without absorption. Therefore, even if the inner strain of the plastic material itself creates some birefringence, thus producing light with the prohibited polarization, it is immediately absorbed inside the polarizer. Absorbing nature of a sheet polarizer makes it virtually insensitive to beam convergence, allowing to work with the cone angles as big as 30°. However, rolling manufacturing technology and softness of the material distort the wavefront to several wavelengths of visible light. For comparison, roughness of an ordinary window glass is about 2 nm. In order to protect the surface from bigger scratches, manufacturers laminate both surfaces of a plastic sheet with protective adhesive films. These films stick firmly to the polarizer sheet, and may not detach even during cutting. Although they are murky and relatively rough, inexperienced user often makes a mistake, leaving them on the polarizer, and afterwards cannot understand why so much scattering comes from it.

In order to improve wavefront quality, plastic polarizer may be placed between two glass disks, usually made from BK7 glass or fused silica, with the gaps from both sides filled with index-matching bonder (Fig. 5.3). Although the polarizing substance remains the same—the dichroic plastic—such polarizers are commonly called dichroic polarizers in order to distinguish them from sheet polarizers.

The best polarizing quality, in terms of transmission, extinction ratio, and power limit, is achieved with prism polarizers made from birefringent crystals, such as calcite (Iceland spar). During almost 200 years of evolution since the invention of the first polarizing prism by William Nicol in 1828, numerous designs were suggested, one better than another. Some became outdated, like the Nicol prism itself, others are still in service. The most popular types are listed in Table 5.2, including the now obsolete Nicol and Foucalt prisms, which, however, are important for understanding the principles. The Nomarski prism, also known as a modified Wollaston prism, is a key component of differential interference contrast (DIC) microscopes, but it is not used in laboratory practice. Nevertheless, it is worth including in our discussion in order to compensate for numerous misleading explanations in the literature. The entire family of prismatic polarizing elements may be classified into two subclasses: polarizers, i.e. the elements for transmitting only one polarization, and polarization separators—elements that spatially separate two orthogonally linearly polarized waves. Since applications for these two subclasses are different, we consider them separately.

The Nicol prism was the first high-quality polarizer. Together with the Foucault prism—probably the first air-spaced polarizer—they form a subclass of slanted prisms, with the incident ray inclined to the front surface. Slanted prisms suffer from lateral displacement *d*, which may become almost intolerable when the polarizing axis should be adjusted by rotating the prism around the coming beam. Also, this feature makes the prisms relatively long. All that makes slanted prisms optically inferior to the Glan-type prisms with perpendicular front surface. However, Nicol and Foucault prisms are much simpler in manufacturing, require much less polishing material and time, and produce much less calcite wastage. The front wedge of the Nicol prism is typically 68°—only around 3° less than in natural geometry of the calcite crystal. It is believed that the very first prism was made by Nicol himself with the natural crystal angle of 71°, and only later the practice of polishing to 68°, which increases the field of view, was introduced, following the great success of the first polarizers. The requirement of wide field of view (about 20° in Nicol prisms) dictates that the angular separation of the ordinary and extraordinary rays inside the crystal should be as big as possible. Although the traditional Nicol prism geometry is not optimized from this point of view, it is actually not far away from the optimum. In order to understand that, we need to revisit the principles of propagation of light waves in birefringent crystals. More detailed mathematical considerations, with computations and Fortran codes, are summarized in Chap.12.

The first thing that should be understood about the birefringent crystals is that optical waves propagate in the directions that are not necessarily perpendicular to the wavefronts. The second rule is that the refractive index *n* of a wave depends on the propagation direction and polarization. In the so-called uniaxial crystals, like quartz, calcite, α-BBO, and YBO_4, there is only one certain direction called the optical axis, which determines optical properties. If polarization vector of a wave

Table 5.2 Polarizing prisms[a]

Type	Material	Spectral range (nm)	Scheme	Spacer
Polarizers				
Nicol	Calcite	350–2300	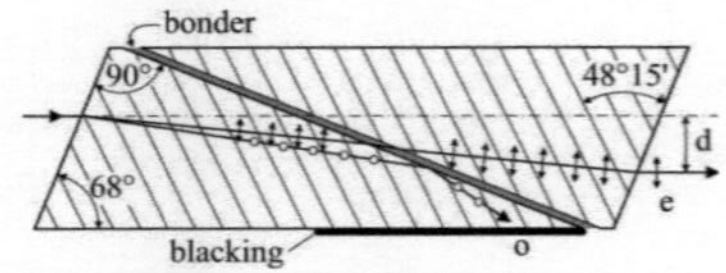	Canada balsam
Foucault	Calcite	350–2300	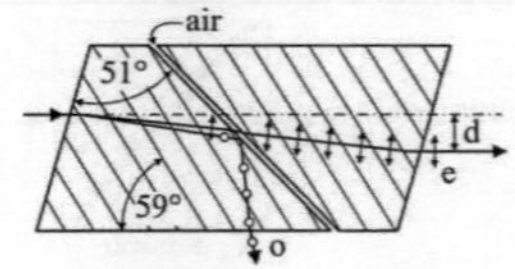	Air
Glan-Taylor (Glan-laser)	α-BBO Calcite YVO_5	190–3500 350–2300 500–4000	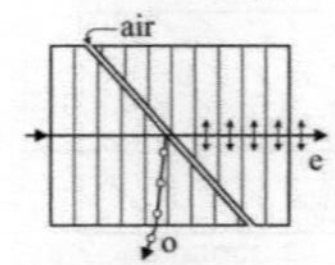	Air
Glan-Thompson	α-BBO Calcite	190–3500 350–2300	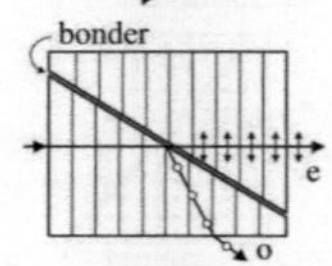	UV bonder

(continued)

Table 5.2 (continued)

Type	Material	Spectral range (nm)	Scheme	Spacer
Glan-Taylor (Brewster angle)	Calcite YVO_5	350–2300 500–4000	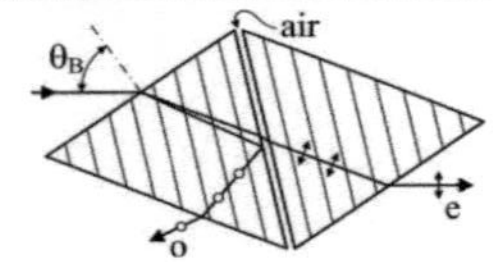	Air
Polarization separators				
Wollaston	α-BBO calcite YVO_5 Quartz	190–3500 350–2300 500–4000 200–2300	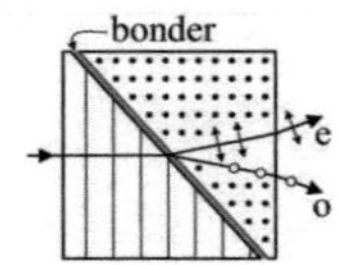	UV bonder
Nomarski	α-BBO Calcite YVO_5	190–3500 350–2300 500–4000	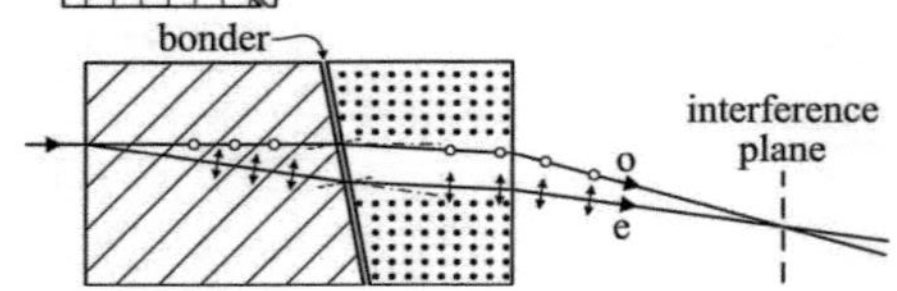	UV bonder
Rochon	α-BBO Calcite YVO_5 Quartz	190–3500 350–2300 500–4000 200–2300	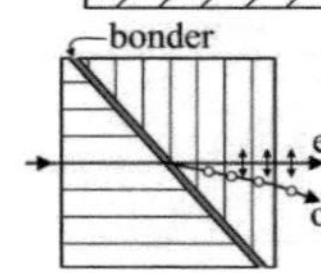	UV bonder

(continued)

Table 5.2 (continued)

Type	Material	Spectral range (nm)	Scheme	Spacer
Beam-splitting cube	Glass	350–2000	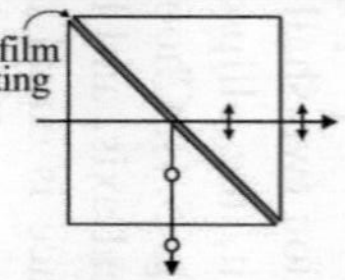	Multilayer coating

[a] Hatching and black dots indicate directions of optical axes in the crystals. Open circles and double arrows show orthogonal polarization directions. In the schemes, the rays are shown for calcite ($CaCO_3$) and α-BBO (α-$Ba_2B_2O_4$)—negative uniaxial crystals. For yttrium orthovanadate YVO_4, a positive uniaxial crystal, extraordinary and ordinary rays swap. Calcite is preferable when low scattering is needed. To make the picture clear, Nomarski prism is shown not to scale: actually, it is flat and wide

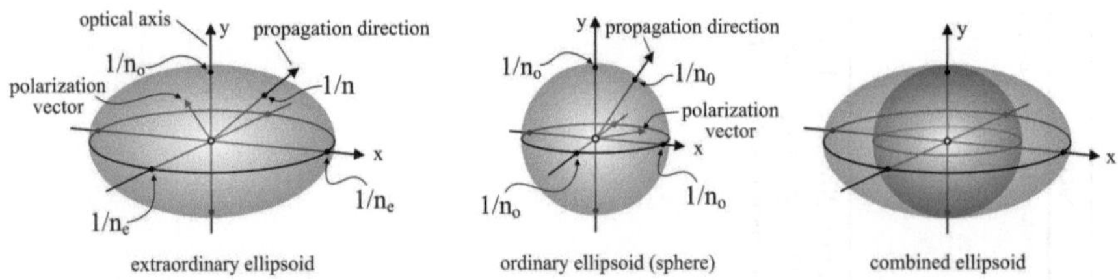

Fig. 5.4 Ellipsoid of refractive indices for uniaxial negative crystal ($n_o > n_e$)

is perpendicular to this axis, then the wave propagates like in isotropic medium, and its wavefront is always perpendicular to propagation direction. This wave is called the ordinary wave, and its refractive index n_o does not depend on propagation direction. The wave whose polarization vector is not perpendicular to the optical axis, propagates differently: its refractive index depends on propagation direction, and the wavefront is not perpendicular to this direction (we shall see it later). Such a wave is called the extraordinary wave. For it, the ellipsoid of refractive indices may be introduced in the following way (Fig. 5.4). Choose the Cartesian system of coordinates with the y axis along the optical axis, and form a surface such that the radius-vector to each point on this surface is equal to $1/n$, where n is the refractive index for the extraordinary wave, traveling in this direction. Clearly, the radius-vector is proportional to the speed of the wave in this particular direction—a feature that later will determine the procedure of finding propagation directions for extraordinary waves. Since optical waves are transversal, polarization vector is always perpendicular to propagation direction. As such, all the waves propagating along the optical axis will have the same refractive index n_o, whatever their polarization is. Consequently, the section of the ellipsoid along the optical axis will produce an ellipse with semiaxes $1/n_o$ and $1/n_e$, where n_e is the refractive index of the extraordinary wave, propagating perpendicular to the optical axis. Any point on the ellipse in the x–y plane satisfies equation

$$n_e^2 x^2 + n_o^2 y^2 = 1.$$

Obviously, with the polarization vector perpendicular to the optical axis, propagation direction may be arbitrary. Thus, we may introduce the ellipsoid of indices for the ordinary wave as well, but this ellipsoid degenerates to the sphere with the radius $1/n_o$. The practice is to draw these two surfaces together, and they touch each other along the optical axis (Fig. 5.4). There are two options: $n_o > n_e$ and $n_o < n_e$, which are commonly referred to as uniaxial negative and uniaxial positive crystals respectively. For example, in calcite, $n_o = 1.66$ and $n_e = 1.49$ at 589 nm. For α-BBO $n_o = 1.65$ and $n_e = 1.53$. These two crystals are uniaxial negative.

Consider now the unpolarized ray, coming to a surface of a uniaxial crystal whose optical axis is tilted to the surface normal and lies in the plane of the drawing (Fig. 5.5). For the s-polarized component, whose polarization vector is

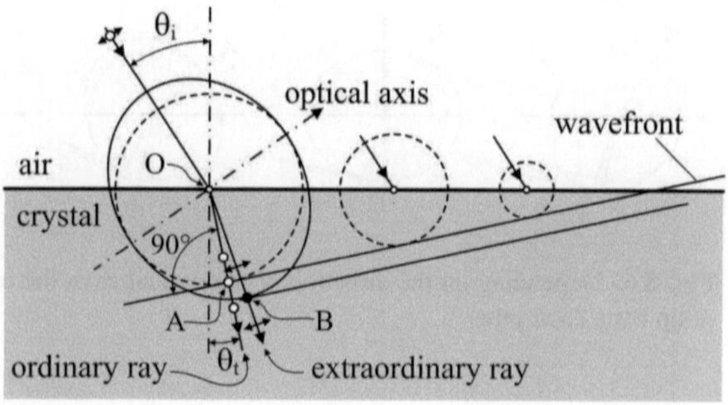

Fig. 5.5 In the ordinary wave, propagation direction is always perpendicular to the wavefront. In the extraordinary wave it is not. The wavefronts of the ordinary and extraordinary waves are approximately parallel

perpendicular to the plane of the drawing and to the optical axis, refractive index does not depend on propagation direction, and is equal to n_o—this is the ordinary wave. Its ellipse of refractive indices is merely a circle shown in the dashed line in Fig. 5.5. For this wave, it is easy to find the direction of the refracted wave, using the Snell law:

$$\sin\theta_i = n_o \sin\theta_t.$$

For the extraordinary wave, however, whose polarization vector lies in the plane of incidence, refractive index depends on the propagation direction, and this direction, in turn, depends on the refractive index. Thus, it is necessary to compose a system of two equations—for the directional angle and for the refractive index—and solve it. Leaving the exact mathematics to Chap.12, we are going to show how to solve this problem qualitatively on the basis of the Huygens principle. According to it, every point on the interface acts as a point-like source of secondary waves with the phases equal to the phases of the incident wave (Fig. 5.5). These secondary waves form a plane wavefront inside the crystal, which touches the circles, coming from the secondary waves in the crystal. It is important that the ellipse of refractive indices is defined in such a way that the section OA is proportional to the speed of the wave in the crystal, i.e. to $1/n$. Therefore, the tangent to the ellipse (circle for the ordinary wave) of the refractive indices coincides with the wavefront. The same holds true for the extraordinary wave.

We have already mentioned the two rules that should be used to understand propagation of light in crystals. The third rule is that the wavefronts are always refracted according to the Snell law, even though the energy flow (direction of the ray itself) may not. In our case, this means that the wavefront of the extraordinary wave will be approximately the same as that of the ordinary wave, because their refractive indices differ insignificantly. Although this is a wrong approximation for exact mathematical computations (see Chap.12), it is a perfect assumption for qualitative considerations. For us, it makes a guideline: draw the tangent to the extraordinary ellipse in Fig. 5.5 parallel to the wavefront of the ordinary wave. The OB is proportional to the speed of the extraordinary wave, and consequently, this is the direction of the extraordinary ray (approximately).

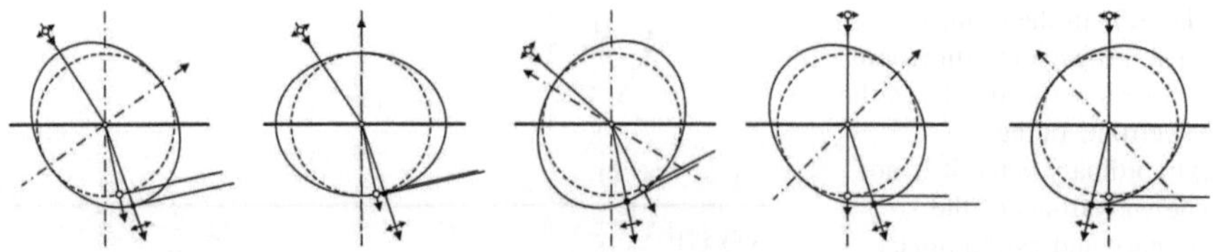

Fig. 5.6 Depending on the direction of the optical axis, the ordinary and extraordinary rays may jump over each other

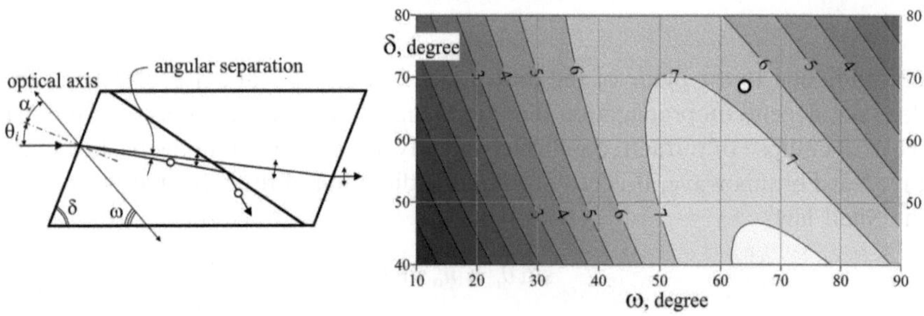

Fig. 5.7 Two-dimensional map of angular separation as a function of the optical axis tilt ω and front face angle δ. *Digits* on the map mark the levels of constant angular separation (degrees). The calcite refractive indices $n_o = 1.6584$, $n_e = 1.4864$. *White circle marks* the traditional design, giving the value for angular separation about 6.5°

Geometrical technique described above is useful for qualitative analysis of rays in birefringent crystals, particularly for understanding mutual orientation of the ordinary and extraordinary rays (Fig. 5.6).

Now, it is time to recall that theoretical discussion above was inspired by the question how big is angular separation between the ordinary and extraordinary rays in the traditional design of the Nicol prism, with the front face at $\delta = 68°$ to the base and the optical axis at 48.25° (48°15′) to the front face. Figure 5.7 gives the answer.

Smaller angles δ would give bigger angular separation, but bigger displacement and distortion of the field of view as well. This is the reason why the Nicol prisms are nowadays out of use, superseded by the Glan-type prisms.

The Glan-type prisms are composed of two right-angled prisms, separated on their long faces either by air gap (Glan-Taylor or Glan-laser) or by cement (Glan-Thompson), with the optical axis being parallel to the front face. This type of prisms uses the difference in refractive indices of the ordinary and extraordinary waves to separate polarizations on reflection, whereas the slanted prisms like Nicol and Foucault use angular separation. Since the optical axis is perpendicular to input rays, the difference between refractive indices of the ordinary and extraordinary rays is a maximum, and equals $n_o - n_e$ (Fig. 5.4). Consequently, a wider field of view or a shorter prism is equally possible, and polarization uniformity over the field of view is better. Typical field of view of a Glan-Taylor prism is about 8°. In calcite, for example, $n_o - n_e$, and total internal reflection occurs to the ordinary ray polarized perpendicularly to the plane of incidence (s-polarization). The extraordinary ray (p-polarization) is partially transmitted and also partially reflected. Therefore, totally reflected s-polarized ray is mixed with some portion of the p-polarized ray, so that the reflected ray is not totally polarized. The transmitted p-polarized extraordinary ray is 100 % polarized.

But the biggest practical convenience of the Glan-type prisms is only hardly noticeable displacement of rays. Ideally, there must be no displacement at all, but manufacturing tolerances in alignment of optical axes in the two halves of the prism make the output ray to rotate when the prism is turned around the line of sight. However, the radius of rotation is not big, practically less than 1 mm. The output ray may possibly deviate angularly, when the prism is rotated, with the normal tolerance being about 1 mrad. This defect is mostly caused by the mismatch of prismatic angles and misalignment of the two halves of the prism.

The Glan-Taylor prism uses air spacing with maximum gradient of refractive indices on the interface $n_o - 1$. The air gap is difficult to assemble, preserving parallelicity of the surfaces. An easier way is to cement these surfaces together, which is done in the Glan-Thompson design. However, in order to obtain total internal reflection on the crystal-cement interface, which has smaller gradient of refractive indices, the cut must be made at bigger angle of incidence. Therefore, Glan-Thompson prisms are longer.

The Glan-Taylor prism, with the cut made close to the Brewster angle, reduces unwanted reflection of the p-polarized light at the expense of smaller field of view. This gives even better transmission, which is important in high-power laser applications. Also, the probability of the laser-induced damage of bonder is excluded. As to the narrower field of view, it is less important in laser applications, where the beams are usually highly collimated. Prisms of that type are commonly called the Glan-laser polarizer.

Glan-Taylor and Glan-laser prisms produce some back-reflection from the front and rear faces, cut normally to the laser beam. Antireflection coating may lower this reflection to the value about 1 %, but at the expense of the transmission and probability of break-down in the film. In applications where transmission is important, the so-called Brewster-angle Glan-Taylor polarizer may be used. It repeats the idea of the Nicol prism design, with the front face tilted at the Brewster angle. This feature reduces reflection of the p-polarized beam to infinitesimal levels, providing transmission better than 98 %.

The most important parameter of any polarizer is its extinction ratio, i.e. the ratio of the intensity passed through two parallel polarizes to that of the two crossed ones. Catalogs of optical equipment commonly cite only the scales for the extinction ratio, like «10^4» or «10^6». But what can be the real values? It is not a simple thing to measure directly six orders of magnitude of any signal, preserving the next reliable digit of the reading. If, for example, we use a digital voltmeter with the upper limit of 10 V, then we need to reliably measure the signal of about one microvolt. Compare it to typical industrial electromagnetic noise in the

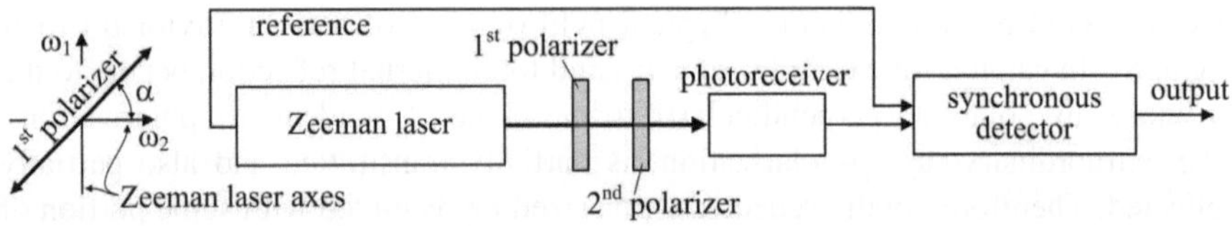

Fig. 5.8 Two polarizers are set to measure extinction ratio. Zeeman laser (Agilent Hewlett-Packard HP5517B; 632 nm; Chap. 2), with frequency split 2.34 MHz, produces sinusoidally modulated intensity through the 1st polarizer directed 45° to the axes of the laser. The 2nd polarizer is set either parallel (*open*) or crossed (*closed*) to the first one. Synchronous detector (Stanford Research SR844; 200 MHz; Chap. 4), with the time constant 1 s, measures the signal on the reference from the laser. The bandwidth of the photoreceiver is not important because modulation is sinusoidal. In the open position, the photoreceiver gain is set to produce 1 V output signal on all the types of polarizers, compensating for their different transmission

laboratory of about millivolt level. Therefore, usually such measurements are made indirectly: the intensity of the light source is attenuated by an optical filter, measured at parallel (open) polarizers, then filter is removed and the measurement is repeated at crossed polarizers. But no one knows the exact attenuation of the optical filter, because to measure it we again need direct measurements over six orders of magnitude. Nevertheless, the solution does exist, based on what was explained in Chap. 4 as synchronous detection. With high-frequency modulation, the industrial noise can be left behind the narrow-pass filter, making microvolt signals quite measurable. It sounds simple, but still some practical precautions should be taken. First, modulated light source is needed. The modulation should be sinusoidal, in order to detect and correct any possible saturation in photoreceiver and avoid influence of the amplifier finite bandwidth on the shape of the signal. The light beam must be collimated in order to exclude influence of the polarizer finite field of view. Light intensity of the collimated beam must be of the milliwatt scale, making the one millionth part of it detectable by a photodetector. And finally, the intensity should be stabilized in time, thus making second-long measurement reliable. A possible experimental arrangement, satisfying these conditions, is shown in Fig. 5.8 and explained in the caption. The 1st polarizer does not require precise rotational motion because its purpose is only to produce sinusoidal modulation. The amplitude of sinusoidal modulation is proportional to the product of the projections of the laser axes onto the axis of the 1st polarizer:

$$\cos(\alpha)\cdot\sin(\alpha)=\frac{1}{2}\sin(2\alpha).$$

It is maximized when $\alpha = 45°$, and approach to this position is smooth, permitting coarse angular adjustment (Fig. 5.9). The 2nd polarizer, in the open position, also does not require precise rotation, because, according to the Malus law, the intensity through both of them is proportional to $\cos^2\beta$, which maximizes relatively slowly as β approaches zero. However, for approaching the closed position, fine rotation is needed, because at this point relative variations are big as the signal amplitude minimizes, coming to zero. Any kind of lever-driven micrometric

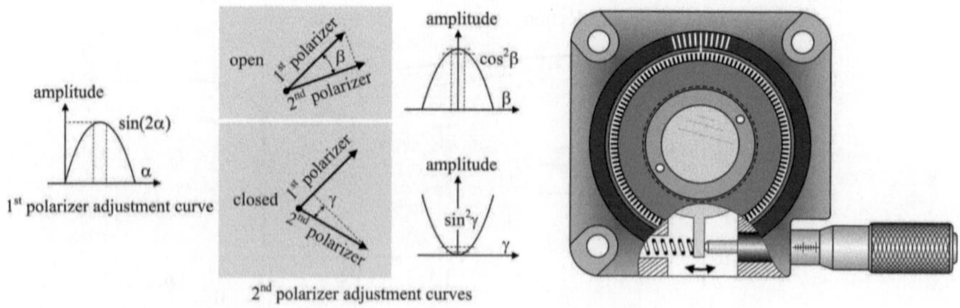

Fig. 5.9 Fine adjustment to closed position with crossed polarizers requires precise rotation

Table 5.3 Direct measurements of the extinction ratio

Polarizer	Open position signal (V)	Closed position signal (μV)	Extinction ratio
Sheet polarizer	1.0	25.5	4×10^4
Dichroic	1.0	3.5	2.9×10^5
Glan-Taylor	1.0	1.1	9.1×10^5

rotation stage may serve well for this application (Fig. 5.9). Results are summarized in Table 5.3 and do not require additional comments.

The shape of a crystal polarizer may be important in practice. Manufacturers offer two types: cylindrical and rectangular (Fig. 5.1). The cylindrical shape is the only choice when axial rotation is needed, because the rectangular one can never be accurately fixed in a rotating holder. On the contrary, when just vertical or horizontal polarization is needed, the rectangular shape is more convenient because polarization direction is already fixed relative to the bottom surface, and it can be easily installed on any flat platform.

5.2 Polarization Separators

The function of polarizers is to stop one polarization component and transmit the opposite one. However, numerous applications require both components to be used, but separately. This is done by means of polarization separators. The simplest, least expensive, and most widely used is the polarizing beam-splitting cube (Table 5.2). It is always designed to transmit the p-component (parallel to the plane of incidence) and reflect the s-component. The reason for this is stronger reflection of the s-component relative to the p-component at a single interface between two optical materials (Fig. 5.10). Therefore, designing the multilayer reflective structure, it is easier to obtain higher reflectivity for the s-component and higher transmission for the p-component. At one specific wavelength, the separating ratio may be as high as 10^2 with the transmittance in both the channels around 95 %.

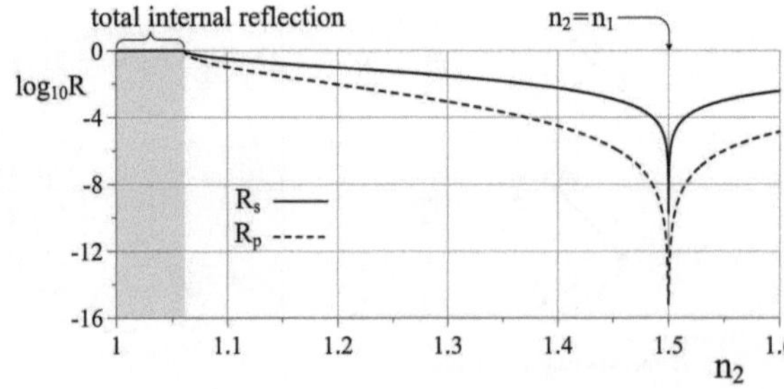

Fig. 5.10 Energy reflection coefficients R_s and R_p for the s- and p-polarized components at the interface between the glass (refractive index $n_1 = 1.5$) and another optical material with the refractive index n_2. 45° angle of incidence. For all the practical values of n_2, $R_s > R_p$

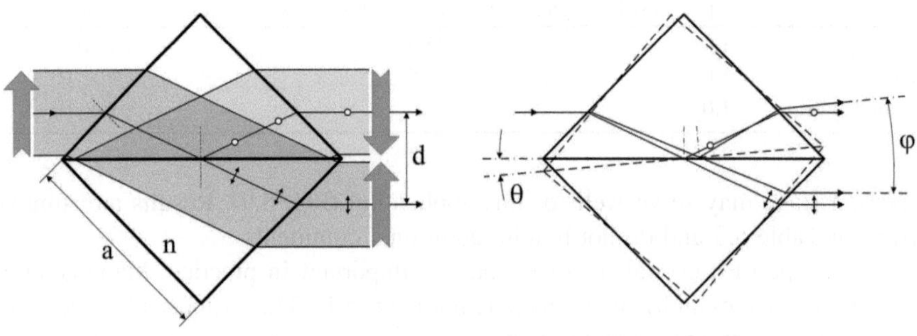

Fig. 5.11 Polarizing beam-splitting rhomb. The output beams are spatially separated by $d = \frac{a}{\sqrt{2}}\left[1 - (2n^2 - 1)^{-1/2}\right]$, where n is the refractive index. Adjusting accurately the input angle θ, it is possible to control angular separation φ of the output beams

The polarizing beam-splitting cube is made of two cemented right angle prisms. The multilayer coating—the key element responsible for beam separation—is deposited in vacuum on one of them. This particular prism, with the multilayer coating on it, should be used as the optical entrance to the cube, and is always marked by a black dot on it. If another prism is used as the entrance, then the multilayer coating works not exactly as it was designed because the wave comes to it through the cement and not through the glass. Although refractive indices of the cement (nowadays a synthetic bonder) and the glass are close, some small difference does exist, which may slightly change the performance. Inexperienced users make this mistake with 50 % probability, since they do not pay attention to any marks on the cube.

A modification of a polarizing beam-splitting cube, sometimes strangely called the energy separator cube, is shown in Fig. 5.11. It is useful when the separated beams must propagate either parallel, like in industrial interferometers that are analyzed in Chap. 6, or slightly divergent, like in various types of differential microscopes and surface profilers. The possibility of controlling separation angle is

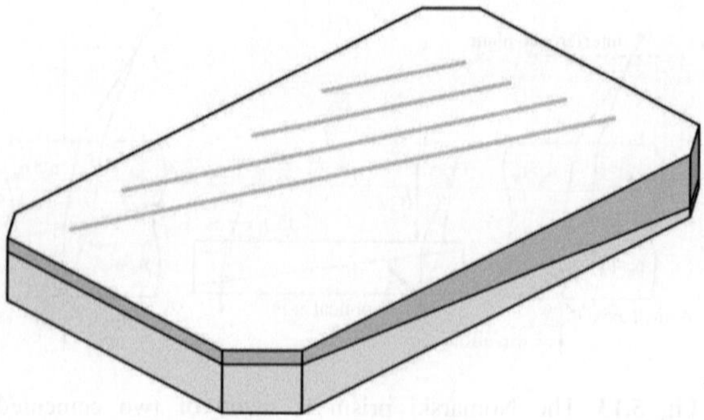

Fig. 5.12 The Nomarski prism is designed to be installed in a microscope, in the illumination beam below the sample. Therefore, it is usually made as a plate

a very useful feature for microscopy applications, unavailable in polarization separators based on crystals. Among the latter, the most commonly used is the Wollaston prism (Table 5.2), which deflects the orthogonal polarizations to the opposite sides from the input beam axis. Separation angle differs, depending on the material: 2° for quartz, 16° for calcite and α-BBO, and 20° for yttrium orthovanadate. The Rochon prism does essentially the same job, but deflects only one ray. Therefore, comparing to the Wollaston prism, angular separation is roughly two times smaller: 1° in quartz, 8° in calcite and 6° in yttrium orthovanadate.

The Nomarski prism looks more like a plate (Fig. 5.12). It is a very unusual type of a polarization serparating prism, designed relatively late, in the middle of the twentieth century, specifically for microscopy applications to produce differential interference contrast (DIC). Like in the Nicol prism, the optical axis of the crystal is neither perpendicular nor parallel to the input ray, which often makes it difficult to understand the principle of operation. The idea of the Nomarski concept is to first spatially separate orthogonally polarized rays, letting them pass through separate areas of the sample, and then combine them again in the image plane, thus producing interference pattern. In the Wollaston and Rochon prisms, oppositely polarized rays deviate from the prism center and never intersect again. In the Nomarski combination, these rays converge outside the prism to the interference plane, revealing unusual phenomenology, depending on lateral position of the input beam and inclination angle of the optical axis of the crystal (Fig. 5.13).

Even with the tilted optical axis in the lower wedge, not every Nomarski combination produces convergent rays: some produce divergent (Fig. 5.14).

5.3 Phase Elements

Phase elements are used to change polarization state of the polarized beams. For that, they introduce certain phase difference in the two orthogonal polarizations. The first in the list of phase elements are the waveplates (Fig. 5.15).

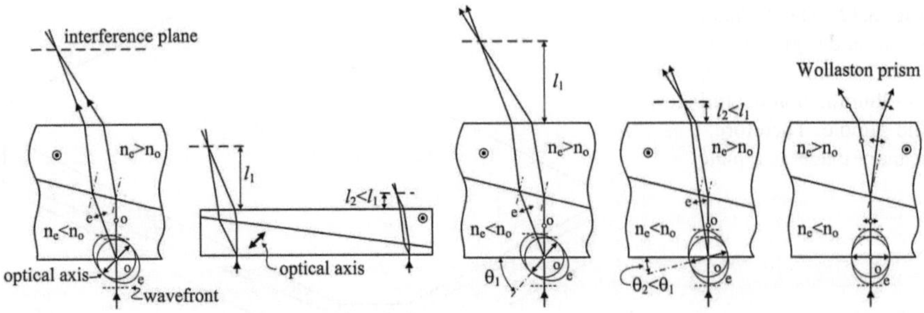

Fig. 5.13 The Nomarski prism is made of two cemented flat wedges cut from uniaxial birefringent crystals such as calcite, for instance. In it, ordinary and extraordinary rays experience cumbersome transformations, depending on the direction of the optical axis of the lower wedge and lateral position of the input ray along the front surface. When the optical axis is parallel to the front surface, the Nomarski prism transforms to the Wollaston prism. This explains its another name: the modified Wollaston prism. In the *upper wedge*, the *circled black dot* shows direction of the optical axis, which remains unchanged in all the transformations

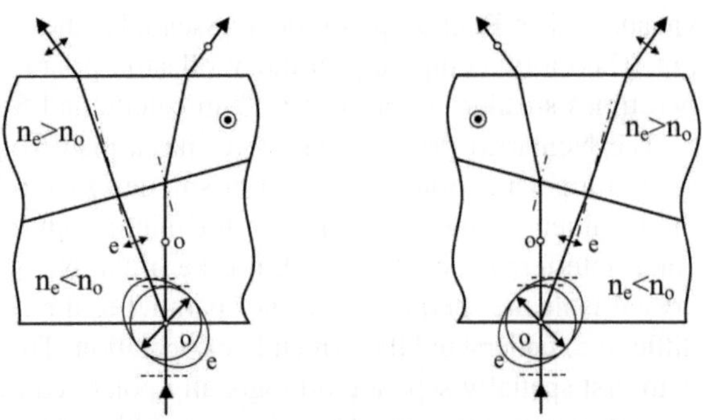

Fig. 5.14 In the Nomarski prism, certain orientations of the optical axis and the wedge produce divergent designs

Phase plates are made of birefringent crystals with different refractive indices n_1 and n_2 along two orthogonal axes (Fig. 5.16). Suppose the input wave $a \cdot \cos \omega t$ is linearly polarized, making 45° with the optical axis of a waveplate. Its projections onto the crystal axes are

$$\frac{a}{\sqrt{2}} \begin{bmatrix} \cos \omega t \\ \cos \omega t \end{bmatrix}.$$

After the wave plate, these projections acquire different phases $2\pi n_1 l/\lambda$ and $2\pi n_2 l/\lambda$, where l is the thickness of the waveplate and λ is the wavelength. Only their difference matters:

$$\varphi = \frac{2\pi}{\lambda}(n_1 - n_2)\, l.$$

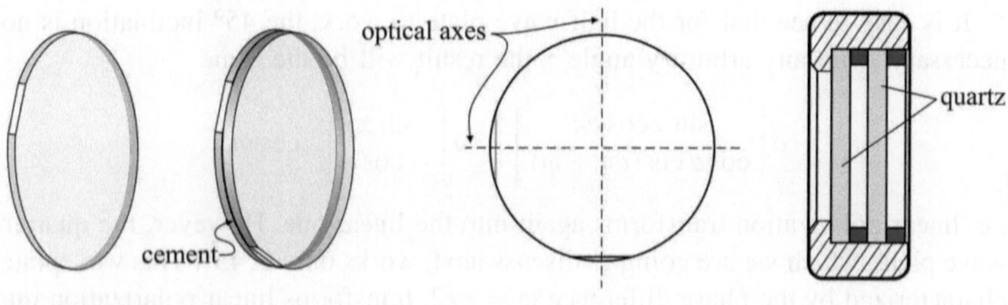

Fig. 5.15 Waveplates: the multiple order (*at left*) and the first-order (*in the middle*). *Side cuts* indicate the optical axis. For high-power applications, when the bonder between the two plates may be damaged, the air-spaced first-order assemblies are available on the market (*at right*)

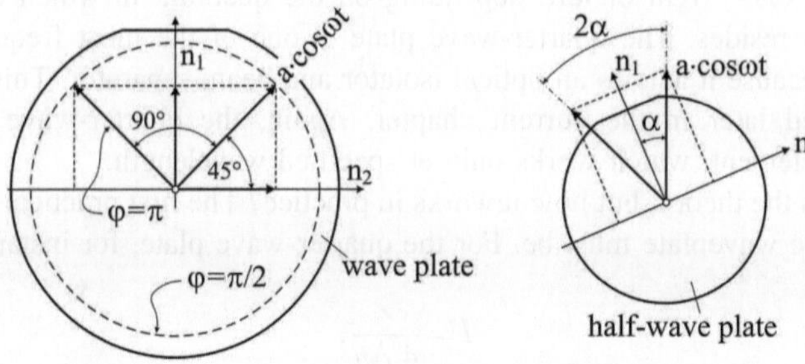

Fig. 5.16 Depending on the phase difference, the waveplate transforms linear polarization into either a circular (*quarter-wave plate*) or a linear polarizations (*half-wave plate*). When the half-wave plate rotates around the input ray, the output linear polarization rotates twice as fast (*at right*)

In the same coordinates, the output wave becomes

$$\frac{a}{\sqrt{2}}\begin{bmatrix}\cos\omega t\\ \cos(\omega t+\varphi)\end{bmatrix}.$$

With the thickness l polished to satisfy $\varphi = \pi$, the output wave continues to be a linearly polarized wave

$$\frac{a}{\sqrt{2}}\begin{bmatrix}\cos\omega t\\ -\cos\omega t\end{bmatrix},$$

with the polarization turned by $2 \times 45° = 90°$. Such waveplates are called the half-wave plates, as the phase difference is equal to half of the wavelength. It is essential that this effect takes place only for one particular wavelength, for which $\varphi = \pi$. All the wave plates are designed for a specific spectral line, like for the helium-neon laser $\lambda = 633$ nm, for instance. It is a common mistake to use a waveplate of unknown design wavelength, because the result may be incomprehensible. Thus, all the waveplates are chromatic. The only achromatic phase element is the Fresnel rhomb, or its analog the Mooney prism, which will be discussed below.

It is easy to see that for the half-wave plate to work, the 45° inclination is not necessary: with any arbitrary angle α the result will be the same

$$a\begin{bmatrix} \sin\alpha\cos\omega t \\ \cos\alpha\cos(\omega t+\varphi) \end{bmatrix} = a\begin{bmatrix} \sin\alpha \\ -\cos\alpha \end{bmatrix}\cdot\cos\omega t,$$

i.e. linear polarization transforms again into the linear one. However, the quarter-wave plate, which we are going to discuss next, works only at 45°. This waveplate, characterized by the phase difference $\varphi = \pi/2$, transforms linear polarization into

$$\frac{a}{\sqrt{2}}\begin{bmatrix} \cos\omega t \\ \cos\left(\omega t+\frac{\pi}{2}\right) \end{bmatrix} = \frac{a}{\sqrt{2}}\begin{bmatrix} \cos\omega t \\ -\sin\omega t \end{bmatrix}$$

the circular one—right or left, depending on the quadrant in which the coming polarization resides. The quarter-wave plate is one of the most frequently used elements because it acts as an optical isolator and beam separator. This topic will be discussed later in the current chapter. Again, the quarter-wave plate is a chromatic element, which works only at specified wavelength.

That was the theory, but how it works in practice? The first practical question is how thin the waveplate must be. For the quarter-wave plate, for instance,

$$l = \frac{\lambda}{4\,\Delta n},$$

where $\Delta n = |n_1 - n_2|$ is called the birefringence value. If the quarter-wave plate would be made from calcite with birefringence $\Delta n = 0.17$, then the waveplate thickness should be as small as a couple of wavelengths—obviously unreal solution, even assuming fine polishing on thick glass substrate. For this reason, calcite—a very popular material for crystal polarizers—is never used for waveplates. Here the quartz dominates, with much smaller value of $\Delta n = 0.009$. But even it gives $l \approx 17\mu$—too thin for practical manufacturing. Therefore, cheap waveplates are always polished to an odd multiple M of the minimum thickness

$$l = M\frac{\lambda}{4\,\Delta n}$$

of the order of 1 mm, making the phase difference

$$\varphi = \frac{\pi}{2} + \pi\cdot m,$$

with m—a big integer called the multiplicity factor. Such waveplates are called the multiple-order plates. Mathematically it does not present any problem, but practically it does, because relatively thick optical plate thermally expands, changing the design phase. Typically, the multiple-order quartz plates change optical path difference between the ordinary and extraordinary rays with a thermal coefficient of the order of 1 nm/°C. Later in this section, we shall see how big or small this is, discussing influence of phase errors on ellipticity of polarization. Increased chromaticity is also a problem: only narrow spectral interval of wavelengths near the design wavelength can work.

These problems are solved in more expensive options called the first-order waveplates. In them, two relatively thick quartz plates are turned 90° and cemented together (Fig. 5.15). For quick understanding, it may be assumed that phase differences after the waveplates are exactly

$$\varphi = \pm\frac{\pi}{4} + \pi m,$$

and after stacking they subtract to produce $\varphi = \pi/2$. In practice, however, the exact thicknesses l_1 and l_2 of the plates are not controlled, letting the initial phase differences be arbitrary

$$\varphi_1 = \alpha + \pi m \quad \text{and} \quad \varphi_2 = \beta + \pi m.$$

Both plates are cut from one wider crystal plate in order to ensure the same multiplicity factor m. After cementing, one surface of the stack is carefully polished, making $|\alpha - \beta|$ equal to either $\pi/2$ or π for the quarter- or half-wave plates respectively. Thus, in the first-order waveplates, the phase difference is proportional to $\Delta l = l_1 - l_2 \ll l_{1,2}$ and not to the entire plate thickness l like in the multiple-order options. The same procedure of meeting the entire tolerance for the two waveplates is applied to the air-spaced assembly as well, but they are polished on optical contact with thick glass plate between them and separated afterwards. The phase difference after the first-order waveplate is then

$$\varphi = \frac{2\pi}{\lambda}(n_1 - n_2)\,(l_2 - l_1) = \frac{2\pi}{\lambda}\Delta n\,\Delta l,$$

making the equations for the quarter- and half-wave plates as follows:

$$\frac{2\pi}{\lambda}\Delta n\,\Delta l = \frac{\pi}{2} \quad \text{and} \quad \frac{2\pi}{\lambda}\Delta n\,\Delta l = \pi.$$

Thermal coefficient for the first-order waveplates is about 0.01 nm/°C, which makes them practically insensitive to thermal variations.

The Fresnel rhomb transforms linear polarization into circular (Fig. 5.17). It is an achromatic phase element, i.e. its functionality does not depend on wavelength (to the extent of dispersion). In it, polarization transformation occurs at total internal reflection due to phase difference, introduced between the polarizations parallel (p-polarization) and perpendicular (s-polarization) to the plane of incidence. Understanding of this principle requires some algebraic manipulations.

The Fresnel formulas determine amplitude reflection coefficients r_p and r_s at the interface between glass with refractive index n and air:

$$r_\text{p} = \frac{\cos\theta - n\sqrt{1 - n^2\sin^2\theta}}{\cos\theta + n\sqrt{1 - n^2\sin^2\theta}}; \quad r_\text{s} = \frac{n\cos\theta - \sqrt{1 - n^2\sin^2\theta}}{n\cos\theta + \sqrt{1 - n^2\sin^2\theta}}.$$

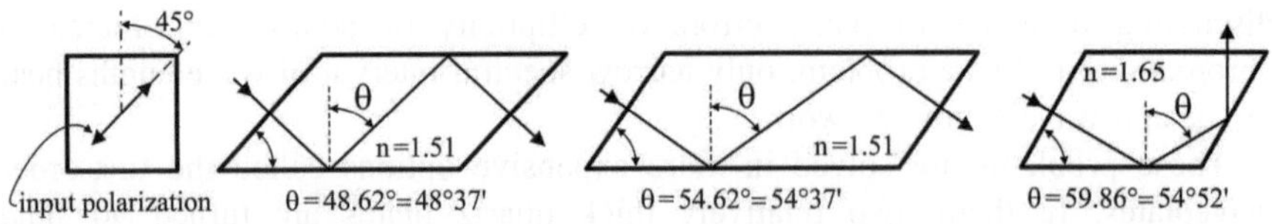

Fig. 5.17 The Fresnel rhombs, satisfying two possible solutions for the angle of incidence. Linear polarization at the input must make 45° to the rhomb in order to produce circular polarization at the output. At right is the modification often called the Mooney rhomb, which uses heavy flint glass with high refractive index and returns the ray at the obtuse angle to the incident one. This geometry also has two solutions for the incident angle, but in practice only one is used, with the bigger angle

Here θ is the incidence angle in glass. Until it is smaller than the angle of total internal reflection determined by the condition $n \sin\theta = 1$, both r_{p} and r_{s} are real, and the phase difference between the incident and reflected waves may only be zero or 180° (inversed sign). Within the angular range of total internal reflection, reflection coefficients become complex

$$r_{\mathrm{p}} = \frac{\cos\theta - i n\sqrt{n^2\sin^2\theta - 1}}{\cos\theta + i n\sqrt{n^2\sin^2\theta - 1}}; \quad r_{\mathrm{s}} = \frac{n\cos\theta - i\sqrt{n^2\sin^2\theta - 1}}{n\cos\theta + i\sqrt{n^2\sin^2\theta - 1}}$$

with $i = \sqrt{-1}$, and any phase shifts may be expected. It is easy to verify that the last string of formulas gives unity moduli of reflection coefficients $|r_{\mathrm{p}}| = |r_{\mathrm{s}}| = 1$, thus complying with the condition of total internal reflection.

Consider now the structure of formulas for r_{p} and r_{s}

$$r = \frac{a - ib}{a + ib} = e^{2i\psi}$$

and its graphical interpretation in Fig. 5.18. The phase shift on reflection φ doubles the phase ψ determined as

$$\psi = \arctan\frac{b}{a}.$$

For p- and s-polarizations we have respectively

$$\psi_p = \arctan\frac{n\sqrt{n^2\sin^2\theta - 1}}{\cos\theta}; \quad \psi_s = \arctan\frac{\sqrt{n^2\sin^2\theta - 1}}{n\cos\theta}$$

For converting the linear polarization into the circular one, the phase shift must equal $\pi/2$:

$$2|\psi_p - \psi_s| = \frac{\pi}{2}.$$

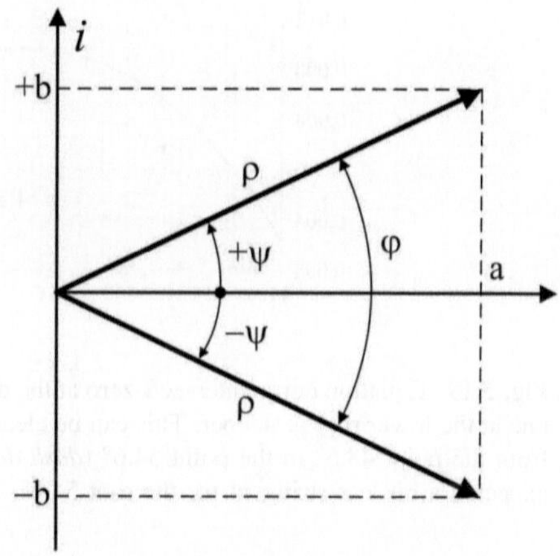

Fig. 5.18 The phase shift on reflection is equal to $\varphi = 2\psi$

The first question is whether or not it is possible to obtain such a big phase difference at one reflection in principle. To answer it, consider

$$\tan(\psi_p - \psi_s) = \frac{\tan\psi_p - \tan\psi_s}{1 + \tan\psi_p \cdot \tan\psi_s} = \frac{\cos\theta\sqrt{n^2\sin^2\theta - 1}}{n\,\sin^2\theta}$$

and find its maximum as a function of θ. Equalizing the derivative to zero, obtain the equation for the extremum:

$$\frac{2 - (n^2+1)\sin^2\theta}{n\sin^3\theta\sqrt{n^2\sin^2\theta - 1}} = 0.$$

The solution is

$$\sin^2\theta_{\max} = \frac{2}{n^2+1}$$

which determines the maximum of the phase difference as

$$\tan(\psi_p - \psi_s)\big|_{\max} = \frac{n^2-1}{2n}, \quad 2(\psi_p - \psi_s)\big|_{\max} = 2\arctan\frac{n^2-1}{2n}.$$

The most common glass in optical industry is the BK7 with $n = 1.51$, which gives the maximum phase difference only $45.94° = 45°56'$—two times less than needed. Fresnel solved this problem by applying two consecutive reflections, each one introducing phase difference exactly 45°. For that, the angle of incidence θ must be slightly less than the $\theta_{\max}$. To find the exact value of θ, we have to solve the equation

$$\tan(\psi_p - \psi_s) = \tan\left(\frac{45°}{2}\right),$$

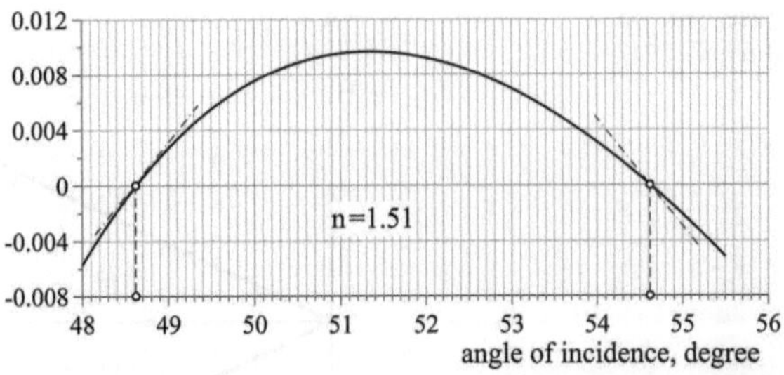

Fig. 5.19 Equation curve intersects zero at the points 48.6° and 54.6° with different tangents: the one at the lower root is steeper. This can be clearly seen by dragging the mirror-reflected tangent from the point 48.6° to the point 54.6° (*dash-dotted lines*). As such, angular tolerances may be expected a bit less stringent for the root 54.6°

which transforms to a more convenient form

$$\frac{\sqrt{(n^2x-1)(1-x)}}{nx}=t$$

with new variables $t \equiv \tan(45°/2) \approx 0.41$ and $x \equiv \sin^2\theta$. The solutions are

$$x_{1,2}=\frac{n^2+1\pm\sqrt{n^4+1-2n^2(2t^2+1)}}{2n^2(t^2+1)}; \quad \theta_{1,2}=\arcsin\sqrt{x_{1,2}}.$$

For the BK7 glass with $n = 1.51$, $\theta_1 = 48.62° = 48°37'$, and $\theta_2 = 54.62° = 54°37'$. The doubled 37′ is not a misprint but a coincidence. For the heavy flint glass with $n = 1.65$ that is used in the Mooney modification (Fig. 5.17), $\theta_1 = 40.35°$ and $\theta_2 = 59.86°$.

Are the both roots equally good for practice? From purely dimensional point of view (Fig. 5.17), the smaller angle makes better impression because the size of the entire rhomb is smaller. However, the size is not a key factor, and from the point of view of functional performance the bigger angle, i.e. 54°37′, is preferable. There are two reasons for that: it is farer from the total internal reflection threshold (42°), and angular tolerances for this root are less stringent. This is explained in Fig. 5.19, showing the curve of the equation function

$$\cos\theta\sqrt{n^2\sin^2\theta-1}-0.41\cdot n\sin^2\theta.$$

The biggest disadvantage of the Fresnel rhomb in its classic form shown in Fig. 5.17 is the displacement of the output ray, which could probably make this optical element obsolete, if it were not for its achromaticity. The achromatic advantage over waveplates is so important that even additional complexity and higher manufacturing price can be paid for the designs without displacement (Fig. 5.20).

The single Fresnel rhomb performs the same function as the quarter-wave plate, but it does not suffer from chromaticity. To extend this very useful advantage to

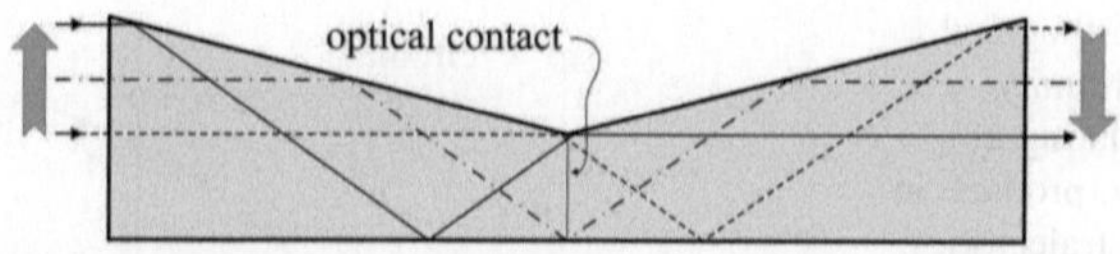

Fig. 5.20 The so-called K-prism is achromatic and serves as the quarter-wave plate without displacement of the outer beam. The two parts act exactly like in the Fresnel rhomb, with the same bottom angle of incidence 54.6° and two smaller angles of incidence on the upper surface. Each of the two reflections on the upper surface introduces only half of the phase shift on the bottom surface, which corresponds to the angle of incidence 76.02° for the BK7 glass. The two parts are put on optical contact with very sharp edge at the bottom, which makes the seam barely visible. Optical contact (bonding) is a very reliable technique for permanent connection of two flat (conformal in general) optical elements made of the same glass: due to interatomic forces between identical materials, the adhesion is strong to keep the parts together. The advantage of optical bonding is the higher power that can be transmitted through the interface

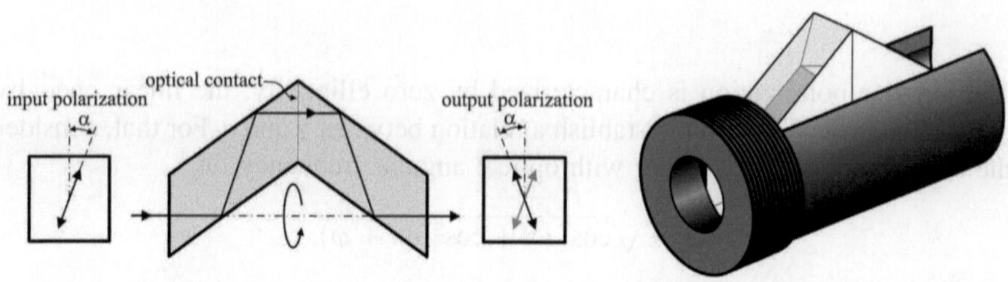

Fig. 5.21 Polarization rotator. Combination of two Fresnel rhombs acts as a half-wave plate. Although looking bulky, such an element has another advantage over an ordinary Fresnel rhomb: the non-displaced output. Therefore, it may be conveniently rotated around the ray, for adjustment purposes. The entire assembly is made from two rhombs, connected on the optical contact. The holder that can be inserted into various standardized rotation stages is a plus (*at right*)

the class of half-wave plates, a combination of two rhombs may be used as in Fig. 5.21. Since this optical element does not look like a plate, it is commonly called polarization rotator, emphasizing its basic application.

Now, it is time to compare spectral limitations of phase elements numerically. For certainty, consider quarter-wave elements, i.e. multiple- and first-order waveplates and the Fresnel rhomb. All of them are designed to introduce the phase difference exactly $\varphi_0 = \pi/2$ at a certain wavelength λ_0. If the wavelength changes slightly by $\Delta\lambda = \lambda - \lambda_0$, then the phase difference also changes to $\varphi = \varphi_0 + \varepsilon$, making the circular polarization elliptical (Fig. 5.22). The question is how far the new elliptical trajectory differs from the initial circular one. This difference may be evaluated numerically in terms of the parameter called ellipticity:

$$\chi = \frac{r_1 - r_2}{r_1} = 1 - \frac{r_2}{r_1}.$$

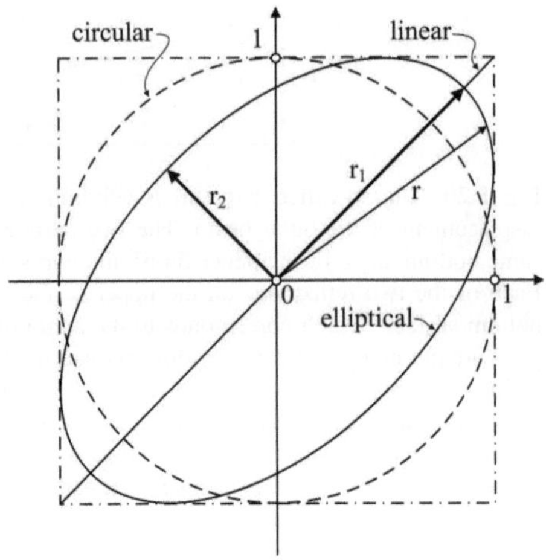

Fig. 5.22 Two orthogonal polarizations with unity amplitudes, oscillating at optical frequency, produce in general elliptical trajectories, which degenerate into either circular or linear, depending on the phase difference φ

The circular polarization is characterized by zero ellipticity, the linear one—by unity ellipticity. We need to establish a relation between χ and ε. For that, consider the radius-vector r, oscillating with optical angular frequency ω:

$$r = \sqrt{\cos^2 \omega t + \cos^2(\omega t + \varphi)},$$

and find its extrema, which are r_1 and r_2. For better tracking the computations, define $x = \omega t$. Then, equalizing the derivative over x to zero, we find the equation:

$$\sin 2x = -\sin(2x + 2\varphi) \quad \text{or} \quad \tan(2x) = -\frac{\sin 2\varphi}{1 + \cos 2\varphi}.$$

This single equation gives two roots

$$-\frac{\pi}{2} - \delta \quad \text{and} \quad +\frac{\pi}{2} - \delta,$$

as it is explained in Fig. 5.23. Since at the extrema $\sin 2x = -\sin(2x + 2\varphi)$, we may apply some trivial trigonometric manipulations and write the radius-vector at these two points as

$$r_{1,2} = \sqrt{1 + \cos 2x_{1,2}}\,,$$

where

$$2x_1 = -\frac{\pi}{2} - \delta \quad \text{and} \quad 2x_2 = +\frac{\pi}{2} - \delta.$$

This leads to

$$r_{1,2} = \sqrt{1 \pm \sin \delta}.$$

Fig. 5.23 The equation with tangent gives two roots: one slightly less than $-\pi/2$ and another slightly less than $+\pi/2$

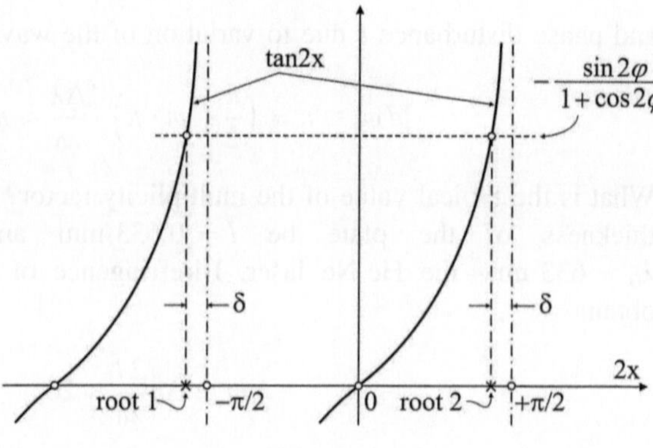

Now, we need to express δ in terms of the initial phase disturbance ε. For that, we use the smallness of ε:

$$\sin 2\varphi = \sin(\pi + 2\varepsilon) \approx -2\varepsilon, \quad \text{and} \quad \cos 2\varphi = \cos(\pi + 2\varepsilon) \approx -1 + 2\varepsilon^2.$$

With this,

$$\tan 2x_1 \approx -\frac{-2\varepsilon}{1 - 1 + 2\varepsilon^2} = \frac{1}{\varepsilon},$$

but on the other hand, using definition of δ,

$$\tan 2x_1 = \frac{\sin\left(-\frac{\pi}{2} - \delta\right)}{\cos\left(-\frac{\pi}{2} - \delta\right)} \approx \frac{1}{\sin\delta}, \quad \tan 2x_2 = \frac{\sin\left(+\frac{\pi}{2} - \delta\right)}{\cos\left(+\frac{\pi}{2} - \delta\right)} \approx \frac{1}{\sin\delta}.$$

Therefore, $\sin\delta = \varepsilon$ and

$$r_{1,2} = \sqrt{1 \pm \varepsilon}.$$

Ellipticity is then

$$\chi = 1 - \frac{r_2}{r_1} \approx \varepsilon.$$

Now we can compare various phase elements practically, understanding the physical meaning of what we compare. What may be considered as good circular polarization? When we rotate a good polarizer, and can hardly see any variations in the transmitted intensity. In a usual experiment, 10 % variation is slightly noticeable, so that we can assume tolerable ellipticity as $\chi = 10$ %. With the multiple-order quarter-wave plate, whose multiplicity factor is m,

$$\varphi = \frac{2\pi}{\lambda_0}\Delta n \cdot l = \frac{\pi}{2} + m \cdot \pi,$$

and phase disturbance ε due to variation of the wavelength $\Delta\lambda$ is

$$|d\varphi| \equiv \varepsilon = \left(\frac{\pi}{2} + m \cdot \pi\right) \cdot \frac{\Delta\lambda}{\lambda_0} \approx m \cdot \pi \cdot \frac{\Delta\lambda}{\lambda_0}.$$

What is the typical value of the multiplicity factor? Just for easy calculations, let thickness of the plate be $l = 0.633$ mm and the design wavelength $\lambda_0 = 633$ nm—the He-Ne laser. Birefringence of quartz $\Delta n = 0.009$, and we obtain

$$m \approx \Delta n \frac{2l}{\lambda_0} \sim 20.$$

Therefore, since $\chi \approx \varepsilon$, relative monochromaticity is

$$\frac{\Delta\lambda}{\lambda_0} = \frac{\chi}{\pi m} \sim 1.5 \times 10^{-3}.$$

The tolerable wavelength deviation for $\lambda_0 = 633$ nm is thus $\Delta\lambda \sim 1$ nm—a very small value comparable with spectral resolution of a spectrometer.

For the first-order quarter-wave plate,

$$\varphi = \frac{2\pi}{\lambda_0} \Delta n \cdot \Delta l = \frac{\pi}{2},$$

and

$$|d\varphi| \equiv \varepsilon = \frac{\pi}{2} \frac{\Delta\lambda}{\lambda_0}.$$

With the same values of the parameters,

$$\frac{\Delta\lambda}{\lambda_0} = \frac{2}{\pi} \chi \sim 6 \times 10^{-2}, \quad \Delta\lambda \sim 40\,\text{nm}$$

much less stringent tolerance than for the multiple-order waveplate, but still restrictive even for laser diodes.

Returning now to thermal stability of waveplates, we can better understand its consequences. How big or small are the temperature coefficients 1 and 0.01 nm/°C, cited earlier for the multiple- and first-order waveplates? For instance, 5 °C variation of ambient temperature causes 5 nm optical path error in a multiple-order quartz plate, which is equal to $\varepsilon = 0.05$ or $\chi = 5$ % at $\lambda = 633$ nm—a value that can be sensed only in rather accurate measurements. As to the first-order waveplate, it may then be considered as thermally insensitive.

Talking about phase errors, it is worth mentioning some tricks with tilting the waveplates. Consider ellipsoids of refractive indices as they are positioned in the waveplate (Fig. 5.24). Tilts may be done either around optical axis or around the perpendicular axis. In both cases, geometrical thickness of the plate increases. Phase difference, however, behaves differently. When the axis of rotation

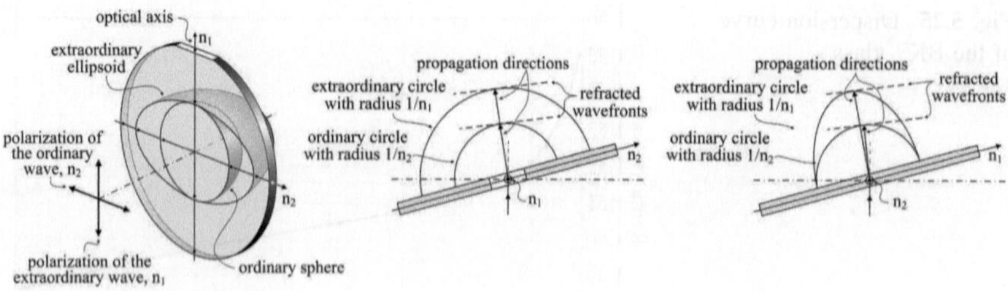

Fig. 5.24 Two types of tilting that may affect phase difference: around the optical axis (*in the middle*) and around perpendicular axis (*at right*). The results are opposite

coincides with the optical axis, propagation directions and refractive indices of both the extraordinary and ordinary rays inside the crystal remain the same. Therefore, phase difference only increases. When the axis of rotation is perpendicular to the optical axis, both propagation direction and refractive index of the extraordinary wave change: in the negative crystals refractive index increases, in the positive—decreases. In both the cases, birefringence $\Delta n = |n_1 - n_2|$ decreases, so that the phase difference may either increase or decrease, depending on the crystal. Quartz is the positive uni-axial crystal, and the combined effect for it is such that the phase difference decreases. Thus, the rule of thumb sounds like this: for quartz waveplates, tilts around the optical axis increase the phase difference, and around the perpendicular axis—decrease it. In the latter case, also some very small angular split of the extraordinary and ordinary rays does occur.

Theoretically, all this sounds like we can tune the waveplate, tilting it accurately around one or another axis, which would be great thing for fine adjustment. However, we must not forget that in reality the optical axis of a waveplate—either the quarter—or the half-wave—always makes some angle to verticality, and this angle also must be adjusted. Thus, the entire two-axis angular stage would be too cumbersome and bulky for simple laboratory applications. Therefore, the aforementioned effects should be considered more like warning to avoid any significant tilts during mounting the waveplate.

Finally, we address spectral tolerance of the Fresnel rhomb. Its governing equation at single reflection is

$$\tan\left(\frac{\varphi}{4}\right) = \frac{\cos\theta\sqrt{n^2\sin^2\theta - 1}}{n\sin^2\theta},$$

where φ is the total output phase shift and $\theta = 54.62°$ is the angle of incidence inside the prism. The factor 1.4 stands because it is only one reflection of the total two. According to the logic of our approach, we must give perturbation $\varepsilon = \chi$ to the phase difference $\varphi = \pi/2 \pm \varepsilon$ and compute from the governing equation new values for the refractive index n

$$n = \left[\sin\theta\sqrt{1 - t^2\tan^2\theta}\right]^{-1}, \quad t \equiv \tan\left(\frac{\pi}{8} \pm \frac{\varepsilon}{4}\right),$$

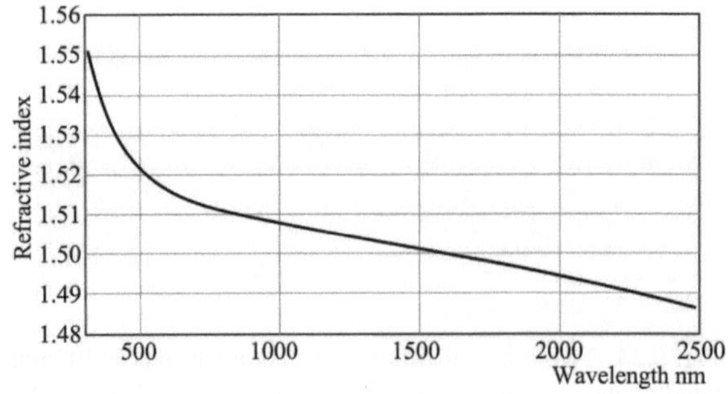

Fig. 5.25 Dispersion curve of the BK7 glass

which will differ slightly from the design value $n = 1.51$ of the BK7 glass. Then, looking at the dispersion curve of the BK7 glass (Fig. 5.25), determine the wavelengths, corresponding to the new refractive indices. This will give the spectral range, in which the Fresnel rhomb introduces less than $\chi = 10\ \%$ ellipticity. Computations give $n_{\max} = 1.57$ for $\varepsilon = +0.1$ and $n_{\min} = 1.46$ for $\varepsilon = -0.1$. These values are even beyond the measured range of the refractive index of the glass, which means that at least in visible domain 10 % ellipticity is guaranteed.

5.4 Compensators

Compensators are variable waveplates, commonly used to correct polarization state of collimated beams. For example, when highly polarized laser beam passes through a system of optical elements or reflects from mirror-like surfaces, its original polarization acquires some ellipticity, which may present a hindrance for subsequent analysis. Compensators readjust such weakly elliptical polarization into linear polarization again. The two most frequently used compensators are the Babinet-Soleil compensator (Fig. 5.26) and an ordinary quarter-wave plate (Fig. 5.15). However, the Babinet-Soleil compensator is much more versatile and accurate instrument that can be used also for precise transformations and analysis of the input polarization, to introduce known phase shifts in orthogonally polarized components, to test fixed waveplates, and for other applications.

Without exceptions, the Babinet-Soleil compensators cover at least one wavelength of optical path difference in visible domain. Thus, it is easy to evaluate the angle of quartz wedge needed for that. With quartz birefringence $\Delta n = 0.009$, one wavelength optical path difference is gained through $\lambda/\Delta n \approx 70\,\mu$ of the crystal at the wavelength of He-Ne laser $\lambda = 0.63\ \mu$. Assuming comfortable micrometer screw incursion about 10 mm, it gives the wedge angle $\sim 70\ \mu/10\ \text{mm} = 0.4°$.

The Babinet-Soleil compensator and the quarter-wave plate compensate for the ellipticity differently. In order to understand that, some brief introductory

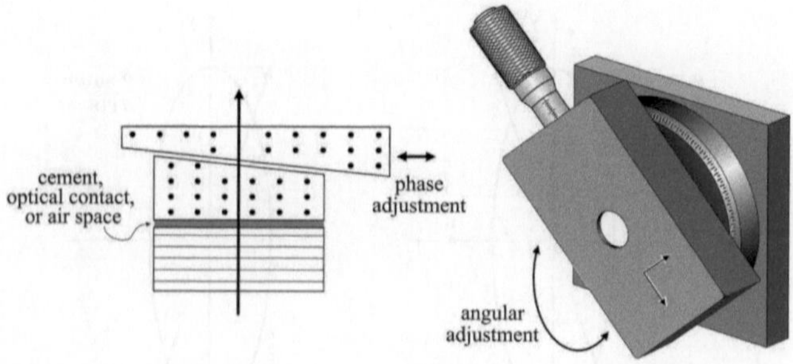

Fig. 5.26 The Babinet-Soleil compensator is physically a first-order waveplate, in which one of the birefringent plates is composed of two matching wedges, moving along each other (*at left*). *Horizontal hatching* and *dots* indicate optical axes in the cross section. The fixed wedge is always shorter than the moving one. The angle of the wedge is greatly exaggerated. *Arrows* on the front surface of the compensator always show orientation of optical axes (*at right*)

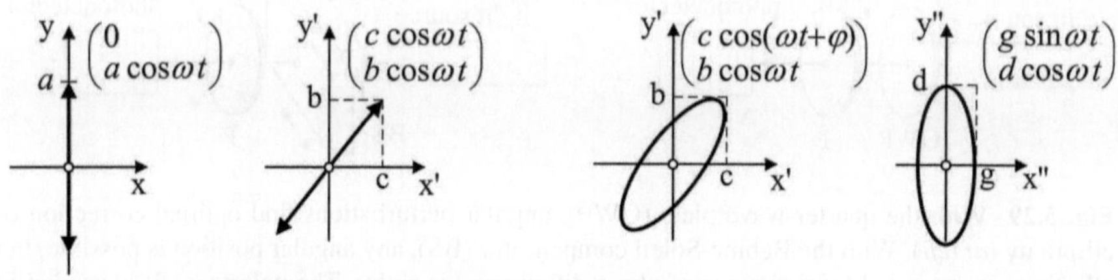

Fig. 5.27 Two Cartesian system of coordinates (x, y) and (x', y'), rotated one relative to another, present differently linear polarization of the same wave with optical frequency ω (*at left*). Differently will be presented the elliptical polarization as well (*at right*)

comments are needed, explaining various forms for presenting elliptical polarization. To begin with, the pure linear polarization can be presented differently in different system of coordinates, as shown in Fig. 5.27. Initial linear polarization in the rotated system (x', y') is characterized by new parameters b, c. If a small phase shift φ is added to the x' component, then linear polarization becomes tilted ellipse, confined within the rectangle with semi-sides b and c. But the trajectory is still an ellipse, and as such, it can be presented in a traditional form

$$\begin{pmatrix} g \sin \omega t \\ d \cos \omega t \end{pmatrix},$$

if rotated vertically. If an optical beam with such a polarization passes through the quarter-wave plate, whose optical axes are directed exactly along (x'', y''), then it will be converted into the linear polarization directed diagonally in the rectangle with semi-sides d and g (Fig. 5.28). The only problem is that the directions of both the axes (x'', y'') and the resultant polarization vector in this system are unknown: the trial-and-error procedure should be applied. The search begins with inserting a polarizer

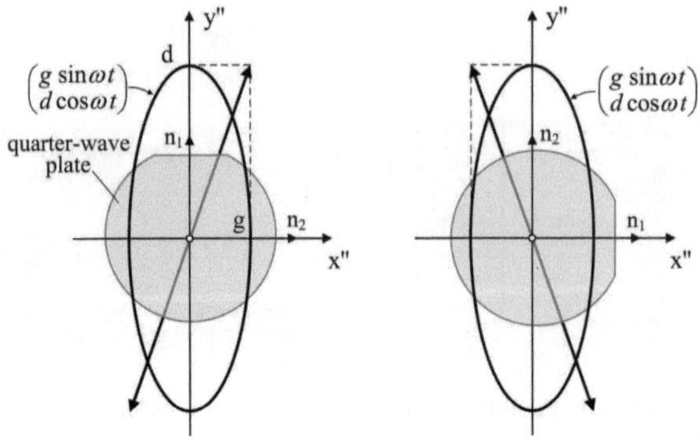

Fig. 5.28 The quarter-wave plate with its axes, coinciding with the axes of the ellipse, converts any elliptical polarization into the linear one, but the direction of the latter cannot be predicted

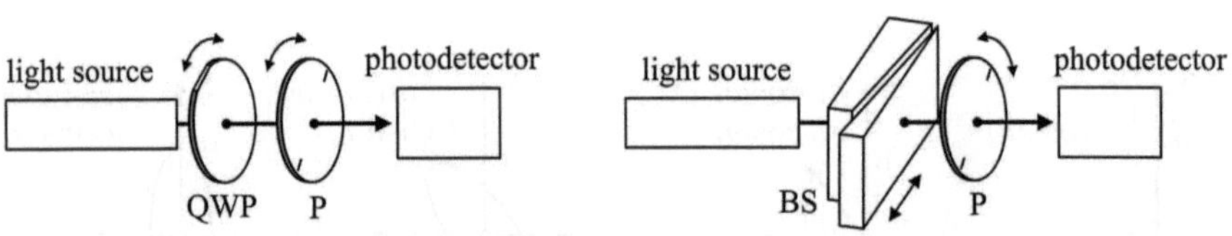

Fig. 5.29 With the quarter-wave plate (QWP), angular perturbations find optimal correction of ellipticity (*at left*). With the Babine-Soleil compensator (BS), any angular position is possible, but ellipticity is corrected by variations of phase difference (*at right*). The polarizer (P) is needed in both cases to find the optimum

and finding its angular position of minimum transmitted intensity (Fig. 5.29). This coarsely identifies direction of the longer axis of the ellipse (perpendicular to the polarizer axis). Then insert the quarter-wave plate between the light source and the polarizer, with its optical axis approximately along the polarization ellipse. Adjust the polarizer to the minimum intensity again. Give the quarter-wave plate some rotational variation and find the minimum of intensity. Then rock the polarizer, and so on, until the minimum-minimorum is found. In this position, ellipticity of the beam is corrected, but the direction of the final linear polarization may be different from the initial expectation.

Using the Babine-Soleil compensator, we work in the (x', y') system of coordinates, and care only about the phase difference φ. But since the angle of rotation between the (x, y) and (x', y') systems may be arbitrary, it means that we do not care about angular orientation of the compensator: we may position it as comfortable for us. This is the difference between the quarter-wave plate and Babine-Soleil compensation: the first one has fixed phase difference (quarter of wave) and must be rotated, whereas the second one is adjustable and, therefore, may be kept in place. The compensating procedure with the Babine-Soleil compensator is explained in Fig. 5.30. In the system of coordinates (x', y'), whose axes are along optical axes of the compensator, the ellipticity is presented as a vector

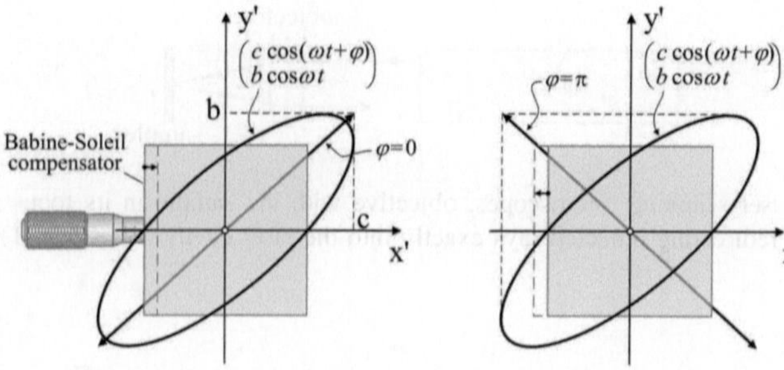

Fig. 5.30 Whatever orientation of the Cartesian system (x', y') is, it is always possible to find parameters b, c, and φ, fitting any ellipse in this particular system of coordinates. Parameter φ has the physical meaning of the phase difference between projections of the running vector onto the axes. The Babine-Soleil compensator is designed to make this difference zero. With zero φ, the ellipse collapses to a diagonal line, i.e. linear polarization. Since the range of the available phase differences exceeds 2π, not only $\varphi = 0$ but also $\varphi = \pi$ may convert the ellipse into the line (*at right*)

$$\begin{pmatrix} c\,\cos(\omega\,t + \varphi) \\ b\cos\omega\,t \end{pmatrix},$$

and the only thing we have to do is to move the prisms so as to compensate for the phase difference φ. Since the Babine-Soleil compensator always covers the phase difference of at least one wavelength, it always can be done. Exact position of the micrometer screw when it happens and direction of the compensated linear polarization cannot be predicted because we do not know directions of (x', y'). But we know that somewhere within the total incursion of the micrometer screw it happens. Thus, we must locate position of total compensation, and the polarizer is the sensor for that (Fig. 5.29). Again, giving consecutive perturbations to the micrometer screw and the polarizer angle, we find the minimum transmitted intensity—the point of total compensation.

5.5 Isolators

Stability of lasers may suffer from back-reflected beams. Mostly, it does not happen because for an optical beam to penetrate the laser cavity, its propagation vector must be strictly parallel to cavity axis, and they are not parallel due to random tilts of all the reflecting surfaces. However, there are very important practical applications where it happens by design, and therefore isolation becomes vital (Fig. 5.31). When the laser beam is polarized (linearly or circularly), isolation of two types may be used (Fig. 5.32).

The first and the simplest one is the waveplate isolator. It is called so because the key component of it is the quarter-wave plate combined with a polarizer. Now, it is very important to understand that the quarter-wave plate is a reciprocal element, which means that it acts equally on the beam going right and on the beam

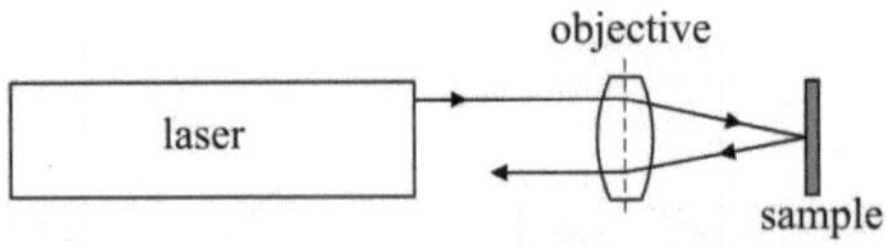

Fig. 5.31 In laser scanning microscopes, objective with the sample in its focus make perfect retro-reflector, redirecting reflected rays exactly into the laser cavity (see Chap. 1)

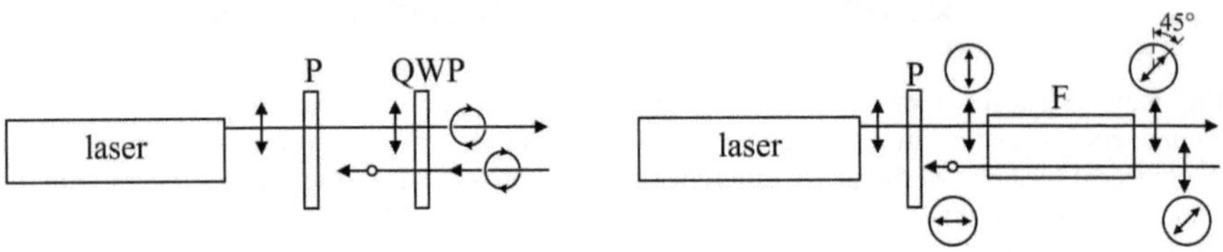

Fig. 5.32 Linearly polarized beam can be isolated by means of a waveplate (*at left*) or Faraday rotator (*at right*). *Double arrows* and *open dots* show direction of linear polarization. For certainty of the figure, initial polarization is assumed vertical. Polarizer (P) transmits vertical and blocks horizontal polarizations. Quarter-wave plate (QWP) converts linear polarization to the circular one. Note that the reflected beam is always of an opposite circular polarization relative to the input one, due to boundary conditions on the reflecting surface. In the Faraday rotator (F), parameters are adjusted to rotate linear polarization by 45°. Doing this twice, the isolator converts polarization of the reflected beam to 90°, and the polarizer blocks it on re-entering the laser

going left (the Faraday element, which will be discussed below, is a non-reciprocal element). The reason why linear polarization, coming from left to right, converts to the opposite after the wave passes the plate from right to left is that circular polarization changes to the opposite upon reflection. The latter becomes clear, considering boundary conditions on the reflecting surface: the electrical field preserves its direction, while the propagation vector changes to the opposite.

In the Faraday rotator, linear polarization continuously rotates in one direction as the wave propagates through the optically-active medium. Unlike the case of birefringent waveplates, linear polarization does not undergo transformations into elliptical and back, but remains linear on the entire way. The Faraday rotation is a non-reciprocal effect, meaning that the wave, propagating in the opposite direction, experiences opposite direction of rotation. The physical reason for that is the so-called Faraday effect, when axially applied magnetic field makes one circular polarization propagate faster than the opposite one. Consider linearly polarized wave with electrical field vector $E = 2a\cos\omega t$ along y axis:

$$2a\begin{pmatrix} 0 \\ \cos\omega t \end{pmatrix}.$$

This vector can be combined as a sum of two virtual vectors, rotating in the opposite directions:

$$a\begin{pmatrix}\sin\omega t\\ \cos\omega t\end{pmatrix} \quad \text{and} \quad a\begin{pmatrix}-\sin\omega t\\ \cos\omega t\end{pmatrix}.$$

Assume as a hypothesis (we shall prove it later) that in the presence of magnetic field these two virtual circularly polarized wave propagate with different refractive indices n_r and n_l, with the subscripts meaning «right» and «left» polarization. Then, after the distance L, these two vectors acquire phase shifts proportional to n_r and n_l, and recombine to the sum

$$\begin{aligned} a&\begin{pmatrix}\sin(\omega t - \frac{2\pi}{\lambda}n_l L) - \sin(\omega t - \frac{2\pi}{\lambda}n_r L)\\ \cos(\omega t - \frac{2\pi}{\lambda}n_l L) + \cos(\omega t - \frac{2\pi}{\lambda}n_r L)\end{pmatrix}\\ &= 2a\begin{pmatrix}\sin\frac{\pi L}{\lambda}\Delta n\\ \cos\frac{\pi L}{\lambda}\Delta n\end{pmatrix}\cdot\cos\left[\omega t - \frac{\pi L}{\lambda}(n_r + n_l)\right]\end{aligned}$$

with $\Delta n = n_r - n_l$ and λ being approximately the same wavelength for both the virtual waves. This is a linearly polarized wave, being rotated by the angle

$$\alpha = \frac{\pi L}{\lambda}\Delta n$$

against its initial vertical direction. The difference between refractive indices Δn is proportional to magnetic field, and the following shows why.

Within the classical model, magnetic field B causes the so-called Larmor precession of electrons with angular frequency

$$\Omega = \frac{e}{2m}B,$$

where e and m are the electron charge and mass. When the vector of the magnetic field is along propagation direction of light, the right- and left-circularly polarized waves interact with the electrons at different frequencies $\omega \pm \Omega$, with the negative sign corresponding to the right-polarized wave (electrons are negative and precess in the opposite direction). Dispersion of the refractive index causes the difference between n_r and n_l:

$$n_r = n(\omega + \Omega) = n + \frac{dn}{d\omega}\Omega = n + \frac{dn}{d\lambda}\frac{d\lambda}{d\omega}\Omega = n - \frac{\lambda^2 e}{4\pi mc}\cdot\frac{dn}{d\lambda}\cdot B,$$

$$n_l = n + \frac{\lambda^2 e}{4\pi mc}\cdot\frac{dn}{d\lambda}\cdot B$$

so that

$$\Delta n = -\frac{\lambda^2 e}{2\pi mc}\cdot\frac{dn}{d\lambda}\cdot B.$$

The angle of rotation per unit length is then defined by the so-called Becquerel formula:

$$\frac{\alpha}{L} = -\frac{\lambda e}{2mc}\cdot\frac{dn}{d\lambda}\cdot B,$$

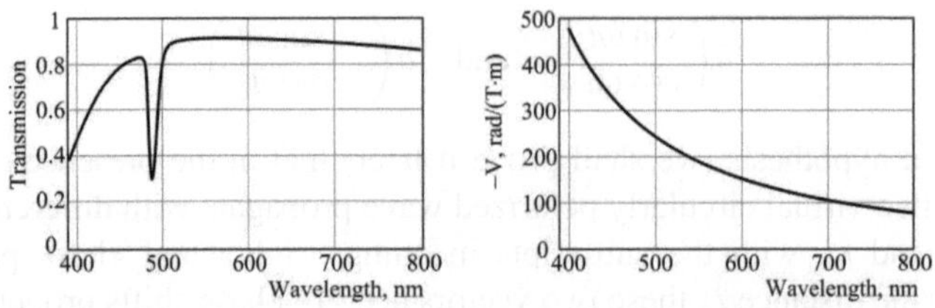

Fig. 5.33 The Verdet constant V of TGG decreases rapidly from short to long wavelengths: $K = 4.45 \times 10^7$ rad nm^2/(T·m); $\lambda_1 = 258.2$ nm. Transmission displays another trend with deep absorption band below 500 nm

and the total angle of rotation

$$\alpha = V \cdot BL,$$

with the proportionality coefficient

$$V = -\frac{\lambda e}{2mc} \cdot \frac{dn}{d\lambda}$$

called the Verdet constant. From these formulas, it follows that the Faraday rotation is stronger in highly dispersive materials. The Verdet constant is rapidly varying function of wavelength, a good analytical approximation of which is

$$V(\lambda) = \frac{K}{\lambda_1^2 - \lambda^2},$$

λ_1 being the parameter. The highest Verdet constant is observed in the terbium gallium garnet $Tb_3Ga_5O_{12}$ (TGG): -134 rad/(T·m) at 633 nm (Fig. 5.33). The negative sign means that polarization rotates counter-clockwise when propagating along the direction of the magnetic field. The sign defines direction of polarization rotation. Very high magneto-optical activity also found in special sorts of terbium-doped flint glass called MOS-4 and MOS-10 (73 and 87 rad/(T·m) respectively), but TGG is preferable, owing to better transparency and power resistance. Even with that big Verdet constant, Faraday isolators require very strong magnetic field around several Tesla in order to obtain required 45° rotation in relatively short active rods of about 1 cm long. Such strong fields are created by permanent rare-earth magnets like neodymium iron boron, for instance. Magnets are made in a form of thick cylinders with narrow axial channel, where the active optical rod is inserted (Fig. 5.34). To minimize magnetic dissipation the channel must be narrow, that is the reason why optical apertures of Faraday isolators are usually small: 2–3 mm. Bigger diameters are also available, but for disproportional price.

The basic concept of the Faraday isolator, shown in Fig. 5.32, assumes only one polarizer, but the advanced versions use two (Fig. 5.34), and this is deemed the first advantage of the Faraday scheme over the waveplate concept: better isolation—typically $\sim 10^3$–10^4. As we have already seen in the previous section,

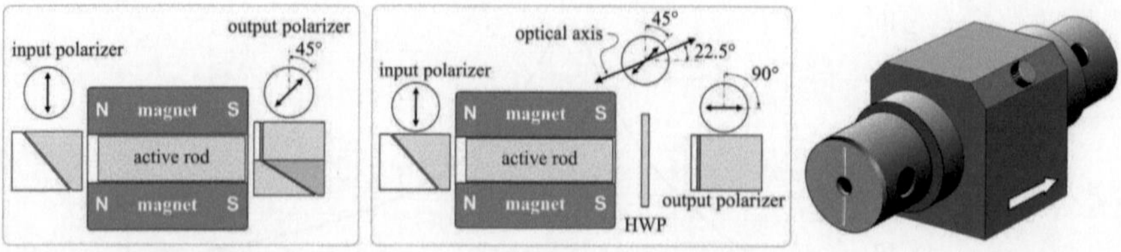

Fig. 5.34 Two options of the Faraday isolator are available for the user convenience: with 45° and horizontal output polarizations. Final adjustment of the rotation angle to exactly 45° is made by fine positioning of the active rod inside the magnet. The half-wave plate (HWP) doubles the 45°–22.5° to final 90° direction of the output polarization. The *arrow* on the base clamp always shows direction of the input beam. The screw on the clamp may be released to rotate the isolator to a proper input polarization usually indicated by a *line* on the front surface. Direction of polarization rotation is shown for positive Verdet constant

waveplates are vulnerable to temperature and tilts, and the reflected wave may suffer from ellipticity. Therefore, in the waveplate isolator, conversion of the reflected wave into linear polarization may not be complete, and some portion of the reflected beam may still pass through the polarizer. The second advantage is that circular polarization is unusable in certain situations, while linear polarization of the Faraday scheme is acceptable without exceptions. However, even though isolation of the Faraday devices may be higher, the waveplate isolators are cheaper, lighter, have no practical limitations in beam size, and can not only stop the reflected wave but also separate it from the direct beam—a very important feature used in interferometry (see Chap. 6).

Isolators described above suffer from chromaticity, i.e. good isolation can be achieved only within relatively narrow spectral interval. For the waveplate isolators, this follows from the analysis that has been done above for quarter-wave plates. For the Faraday devices, some additional insight is needed. Consider Faraday isolators with two polarizers as shown in Fig. 5.34. Isolation is determined as the ratio of the reflected optical power I_0, coming back to the isolator, to the portion of it I that has passed through the input polarizer. According to the Malus law, with a perfect polarizer set at the angle θ relative to polarization of the wave

$$I = I_0 \cos^2 \theta,$$

so that the isolation is

$$\frac{I_0}{I} = \frac{1}{\cos^2 \theta}.$$

In this formula, the angle $\theta = \pi/4 + \alpha$ is composed of the fixed angle $\pi/4$ of the output polarizer and the angle of polarization rotation α only on the way back. It does not depend on the angle of rotation on the direct way because the output polarizer fixes the angle at which the back-reflected wave passes through it. Therefore, variations of the rotation angle on the direct way can only affect

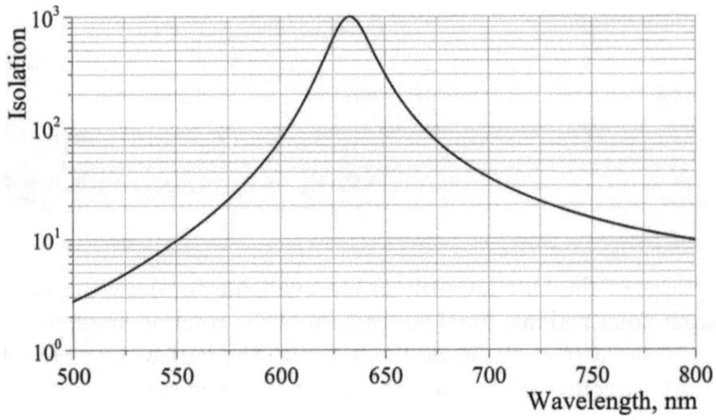

Fig. 5.35 Theoretical spectral performance of a hypothetical Faraday isolator made of TGG and designed for 633 nm. Damping parameter $p = 10^{-3}$

transmission, and not the isolation. This is the reason why the scheme with two polarizers is advantageous.

Suppose now that the wavelength changed from the design value λ_0 to λ. Then the Verdet constant changes from V_0 to $V(\lambda)$, and so does the angle θ:

$$\theta = \frac{\pi}{2} - \frac{\pi}{4}\left(1 - \frac{V(\lambda)}{V_0}\right),$$

making isolation smaller:

$$\frac{I_0}{I} = \sin^{-2}\left[\frac{\pi}{4}\left(1 - \frac{V(\lambda)}{V_0}\right)\right].$$

This formula gives infinity value at the design wavelength $\lambda = \lambda_0$, which is obviously impractical and can be corrected by a small empirical damping parameter p in the denominator (Fig. 5.35):

$$\frac{I_0}{I} = \frac{1}{\sin^2\left[\frac{\pi}{4}\left(1 - \frac{V(\lambda)}{V_0}\right)\right] + p}.$$

Numerically, parameter p is none other than the reciprocal of the maximum isolation. Note, that in this approximation, spectral characteristic depends only on dispersion of the Verdet constant and does not depend on the length of the crystal. Figure 5.35 shows that spectral offset of only 20 nm causes tenfold drop of isolation, which presents certain discomfort to both the users and manufacturers, who need to quickly modify the device for a particular wavelength. In order to ease this practical problem and extend spectral performance of a single device, manufacturers offer the so-called broadband adjustable Faraday isolators, covering up to 200 nm of spectral range. The idea is to use non-uniformity of magnetic field inside the channel of the magnet, axially adjusting the active rod inside it (Fig. 5.36). Since the angle of rotation builds up cumulatively, specific local variations of the magnetic field are not important.

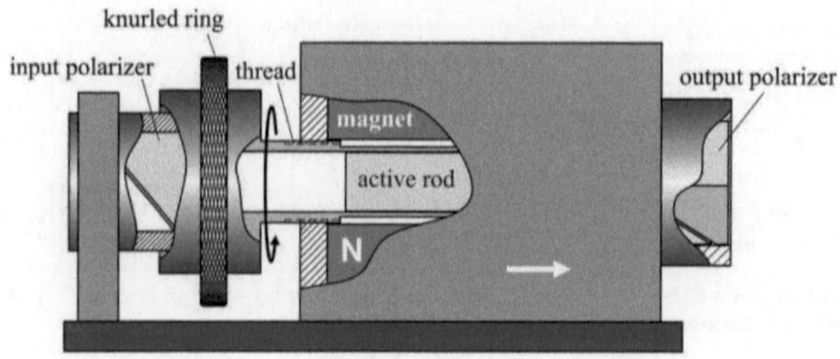

Fig. 5.36 In broad-band adjustable Faraday isolators, the user screws the active rod in or out of the magnet, thus changing cumulative angle of rotation, making it equal to 45° only for a particular wavelength of interest. Directions of polarizers are shown for positive Verdet constant

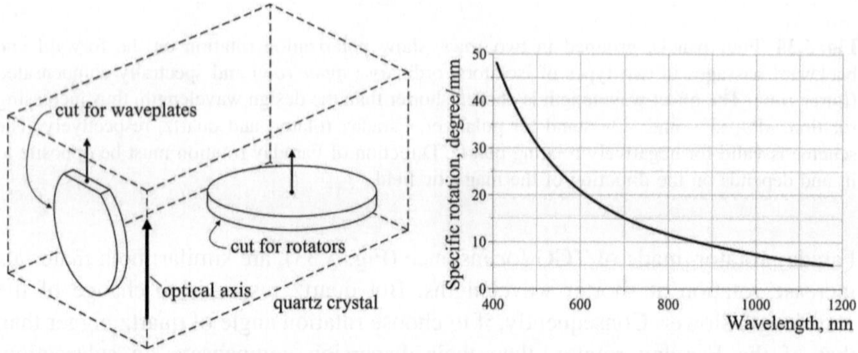

Fig. 5.37 Quartz exists in two forms; one form, called Dextro-Rotary quartz (D-quartz), rotates polarized light in a clockwise direction. Laevo-Rotary quartz (L-quartz) rotates in a counter-clockwise direction. Fused quartz is not optically active. Dispersion curve of a quartz rotator is similar to that of the Faraday rotator, the TGG for instance

However, the problem of narrow spectral response remains even in adjustable versions of the Faraday isolators, making them inapplicable when spectrally wide optical beams are used. In that case, the fixed broadband isolators should be used. The idea of this kind of a device is to compensate for dispersion of the non-reciprocal Faraday rotation by means of reciprocal optical rotation, like that of the quartz. Quartz is not only a uniaxial birefringent crystal but also an optical rotator—a solid that rotates direction of linearly polarized optical wave even without external magnetic field (Fig. 5.37). Such crystals are called optically active crystals, and optical activity of quartz is very big: specific rotation of quartz, i.e. angle of rotation per unit length, is 18.7°/mm at 633 nm. Unlike the Faraday effect, optical activity is reciprocal: change of propagation direction changes direction of rotation. This is why optically active rotators alone cannot serve for isolation, but in combination with non-reciprocal Faraday rotators they can compensate for the Verdet constant dispersion, thus increasing spectral range. The scheme in Fig. 5.38 shows how it can be done. Dispersion curves of the quartz rotator (Fig. 5.37) and

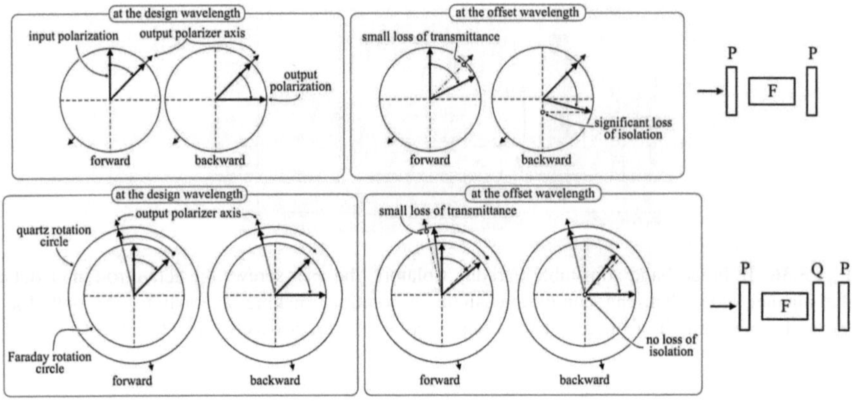

Fig. 5.38 Four panels, grouped in two *rows*, show polarization rotation on the forward and backward passages in two types of isolators: ordinary (*upper row*) and spectrally compensated (*lower row*). The offset wavelength is chosen shorter than the design wavelength, thus increasing rotation. «P», «F», and «Q» stand for polarizer, Faraday rotator, and quartz, respectively. The scheme is valid for negatively rotating quartz. Direction of Faraday rotation must be opposite to it, and depends on the direction of the magnetic field

Faraday rotator, made of TGG for instance (Fig. 5.33), are similar: both materials increase rotation at shorter wavelengths. But quartz responds to change of the wavelength slower. Consequently, if to choose rotation angle of quartz bigger than that of the Faraday rotator, then their dispersion compensate on subtraction. Typically, optimal quartz plate thickness is close to 4.1 mm. Careful compensation may produce as wide spectral range as 80 nm around central wavelength 800 nm, decreasing isolation only two times at the ends of this interval.

Eventually, we are going to address the case of isolation of non-polarized beams. This problem became essential with the spread of Zeeman lasers (Chap. 2) in semiconductor industry and fiber-optic communication systems. The output beam of a Zeeman laser consists of two orthogonal polarizations, thus making the aforementioned isolators irrelevant. As to the fiber optic, ordinary fibers do not preserve certain polarization, making the output beam randomly polarized. The solution was found with the devices called polarization-independent Faraday isolators. There are two types of such devices, using either birefringent crystals or polarizing beam-splitting cubes, that are shown in Fig. 5.39. The figure with captions is self-explanatory. When a wave comes at normal incidence to the surface of a birefringent crystal whose optical axis is inclined to the surface, ordinary and extraordinary waves split angularly (Sect. 5.1), with the extraordinary wave polarized in the plane of incidence. The ordinary wave does not change direction. When the polarizing cube is used, it is always made to transmit p-polarization (Sect. 5.2), i.e. with electrical vector parallel to the plane of incidence.

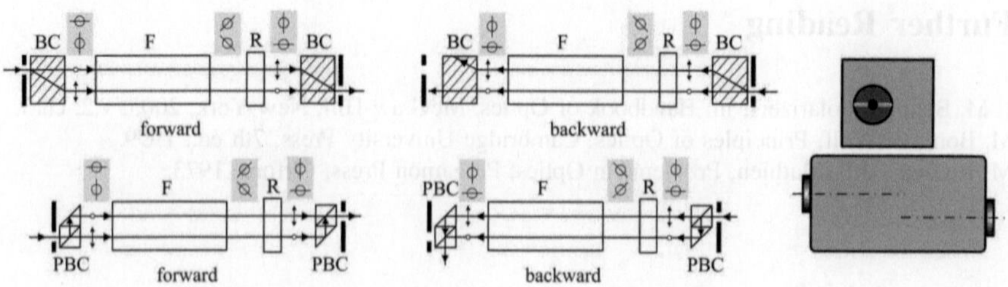

Fig. 5.39 In polarization-independent isolators, two polarizations are initially split, and then recombined to produce a single beam. On the way back, non-reciprocal Faraday rotator (F) splits these waves again, but they cannot recombine and are blocked due to angular or spatial separation. Polarization-independent isolators can be easily recognized by slightly shifted positions of input and output apertures. Hatching indicates direction of optical axis in birefringent crystals (BC). Reciprocal rotator (R) changes polarization by 90°. Either half-wave plate or quartz optically active rotator may be used for this purpose. PBC stands for polarizing beam splitter. *Open dots* and *double arrows* indicate polarizations

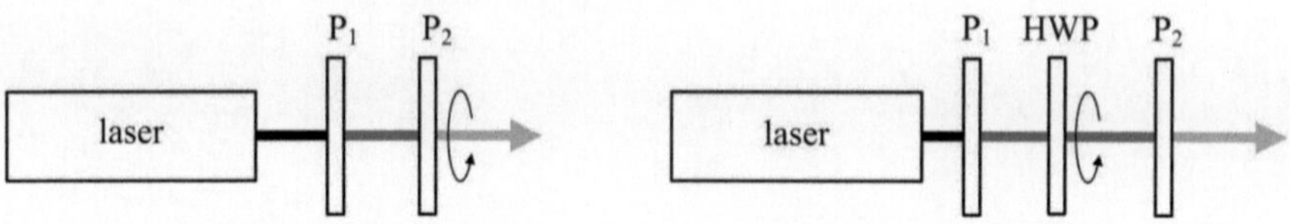

Fig. 5.40 The simplest scheme of a variable attenuator contains only two polarizers P_1 and P_2, one mounted in a rotation stage. Another option contains additionally the half-wave plate HWP in a rotation stage

5.6 Variable Attenuators

One of the simplest applications of polarizers in laboratory practice is variable attenuator (Fig. 5.40).

At least one rotating holder is needed to arrange this scheme. According to Malus law, intensity changes as cosine squared of the rotation angle. In another arrangement that may be used when the polarizer cannot be rotated conveniently (Nikol prism, for instance), a half-wave plate is inserted in between two polarizers. Smooth variation of intensity is a very important feature.

List of Common Mistakes

- forget to remove protective foil from the sheet polarizer before using it;
- ignore the entrance mark on a beam-splitting cube;
- the wave plate designed for a particular wavelength is used at a different wavelength.

Further Reading

J. M. Bennett, Polarizers, in: Handbook of Optics, McGraw-Hill, New-York, 2005, v.2, ch.3.
M. Born, E. Wolf, Principles of Optics, Cambridge University Press, 7th ed., 1999.
M. Rouseau, J.P. Mathieu, Problems in Optics, Pergamon Press, Oxford, 1973.

Chapter 6
Interferometers

Do not be scared: interferometers are reliable instruments when you are aware of very few know-how.

Abstract Among tremendous variety of geometries that interferometry as a science acquired during centuries, this chapter focuses on only few selected configurations that became practical standards. Section one introduces indispensable plane-wave Michelson configuration, and discusses effects of polarization, beamsplitter transmittance, and tilts of the mirrors on interference pattern. In simple vector interpretation, these effects become clear, and optimal conditions for fringe contrast, spatial frequency, and beamsplitter ratio are established. The most frequently used fringe counting technique is introduced here in generalized configuration, later explained in full detail in the next section, considering heterodyne interferometry. Analytical formula is derived for the signal variation as a function of mirrors angular misalignment. Fundamental property of invariance of the interference integral is explained. From these theoretical basics, practical rules of aligning Michelson interferometer become straightforward, and pictures of real interference patterns in various consequitive phases of alignment are presented. The effect of temporal coherence on visibility of fringes is analyzed rigorously in analytical form, introducing the visibility function. Experimental results presented in a series of images recorded in spatial steps of ten micrometers corroborate theoretical conclusion. This effect has clear influence on the design of the mechanism for adjustment of mirrors, and the best solution is explained. The solution that does not require angular adjustment of mirrors implements corner cube reflectors—the optical element explained in Chap. 1. The absence of interference fringes in this type of the interferometer is portrayed in a series of experimental pictures that can hardly be found elsewhere. Such pictures are better observed with telecentric lenses (Chap. 1), and important practical know-how is explained. Chapter 1 already briefly introduced polarization properties of a corner cube, and in this section, these effects are presented in a series of impressive experimental pictures. Practical recommendations of how to avoid images of chamfers of the cube corners in the field of view may be useful. The second section gives detailed understanding of functionality and geometry of the heterodyne interferometers—devices widely used in industry to control nano-scale positioning. The self-explanatory figure and real oscilloscope traces give necessary impression about this technology. Various configurations are possible with standard high-quality

V. Protopopov, *Practical Opto-Electronics*, Springer Series in Optical Sciences 184,
DOI: 10.1007/978-3-319-04513-9_6,

modules available on the market. Some scientific applications require another peculiar configuration that may be composed with the help of the energy separator cube explained in Chap. 1. The shear interferometer for controlling parallelicity of laser beams is the most simple and common device in optical laboratory, but its theory is rather complicated and can be understood fully only with the help of detailed three-dimensional mathematical analysis and computer simulation that are left for the Chap. 12. Without experimental pictures presented in the third section it is not easy to realize this technique, but when the idea is grasped it is only a pleasure to use shear interferometer for perfect adjustment of laser collimators. Besides, with the help of theoretical analysis presented in this section, it turns out to be possible to estimate aberrations. However, this approach requires three-dimensional numerical ray-tracing routines that are summarized in Chap. 12. Imaging interferometers, freely available in the form of Mireau and Michelson interference objectives, are much more than an ordinary microscope objectives, and require special practice to handle them in order to avoid common mistakes. All that is described in the fourth section, starting from the generalized concept of an imaging interferometer and theoretical requirements for image contrast. A series of experimental interference patterns obtained on images of incoherent tungsten halogen lamp (Chap. 2) clearly explain these requirements. The newly purchased interference objective never works on the microscope without special, often tedious alignment. This process is carefully explained and the explanation is supported by detailed figures and photographs. The last section of this chapter is devoted to spectral interferometry. It starts with clear graphical explanation of the concept of this simple and reliable technique, followed by experimental spectrum obtained with very primitive experimental means, always available in any optical laboratory. Interference oscillations in the spectrum of a broad-band light source (Chap. 2) are clearly connected with the optical path difference of a transparent stratified sample. The fast Fourier transform performed on the spectrum can compute absolute values of thicknesses of layers if some non-trivial mathematical transformations are made. Explicit mathematical formulas are given.

6.1 Plane-Wave Interferometers

Among all the variety of sophisticated optical configurations and techniques that interferometry accumulated during centuries, we shall discuss only very few types of practical schemes that can be handled by an average user with the help of common components available on the market. The basic scheme for that is the plane-wave interferometer of the Michelson type (Fig. 6.1).

First, consider monochromatic light with optical frequency ω and wavelength λ. After splitting, the two partial beams produce complex waves $\vec{E}_1$ and $\vec{E}_2$ in the arbitrary chosen interference plane xy with vector $\vec{r}$ in it:

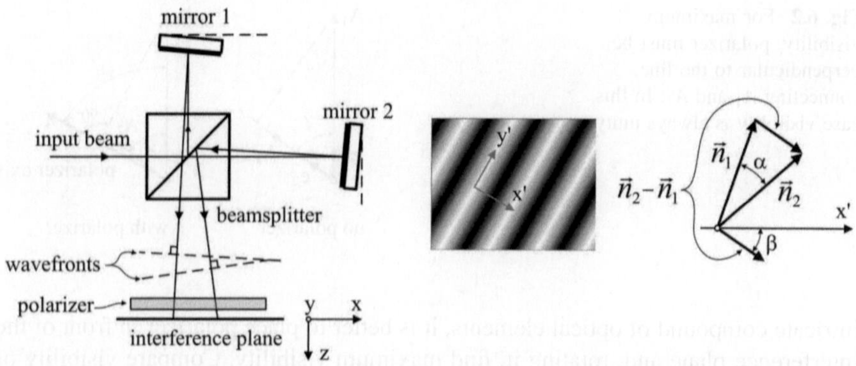

Fig. 6.1 In the plane-wave interferometers, the input beam is commonly the laser beam. Two mirrors may be tilted with respect to each other. Properly oriented polarizer may increase visibility of fringes in case of different polarization of the interfering waves

$$\vec{E}_1 = \vec{e}_1 A_1 \exp(i\omega t - i\frac{2\pi}{\lambda}\vec{n}_1\vec{r} - i\frac{2\pi}{\lambda}L_1),$$
$$\vec{E}_2 = \vec{e}_2 A_2 \exp\left(i\omega t - i\frac{2\pi}{\lambda}\vec{n}_2\vec{r} - i\frac{2\pi}{\lambda}L_2\right),$$

where $i = \sqrt{-1}$, $\vec{e}$ are the unity polarization vectors, A—amplitudes, $\vec{n}$—normal vectors of the wavefronts, and L are the optical paths (accounting for the refractive indices) that the two beams have passed on their ways. We are interested in the intensity of the combined field $\vec{E}$ because photodetectors respond to intensity of light:

$$I = \left|\vec{E}\right|^2 = \left|\vec{E}_1 + \vec{E}_2\right|^2 = I_1 + I_2 + 2(\vec{e}_1\vec{e}_2)\sqrt{I_1 I_2}\cos\left(\frac{2\pi}{\lambda}x'2\sin\frac{\alpha}{2}\cos\beta + \varphi\right),$$

where $I_{1,2} = A_{1,2}^2$—intensities of the partial beams, $\varphi = \frac{2\pi}{\lambda}(L_2 - L_1)$, α is the angle between $\vec{n}_1$ and $\vec{n}_2$, β is the angle between the differential normal $\vec{n}_2 - \vec{n}_1$ and the interference plane, and the axis x' is directed along the projection of the differential normal. The entire interference picture looks like tilted stripes of dark and light areas, which are called fringes. The so-called visibility of fringes

$$v = \left|\frac{2(\vec{e}_1\vec{e}_2)\sqrt{I_1 I_2}}{I_1 + I_2}\right|$$

characterizes modulation depth of this picture. When the two partial beams are orthogonally polarized, the scalar product $(\vec{e}_1\vec{e}_2) = 0$, and there are no fringes. Thus, the first condition for good visibility of fringes is the same polarizations of partial waves. It is a common mistake to combine two waves with different or unknown polarizations, because spatially modulated component of I may be hardly visible on the background of the sum $I_1 + I_2$. When polarization composition of interfering waves is unknown, for example, when one of the beams passes through

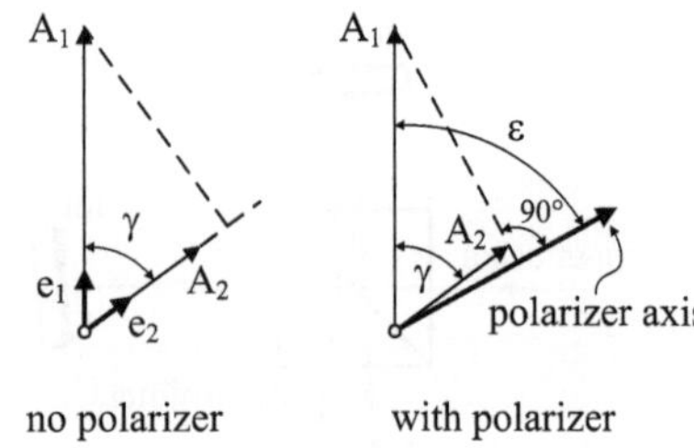

Fig. 6.2 For maximum visibility, polarizer must be perpendicular to the line, connecting A_1 and A_2. In this case visibility is always unity

intricate compound of optical elements, it is better to place polarizer in front of the interference plane and, rotating it, find maximum visibility. Compare visibility of fringes with polarizer and without it (Fig. 6.2). Without polarizer,

$$v = \left| \frac{2\sqrt{I_1 I_2}}{I_1 + I_2} \cos\gamma \right|$$

reaches its maximum $v = 1$ only when $|\cos\gamma| = 1$ and $I_1 = I_2$. In all the other cases, $v < 1$. With the polarizer, making angle ε with vector A_1, the Malus law gives

$$v = \left| \frac{2\sqrt{I_1 I_2} \cos\varepsilon \cdot \cos(\gamma - \varepsilon)}{I_1 \cos^2\varepsilon + I_2 \cos^2(\gamma - \varepsilon)} \right| = \left| \frac{2\sqrt{I_1 I_2} \cdot p}{I_1 p^2 + I_2} \right|, \quad p = \frac{\cos\varepsilon}{\cos(\gamma - \varepsilon)}.$$

Finding maximum over p by differentiating, we obtain

$$p_{\max} = \pm\sqrt{\frac{I_2}{I_1}} = \pm\frac{A_2}{A_1}, \quad A_1 \cos\varepsilon = A_2 \cos(\gamma - \varepsilon)$$

which means that projections of A_1 and A_2 on the polarizer axis are equal. At this condition, $v = 1$. Thus, proper adjustment of polarizer always gives maximum visibility of fringes.

If not the visibility v but the amplitude of modulation $a = 2(\vec{e}_1 \vec{e}_2)\sqrt{I_1 I_2} = 2\sqrt{I_1 I_2}\cos\gamma$ is important, like in heterodyning, then polarizer also helps to maximize the signal. With the notations of Fig. 6.2, $a = 2\sqrt{I_1 I_2}\cos\varepsilon \cdot \cos(\gamma - \varepsilon)$, with its maximum being at $\varepsilon = \gamma/2$:

$$a_{\max} = 2\sqrt{I_1 I_2}\cos^2\frac{\gamma}{2} = \sqrt{I_1 I_2}(1 + \cos\gamma) \geq 2\sqrt{I_1 I_2}\cos\gamma.$$

It means that the amplitude of modulation with properly adjusted polarizer is always bigger than that without a polarizer.

Polarizer is not needed when polarization of both the interfering waves is exactly the same, i.e. $\gamma = 0$. Then

$$v = \frac{2\sqrt{I_1 I_2}}{I_1 + I_2},$$

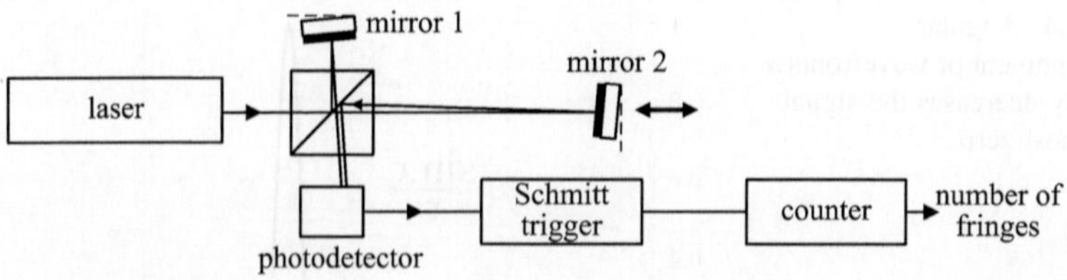

Fig. 6.3 The Michelson interferometer in the fringe-counting mode. The Schmitt trigger is an electronic circuit that abruptly changes its output from logical «0» to logical «1» when the input signal crosses the preset threshold

and the question is how visibility depends on the balance between intensities I_1 and I_2 of the interfering beams. It is easy to show that the maximum is at $I_1 = I_2$, and $v_{max} = 1$.

The next question is how visibility of fringes depends on the transmission and reflection coefficients T and R of the beamsplitter. The answer may be somewhat surprising: it does not. Indeed, assuming 100 % reflectivity of mirrors and the absence of absorption in the beamsplitter, $I_1 = I \cdot R \cdot T$ and $I_2 = I \cdot T \cdot R$, which means that $I_1 = I_2$ and $v = 1$. However, whereas visibility does not depend on the beamsplitting proportion, the amplitude of modulation does:

$$a = 2\sqrt{I_1 I_2} = 2IRT = 2IR(1 - R).$$

Obviously, the amplitude reaches its maximum when $R = T = ½$. The amplitude is important because it determines signal-to-noise ratio of measurements, and as such should be kept as big as possible. Therefore, for interferometers, 50 % beamsplitters are always used.

Plane-wave interferometer of the Michelson type is most frequently used for measuring linear displacements or vibrations. The simplest scheme for that is shown in Fig. 6.3. The laser is supposed to provide linear polarization, so that we may not worry about visibility of fringes: it is unity. But what we do worry is the adjustment of mirrors to provide maximum amplitude of the photodetector signal. The mirror 2, which is actually the sensor that is fixed on either the translational stage or vibrating part to be measured, may be considered as only coarsely adjustable so that only to return the laser beam back on the beamsplitter. As to the mirror 1, it is supposed to be finely adjustable, accurate enough to align the wavefront with the wavelength precision. Mechanical details of such aligners are outlined in Chap. 10. Interference results in fringes on the photodetector sensitive surface:

$$I = I_1 + I_2 + 2\sqrt{I_1 I_2}\cos\left(\frac{2\pi}{\lambda}x'2\sin\frac{\alpha}{2}\cos\beta + \varphi\right),$$

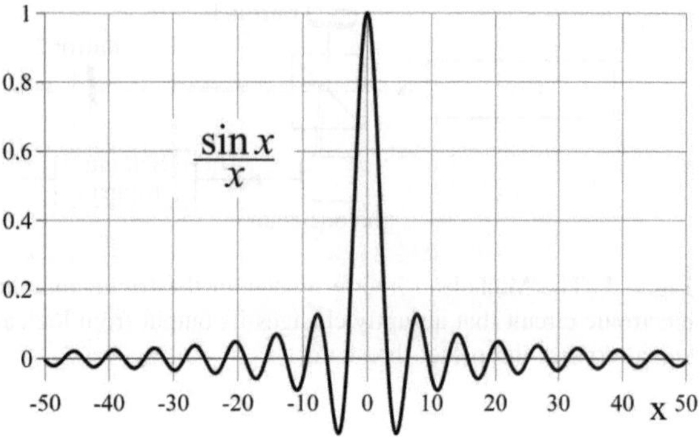

Fig. 6.4 Angular misalignment of wavefronts α quickly decreases the signal to almost zero

where the phase

$$\varphi = \frac{2\pi}{\lambda}(L_2 - L_1)$$

is the informative parameter to be determined in order to measure $L_2 - L_1$. The variable part of the electrical signal is then proportional to the integral over the photodetector sensitive area $S = b \times b$, which we assume to be a square:

$$\int_{-b/2}^{b/2}\int_{-b/2}^{b/2} \cos\left(\frac{2\pi}{\lambda}x'w + \varphi\right) dx'\, dy' = b^2 \cos\varphi \cdot \left[\frac{\sin\left(\frac{\pi b w}{\lambda}\right)}{\frac{\pi b w}{\lambda}}\right] \equiv b^2 \cos\varphi \cdot f(w),$$

$$w = 2\sin\frac{\alpha}{2}\cos\beta.$$

The first conclusion is that the electrical signal oscillates as $\cos\varphi$ as the optical path L_2 to the mirror 2 changes. Counting these oscillations, it is possible to measure variations of L_2 in increments of the wavelength λ. This is how the simplest position-sensitive interferometer works.

The second conclusion is that the amplitude is the sharply falling function (Fig. 6.4) of the misalignment angle α:

$$f(w) \approx \frac{\sin\left(\frac{\pi b \alpha}{\lambda}\right)}{\frac{\pi b \alpha}{\lambda}}$$

for small α and β (Fig. 6.1). Central lobe of this function is located between $-\pi$ and π, so that angular misalignment α must be

$$\alpha < \frac{\lambda}{b},$$

which is none other than the diffraction angle on the photodetector aperture b or the beam diameter, whichever is smaller. In order to mitigate this requirement, some would suggest to use smaller photodetector. But this leads to drop of the signal as b^2. Others would suggest to use a lens to focus the entire optical power

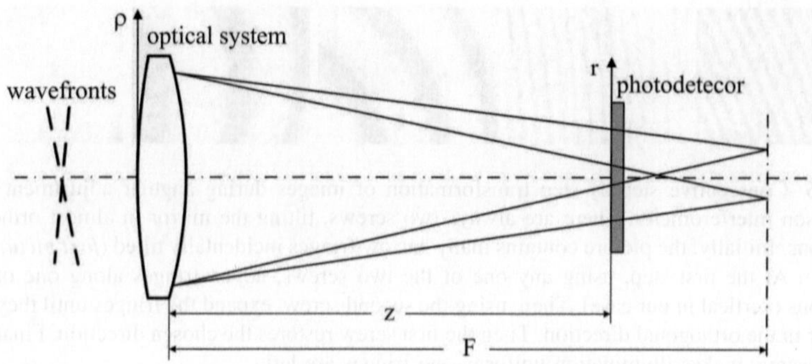

Fig. 6.5 Generalized optical system transfers waves from its input plane $\vec{\rho}$ into the plane of a photodetector $\vec{r}$

onto a small photodetector, and this virtual suggestion inspires another very important conclusion: no lens can help. Indeed, let the optical system be focusing the two interfering beams onto a photodetector (Fig. 6.5). Whatever the focal length F of the lens and separation z from it are, the optical fields on the photodetector are determined by the Fresnel transform:

$$E_{1,2}(\vec{r}) = C_0 e^{iC_1 r^2} \int E_{1,2}(\vec{\rho})\, e^{iC_2\rho^2 + iC_3 \vec{\rho}\vec{r}}\, d^2\rho,$$

where C_j are unimportant constants, depending on z, F, and wavelength λ. If the photodetector fully intercepts the beams, then the amplitude of the modulated electrical signal is proportional to integral over the entire plane $\vec{r}$:

$$\int E_1(\vec{r})E_2^*(\vec{r})d^2r = |C_0|^2 \iiint E_1(\vec{\rho}')E_2^*(\vec{\rho})e^{iC_2\left(\rho'^2-\rho^2\right)+iC_3(\vec{\rho}'-\vec{\rho})\vec{r}}d^2r\, d^2\rho'\, d^2\rho.$$

The asterisk denotes complex conjugate. Integration over $\vec{r}$ gives the Dirac delta-function $\delta(\vec{\rho}\prime - \vec{\rho})$, which subsequently eliminates all the other variables except the last one, say $\vec{\rho}$, simplifying to

$$\int E_1(\vec{r})\, E_2^*(\vec{r})\, d^2r = const \cdot \int E_1(\vec{\rho})\, E_2^*(\vec{\rho})\, d^2\rho.$$

Applying energy conservation law to the case when the two waves are identical, we find that the constant must be unity, and finally

$$\int E_1(\vec{r})\, E_2^*(\vec{r})\, d^2r = \int E_1(\vec{\rho})\, E_2^*(\vec{\rho})\, d^2\rho.$$

This means that whatever the optical system is used to focus interfering beams onto the photodetector, the result will be the same as without focusing. This relation holds true for any optical system, including free space. Therefore, another conclusion is that photodetector signal does not depend on how far from the beamsplitter it is located, if only the entire beam is intercepted.

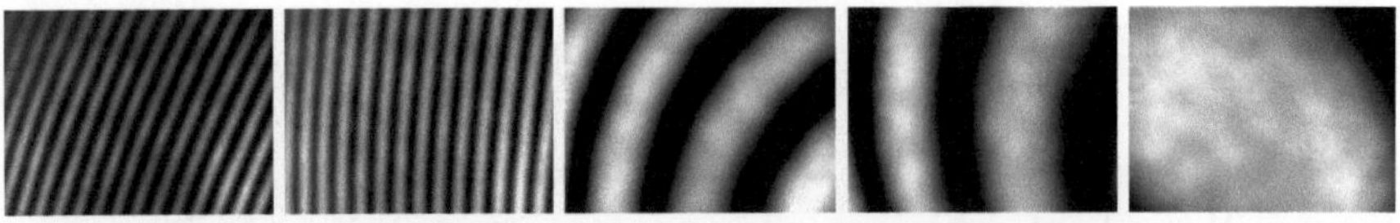

Fig. 6.6 Consecutive step-by-step transformation of images during angular adjustment of the Michelson interferometer. There are always two screws, tilting the mirror in almost orthogonal directions. Initially, the picture contains many narrow fringes incidentally tilted (*first picture from the left*). At the first step, using any one of the two screws, adjust fringes along one of these directions (vertical in our case). Then, using the second screw, expand the fringes until they begin to shear in the orthogonal direction. Then the first screw restores the chosen direction. Finally, the second screw makes illumination uniform—no fringes are left

The argument of the function *f* can be rewritten in another form, introducing the number N of fringes within the sensitive area. The space l between two adjacent fringes is determined by the condition $l\alpha = \lambda$, so that

$$\frac{b\alpha}{\lambda} = \frac{b}{l} = N.$$

Ideally, $\alpha = 0$ and there are no fringes on the photodetector. Thus, the aim of the angular alignment is to obtain zero number of fringes within the photodetector area. Practically, it cannot be done with only single-channel photodetector, and always an imaging detector like CCD (charge coupled device) camera must be used to obtain good result. Usually, only one fine adjustment is needed—on the fixed mirror that we labeled as mirror 1 in Fig. 6.3. As to the moving mirror 2, it always must be as light and small as possible, thus allowing only coarse adjustment necessary to reflect the beam back onto the beamsplitter. Leaving design details of advanced adjustment mechanisms to Chap. 10, consider only typical images that the user will see when the CCD camera is coupled to the output of the interferometer instead of the single-channel photodetector (Fig. 6.6). Finally, only uniform illumination must be seen on the screen. All the pictures in Fig. 6.6 show some sphericity—circular shape of fringes. This is the consequence of collimation errors of the light source. Below, in Sect. 6.3, it will be explained how collimation can be easily adjusted with the help of shear-plate interferometers.

However, the situation is such simple only for highly coherent source, like helium-neon lasers. For other light sources, including even laser diodes, finite coherence length (finite spectral width) creates significant practical problems, which we are going to discuss next. Assume the ideal case of perfectly matched wavefronts of the two interfereing waves at a single wavelength λ. Then the fringe pattern

$$1 + \cos\left(\frac{2\pi}{\lambda} x' 2 \sin\frac{\alpha}{2} \cos\beta + \varphi\right)$$

is 100 % modulated. However, if there are many spectral components uniformly distributed in the spectral interval $\Delta\lambda$ between $\lambda - \Delta\lambda/2$ and $\lambda + \Delta\lambda/2$, then this pattern will be averaged over this spectral interval, giving

$$1 + \frac{1}{\Delta\lambda} \int\limits_{\lambda-\Delta\lambda/2}^{\lambda+\Delta\lambda/2} \cos\left[\frac{2\pi}{\lambda} x' 2 \sin\frac{\alpha}{2} \cos\beta + \frac{2\pi}{\lambda}(L_2 - L_1)\right] d\lambda.$$

Inside the cosine, isolate the term

$$\Phi = 2\pi x' 2 \sin\frac{\alpha}{2} \cos\beta + 2\pi(L_2 - L_1)$$

that does not contain λ, introduce new variable $u = \Phi/\lambda$, and integrate, understanding that relative variation of u is small inside the interval, so that

$$\frac{1}{u^2} \approx \frac{\Phi^2}{\lambda^2}.$$

Then the average fringe pattern reduces to

$$1 + \frac{\lambda^2}{\Phi\Delta\lambda}\left[\sin\left(\frac{\Phi}{\lambda - \Delta\lambda/2}\right) - \sin\left(\frac{\Phi}{\lambda + \Delta\lambda/2}\right)\right].$$

Using standard expansion for small ε

$$\frac{1}{1 \pm \varepsilon} \approx 1 \mp \varepsilon,$$

and applying trigonometric identities, we obtain the final formula for the fringe pattern:

$$1 + v \cdot \cos\left(\frac{2\pi}{\lambda} x' 2 \sin\frac{\alpha}{2} \cos\beta + \varphi\right), \quad v = \frac{\sin\left(\frac{\Phi\Delta\lambda}{2\lambda^2}\right)}{\frac{\Phi\Delta\lambda}{2\lambda^2}}.$$

Thus, the visibility of fringes v is not unity any more, and we already know the function that determines its value (Fig. 6.4): it only differs from zero when its argument is less than π. For the case of perfect angular alignment, when $\alpha = 0$, this formula imposes limitation on optical path difference to the mirrors:

$$|L_2 - L_1| < \frac{\lambda^2}{\Delta\lambda} \equiv l_c.$$

Parameter l_c is commonly understood as the coherence length. From Chap. 2, we know that for helium-neon lasers coherence length may be of a kilometer scale, thus presenting no practical limitations. However, for other available sources the

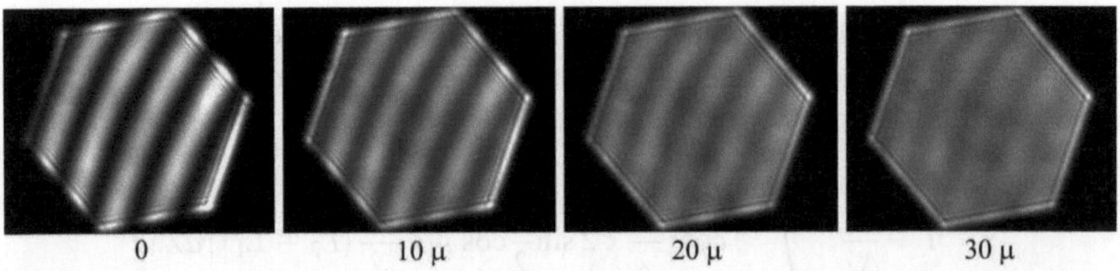

Fig. 6.7 Fringe pattern obtain in the Michelson interferometer powered by laser diode with the wavelength 650 nm. Mirror travel relative to zero path difference is indicated below each picture. The dark hexagonal frame is imposed by an iris diaphragm in front of the CCD camera

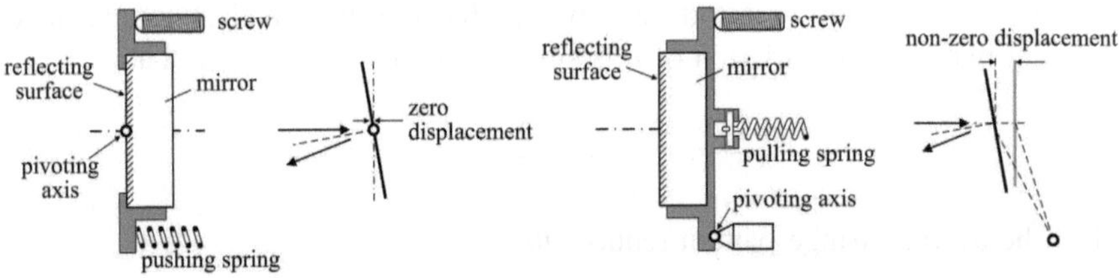

Fig. 6.8 For angular adjustment of wide-band interferometers, the pivoting center of the mirror must lie in its reflecting surface (*at left*). Standard modules do not satisfy this requirement (*at right*), introducing additionally optical path difference during angular adjustment. This significantly complicates adjustment

limitation is severe. For instance, spectral width of laser diode radiation in visible domain is about 5 nm, which gives only 80 μ optical path difference or 40 μ mirror travel. If inequality between L_2 and L_1 is bigger than no fringes will be visible. Figure 6.7 gives visual perception of this phenomenon.

The aforementioned phenomenon makes adjustment of wide-band interferometers very time-consuming, especially when they are designed improperly. The biggest mistake is to use standard mirror-adjustment units available in general-purpose optical kits. Figure 6.8 exemplifies right and wrong designs. When the new interferometer is assembled and adjustment started, there are two independent parameters to be fit: mirror angle and the optical path difference between the mirrors. Probability of initial zero path difference is infinitesimal, therefore whatever the angular adjustment is, no fringes will be visible, and all the efforts will be in vain. The solution is to split adjustment process in two: first, use helium-neon laser and roughly adjust angular position of the mirror to observe any fringes, and only after that use wide-band source to find position of the second (moving) mirror at which fringe contrast is maximal. Have in mind that probability of perfect start-up angular alignment is as small as that of initial zero path difference, therefore initial spatial frequency of fringes is always so big, that it is even beyond

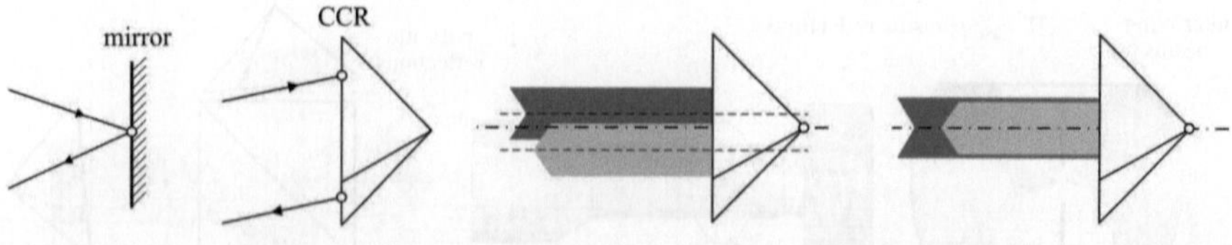

Fig. 6.9 Using CCRs in the interferometers, angular adjustment is not needed as the reflected beam is directed exactly backwards by the principle of operation. Instead, lateral adjustment should be made in order to place the reflector apex on optical axis. This adjustment is on a macroscopic scale, meaning that single-wavelength precision is not necessary and visual positioning is quite sufficient. Note that optical paths inside the CCR are the same for all the rays, and for the design purpose may be estimated as the path to its vertex

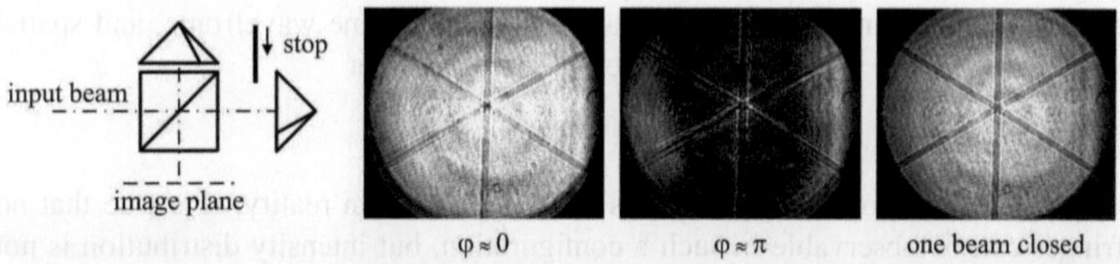

Fig. 6.10 With corner cube reflectors (CCRs) in both shoulders of the interferometer, no angular adjustment is needed, and no fringes can be observed in the output plane—exactly what is needed for industrial applications. Changes of optical path differences result in variations of total phase φ, causing average brightness to go up and down, preserving at the same time the image structure. With one beam shut, a steady cross section image of a reflected beam can be seen, showing six star-like geometry. In the particular case shown in pictures, not only the vertices but also the edges of the two CCRs were approximately superimposed by axial rotation. Diameter of the beam is 15 mm

spatial resolution of the CCD camera, leaving no chance to detect interference even moving the second mirror.

Abrupt vanishing of fringes, when optical path difference exceeds coherent length, is used in imaging interferometers to construct three-dimensional pictures of object surfaces. This type of interferometers will be analyzed in Sect. 6.4.

Numerous industrial applications of heterodyne interferometers, which will be considered in the next section, cannot rely on tedious manual adjustments like those explained above. In order to avoid it, corner cube reflectors (CCRs) can be used instead of mirrors (Chap. 1). However, CCR does not act exactly like a mirror in a sense that the outgoing ray does not emerge at the same point where the incoming ray intersects the front surface (Fig. 6.9). Since every ray inside CCR travels the same optical path, the reflection of a plane wave would be the plane wave again, propagating exactly backwards, if the CCR has exactly 90° geometry. In this ideal case, the interferometer equipped with two CCRs (Fig. 6.10) would never show any fringes within the sectors of constant polarization (see Chap. 1)

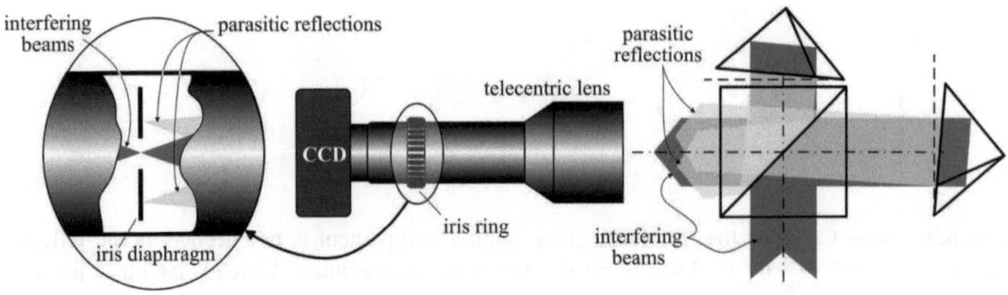

Fig. 6.11 In a telecentric lens, iris diaphragm is installed in the Fourier plane, where parallel input beams focus (Chap. 1). Properly narrowed, iris blocks reflections, transmitting only the interfering waves

because the two interfering waves have exactly the same wavefronts, and spatial distribution of the intensity would be merely a constant

$$I(\overrightarrow{r}) = I \cdot (1 + \cos \varphi).$$

varying in time, following variations of the phase φ. In reality, it is true that no fringes can be observable in such a configuration, but intensity distribution is not exactly a constant, suffering from imperfections of the laser beam, displaying chamfers of the CCR (inevitable result of manufacturing), dust on laser window, etc.

There are some practical tricks of how to better observe interference pattern, avoiding parasitic reflections from flat surfaces of beam-splitting cube and CCRs. Telecentric lens with adjustable iris diaphragm and large enough input aperture is a very convenient instrument for this (Fig. 6.11).

In general, CCR transforms linear polarization of the input beam into elliptical because phases of orthogonal polarizations change in total internal reflection (Chaps. 1 and 5). But how severe consequences should we expect in practice? The answer is encouraging: in interferometry, depolarization caused by corner cube reflectors (CCRs) may be neglected. Figure 6.12 gives visual impression of what happens when polarizer (serves as an analyzer in this case) is installed in the output beam of the interferometer.

We return now to the problem of measuring displacements as outlined in Fig. 6.3, and must admit that with corner cube reflectors (CCRs), here is no big sense to analyze the entire CCR area since there are no fringes in it. As such, the laser beam may be made as narrow as only fit into one of the six sectors of a CCR. Then chamfered edges of the pyramid do not cross the returned beam, which is an obvious advantage. This optical scheme is shown in Fig. 6.13.

The fringe-counting technique that was discussed in context of the Fig. 6.3 makes it possible to measure displacements with errors about one wavelength, i.e. ± 0.6 micron. This is not enough for many applications. The breakthrough was done with phase-measuring technology based on heterodyne interferometry, which is discussed in the next section.

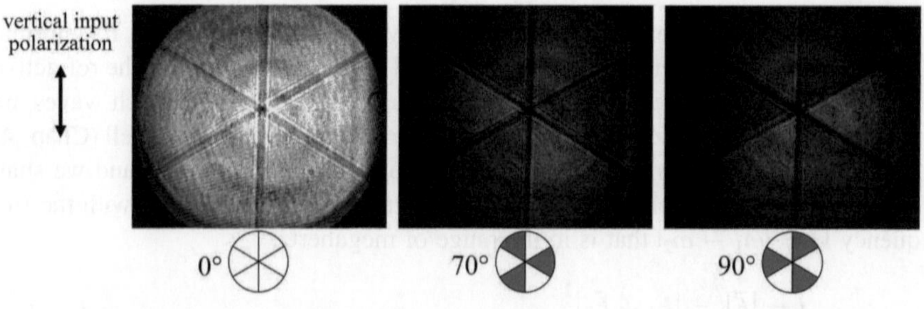

Fig. 6.12 Interference patterns at the output of the Michelson interferometer with two CCRs in its shoulders. With linear (vertical in this particular case) input polarization, no significant change to the interference pattern occurs when the analyzing polarizer is set in the same direction (*at left*). Some minor part of total intensity does infiltrate into the orthogonal polarization, but it hardly noticeable visually (photos *in the middle* and *at right*). The numbers show the angle that the analyzer makes with the vertical direction. Circular diagrams emphasize difference of illumination in the sectors, which may not be properly seen in print. Digital contrast enhancement was not applied to the last two photos in order to preserve original balance of illumination with the left picture

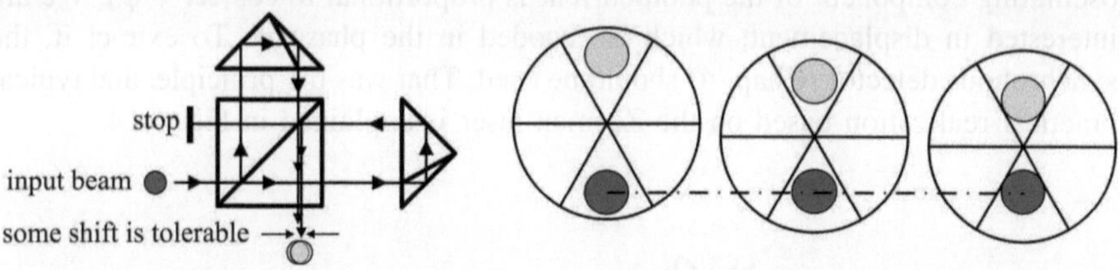

Fig. 6.13 Narrow laser beam can be fitted into only one sector of a CCR. Proper superposition of interfering beams is accomplished by lateral adjustment of CCRs. Wavefront matching is guaranteed by design, and does not depend on tilts of CCRs. Thus, modulated component of the interference signal depends only on the overlapping area of the two beams. Practically, as big shift as 25 % of the beam diameter is tolerable

6.2 Heterodyne Interferometers

In heterodyne interferometers, the two interfering waves have slightly different frequencies ω_1 and ω_2, so close to each other that their wavelengths λ_1 and λ_2 may be considered equal to λ:

$$\vec{E}_1 = \vec{e}_1 A_1 \exp\left(i\omega_1 t - i\frac{2\pi}{\lambda}\vec{n}_1\vec{r} - i\frac{2\pi}{\lambda}L_1 \right),$$
$$\vec{E}_2 = \vec{e}_2 A_2 \exp\left(i\omega_2 t - i\frac{2\pi}{\lambda}\vec{n}_2\vec{r} - i\frac{2\pi}{\lambda}L_2 \right),$$

where $i = \sqrt{-1}$, $\vec{e}$ are the unity polarization vectors, A - amplitudes, $\vec{n}$—normal vectors of the wavefronts, and L are the optical paths (accounting for the refractive indices) that the two beams have passed on their ways. Typically, such waves are produced either by the Zeeman laser (Chap. 2) or by an acousto-optic cell (Chap. 4) in one of the beams. In these two cases, the optical schemes differ, and we shall consider them separately. The interference pattern is now oscillating with the frequency $\Omega = |\omega_1 - \omega_2|$ that is in the range of megahertz:

$$\begin{aligned} I &= \left|\vec{E}\right|^2 = \left|\vec{E}_1 + \vec{E}_2\right|^2 \\ &= I_1 + I_2 + 2(\vec{e}_1\vec{e}_2)\sqrt{I_1 I_2}\cos\left(\Omega t + \frac{2\pi}{\lambda}x'2\sin\frac{\alpha}{2}\cos\beta + \varphi\right), \end{aligned}$$

where $I_{1,2} = A_{1,2}^2$—intensities of the partial beams, $\varphi = \frac{2\pi}{\lambda}(L_2 - L_1)$, α is the angle between $\vec{n}_1$ and $\vec{n}_2$, β is the angle between the differential normal $\vec{n}_2 - \vec{n}_1$ and the interference plane, and the axis x' is directed along the differential normal projection. All that has been told in Sect. 6.1, regarding matching of wavefronts, remains true in this case as well. Adjustment of mirrors or usage of corner cube reflectors (CCRs) makes interfering wavefronts parallel, i.e. $\alpha = 0$, and the oscillating component of the photocurrent is proportional to $\cos(\Omega t + \varphi)$. We are interested in displacement, which is encoded in the phase φ. To extract it, the synchronous detector (Chap. 4) should be used. That was the principle, and typical practical realization based on the Zeeman laser is explained in Fig. 6.14.

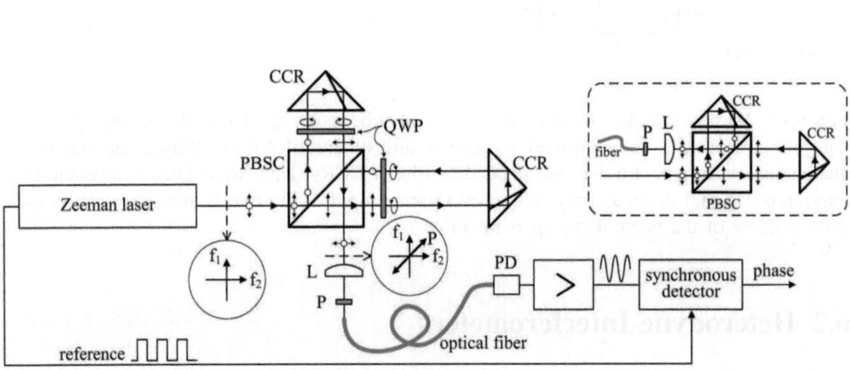

Fig. 6.14 Polarizing beam-splitting cube (PBSC) separates two orthogonal polarizations with cycling frequencies f_1 and f_2. After passing two consecutive times through the quarter-wave plates (QWP), their polarizations change to orthogonal, enabling the PBSC to recombine them with practically no loss of intensity. In order to make these waves interfere, polarizer P is set 45° relative to their directions. After passing through the polarizer, the two initially independent waves become one linearly polarized wave with its intensity oscillating at frequency f_1–f_2. It means, that from now on, any depolarization will not affect intensity modulation. Therefore, any optical fibers, even those that does not preserve polarization (see Chap. 7), may be used to transmit this modulation to photodetector (PD). The lens L focuses the plane wave into the fiber core. Amplified sinusoidal electrical signal from the PD feeds synchronous detector (see Chap. 4) to produce the output signal proportional to the phase. The option shown in the dashed frame does not require QWPs but is less convenient, because it leaves less space for the fiber-optic pick-up

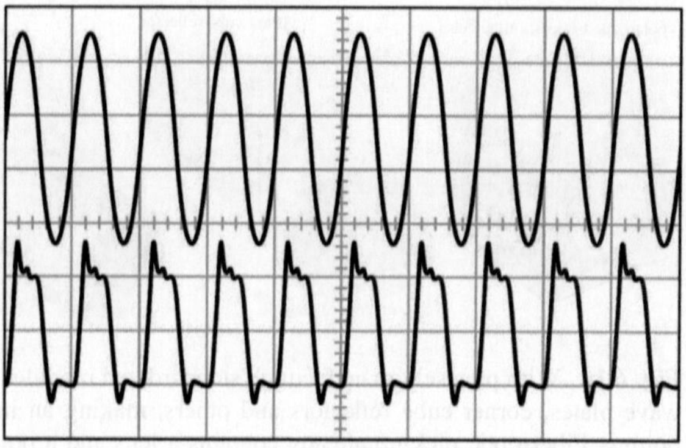

Fig. 6.15 Interference signal trace (*above*) in comparison with the reference (*below*). The carrier frequency f_1–f_2 is typically 2–3 MHz (Chap. 2)

Possibly only one question can arise after examining this figure: why the polarizer is placed after the focusing lens, i.e. in the convergent beam, and not in front of the lens into the parallel beam? The question is natural because from Chap. 5 we know that high-quality polarizers are vulnerable to divergence. The answer is: the cost. The entire system does not suffer from the lack of power provided by the Zeeman laser, therefore small loss of modulated signal that arises from convergence of the beam, passing through the polarizer, is quite tolerable. On the other hand, additional cost associated with the larger size of the polarizer would be intolerable in mass production. Therefore, the polarizer in this system is usually a tiny square of glass-laminated polarizing plastic of about 2 × 2 mm in size, placed just in front of the optical fiber.

A typical pair of electrical signals that come to the inputs of synchronous detector is shown in Fig. 6.15. The interference signal is (must be) always a fairly good sinusoid, if it is not corrupted by saturation inside the amplifier. As to the reference signal, it always shows considerable distortion and uncertified delay that occurs on the way through long cables that may be unmatched in impedance. Nevertheless, this distortion does not affect phase measurements because initial calibration is always done beforehand. At frequencies in megahertz range, phase can be measured with 0.01° precision, which gives potential accuracy of measuring displacements of the order of 0.02 nm. However, this limit is never reached because of other factors like temperature instability, variations of laser frequency, incomplete polarization separation, mismatch of wavefronts, etc. Practically, 2 nm resolution can be reliably obtained with interferometers shown in Fig. 6.14. Such systems are widely used in industry, and all the parts are designed as independent modules for quick interchange to produce easily reconfigurable interferometric systems. Some examples, available on the market, are portrayed in Fig. 6.16.

When extremely high precision of measuring displacements is needed, on the scale of Angstroms, then incomplete separation of frequencies ω_1 and ω_2 in the scheme of Fig. 6.14 becomes a limiting factor. This obstacle may be overcome with the help of two separate sources with frequencies ω_1 and ω_2, created by acousto-optic modulators (Chap. 4). Zeeman laser—such a convenient tool in

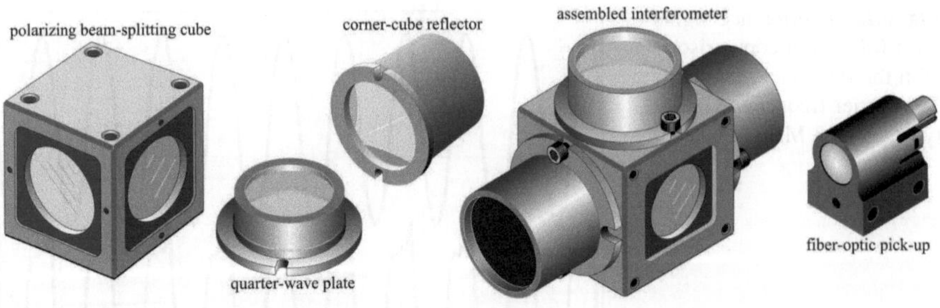

Fig. 6.16 With precisely manufactured standardized modules like beam-splitting cubes, quarter-wave plates, corner cube reflectors and others, making an interferometer becomes an issue of bolting. Fiber-optic pick-up already contains a lens and a polarizer directed 45° to the base

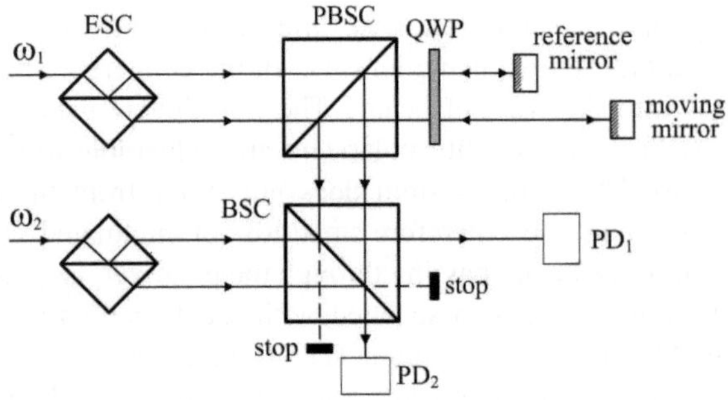

Fig. 6.17 Laser beams with equal linear polarizations and different frequencies ω_1 and ω_2 are generated separately, making polarization separation unnecessary. However, for better energy efficiency, the polarizing beam-splitting cube (PBSC) is used in combination with the quarter-wave plate (QWP). The mixing cube is non-polarizing (BSC). The energy separator cubes (ESC, Chap. 1) split the beams laterally without change of polarization. The first photodetector (PD_1) measures signal from the moving mirror, and the second one (PD_2)—from the reference mirror, providing the reference signal for synchronous detection

simpler experiments—cannot be used in this case, and reference signal must be generated in a separate channel as shown in Fig. 6.17.

6.3 Shear-Plate Interferometer

Shear-plate interferometer is probably the simplest in use and, at the same time, very efficient tool for adjustment of collimators (Fig. 6.18). Parallel laser beam is always needed in the laboratory, and there must be some instrument to measure how parallel the beam is. This job can be easily done by a shear-plate interferometer. Nowadays, all the shear-plate interferometers work with a small wedge in

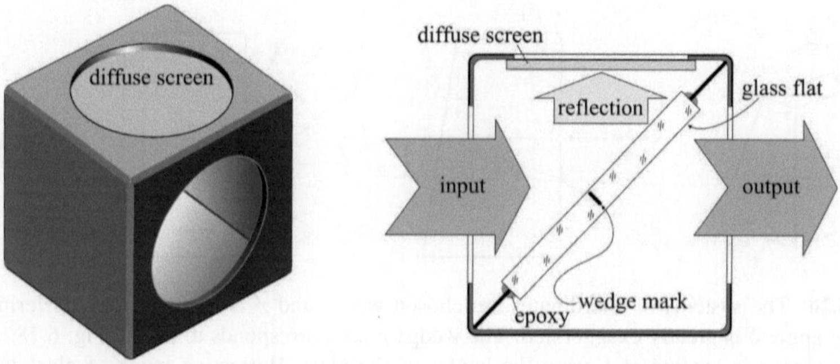

Fig. 6.18 Shear-plate (or shearing) interferometers are available in a variety of sizes, designed to work with laser beams from 1 to 10 mm in diameter. They contain only two optical elements: an optical flat plate with a tiny wedge of about tens of arc seconds, and a diffuse screen to observe interference pattern

Fig. 6.19 Perfect wavefront is spherical, corresponding to plane wave when $R_0 = \infty$

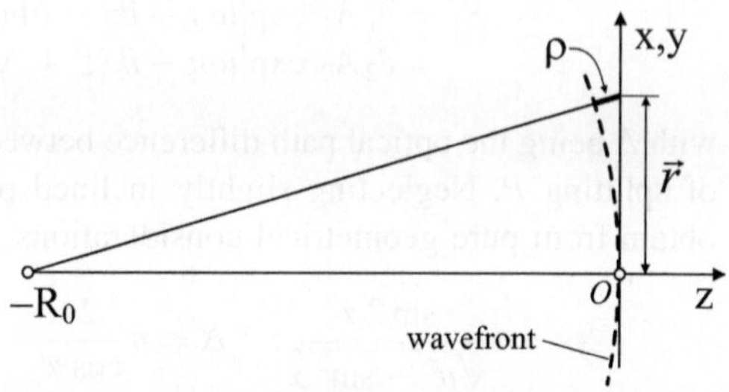

the common optical flat plate in order to make interference fringes visible even in perfectly plane wave. When the beam from the adjustable collimator reflects from two surfaces of the glass flat, it produces interference pattern on the diffuse screen. Perfectly parallel beam produces straight fringes parallel to optical axis of the beam. Any sphericity or other aberrations produce curved fringes directed differently. No special adjustment is needed except for the coarse positioning of the box in order to bring fringes roughly in the middle of the screen. This is both the idea and instruction for operation in one. However, despite exceptional simplicity of the design and operation, the physical background is rather complicated and needs some mathematics to understand the subject.

First, consider the beams without aberrations. Then the wavefront is spherical with the radius of curvature R (Fig. 6.19). In a spherical wave, originating at $z = -R_0$ with the wavelength λ, the phase Φ is the function of x, y, z:

$$\Phi(x,y,z) = \frac{2\pi}{\lambda}\cdot\rho = \frac{2\pi}{\lambda}\cdot\left[\sqrt{R^2+r^2}-R\right] \approx \frac{2\pi}{\lambda}\cdot\frac{r^2}{2R} \equiv \frac{k}{2R}r^2$$

with $k = 2\pi/\lambda$ being the wave number. In our notations, obviously $R = R_0 + z$.

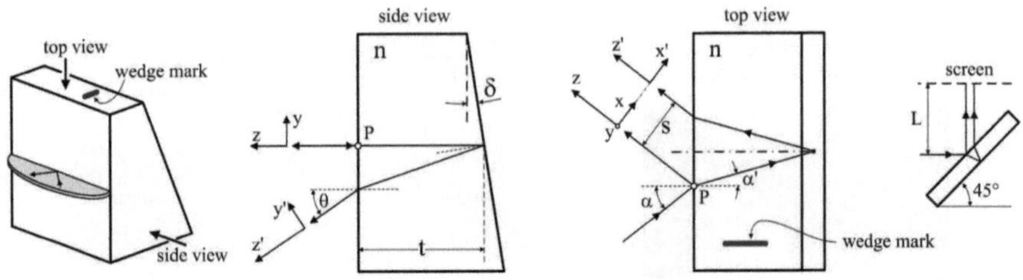

Fig. 6.20 The systems of coordinates are chosen with z and z' axes along the interfering rays. Wedge angle δ is greatly exaggerated. The wedge mark corresponds to that in Fig. 6.18, and the diffuse screen is located at L from the center of the plate. Refractive index of glass is n. For practical reasons, angle of incidence α is always 45°

Now it is time to introduce systems of coordinates associated with the interfering waves (Fig. 6.20). Their origins lie in the same plane perpendicular to the reflected rays. With it, the interfering waves are

$$\vec{E}_1 = \vec{e}_1 A_1 \exp[i\omega t - ikz - i\Phi(x,y,z)],\ \vec{E}_2 \\ = \vec{e}_2 A_2 \exp[i\omega t - ik(z' + \Delta) - i\Phi(x', y', z' + \Delta)],$$

with Δ being the optical path difference between the waves acquired after the point of splitting P. Neglecting slightly inclined path due to the wedge, it is easy to obtain from pure geometrical considerations

$$s = t\frac{\sin 2\alpha}{\sqrt{n^2 - \sin^2\alpha}};\quad \Delta = n\frac{2t}{\cos\alpha'} - s\tan\alpha = 2t\sqrt{n^2 - \sin^2\alpha}.$$

Polarization does not change after reflection (it changes only after total internal reflection, see Chap. 5) and the amplitudes remain practically the same, so we have to watch only phase variations in the interference pattern. Using standard trigonometric identity, it may be written as

$$\sin^2\left\{\frac{1}{2}[k(z - z' - \Delta) + \Phi(x,y,z) - \Phi(x', y', z' + \Delta)]\right\}.$$

The most complicated thing now is to establish a relation between x, y, z and x', y', z'. Leaving detailed mathematics to Chap. 12, we write down only the final approximation that holds true for $\delta \ll 1$:

$$\theta \approx 2\delta\sqrt{n^2 - \sin^2\alpha},\quad \begin{cases} x' = x - s \\ y' = y + \theta z \\ z' = -\theta y + z \end{cases}.$$

This transformation must be substituted into Φ. Since we analyze fringes in the plane of the diffuse screen located at $L \sim 5$ cm from the center of the plate (Fig. 6.20), we must put $z = L$ and $R = R_0 + L$. Then $y' = y + \theta L$. However, with the wedge angle δ being small, the angle θ is also small, of the order of 10^{-4} radian, so that the product $\theta L \sim 5$ micron is much smaller than any

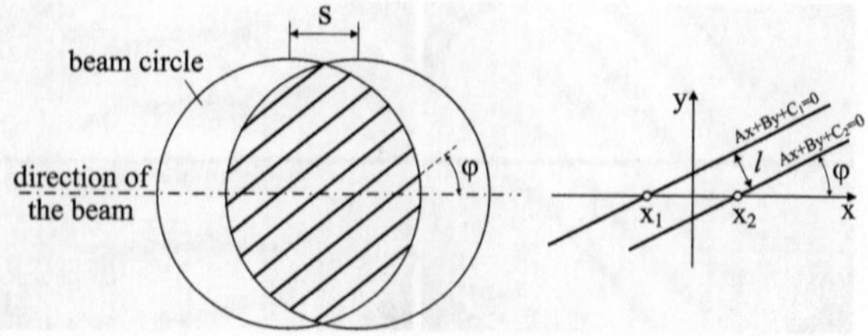

Fig. 6.21 Interference pattern is formed in the middle of two overlapping circles. If the laser beam has no aberrations, the pattern is composed of *straight lines*. Inclination angle φ tells about parallelicity of the beam

resolvable fringe, and may be dropped. The same approximation may be accepted for z', simplifying transformation to

$$\begin{cases} x' = x - s \\ y' = y \\ z' = z = L \end{cases}.$$

Typical range of wavefront radii R_0, that may be expected from the collimator, is between 10 m and 1 km. Therefore, $L \ll R_0$ and $R \approx R_0$. As a result, the interference pattern becomes

$$\sin^2\left[\frac{k}{2}(y\,\theta + x\frac{s}{R} - \Delta)\right].$$

The maximum and minimum intensity of the pattern are defined by the conditions

$$\frac{k}{2}(y\,\theta + x\frac{s}{R} - \Delta) = \frac{\pi}{2} + \pi\,m; \quad \frac{k}{2}(y\,\theta + x\frac{s}{R} - \Delta) = \pi\,m;$$

which represent a series of parallel lines—fringes (Fig. 6.21). Two parallel lines can be presented in the general form

$$A\,x + B\,y + C_1 = 0 \text{ and } A\,x + B\,y + C_2 = 0$$

Inclination angle φ of each line satisfies the relation

$$\sin\varphi = \frac{A}{\sqrt{A^2 + B^2}}.$$

Separation l is

$$l = (x_2 - x_1)\cdot\sin\varphi = \frac{|C_2 - C_1|}{\sqrt{A^2 + B^2}}.$$

In our case

$$A = \frac{k s}{2R}; \quad B = \frac{k\theta}{2}; |C_2 - C_1| = \pi;$$

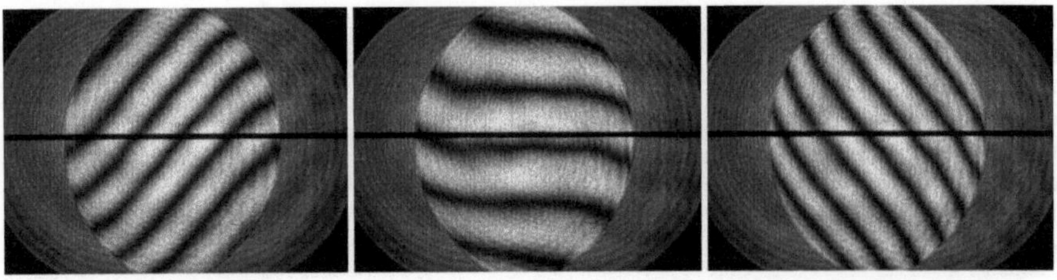

Fig. 6.22 Adjustment of collimator. The output beam passes through the stages of divergence, parallelicity, and convergence (from *left* to *right*). Not exactly straight lines of interference pattern says about aberrations. The diffuse screen always has axial black line in the middle of it for better visual assessment of parallelicity. Beam diameter is 25 mm

which leads to

$$\sin\varphi = \frac{s}{\sqrt{s^2 + R^2\theta^2}}; \quad l = \frac{\lambda}{\sqrt{\left(\frac{s}{R}\right)^2 + \theta^2}}.$$

These are two basic formulas, explaining performance of the shear-plate interferometer. First, consider the inclination angle φ. Without the wedge, i.e. when $\theta = 0$, $\sin\varphi = 1$ and $\varphi = \pi/2$, meaning that fringes are always perpendicular to the axis of the laser beam. In this case, separation of fringes becomes

$$l = \frac{\lambda R}{s},$$

tending to infinity as $R \to \infty$. This is what happens when parallel glass flat, also a shearing interferometer, is illuminated with laser light. Beginning from some value of R, the user stops seeing any fringes at all, and the adjustment process actually stops. Alternatively, in the wedge-type interferometer $\theta \neq 0$, fringes do not vanish with $R \to \infty$, and their spacing quickly becomes a finite value $l = \lambda/\theta$. For example, if the diameter of the laser beam is 5 cm, and the convenient interference pattern contains, say, 5 fringes separated by 1 cm, then $\theta \sim 10$ arc seconds and the wedge angle δ must be roughly two times smaller: ~ 5 arc seconds. This is the reason why manufacturers make the wedge angle dependent on the beam diameter D. This tendency may be roughly estimated as

$$\theta \approx \frac{5\lambda}{D}.$$

In the wedge-type interferometers, fringes do not vanish but change their tilt as $R \to \infty$: from almost vertical when $R\theta \ll s$ ($\varphi \approx \pi/2$) to almost horizontal when $R\theta \gg s$ ($\varphi \approx 0$). The aim of adjustment is then to make fringes horizontal. When the beam transforms from a convergent to a divergent one, R changes sign, and so does the inclination angle. Figure 6.22 shows how all this works in reality.

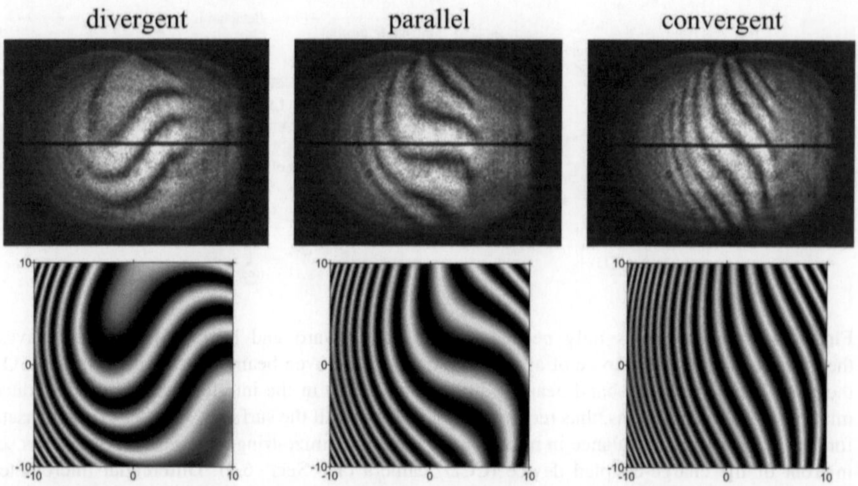

Fig. 6.23 *The upper line of pictures* are real fringe patterns obtained on inexpensive variable collimator of the Galilean type with relatively big spherical aberration. They are compared with simulated maps (*bottom pictures*). Spherical aberration coefficient *c* was chosen to produce the best qualitative agreement with the experiment

The results above show that the shear-plate interferometer not only clearly identifies parallelicity of the beam, but also displays aberrations as any variation from straight lines. In this context, it is interesting to know whether or not the theory developed above can predict fringe patterns of aberrated beams. For example, consider the influence of only spherical aberration alone, which can be described as the term of the fourth power in the phase function:

$$\Phi(x, y, z) = \frac{k}{2R} r^2 + c \cdot r^4.$$

Chapter 12 explains how to make simulation codes, using ray-tracing procedures, that can compute such problems numerically. Here, we only present the results of this simulation, comparing them to experimental pictures (Fig. 6.23). Good qualitative agreement between simulation and experiment makes it possible, in principle, to estimate contribution of each aberration by comparing numerical pictures with the real ones.

6.4 Imaging Interferometers

Plane-wave interferometers, discussed in the previous sections, deal with the so-called non-localized fringes—the interference pattern that preserves its contrast in any plane of observation. Imaging interferometers produce the so-called localized fringes—the interference pattern that can be observed with maximum contrast only

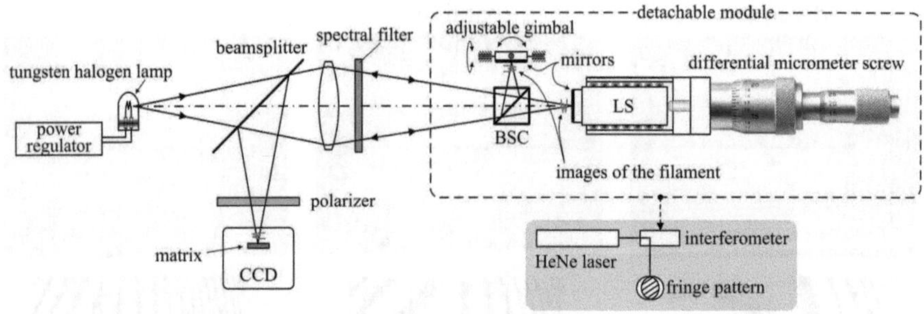

Fig. 6.24 Beamsplitter is only needed to separate forward and backward traveling waves, therefore it does not need to be of a very high quality, and even beam-splitting plate suffices. On the other hand, the broad-band beam-splitting cube (BSC) in the interferometer must introduce minimum phase aberrations, thus requiring high quality of all the surfaces. In order to compensate for any polarization misbalance in returning rays and maximize fringe contrast, a polarizer is set in front of the charge-coupled device (CCD) camera (see Sect 6.1). Differential micrometer screw has usually 10 micron graduation on the coarse dial and 0.5 micron on the fine one. It moves fine linear ball-bearing stage (LS), to which the mirror is attached (fast-hardening epoxy does well)

in one definite plane. For this reason, handling imaging interferometers is much more tricky process, requiring certain know-how that will be explained below.

Imaging interferometers are mostly used to measure height distribution on sample surfaces, i.e. surface relief. For that, they produce images of sample surfaces with interference pattern on them. Analysis of this pattern then allows to extract information about height of the features. Interference pattern can be obtained from any object, not necessarily coherent, even the filament of a tungsten halogen lamp. To begin with, consider a simplified scheme shown in Fig. 6.24 with all the elements that are required for its functionality in practice. First of all, the interferometer must be assembled on a detachable frame in order to adjust wavefronts of interfering waves separately, using high-coherence helium-neon laser. Its longitudinal coherence always exceeds any possible initial optical path difference in the interferometer, making fringes always visible. Typically, from one to five fringes may be left in the pattern in order to easily visually detect zero-difference, adjusting lateral position of the moving mirror in white light.

The next step is to install the interferometer module in white light and, gently moving the mirror, adjust zero optical path difference between the shoulders. For that, it is better to install a narrow-pass interference filter to extend initial fringe-visibility range. Typically, 10–15 nm spectral width will give about 50 micron visibility travel range on the micrometer. Finding the maximum visibility position is a tedious procedure, requiring patience. It is better to use differential micrometer screw that has two dials: the coarse and the fine. First moment of fringe beatings can be found, using the coarse screw, and adjustment to maximum should be done by the fine screw. The result is shown in Fig. 6.25. The «in focus» set of pictures corresponds to situation when the image of a tungsten filament is sharply projected

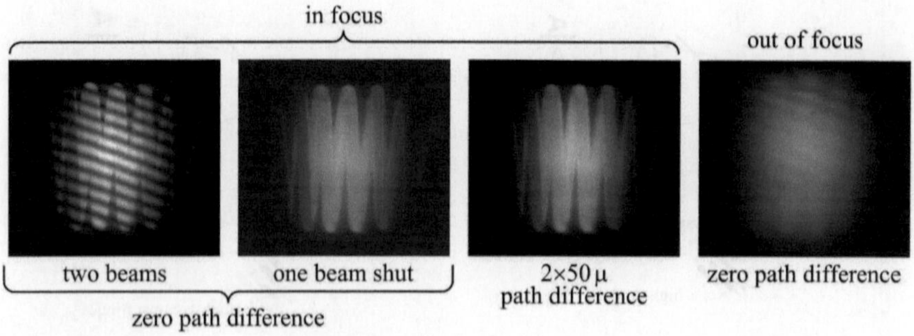

Fig. 6.25 Zero optical path difference corresponds to maximum fringe contrast. Green interference filter with 15 nm full spectral width at half maximum makes coherence length smaller than 100 μ, destructing any visible interference at this optical path difference. But even with zero optical path difference, displacement of the CCD camera from the image plane makes fringes barely visible (*right picture*)

onto the mirrors of the interferometer, and the reflected rays are also focused sharply on the CCD sensitive area (matrix). Interference fringes are well seen within the contour of the filament when two interfering beams are open, and vanish when either one beam is shut or optical path difference far exceeds coherence length of light. Everything looks consistent with general phenomenology of interference that was presented in the previous sections of the current chapter. But here is the new question: if the filament is a three-dimensional macroscopic structure extended to millimeters in depth, and spacing between two adjacent fringes corresponds to optical path difference equal to one wavelength of green light that is half a micrometer, than why all the fringes are straight lines? The answer is composed of two independent statements. First, the interference pattern in the plane of CCD matrix is not produced by rays, coming from the lamp filament—those are spatially incoherent and cannot produce any interference. Second, interference pattern is produced by two images of the two interferometer mirrors, which are mutually coherent unless optical path difference between them is bigger than coherence length of light. The mirrors are flat and slightly tilted relative to one another by the angle $\vec{\theta}$ (we ascribe direction of inclination to the angle, making it a vector), so that the optical fields over the mirror surfaces $\vec{r}$, split identically by the beamsplitting cube, are produced by equal time-independent intensity distributions $I(\vec{r})$ and time-dependent phase $\psi(\vec{r},t)$and polarization $\vec{e}(\vec{r},t)$ distributions:

$$\vec{E}_1 = \vec{e}(\vec{r},t)\cdot\sqrt{I(\vec{r})}\exp[i\,\psi(\vec{r},t)],$$
$$\vec{E}_2 = \vec{e}(\vec{r},t)\cdot\sqrt{I(\vec{r})}\exp\left[i\,\psi(\vec{r},t)+i\frac{2\pi}{\lambda}2\,\vec{\theta}\,\vec{r}\right].$$

Note the factor of two with θ, as the ray turns two times the tilt angle. Polarization direction is the same in both the fields, making $\vec{e}\vec{e}=1$ all the time in every point $\vec{r}$, and phase terms cancel each other as a result of complex conjugation.

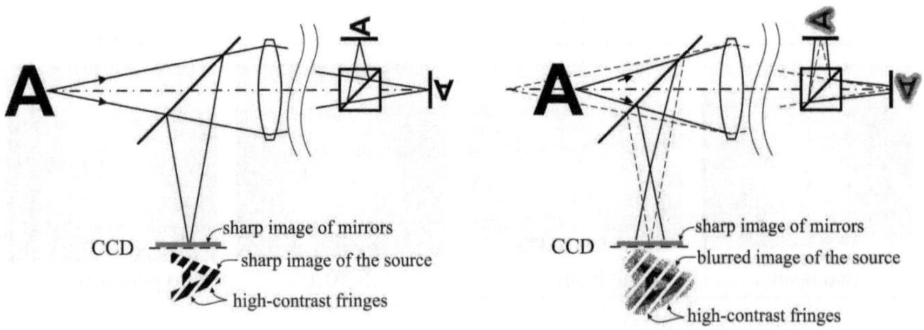

Fig. 6.26 With sharp images of the interferometer mirrors, any picture $I(\overrightarrow{r})$ in the image plane will be covered by high-contrast interference fringes

Then the interfering term $2\mathrm{Re}\vec{E}_1\,\vec{E}_2^*$ (asterisk denotes complex conjugate) is devoid of any time dependence:

$$2\mathrm{Re}\vec{E}_1\,\vec{E}_2^* = 2\,I(\vec{r})\,\cos\left(\frac{4\pi}{\lambda}\,\vec{\theta}\,\vec{r}\right).$$

This is the stable interference pattern, modulated by $I(\overrightarrow{r})$, which is none other than the incoherent image of the lamp filament, which was projected onto the interferometer mirrors by initial concept of the Fig. 6.24.

The above explanation initiates a new question: was it necessary to focus the image of the lamp into the plane of interferometer mirrors? The answer is: definitely not. It was done merely for demonstration purposes, and Fig. 6.26 explains what happens otherwise. But the next question is crucial: is it necessary to focus the images of the mirrors in the plane of the CCD matrix? The answer is: definitely yes. For explanation, consider simplified optical scheme in Fig. 6.27, where beamsplitters are virtually removed for clarity. Suppose the mirrors are perfectly aligned parallel to each other with zero optical path difference between them. Then their reflecting surfaces virtually superimpose, as in the left picture. Let any ray be coming from the source to an arbitrary point A at the mirrors. Then the rays reflected from the two mirrors superimpose from point A to infinity, remaining coherent on their entire ways because these ways are exactly the same. It means that high-contrast interference pattern would be visible in any plane behind the lens, not necessarily in the image plane. «Would be» because with parallel mirrors, there are no fringes in the pattern, and the pattern itself is a uniform illumination. Every time we want a system of interference fringes be visible, for example, to visualize surface relief, we need to create a tilt angle between the mirrors. Let this angle be $\theta/2$. Then the rays reflected from the same point A will be separated by the angle θ, as in the right picture of Fig. 6.27. The rays intersect only in the image plane at the conjugated point A'. According to Descartes principle, the elapsed time is the same for all the rays exiting from A and coming to A', thus preserving temporal coherence. Therefore, interference fringes will always be visible in the image plane. In the other planes where CCD matrix can be shifted to, the rays

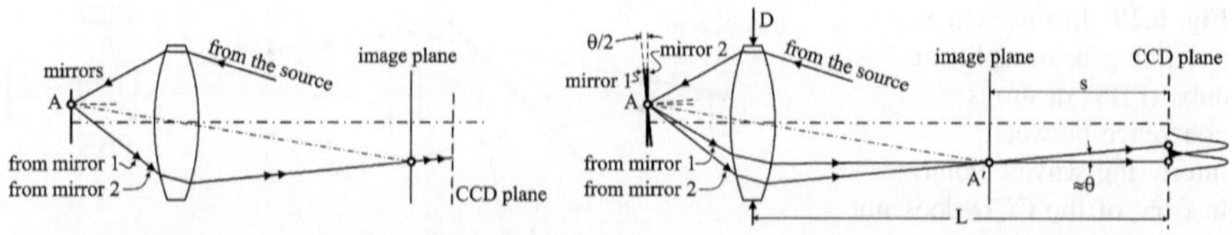

Fig. 6.27 With parallel mirrors, interference pattern is not localized in the image plane as the two rays reflected from the mirrors superimpose everywhere (*left scheme*). With the angular tilt between the mirrors, fringes are localized in the image plane, with the point A' being conjugated to A (*right scheme*)

Fig. 6.28 The Airy function determines diffraction spread *d*

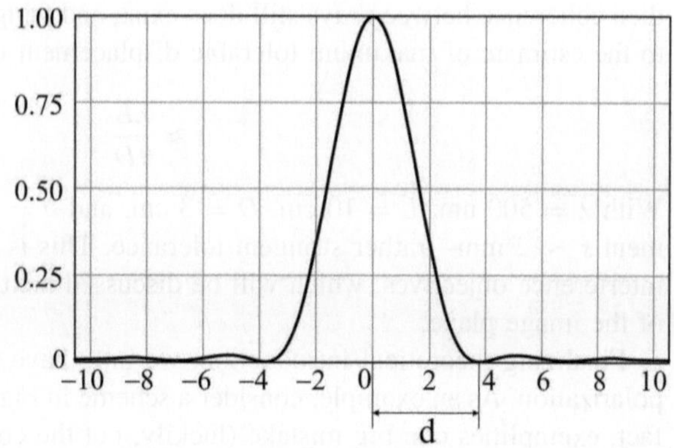

come to different points, and interference is impossible in geometrical approximation. This is the principle of localized interference fringes, exemplifying by the last picture in Fig. 6.25.

However, diffraction on the lens aperture of the diameter D produces finite spread d of the rays in the plane at distance L from the lens. This spread is determined by the well-known Airy function (Fig. 6.28):

$$\left[2\frac{J_1(x)}{x}\right]^2,$$

with $x = \frac{\pi D}{\lambda L} r$, where J_1 is the Bessel function of the first order of the first kind. Its first zero is 3.83, which leads to the widely known relation

$$d = 1.22\frac{\lambda L}{D}.$$

This is not the width of the image that the point A forms in the plane of observation—the latter is determined by divergence of the cone of rays and is much bigger than d. The diffraction spread d determines radius of the area where spatial coherence is preserved even in images of spatially incoherent objects like the filament of an incandescent lamp. Thus, if geometrical separation $s \cdot \theta$ is about d or less,

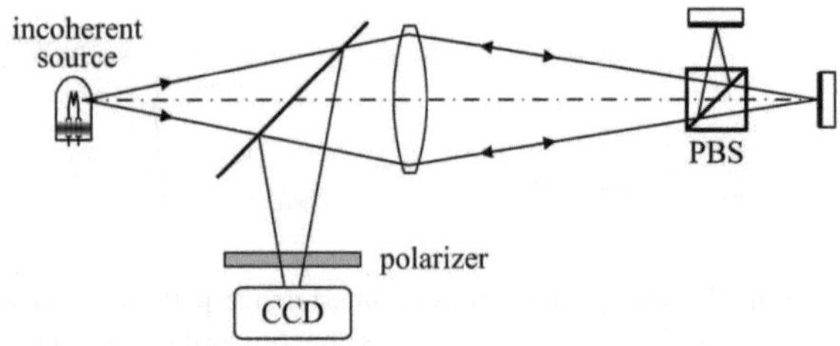

Fig. 6.29 In this scheme, polarizing beam-splitting cube (PBS) destroys coherence between interfering waves. Polarizer in front of the CCD does not help

$$s \cdot \theta \leq d$$

then coherence between rays still does exist, and fringes can be visible. This leads to the estimate of maximum tolerable displacement of the observation plane:

$$s \approx \frac{\lambda L}{\theta D}.$$

With $\lambda = 500$ nm, $L = 10$ cm, $D = 3$ cm, and $\theta = 1$ mrad, maximum displacement $s \sim 2$ mm—rather stringent tolerance. This is why practical application of interference objectives, which will be discussed next, requires careful adjustment of the image plane.

Finalizing theoretical introduction, we only have to make some comments on polarization. As an example, consider a scheme in Fig. 6.29 that looks good, but in fact, exemplifies one big mistake (luckily, not the common one). Considering all that has been told in Sect. 6.1, the scheme looks very promising: polarizing beamsplitter (PBS) returns back 100 % of both the s-polarized (perpendicular to the plane of incidence) and p-polarized components, thus being more energy efficient than the non-polarizing splitter. After coming to the CCD, the orthogonally polarized waves are coupled by the polarizer to produce interference pattern. However, this consideration contains two mistakes. First, energy efficiency would be the same as with the non-polarizing splitter because the polarizer halves total intensity. Second, such a scheme does not produce interference between the two waves reflected from the mirrors because they are produced separately: one by s-polarized component and the second—by the p-polarized. Although these two components originate at the same radiating point on the source, they are statistically independent, thus being totally incoherent with respect to each other. As such, no interference pattern will be observed.

Manufacturers of optical equipment offer imaging interferometers as sealed interference objectives with standard flanges, thus greatly simplifying process of assembling the entire imaging system. Two types of objectives are available: the Mireau and the Michelson types (Fig. 6.30). The object itself plays the role of the second mirror of the interferometer, and Fig. 6.31 shows how the imaging system should be organized. This scheme is essentially a microscope with upper illumination. Some microscopes, particularly in biological applications, use bottom illumination to study transparent specimen. Interference objectives cannot produce interference pattern in bottom illumination by design: there cannot be reflection

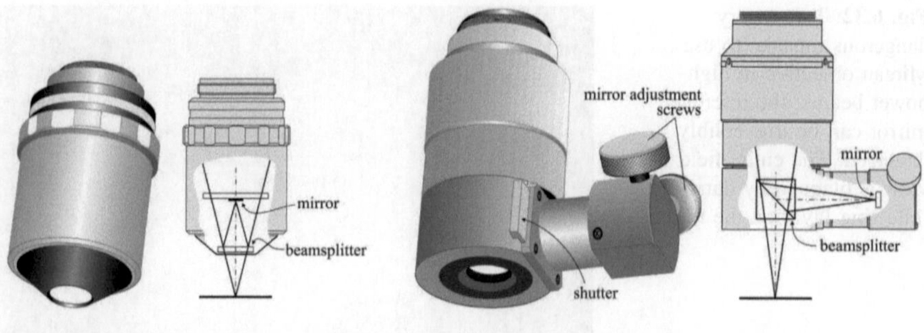

Fig. 6.30 The Mireau (*at left*) and Michelson (*at right*) interference objectives. The two knurled screws in the Michelson objective are used for angular adjustment of the mirror

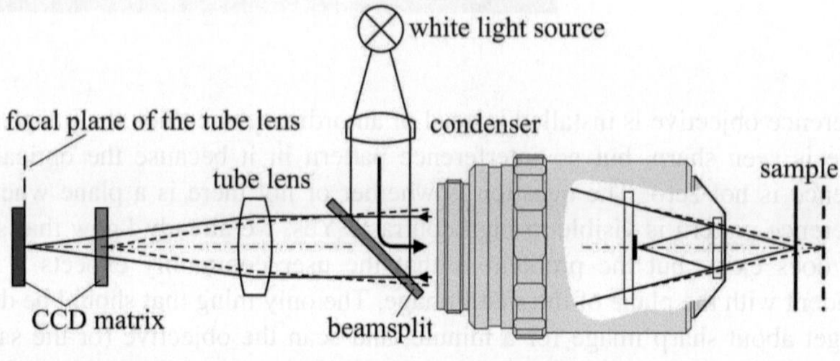

Fig. 6.31 Nowadays, all the interference objectives are infinity-corrected, meaning that they introduce minimum aberrations with parallel output bundle of rays (see Chap. 1). Without exception, focal plane virtually coincides with the mirror, thus requiring that the sample be positioned in the focal plane in order to observe interference pattern (zero optical path difference condition). The system is shown horizontally only to fit the page format. Mostly, the vertical orientation is used

from the reference mirror when illumination comes from the sample. However, some users try to do that, and it is a common mistake.

Another common and frequently made mistake is to work with unadjusted position of the CCD camera. Dashed lines in Fig. 6.31 explain what happens in this case. Consider a situation when the CCD matrix is shifted from the image plane of the tube lens. It is a quite common thing because the tube, containing the lens, always has an adjustment ring. This ring is designed for good purpose—to precisely position CCD matrix in the focal plane. But users do not know the purpose of this ring and how to use it, and after turning it several times and observing no effect in the image through an ordinary objective, just leave it in peace «as is». The sharpness of the image does not change noticeably even if the CCD sensor is significantly shifted from the focal plane of the tube lens, because an ordinary objective is designed well, and as such, delivers sharp image even in converging output bundle of rays (dashed lines in Fig. 6.31). But the position of the sample is now shifted from the focal plane of the objective. Therefore, if the

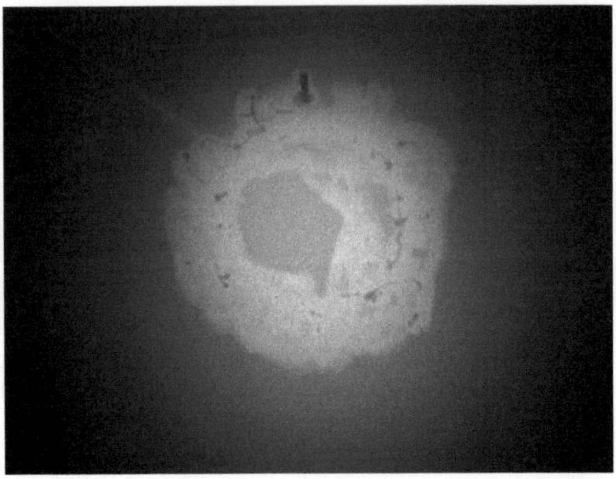

Fig. 6.32 It is a very dangerous mistake to use Mireau objective in high-power beams: the reference mirror can be irreversibly damaged. The entire field of view is plagued by damaged reflecting layer on the mirror

interference objective is installed instead of an ordinary one, then the image of the sample is seen sharp, but no interference pattern in it because the optical path difference is not zero. The question is whether or not there is a plane where the interference pattern is visible in high contrast? Yes, we already know that such a plane does exist, but the problem is that the user commonly expects it being coincident with the plane of the sharp image. The only thing that should be done is to forget about sharp image for a minute, and scan the objective (or the sample, whatever is moving) up and down (to and fro in geometry of Fig. 6.31) to find the fringes. To facilitate easy finding, it is better to install narrow-pass interference filter in the illumination beam. When the fringes are found, keep the sample in this position, and adjust position of the CCD camera for the sharp image. Now, the entire system is adjusted properly.

Optical scheme of the Mireau objective is very compact and uses thin plates for both the beamsplitter and mirror. As such, the numerical aperture may be large, from 0.3 to 0.5, and this is the biggest advantage of the Mireau objective over the Michelson type. When high magnification is needed, the Mireau objective is the only choice. The weakness of this objective is its inability to align the mirror along the sample surface. This results in uncontrollable spatial frequency and direction of fringes, depending on how big the tilt angle between the mirror and the sample is. Another weakness is its permanently connected mirror that cannot be shut in order to remove fringes when necessary. Alternatively, in the Michelson objectives, bulky beam-splitting cube limits numerical apertures typically to 0.075. However, this design includes manual adjustment of the mirror tilt and a shutter—a useful feature, especially when high-power laser beam must be transmitted. Vulnerability of interference objectives to high-power is critical: the light, focused to the sample, is focused onto the mirror as well, jeopardizing integrity of the reflecting layer. The Mireau objective is especially critical because in it, the reference beam cannot be shut. As a warning, Fig. 6.32 exemplifies what may happen to Mireau objective with high-power laser beam.

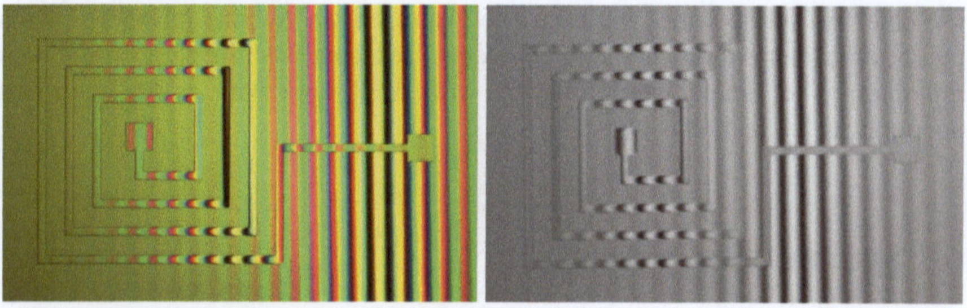

Fig. 6.33 Deep trench etching in silicon as viewed through the Michelson objective. The colour (*at left*) and black-and-white (*at right*) CCD cameras. Fringes are adjusted vertically for better visual perception. The halved number of fringes between black strips gives the depth of etching in wavelengths: 0.5 × 8× 0.6 μ = 2.4 μ

Fig. 6.34 The Linnik imaging interferometer uses ordinary objectives

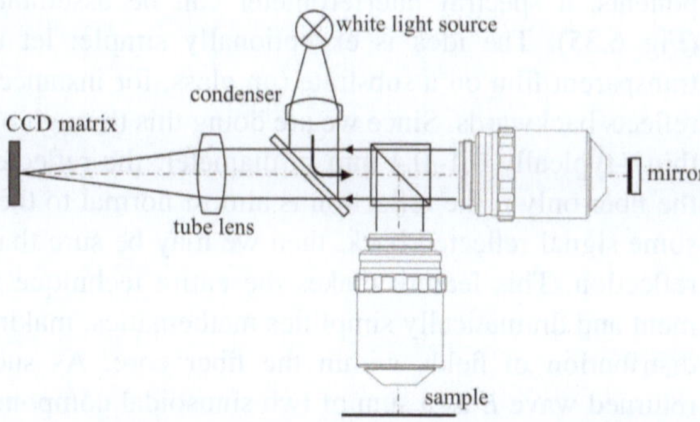

While the Mireau objective is always ready for use, the Michelson objective needs adjusting the mirror in order to make spatial frequency of fringes low enough to be visible. For that, high-coherence helium-neon laser must be coupled to the illumination port instead of the white light source. The easiest way to do this is to use fiber optics (Chap. 7). It was already explained above in this section that temporal coherence of the helium-neon laser radiation is so high that initial optical path difference between the sample and the mirror does not decrease fringe contrast, making visual adjustment easy. After the mirror tilt is adjusted, switch the illumination port to the white light and find position of the best fringe contrast. It was already explained that, in this position, image of the sample must also be the sharpest. Now, if your CCD camera captures colour images, then the typical image looks like the one portrayed in Fig. 6.33.

Factory-made interference objectives of the Mireau and Michelson types are very simple and reliable tools to quickly create an imaging interferometer. In those cases when some special functions are required, like fringe adjustment in imaging with high numerical aperture, then the so-called Linnik scheme may be recommended (Fig. 6.34). The strong feature of this scheme is that, in interference, aberrations of the two objectives compensate each other. Therefore, it is better to use objective of the same type from one manufacturer. Another advantage is that the beamsplitter works in parallel beams, reducing additional aberrations to minimum. The Linnik interferometers work well even with as big numerical apertures as 0.8 and above.

6.5 Spectral Interferometry

Spectral interferometry is a very simple and reliable tool for measuring thickness of thin films. This technique gained its popularity after commercialization of compact fiber-optic spectrometers—relatively inexpensive instruments, connected to a personal computer and easily operated through simple programs (Chap. 9). Considering spectrometers and fiber-optic accessories (Chap. 7) as given components, a spectral interferometer can be assembled almost on a dinner table (Fig. 6.35). The idea is exceptionally simple: let us send a wave towards the transparent film on a substrate (on glass, for instance), and receive something that reflects backwards. Since we are doing this through optical fiber, which is optically thin—typically 0.1–0.4 mm in diameter, the reflected wave can return back into the fiber only if the reflection is almost normal to the surface. Therefore, if we get some signal reflected back, then we may be sure that there was normal-incidence reflection. This feature makes the entire technique insensitive to angular adjustment and dramatically simplifies mathematics, making it possible to ignore spatial distribution of fields within the fiber core. As such, we may write down the returned wave E as a sum of two sinusoidal components with different amplitudes A_1 and A_2 and the phase shift between them:

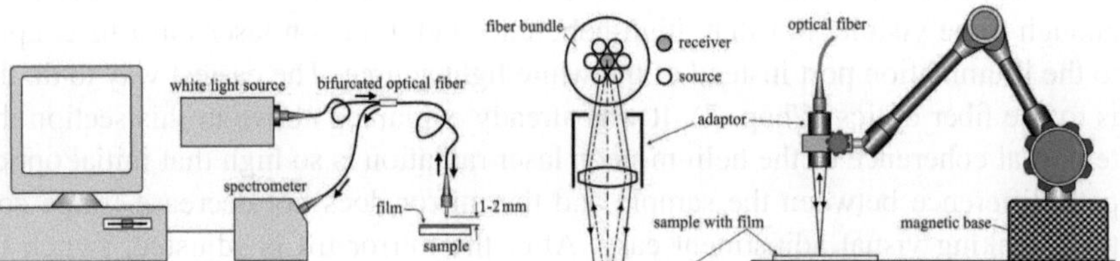

Fig. 6.35 Spectral interferometry requires minimum components: white-light source with fiber-optic adaptor, compact spectrometer connected to a computer, and bidirectional (bifurcated) optical fiber (scheme *at left*; for fiber-optic details see Chap. 7). In order to get strong reflected signal, keep the end of the fiber roughly 1–2 mm above the surface: it optimizes the angle at which the peripheral fibers in the bundle relay the light to the central core, connected to spectrometer. Almost professional system can be obtained by adding a focusing fiber-optic adaptor with a lens (*in the middle*). The end of the fiber and the surface of the sample must be roughly conjugated relative to the lens, which means that the lens must project the image of the fiber onto the surface of the sample. To keep the adaptor in place, a standard triple-jointed arm with magnetic base on steel plate finalizes the system (*at right*)

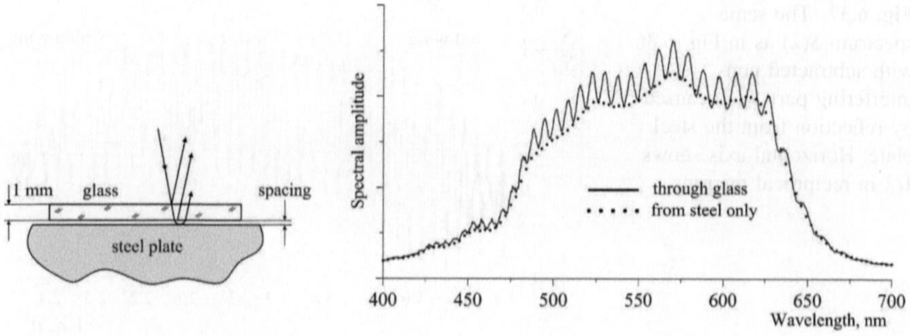

Fig. 6.36 Spectrum of white light on reflection from cover glass placed onto a steel plate. The *dotted curve* shows the spectrum without glass

$$E = A_1 \cos \omega t + A_2 \cos\left(\omega t + \frac{2\pi}{\lambda} 2\,hn\right).$$

The first component is reflected from the top of the film of thickness h and refractive index n, and the second—from its bottom, making the optical path difference equal to 2 hn. Photodetector responds to intensity E^2, therefore the electrical signal is proportional to

$$1 + \gamma \cos\left(\frac{2\pi}{\lambda} 2\,hn\right),$$

where γ is the contrast of the interference pattern, depending on A_1 and A_2. This holds true for every wavelength λ, and the spectrometer gives the signal for each λ separately, i.e. the spectrum. Therefore, the spectrum as a function of λ is modulated as $\cos(4\pi\, hn/\lambda)$. The first thing that must be emphasized is that this modulation is not periodic as a function of wavelength λ, but is periodic as a function of $1/\lambda$. Therefore, in order to determine film thickness h, the Fourier transform must be applied to the measured spectrum as a function of $x = 1/\lambda$.

Next, consider the simplest experiment shown in Fig. 6.36. Clearly visible modulation is the result of interference because when the glass is removed there is no modulation. But the question is which waves produce this modulation? Indeed, there are three waves reflected from the steel base, from the bottom of the glass, and from the top of the glass. To answer this question, we have to explicitly determine spectral spacing of modulation $\Delta\lambda$:

$$\frac{2\pi}{\lambda_1} 2\,hn - \frac{2\pi}{\lambda_2} 2\,hn = 2\pi; \qquad \Delta\lambda \approx \frac{\lambda^2}{2hn}.$$

At $\lambda = 500$ nm, the glass of $h = 1$ mm and refractive index $n = 1.5$ produces spacing $\Delta\lambda \sim 0.1$ nm—far below spectral resolution of a compact spectrometer that is typically $\sim$1 nm. Thus, interference between the waves reflected from

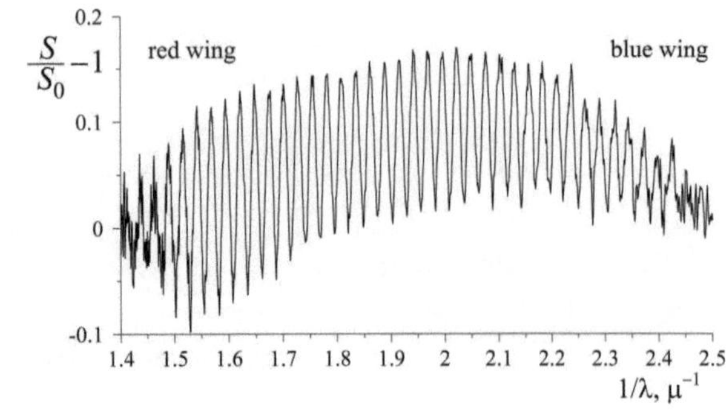

Fig. 6.37 The same spectrum $S(\lambda)$ as in Fig. 6.36 with subtracted non-interfering part $S_0(\lambda)$ caused by reflection from the steel plate. Horizontal axis shows $1/\lambda$ in reciprocal microns

opposite surfaces of glass does exist, but it cannot be seen due to finite spectral resolution of the spectrometer. In general, spectral resolution δ of the spectrometer imposes the upper limit for the film thickness $h_{\max}$ that can be measured around the wavelength λ:

$$h_{\max} \approx \frac{\lambda^2}{2\,n\,\delta}.$$

What we do see in Fig. 6.36 is the interference between the waves reflected from the steel plate and bottom surface of glass. There is about six oscillations in between 525 nm and 575 nm, which gives the estimate $\Delta\lambda \approx 8$ nm and $h \approx 20$ micron ($n = 1$ for air). It is a reasonable value because average size of a dust particle is about 20 micron, and the gap between the glass and the steel plate was due to dust, presumably.

It is not even necessary to perform a special mathematical analysis on data presented in Fig. 6.36 to notice the difference in periodicity between the blue (short wavelengths) and red (long wavelengths) wings of the spectrum. However, the same data presented as function of $x = 1/\lambda$ show real periodical variations (Fig. 6.37).

To finalize this section, consider some very practical hints on how to analyze more complicated cases, using fast Fourier transform (FFT). The FFT can only be applied to uniformly distributed grids, i.e. to sets of points x_k with constant spacing:

$$x_{k+1} - x_k = const.$$

However, spectrometer produces data that are uniformly distributed in wavelength λ

$$\lambda_{m+1} - \lambda_m \approx const,$$

and even this approximation is not accurate enough (see Chap. 9). Thus, the first task is to interpolate non-uniform grid $1/\lambda_m$ into uniform grid x_k, and there are many standard mathematical libraries that can do this job. The basic idea is to determine uniform grid x_k with as many points as we want, then to compute

$\lambda_k = 1/x_k$, and after that to interpolate spectral points $S(\lambda_k)$, using the spectrum $S(\lambda_m)$ supplied by the spectrometer. Number N of interpolated points λ_k may far exceed the number of measured points λ_m.

Suppose this is done, and we have the set $S(x_k)$. Next, we need to apply some model to the spectrum, for example

$$S(x) = 1 + \sum_{j=1}^{M} \gamma_j \cos\left(2\pi x 2\, h_j n_j\right),$$

which represents a set of M films of thicknesses h_j and refractive indices n_j, producing fringe contrasts γ_j. Our goal is to determine γ_j and $b_j \equiv 2\, h_j n_j$. Note that with spectral interferometry, it is impossible to determine thicknesses h_j and refractive indices n_j separately—only in product.

Now, apply the complex FFT to obtain the resultant set z_m as follows:

$$z_m = \sum_{k=1}^{N} S(x_k) e^{-2\pi i (k-1)(m-1)/N},$$

where N is the size of the grid. This formula shows how the FFT is computed: it works only with dimensionless integer numbers k and m, and does not want to know anything about our dimensional arguments x and h. Therefore, we need to establish a relation between the index m and $b = 2h \cdot n$. Doing this, we expect that the Fourier transform gives us a set of peaks—the harmonics—located at m_j, and each m_j identifies b_j. Relative amplitudes z_j of these peaks correspond to relative contrasts γ_j. Thus, substitute the model for $S(x)$ into the formula for FFT, expand cosine as the sum of complex exponents, and equalize the argument of the exponent to zero—this will show us locations of peaks:

$$x_k b = \frac{k-1}{N}(m-1).$$

Use

$$x_k = \frac{x_{\max} - x_{\min}}{N}(k-1)$$

to obtain

$$b = (m-1)\left(\frac{1}{\lambda_{\min}} - \frac{1}{\lambda_{\max}}\right)^{-1}.$$

Thus, every peak in the FFT, located at m_j, corresponds to a certain film with thickness

$$h_j = \frac{m_j - 1}{2n_j}\left(\frac{1}{\lambda_{\min}} - \frac{1}{\lambda_{\max}}\right)^{-1}.$$

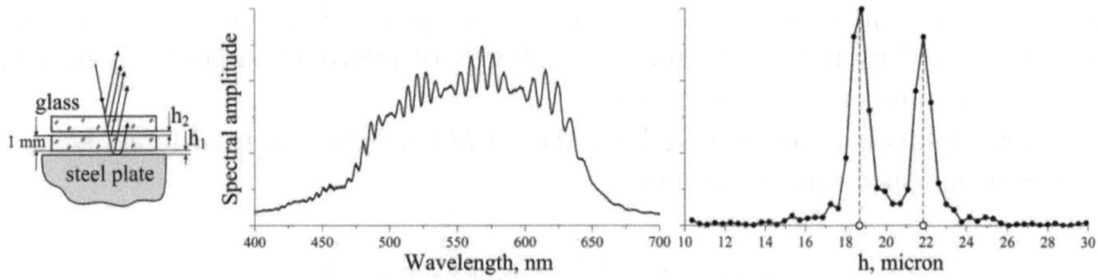

Fig. 6.38 Two air-spaced cover glasses, of 1 mm thickness each, form the reflecting stack (at *left*). Interference in two air spaces h_1 and h_2 produces modulated spectrum (in the *middle*). Horizontal positions of peaks in the FFT curve give the values for h_1 and h_2 (at *right*)

It was already told that we do not know n_j, but for air-spaced gaps situation radically simplifies:

$$h_j = \frac{m_j - 1}{2}\left(\frac{1}{\lambda_{\min}} - \frac{1}{\lambda_{\max}}\right)^{-1}.$$

Thus, scaling the horizontal axis of the FFT according to this formula, we automatically measure thicknesses of all the spaces. Relative amplitudes of the peaks give contrasts γ_j. Typical result is shown in Fig. 6.38. In this experiment, two glass plates were put one on top another, thus forming five reflecting surfaces, of which only four contribute in pairs to visible interference. Spectral amplitude shows more complicated modulation than in Fig. 6.36, and only experienced user can identify it as a beating curve produced by two sinusoids of different frequencies. However, the fast Fourier transform (FFT) not only readily identifies two gaps as two separate maxima, but being calibrated along the horizontal axis, gives the estimates of the spacings as 19 and 22 microns. Their vertical order in the stack cannot be identified exactly, but taking into consideration relative amplitudes, the plausible suggestion is that the higher peak comes from the steel plate.

For spectral interferometry, at least two reflecting interfaces are needed to create interference pattern. This is always the case in thin films. But what about bare vertically patterned surface, with no films on it? In this case, there is only one reflecting surface, therefore no interference occurs. To measure such relieves, the interference objectives of Mireau and Michelson type in combination with fringe-counting algorithms should be used as explained in the preceding section. In this type of objectives, the reference mirror plays the role of the second reflecting interface. As such, with interference objectives, spectral interferometry is also applicable for measuring profiles of bare surfaces. The entire system will not be as simple as before because interference objectives are infinity-corrected, thus requiring the tube lens to focus reflected light on the fiber. Nonetheless, the system will be perfectly working on bare patterned surfaces, measuring the virtual gap between the reference mirror inside the objective and the surface at the point where the light is focused to. To measure the profile of the surface, the object should be scanned along the desired direction. In order to avoid mechanical scanning and increase the speed of measurement, imaging spectrometers may be used as described in Chap. 9.

List of Common Mistakes

- opposite polarization in interfering beams;
- use of polarizing beam-splitting cubes instead of non-polarizing;
- trying to observe interference outside of localization plane;
- bottom illumination for Mireau and Michelson objectives;
- transmitting high-power laser beam through Mireau or Michelson objectives;
- in the white-light interferometer, pivoting axis of the adjustment mirror does not lie in the reflecting plane.

Further Reading

W.H. Steel, Interferometry, Cambridge University Press, 2nd ed., 2009.
M. Born, E. Wolf, Principles of Optics, Cambridge University Press, 7th ed., 1999.
V. Protopopov, Laser Heterodyning, Springer, 2009.
E.L. O'Neill, Introduction to Statistical Optics, Dover, 2003.

Chapter 7
Fiber Optics

A very efficient way to relay optical power from a source to a photodetector.

Abstract This chapter is focused on practical handling of standard optical fiber cables. The first section introduces basics of total internal reflection, concept of numerical aperture, structure of standard optical cables and their specifications. Dimensions and thread type of SMA connectors are summarized as a reference for manufacturing as well as some practical ideas for connecting SMA cables to hand-made opto-electronic equipment. Difference in jacketing is outlined to facilitate right choice. Various standard opto-couplers that can be found on the market are recommended with detailed dimensional specifications of connecting threads. Correct and incorrect connectivity options are explained and summarized in a comprehensive drawing. Spectral transmission of optical fibers may be crucial for ultra-violet applications below 400 nm. Some typical spectral curves show the lower spectral limit where special considerations must be applied. Advantages and disadvantages of plastic optical fibers are summarized and standard connectors explained in a detailed figure. Physics of polarization-maintaining optical fibers of the two standard types is explained in general phenomenology. Multi-fiber bundles are used either for illumination purposes, as it is briefly outlined in the second section, or as imaging elements—rather complicated and expensive optical elements described in the last section. Imaging optical fibers are rather rare, and wide audience does not commonly know their imaging capabilities. Therefore, details of the design and experimental high-magnification images may be useful for understanding performance of these optical elements.

7.1 Fiber Cables

In opto-electronics, with optical arrangement being assembled on an optical table and the electrical part being concentrated around the oscilloscopes and computer often meters away, optical fiber cables provide great flexibility in delivering optical power to a measurement system. Although basic concept of fiber optics is

V. Protopopov, *Practical Opto-Electronics*, Springer Series in Optical Sciences 184,
DOI: 10.1007/978-3-319-04513-9_7, © Springer International Publishing Switzerland 2014

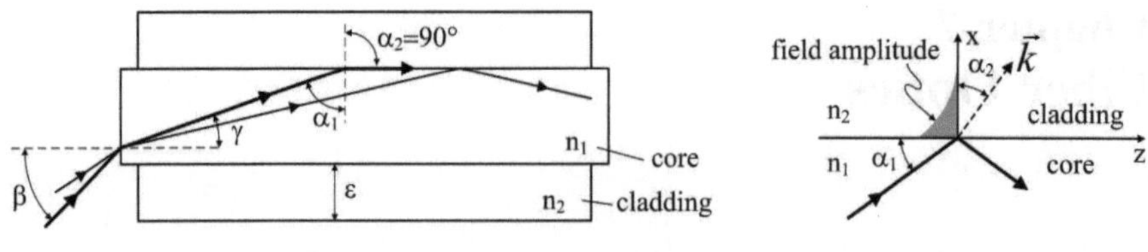

Fig. 7.1 In an optical fiber, rays propagate within the core, being reflected from the cladding. Refractive index of the core n_1 is always bigger than that of the cladding n_2. Inside the cladding, optical field quenches exponentially outwards the core

widely known, some initial introduction is still necessary to understand what particular types of fiber cables are used in laboratory and why. Refractive index of glass $n > 1$ makes it possible to transport optical rays inside the glass rod due to total internal reflection (Fig. 7.1). The Snell law dictates the sine rule for the angles of the incident α_1 and transmitted α_2 rays

$$n_1 \sin \alpha_1 = n_2 \sin \alpha_2.$$

When α_1 is big enough, such that $\sin \alpha_2$ becomes unity, the total internal reflection occurs for the case $n_1 > n_2$.This happens for all the rays with

$$\sin \alpha_1 \geq \frac{n_2}{n_1}.$$

The question is how big can be the acceptance angle β for the rays coming from the air? Since $\sin \beta = n_1 \sin \gamma$ and $\gamma = 90° - \alpha_1$, the following relation follows:

$$\sin \beta = \sqrt{n_1^2 - n_2^2}.$$

This value is commonly called the numerical aperture, and denoted NA. Of course, the sine must not exceed unity, therefore when $n_1^2 - n_2^2 \geq 1$, NA $= 1$, which means that all the rays from hemisphere will be transmitted through the fiber with bigger or smaller efficiency, depending on the reflection from the front surface. One important particular case is the absence of cladding: $n_2 = 1$. With the typical refractive index of glass $n_1 = 1.5$, $n_1^2 - 1 = 1.25$, which means $\beta = 90°$ and NA $= 1$—the highest possible acceptance angle. Why then do all optical fibers contain cladding? Because without it, even smallest amount of dirt or grease on the surface, or merely a contact with another fiber, would create an optical leakage from the core in the spot of a contact. Transmittance through such a fiber would become unreliable. The next question that immediately follows the first one is how thick must be the cladding? The answer requires a deeper insight into the physics of total internal reflection.

According to general laws of electrodynamics, electrical and magnetic fields cannot vanish abruptly on the interface between two mediums: some optical field does exist inside the cladding. But how deep can the field penetrate? At the condition of total internal reflection, propagation vector $\vec{k}$ inside the cladding is imaginary, with its normal projection being

$$k_x = \frac{2\pi n_2}{\lambda} \cos\alpha_2,$$

where λ is the wavelength in vacuum and the cosine is imaginary:

$$\cos\alpha_2 = i\sqrt{\left(\frac{n_1}{n_2}\sin\alpha_1\right)^2 - 1}, \quad i = \sqrt{-1}.$$

In this direction, the field changes as

$$e^{ik_x x} = \exp\left[-\frac{2\pi}{\lambda}\sqrt{n_1^2 \sin^2\alpha_1 - n_2^2}\cdot x\right],$$

decreasing exponentially from the interface. Thus, the penetration depth is of the order of the wavelength, and the cladding thickness ε must be much bigger. In reality, $\varepsilon \sim 20 - 100$ micron, completely blocking optical field from leaking outside the fiber, even in the near-infrared domain.

Rays, entering the fiber at different angles, travel different paths, emerging at the output in different times. For example, consider a straight section of the fiber. Then the two rays, propagating one along the optical axis and the second at the angle γ to it, will pass the paths that differ by a factor of $\cos\gamma$, making propagating times different. Thus, short optical pulse at the input transforms into long pulse at the output. This is the inevitable dispersion of an optical fiber. The question is how big can it be. Since

$$\sin\gamma = \frac{NA}{n_1}$$

may be considered as a small value,

$$\cos\gamma \approx 1 - \frac{NA^2}{2n_1^2}$$

does not differ much from unity. Every t seconds of propagation time result in

$$t \cdot \frac{NA^2}{2n_1^2}$$

delay spread. For instance, with typical value NA = 0.22 and $n_1 = 1.5$, the time spread is not more than 1 %—a relatively small value. However, not the relative value is important but the absolute delay spread. Inside the glass, light propagates with the speed of approximately 2×10^8 km/s. With the 2 m section of the cable, the propagation time is only 10 ns, and the delay spread is only 0.1 ns—hardly noticeable value. But on the longer routes, about 20 km, propagation times

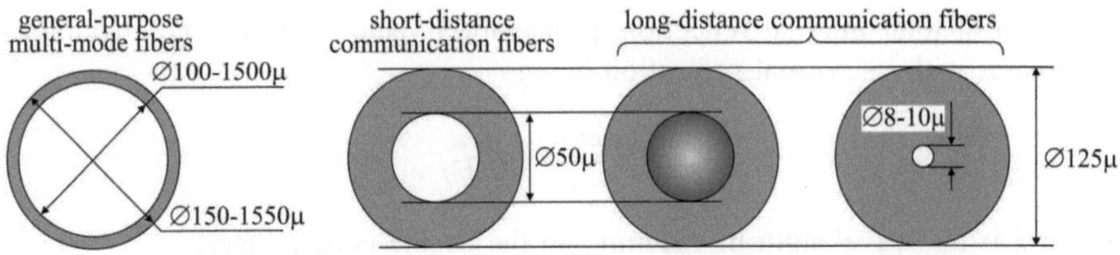

Fig. 7.2 In long-distance communication fibers (*at right*), dispersion is minimized either by narrowing the core to such a size that only one transversal mode can propagate (typically 8–10 μ) or by using parabolic profile of refractive index in the core (graded-index fibers). In the latter case, rays propagate like in a lens, periodically focusing and defocusing, preserving the same propagation time for all the rays. The short-distance communication fibers (*in the middle*) have much higher dispersion than the first two types, but it is tolerable on short distances of the order of 100 m. Diameters of these fibers are standardized: 125 μ cladding and 50 μ core. Diameters of general-purpose fibers (*at left*) are not strictly standardized, ranging from 100 to 1500 μ cores

increases to 100 μs, and the delay spread would be of the microsecond scale—completely unacceptable in high-speed communication systems.

We are talking so much about dispersion in order to clarify the difference between the optical fiber cables used in optical communication systems and those used to only relay optical power within short distances. Whereas in the first case dispersion is the priority, in the second case the priority is efficiency of collecting and relaying the optical power. Therefore, the communication cables are irrelevant to laboratory applications. The proper choice of optical fibers is explained in Fig. 7.2. Fibers with the core diameter less than 50 μare hard to couple to a light source, preserving reasonable efficiency of power transfer. Therefore, general-purpose multi-mode fibers of the core diameter up to 1500 μ(1.5 mm) are commonly used to relay optical signals to photoreceivers. Optical systems, requiring collimated optical beams, may use smaller diameters of about 100 μ.

Both core and cladding are made of one basic material—the fused silica (SiO_2) with $n_1 = 1.46$ at $\lambda = 500$ nm. To create necessary difference in refractive indices, they are doped by other materials. Dopants such as GeO_2 and P_2O_5 increase the refractive index of silica and are suitable for the core. On the other hand, dopants such as B_2O_3 and fluorine (F) decrease the refractive index of silica and are suitable for the cladding. Mostly, the core is not doped and the cladding is. In order to obtain the standard NA = 0.22, the difference between n_1 and n_2 must be

$$n_1 - n_2 \approx \frac{NA^2}{2n_1} = 0.017.$$

Fibers with non-standard numerical apertures, ranging from 0.16 to 0.5, are also available on the market.

Manufacturers offer numerous standard and custom-made fiber cables terminated with connectors. Connector is an essential part of the optical cable that determines flexibility of usage, and therefore should be chosen wisely. In the order of importance, the consideration priorities can be ranked as follows: how easily the

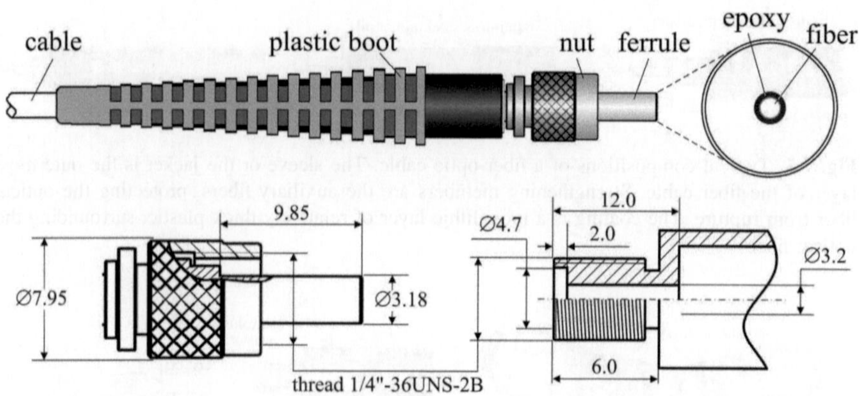

Fig. 7.3 SMA-905 fiber connector

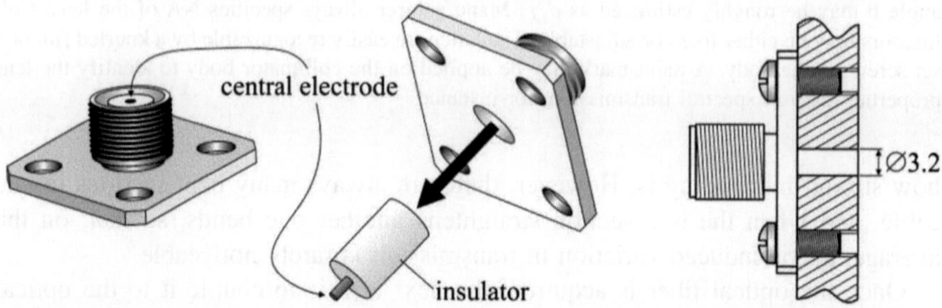

Fig. 7.4 Radio-frequency SMA connector may serve well as a receptacle of the SMA-905 optical fiber

new adaptor can be made for the existing cable; how tightly the fiber can be fixed; availability for large core diameters. The best choice is the SMA-905 connector shown in full detail in Fig. 7.3 together with the dimensions of matching receptacle that can be readily made in a modestly equipped workshop. A very important practical advantage of SMA-905 is that it has a standard threaded nut that fits the conventional SMA connector widely used in radio-frequency electronics (Fig. 7.4). If the receptacle cannot be manufactured, then merely take the SMA chassis connector, press out the central electrode together with the insulator (not much force is needed), and attach the remaining frame to a flat surface with ∅3.2 mm through hole and four M2.5 threaded holes for screws.

Fiber-optic cables are available in various types of sleeves or jackets. Typical structures are shown in Fig. 7.5.

Optical fiber cable is never straight, and the smaller the radius of bending the bigger the angle of incidence on the cladding. Some most tilted rays may run out of the total internal reflection, thus decreasing slightly the optical power delivered to the output. This drop in the transmitted power may reach 5–10 %, depending on

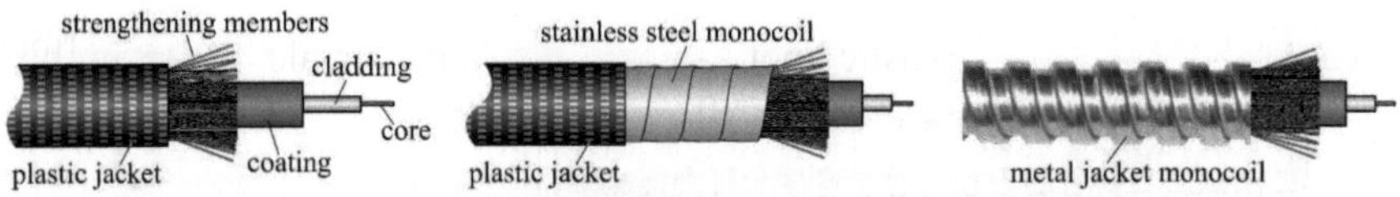

Fig. 7.5 Typical compositions of a fiber-optic cable. The sleeve or the jacket is the outermost layer of the fiber cable. Strengthening members are the auxiliary fibers, protecting the optical fiber from rupture. The coating is a monolithic layer of relatively thick plastic, surrounding the optical fiber

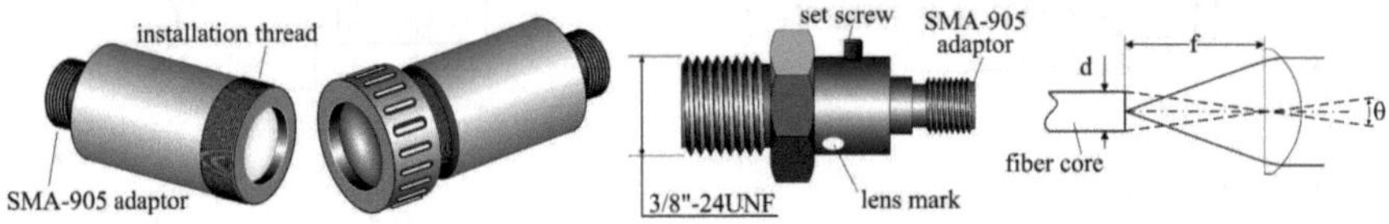

Fig. 7.6 In fiber-optic collimators, aspheric lenses are commonly used (Chap. 1). Divergence angle θ may be roughly estimated as d/f. Manufacturer always specifies NA of the lens. Collimators may be either fixed or adjustable. The latter are easily recognizable by a knurled rim or a set screw on the body. A paint mark may be applied on the collimator body to identify the lens properties such as spectral transmission, for instance

how strong the bending is. However, there are always many bent sections in one cable, and when the one section straightens another one bends, so that, on the average, shape-induced variation in transmission is rarely noticeable.

Once the optical fiber is acquired, the next task is to couple it to the optical beam. This is easily done with numerous standard couplers available from various vendors. Typically, the optical beam is nearly parallel, and then standard fiber-optic collimators suffice (Fig. 7.6). If the light source is point-like, akin the laser diode, then a pair of oppositely placed collimators serves well. Alternatively, a single lens with the source and fiber in conjugated planes solves the problem. Light-emitting diodes are far not the point-like sources (Chap. 2), therefore they cannot be efficiently coupled to a fiber unless the latter is very thick, more than 1 mm in diameter, or a fiber bundle is used (Sect. 7.2). Anyway, numerical apertures (NA) of the lens and the fiber should be equal. If not, then negative aftermaths may be expected (Fig. 7.7).

Fused silica, as an optical material, has excellent transmission from ultra-violet (UV) to near-infrared domains (Fig. 7.8). The same holds true for optical fibers made of it, except that doping of cladding may introduce some extra absorption. Fused silica is more expensive than glass, therefore, when UV transparency below 400 nm is not essential, cheaper glass fibers may be used in visible domain between 400 and 900 nm (Fig. 7.9).

For the central visible domain around 600 nm, plastic optical fibers (POFs) is the least expensive and most convenient choice: large core ∅1 mm, stress-resistant, virtually unbreakable, tolerating as small bending radii as 1 cm. With such a big core, POFs do not require precise alignment for coupling to optical sources,

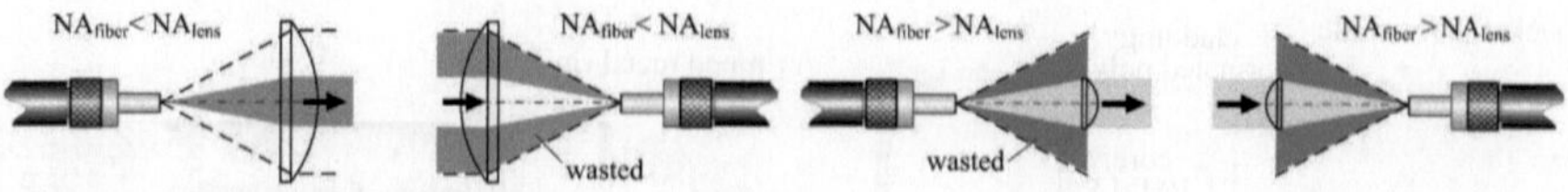

Fig. 7.7 When $NA_{fiber} < NA_{lens}$ on transmission, collimated diameter is less than expected. When $NA_{fiber} < NA_{lens}$ on the receiver end, some fraction of optical power is wasted. When $NA_{fiber} > NA_{lens}$ on transmission, a fraction of optical power is wasted. When $NA_{fiber} > NA_{lens}$ on the receiver end, no bad consequences occur

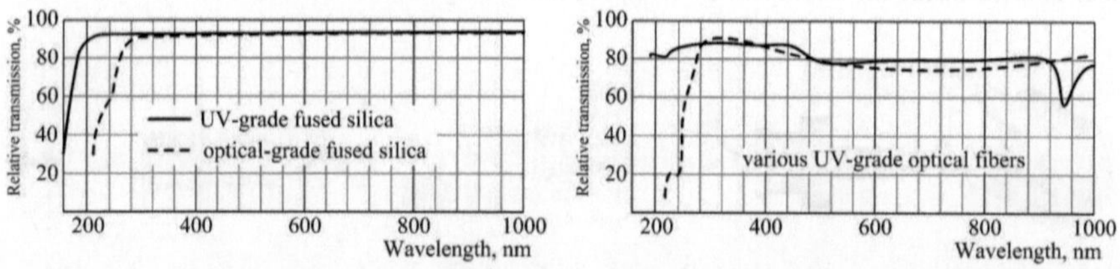

Fig. 7.8 Fused silica (SiO_2) shows exceptionally uniform optical transmission from UV to near-infrared (*at left*). However, optical fibers may exhibit more complicated performance (*at right*)

Fig. 7.9 Spectral transmission of various glass fibers

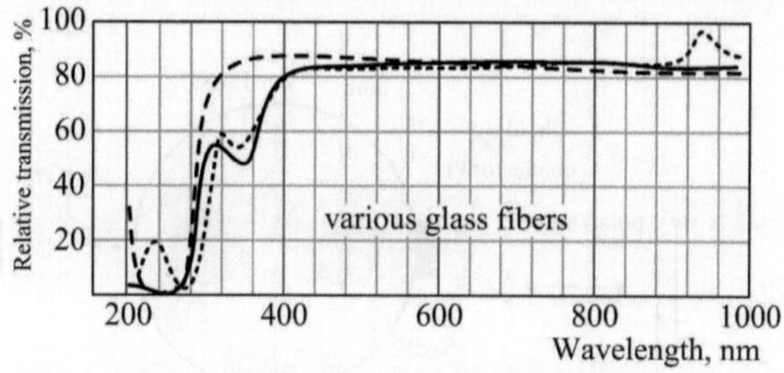

therefore they commonly terminate into much simpler connectors shown in Fig. 7.10, although SMA-905 are also applicable. For POFs, non-adjustable collimators are available from manufacturers of interferometers (Chap. 6) (Fig. 7.11). However, with ∅1 mm core and ≈15 mm focal length of the lens, divergence is rather poor: about several degrees. Therefore, primary purpose of these elements is to deliver light to photoreceivers, and it is better to name them the couplers or pick-ups. It is important to remember, that inside of these couplers there are tiny polarizers, which can be easily removed, giving increase in optical flux by a factor of two.

With nearly random trajectories of rays in optical fiber, it is difficult to expect that polarization of the input light will be preserved at the output, and actually it is not. Polarization extinction ratio at the output of 1 m long cable with ∅600 μ core is about 10, and it rapidly decreases with length. It must be stated explicitly that no optical fiber can maintain arbitrary polarization state from its input to the output. What the so-called polarization-maintaining (PM) optical fibers actually do is only

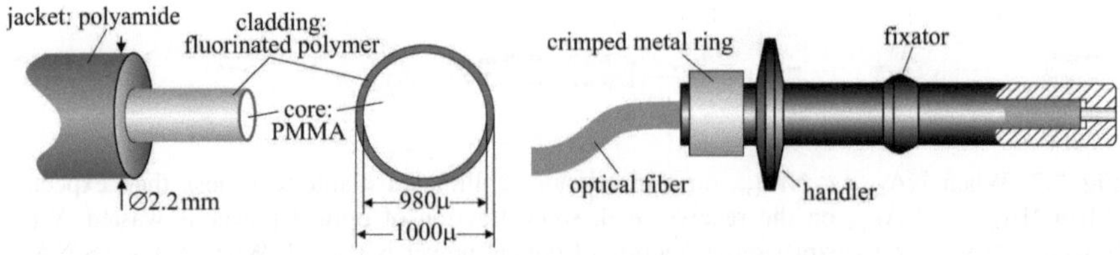

Fig. 7.10 In a POF, the core is made from polymethyl-methacrylate (PMMA) with refractive index 1.49. Connectors are standardized for ∅2.2 mm jacket diameter and ∅1.0 mm core. The fiber is fixed inside the connector by crimping metal ring around the tail

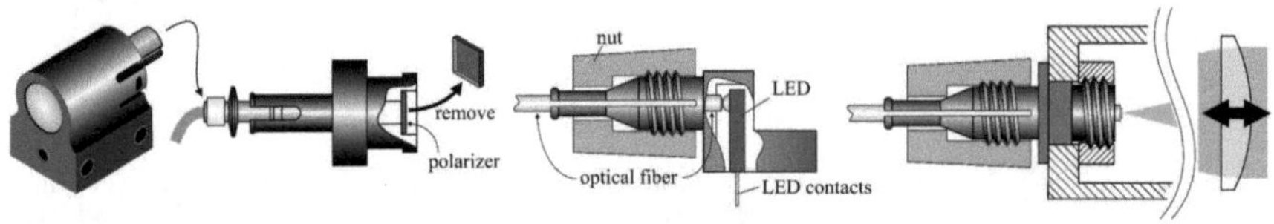

Fig. 7.11 Standard POF couplers (*at left*) are designed to accept connectors shown in Fig. 7.10. The polarizer tile can be easily pressed out. Many adaptors are designed to accept POF without any connector, for instance, the assembly with light-emitting diode (*in the middle*). Some general-purpose adaptors may help to make both the collimators or receivers (*at right*)

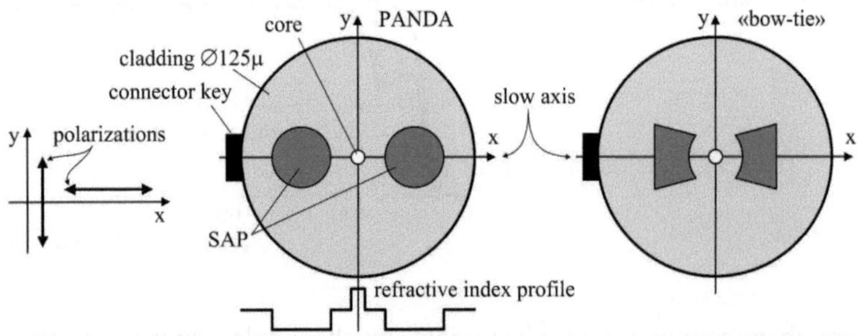

Fig. 7.12 PM fibers create diffraction-enhanced dispersion for the two orthogonally polarized modes along x and y axes, and as such, the core must be of the wavelength scale, typically 5 μ in visible domain. Two types of PM fibers—the PANDA and the «bow-tie»—use the same principle: stress-induced birefringence (Chap. 5). Borosilicate glass ($SiO_2 + B_2O_3$)—material of stress-applying parts (SAP)—has higher thermal expansion than the cladding made from pure silica (SiO_2). SAP shrink more than the cladding during cooling after drawing. It induces high tensile stress in the core (fused silica SiO_2 doped with GeO_2), causing radially-asymmetric birefringence

preserving directions of their two orthogonal modes (Fig. 7.12). The PM fibers are the single-mode optical fibers with radial symmetry intentionally disrupted along orthogonal axes x and y. Thus, these axes define predetermined directions of polarization modes that propagate with different velocities. If polarization of the input light is along one of these axes, then polarization at the output remains the same. The essential parameter of the PM fibers is the so-called cross-talk ratio or

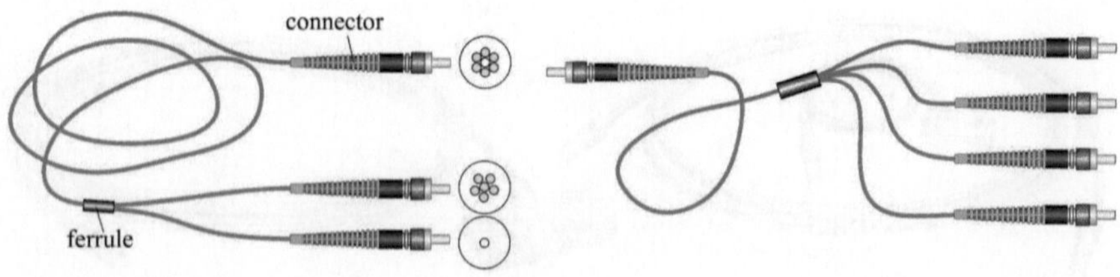

Fig. 7.13 Organization of separate optical fibers inside the bifurcated cable is of a primary importance

the extinction ratio—a value, showing what part of the linearly polarized input power is transferred into the opposite polarization at the output. For all the types of PM fibers, extinction ratio is about 10^2 at 633 nm (HeNe laser wavelength) for the cables 2–5 m long.

If input polarization is set arbitrarily with respect to directions of polarization modes, then the fiber acts as a waveplate (see Chap. 5), transforming polarization along the fiber periodically from linear to elliptical, to orthogonal linear, and so on. With the birefringence $\Delta n = n_x - n_y$ (Chap. 5), longitudinal period of this transformation is equal to

$$L = \frac{\lambda}{\Delta n}.$$

Typically, $\Delta n \sim 10^{-4}$, making $L \sim 2 - 6$ mm in visible domain. The axes x and y are fixed relative to the fiber structure and not in the laboratory system of coordinates, as the fiber may twist. Therefore, the PM fiber cables always have a key on their connectors, associated with directions of polarization modes (Fig. 7.12). Small core diameter of about 5–10 μ makes PM fibers inefficient for coupling to ordinary light sources and impractical for usual applications, leaving them to specialized research areas.

Fiber cables make it easy to deliver optical power from a single source to multiple receivers. For that, the so-called bifurcated cables may be used (Fig. 7.13). Such cables contain several separate optical fibers arranged together at one end and distributed in separate sleeves at other ends. Spatial organization of the cables is regular and in most cases symmetrical. One particular application of bifurcated cables is spectral interferometry, considered in Chap. 6. Manufacturers offer vast variety of cross-sectional geometries, and it is easy to make a mistake, using inappropriate leg for particular application. For example, a very popular type is the hexagonal geometry shown in Fig. 7.13. In it, one leg contains a single fiber and the second leg—six fibers. As such, the single-fiber leg may be used for precise collimation or as a single-mode receiver, while the multiple-fiber end—for illumination. With the core diameter of, say, 100 μ, organization of a cable can hardly be visible by a naked eye, leaving room for mistakes. For instance, using

Fig. 7.14 Fiber bundles are commonly laid inside metal monocoil sleeve in order to protect individual fibers from mechanical damage. Endings may be very different: merely round ferrules (*at left*), ring-shape circular illuminators for microscopy (*at left*, inside), or linear illuminators for line-CCD cameras (*at right*)

six-fibers end for collimation would result in roughly three times bigger divergence than with the single-fiber end, while using single-fiber end for illumination would decrease the total optical flux by six times with respect to the opposite connection. Therefore, always look at the connector tip through a microscope before connecting bifurcated cables.

7.2 Fiber Bundles

Fiber bundles are used for illumination, and contain hundreds of optical fibers placed together without any special organization. The application dictates that numerical aperture of a single fiber should be as big as possible: the more light is collected the better. Optical aperture of a fiber bundle may be as big as ∅10 mm, and one of its endings, specifically the one that is to be connected to a light source, is typically a cylindrical ferrule that is fixed in place by a screw at the light source. However, the opposites endings may take very different shapes, depending on the application (Fig. 7.14). Optical bundles are usually made from glass fibers, thus cutting ultra-violet radiation below 400 nm. On special request, fused silica fibers may be supplied, but at extra cost. Also on special request, individual fibers may be randomized within the bundle, providing better spatial uniformity of illumination. Alternatively, individual fibers may be laid in perfect order at one end relative to another. Such fiber bundles can relay images, as described in the Sect. 7.3.

7.3 Imaging Fibers

Suppose one single long optical fiber is wound many times around a cylinder, grouped tightly at one point, and cut there (Fig. 7.15). Then the fiber bundle is formed, and each single fiber in one end of it is connected exactly to the same

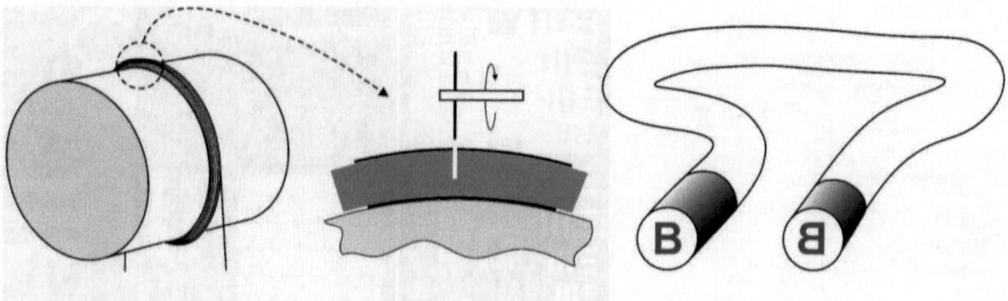

Fig. 7.15 Identical point-to-point connection between the ends of a fiber bundle is the necessary condition for imaging properties

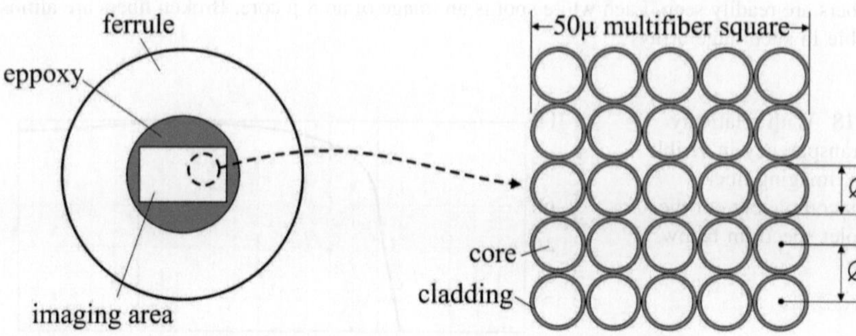

Fig. 7.16 High-quality imaging bundles are composed of great number of regularly packed mutlifibers, also possessing imaging quality

point in the opposite end. Such bundles can relay images from one end to another, and therefore are called the imaging fiber bundles. Another name that is often seen is the coherent fiber bundles, although they have nothing to do with coherence of light.

Although the virtual technology portrayed in Fig. 7.15 is feasible, it does not guarantee high spatial resolution because spacing between adjacent fibers may be different, and they may be packed not to the highest possible density. Therefore, when relatively large imaging area is needed, on the scale of $1 \times 1\ \text{cm}^2$, the bundle is composed of great number of thin tightly packed elementary imaging bundles (multifibers) of about $50 \times 50\ \mu$ area in cross section (Fig. 7.16). Figure 7.17 shows real images obtained through an imaging fiber bundle. Imaging fiber bundles are supposed to work in visible domain and are made from glass. Therefore, prepare for wavelength cut-off at 400 nm (Fig. 7.18).

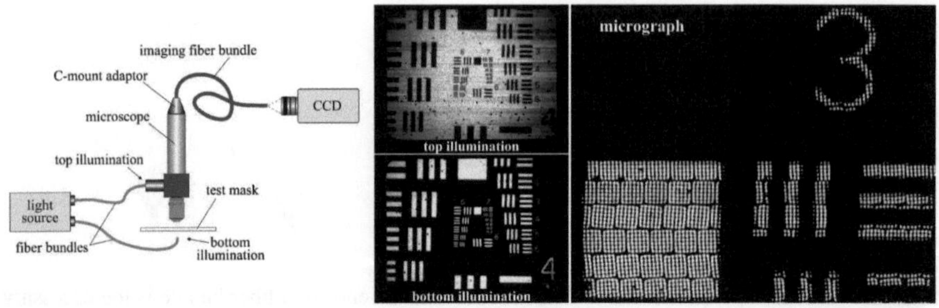

Fig. 7.17 Imaging fiber bundle capabilities. Image area 6.7 × 5 mm^2. The test mask is a chromium-coated glass, producing opposite contrast in upper and bottom illumination. The micrograph of the finest part of the mask pattern clearly shows the structure of the bundle: the multifibers are readily seen. Each white spot is an image of an 8 μ core. Broken fibers are almost inevitable in such huge arrays

Fig. 7.18 With relatively good transparency in visible domain, imaging fiber bundles completely cut the ultraviolet spectrum below 400 nm

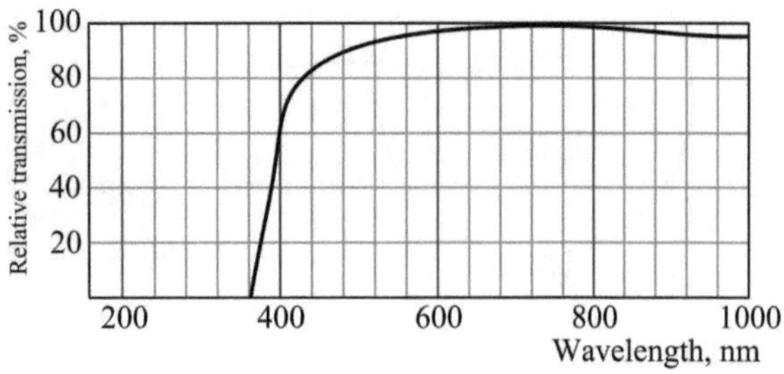

List of Common Mistakes

- mismatch between numerical apertures of the lens and fiber;
- using improper ends in a bifurcated cable.

Further Reading

M. Born, E. Wolf, Principles of Optics, Cambridge University Press, 7th ed., 1999.

Chapter 8
Magneto-Optics

Magneto-optical measurement systems are rare guests in laboratories. However when it happens it is better to be prepared.

Abstract For those who never heard about magneto-optics before, this chapter may serve as a quick guide into this part of optics. Although not very popular in publications, magneto-optics is a solid part of opto-electronics, with applications in microscopy of magnetic domains, inspection of magnetic disks, study of magnetization dynamics of solids, etc. The material of this chapter combines general phenomenology of the magneto-optical Kerr effect, its succinct but not trivial mathematical physics sufficient for making own numerical computations, verified table of magneto-optical coefficients, examples of computations in graphical form, and numerous examples of various practical magneto-optical systems with experimental curves and images. The Sect. 8.1 of the chapter introduces the three types of magneto-optical Kerr effects and explains physical mechanisms that change polarization of light after reflection from magnetized surface. From matrix formalism to vector diagrams, this section comprehensively explains phenomenology of the Kerr effect. The Sect. 8.2 outlines three different types of magneto-optical Kerr systems, compares performance of these technologies, and summarizes advantages and disadvantages in a table. For practice, particular values of magneto-optical coefficients are very important, and the next table summarizes analytical formulas for the three magneto-optical Kerr effects. All the notorious mistakes of the original publications, which have been since then relayed from one publication to another through years, are corrected and checked. This part is followed by an example of a magneto-optical system that can be made from very simple parts described earlier in Chaps. 2, 4 and 5. More sophisticated heterodyne technique is outlined next with explanation of vital practical know-how and experimental curves. The Sect. 8.3 describes magneto-optical imaging systems that are actually the microscopes with numerous options. An unusual combination of a moving diaphragm in the illumination system of a microscope with quadrant photodetector (Chap. 3) makes it possible to switch between various Kerr effects and change magneto-optical contrast, performing image enhancement by digital subtraction of consecutive images. Inspection of magnetic disks, particularly correctness of writing service magnetic tracks, utilizes another optical scheme that takes advantage of rotation: two-dimensional magnetic image can be obtained by only one-dimensional scanning. Typical experimental magneto-optical

V. Protopopov, *Practical Opto-Electronics*, Springer Series in Optical Sciences 184, DOI: 10.1007/978-3-319-04513-9_8,

images are presented, outlining capabilities of different optical technologies. The Sect. 8.4 describes quite a different technique for observing magnetic structures—magneto-optical liquids. It is a very efficient approach suitable when intrusive methods may be applied. Its capabilities become clear from the experimental image presented in the end of chapter.

8.1 Magneto-Optical Kerr Effect

Magneto-optics, except for the Faraday isolators that were considered in detail in Chap. 5, is not a widely known or highly commercialized technique. Nevertheless, it is indispensable when non-intrusive measurements of magnetization or pictures of magnetic structures are needed.

In 1876, Scottish physicist John Kerr observed that linearly polarized light, after reflection from magnetized surface, acquires polarization rotation with respect to the original one, and becomes elliptically polarized. This phenomenon is nowadays commonly referred to as the polar magneto-optical Kerr effect. Two years later he reported the existence of the so-called longitudinal Kerr effect. These two phenomena, together with the third one called the transversal Kerr effect, form what is nowadays known as the magneto-optical Kerr effect.

Magneto-optical Kerr effect may take place only on ferromagnetic materials. Early experiments have shown that the value of the effect is proportional to the magnetization of the sample and not to the external magnetic field. Macroscopically, the magneto-optical Kerr effect can be described in terms of a refractive index tensor, which replaces the ordinary index of refraction. Mathematically, this can be done by introducing the so called gyration vector $\vec{g}$ to the equation, connecting the electric displacement vector $\vec{D}$ with the electric field vector $\vec{E}$:

$$\vec{D} = \varepsilon \vec{E} + i\left[\vec{E}, \vec{g}\right] = \hat{\varepsilon}\vec{E},$$

where $i = \sqrt{-1}$, ε is the ordinary dielectric permittivity of an isotropic media, $\hat{\varepsilon}$ is the tensor of dielectric permittivity that we want to introduce, and square brackets denote the vector product. Thus, the tensor of dielectric permittivity can be written in the form:

$$\hat{\varepsilon} = \begin{pmatrix} \varepsilon & ig_z & -ig_y \\ -ig_z & \varepsilon & ig_x \\ ig_y & -ig_x & \varepsilon \end{pmatrix}.$$

When $\vec{g}$ is a real vector, the dielectric permittivity tensor is Hermitian:

$$\varepsilon_{ik} = \varepsilon_{ki}^{*}.$$

It means that the medium is lossless. However, the magneto-optical Kerr effect is observable only in ferromagnetic medium with losses, so that the gyration vector $\vec{g}$ must be complex. German physicist Woldemar Voigt suggested to account for complexity of the gyration vector by introducing the complex parameter Q, connecting $\vec{g}$ with the magnetization vector $\vec{m}$:

$$\vec{g} = Q \cdot \vec{m}.$$

Then

$$\hat{\varepsilon} = \begin{pmatrix} \varepsilon & iQm_z & -iQm_y \\ -iQm_z & \varepsilon & iQm_x \\ iQm_y & -iQm_x & \varepsilon \end{pmatrix}$$

The microscopic quantum–mechanical theory of magneto-optical phenomena explains the effect as the spin-orbital interaction. Further development of the microscopic theory showed that the phenomenological magnetization directions m_i are none other than directional cosines of the net spin direction in the specimen with respect to coordinate system attached to the optical beam, propagating in the z-direction. In order to realize where the ellipticity of the reflected beam comes from, consider as an example the polar Kerr effect in the simplest geometry of normal incidence shown in Fig. 8.1. In this case, the dielectric permittivity tensor can be written in the form

$$\hat{\varepsilon} = \begin{pmatrix} \varepsilon & q & 0 \\ -q & \varepsilon & 0 \\ 0 & 0 & \varepsilon \end{pmatrix}; q = iQm_z.$$

Off-diagonal elements are complex. The Maxwell equations for the plane wave, propagating in the anisotropic medium, can be written as:

$$\vec{H} = n\left[\vec{k}, \vec{E}\right],$$

$$\vec{D} = -n\left[\vec{k}, \vec{E}\right],$$

where $\vec{k}$ is the unity vector in the direction of propagation (z-direction), and n is the refractive index in the direction of propagation. This leads to

$$\vec{D} = n^2\left\{ \vec{E} - \vec{k}\left(\vec{k} \cdot \vec{E}\right) \right\},$$

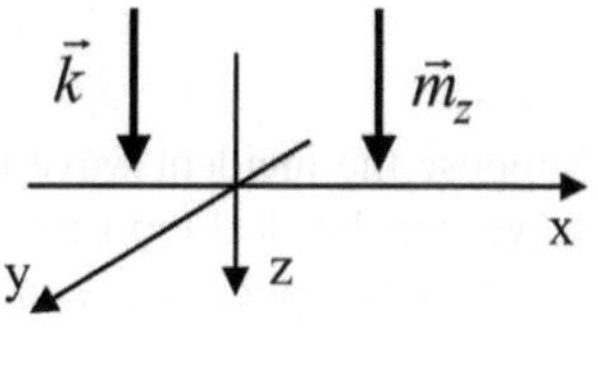

Fig. 8.1 Configuration of the polar Kerr effect with normal incidence of the incoming wave. The magnetized surface lies in the x–y plane

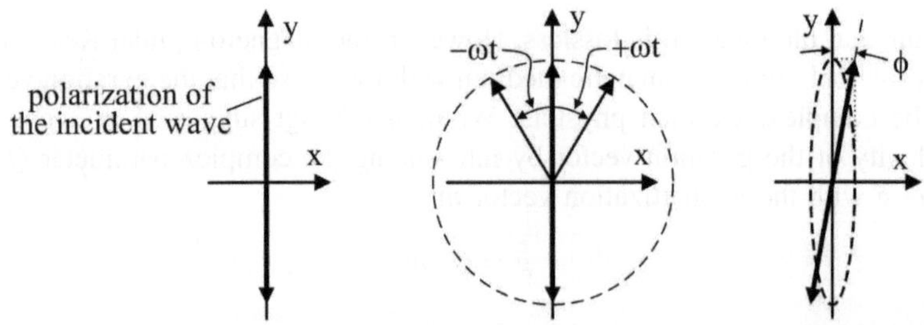

Fig. 8.2 Initially linearly polarized wave (*at left*) may be represented as the sum of the right- and left-circularly polarized waves (*in the middle*). After reflection from magnetized surface, it becomes elliptical (*at right*). Correction of ellipticity with a quarter-wave plate (Chap. 5) results in rotation by the angle ϕ

or, taking into consideration that $\vec{k}$ is perpendicular to $\vec{E}$, to very simple relations:

$$\begin{aligned} &\varepsilon E_x + q E_y \\ &- q E_x + \varepsilon E_y \, . \\ &\varepsilon E_z \end{aligned}$$

For the first two equations to have a nontrivial solution, their determinant must be zero:

$$\left(n^2 - \varepsilon\right)^2 + q^2 = 0.$$

Thus, the medium with the dielectric permittivity tensor $\hat{\varepsilon}$ as above is birefringent with the refractive indices

$$\begin{aligned} n_+^2 &= \varepsilon - iq \\ n_-^2 &= \varepsilon + iq \end{aligned}$$

and the solutions

$$\begin{aligned} E_y^+ &= +i E_x^+ \\ E_y^- &= -i E_x^- \end{aligned},$$

corresponding to right- and left-circularly polarized waves denoted as ≪+≫ and ≪−≫. The right-circularly polarized wave propagates with the refractive index n_+, and the left-circularly polarized wave propagates with the refractive index n_-.

According to the Fresnel formulas, the right- and left-circularly polarized waves will be reflected from the surface with the complex reflection coefficients

$$r_+ = \frac{1 - n_+}{1 + n_+}, \quad r_- = \frac{1 - n_-}{1 + n_-},$$

Suppose the incident wave is linearly polarized in the y–z plane (Fig. 8.2). Such a wave can be decomposed into a sum of the right- and left-circularly polarized waves with optical frequency ω and equal amplitudes:

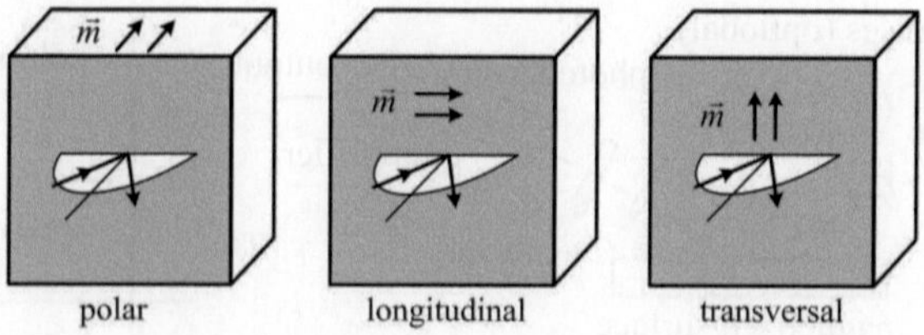

Fig. 8.3 Magneto-optical Kerr effects. $\vec{m}$—magnetization vector

$$\begin{pmatrix} \sin \omega t \\ \cos \omega t \end{pmatrix} + \begin{pmatrix} -\sin \omega t \\ \cos \omega t \end{pmatrix}.$$

Here t is time. After reflection, their amplitudes are no longer equal because the reflection coefficients r_- and r_+ are different. Only to understand the physics, we may assume for simplicity that r_- and r_+ are real values, and write down the reflected wave:

$$r_+ \begin{pmatrix} \sin \omega t \\ \cos \omega t \end{pmatrix} + r_- \begin{pmatrix} -\sin \omega t \\ \cos \omega t \end{pmatrix} = \begin{pmatrix} (r_+ - r_-) \sin \omega t \\ (r_+ + r_-) \cos \omega t \end{pmatrix}.$$

It is an ellipse (Fig. 8.2) with the ratio of the minor axis to the major axis

$$\phi = \frac{r_+ - r_-}{r_+ + r_-} = \frac{n_- - n_+}{1 - n_+ n_-}.$$

This is the physics of the Kerr effect, exemplified in the polar configuration. It must be emphasized again that the Kerr effect phenomenology is different from the Faraday effect: there is no rotation of polarization but only magnetically induced ellipticity. However, elliptically polarized wave may be transformed to linear polarization by means of a quarter-wave plate (Chap. 5). In the same system of coordinates, new polarization will make the angle arctan $\phi \approx \phi$ with the initial one, i.e. with the y axis. The words ≪Kerr rotation angle≫, being frequently met in literature, refer to parameter ϕ.

Summarizing, the three fundamental magneto-optical Kerr effects can be understood from Fig. 8.3. The polar and longitudinal effects result in depolarization. In the transversal configuration, there is no depolarization, but the reflection coefficient depends on magnetization.

8.2 Magneto-Optical Systems

All these effects are commonly very weak: the Kerr rotation angle is about 0.01° and the same is the typical ellipticity (i.e., about 10^{-4}). Therefore, special experimental techniques are necessary to detect such small depolarization. Today,

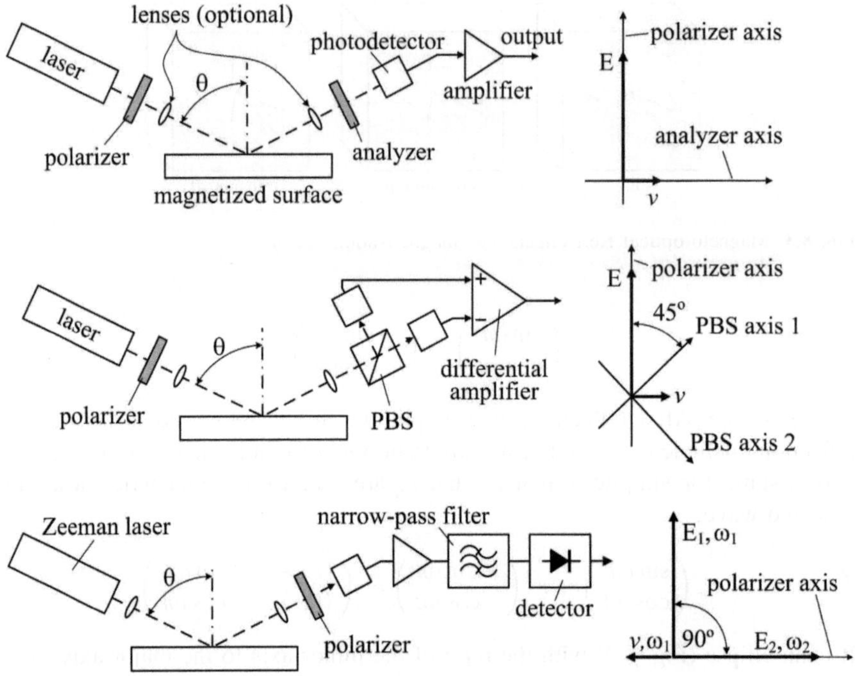

Fig. 8.4 Direct detection (*upper*) and bi-channel homodyne (*in the middle*) schemes use ordinary lasers, whereas the heterodyne scheme (*below*) uses Zeeman two-frequency laser. PBS—polarizing beam-splitting cube. When the direct-detection scheme is used to measure the transversal Kerr effect, the analyzer may be either excluded or redirected along the polarizer

three basic schemes are known for magneto-optic measurements: direct detection, homodyne bi-channel and heterodyne schemes (Fig. 8.4).

The direct detection technique is the most simple in realization. Due to the Kerr effect, the linearly polarized probe wave E generates after reflection the wave $|v| \ll |E|$ of the orthogonal polarization. To reduce both the detector saturation and noise, the analyzer is set cross-oriented with respect to the polarization of the probe beam. Neglecting the influence of the probe beam, the output signal is equal to

$$s = G \cdot |v|^2,$$

where G is the total gain of the detecting electronic circuit. Since it is proportional to $|v|^2$, it does not differentiate between the signs of v, or better to say the phases between E and v. However, the sign of v changes when magnetization vector $\vec{m}$ changes its direction. Therefore, the direct detection technique is insensitive to magnetization direction. The two other techniques—the homodyne and the heterodyne—are both sensitive to magnetization direction.

The homodyne detection technique provides the so-called intrinsic amplification, and, therefore, provides better signal-to-noise ratio than the direct detection method. In this scheme, the output signal is equal to

$$s = G_1 \left| \frac{1}{\sqrt{2}}(E+v) \right|^2 - G_2 \left| \frac{1}{\sqrt{2}}(E-v) \right|^2 ,$$

where G_1 and G_2 are the gains in the two shoulders of the detection module. Assuming $G_1 = G_2 = G$, the output signal equals

$$s = 2\,G \cdot E\,v\,\cos\varphi,$$

where φ is the phase difference between the probe wave and the depolarized Kerr wave. Comparing this formula to the direct detection scheme, we can see that in the homodyne scheme the signal amplitude is $2|E/v| \gg 1$ times bigger, if the phase difference φ is zero. This is the so-called intrinsic amplification. However, the homodyne technique has also two significant disadvantages. The first is the phase dependence of the signal. Indeed, if $\varphi = \pi/2$ then the output signal is zero regardless of the value of the Kerr signal $|v|$. The phase difference φ depends on many factors such as the angle of incidence θ, material type and structure, probe beam polarization. Therefore, it cannot be predicted. The second disadvantage is substantial dependence of the output signal on the inequality of gain coefficients $G_{1,2}$. Suppose $G_1 \neq G_2$. Then

$$s = (G_1 - G_2)\,\frac{1}{2}|E|^2 + 2\,\frac{G_1 + G_2}{2} \cdot E\,v\,\cos\varphi.$$

Since $|E| \gg |v|$, even small gain mismatch substantially influences the output signal.

The heterodyne cross-polarized technique for detection of Kerr depolarization has its own advantages and disadvantages. The key component of this technique is the dual-frequency cross-polarized Zeeman laser (Chap. 2), producing two orthogonally linearly polarized waves $E_1 = a_1 e^{i\omega_1 t}$ and $E_2 = a_2 e^{i\omega_2 t}$ with slightly different frequencies ω_1 and ω_2. Polarization vectors of the both waves are orthogonal to each other $(\vec{e}_1, \vec{e}_2) = 0$, therefore, when entering the photo-detector, these waves produce the output current proportional to the sum of intensities:

$$\left| \vec{e}_1\, a_1\, e^{i\omega_1 t} + \vec{e}_2\, a_2\, e^{i\omega_2 t} \right|^2 = |a_1|^2 + |a_2|^2.$$

However, if a depolarization occurs after reflection, and a third wave with the amplitude v and the polarization vector along $\vec{e}_1$ or $\vec{e}_2$ appears, the situation changes, and an oscillating component appears in the photo-current:

$$\left| \vec{e}_1\, a_1\, e^{i\omega_1 t} + \vec{e}_2\, v\, e^{i\omega_1 t + i\varphi} + \vec{e}_2\, a_2\, e^{i\omega_2 t} \right|^2 = |a_1|^2 + |a_2|^2 + 2 a_2 v \cos[(\omega_1 - \omega_2)\,t + \varphi].$$

After the wave E_1 has generated the orthogonal component with the small amplitude v, it becomes redundant, and has to be blocked in order not to generate additional noise in the photodetector. This can be done by inserting a polarizer in front of the photodetector. Rotating the polarizer, we can use either *s*- or *p*-polarized component of the laser output beam, depending on what type of the Kerr effect we wish to work with. Then the photo-detector signal is proportional to

$$\left|\vec{e}_2\, v\, e^{i\omega_1 t + i\varphi} + \vec{e}_2\, a_2\, e^{i\omega_2 t}\right|^2 = I_2 + 2a_2 v \cos[(\omega_1 - \omega_2)\, t + \varphi],$$

where I_2 is the intensity of the wave E_2. After filtering in the narrow-pass filter, the slowly varying component I_2 vanishes together with low-frequency noise, so that the signal contains only the intermediate frequency signal proportional to the Kerr amplitude v:

$$s = G \cdot 2a_2 v \cos[(\omega_1 - \omega_2)\, t + \varphi],$$

where G is the gain. The final operation performed in the heterodyne scheme is demodulation (detection) of the radio-frequency that leaves only the amplitude of the signal:

$$s = G \cdot 2\, a_2 v = G \cdot 2\, I_2 \frac{v}{a_2}.$$

From this formula, it follows that the heterodyne scheme also provides intrinsic amplification by a factor of $2|a_2/v|$ with respect to the direct detection technique. In this formula, the Kerr amplitude v is originated by the wave E_1. In practice, the orthogonally polarized laser beams have roughly equal intensities, so that $a_1 \approx a_2$. Therefore, the factor v/a_2 practically equals the Kerr reflection coefficient v/a_1.

The biggest disadvantage of the heterodyne scheme is uncontrollable ellipticity of the partial waves that the Zeeman laser generates. Uncontrollable because there is no way to correct polarization of one partial wave without affecting another one. It was explained in Chap. 2 that inherent polarization ratio of helium–neon lasers is about 500:1. Also, we know from Chap. 5 that high-quality crystal polarizers provide extinction ratio up to 10^6. Therefore, in the direct detection and homodyne techniques, polarizer improves polarization purity of the probe beam by several orders of magnitude. In the heterodyne scheme, it cannot be done. Also, additional cost of Zeeman lasers is a significant disadvantage.

Now we can summarize the advantages and disadvantages of the three schemes in the Table 8.1.

Detailed explanation of the detection techniques that was given above is necessary to understand how to compute the Kerr effect magnitudes. When the wave polarized parallel to the plane of incidence (*p*-polarization) comes to the magnetized surface, two orthogonal components generally exist in the wave, outgoing from it: the one that is polarized in the same direction (*p*-component) and the orthogonal one (*s*-component). Thus, we may introduce reflection coefficient r_{pp}, describing the *p*-polarized wave after reflection, and also the coefficient r_{sp} that describes the birth of the orthogonal component polarized in *s*-direction. By convention, the first index refers to what is reflected, and the second one—to the incident polarization. It means that r_{ps} describes the depolarized wave that have

Table 8.1 Comparison of magneto-optic techniques

Scheme	Advantage	Disadvantage
Direct detection	Simplicity: single detector; does not depend on the phase; high polarization ratio	No intrinsic amplification; insensitive to magnetization direction
Bi-channel homodyne	Intrinsic amplification; high polarization ratio; sensitive to magnetization direction	Complexity: two detectors; depends on phase
Heterodyne	Intrinsic amplification; stabilized laser; narrow-pass filtering at intermediate frequency; does not depend on phase; sensitive to magnetization direction	Complexity: Zeeman laser; high cost; low polarization ratio

Table 8.2 Magneto-optical coefficients associated with measuring techniques

Technique	Coefficients
Direct detection	r_{ps}, r_{sp}, or r_{pp}
Bi-channel homodyne	$\mathrm{Re}(r_{sp}r_{pp})$ or $\mathrm{Re}(r_{ps}r_{ss})$
Heterodyne	$\lvert r_{sp}r_{ss}\rvert$ or $\lvert r_{ps}r_{pp}\rvert$

been born in *p*-direction by the incident wave polarized in *s*-direction. Alternatively, r_{ss} is the traditional reflection coefficient of this wave, i.e. without change of polarization. All the aforementioned can be formalized by introducing the matrix of reflection coefficients that connects the vectors of the incident $E^i_{p,s}$ and reflected $E^r_{p,s}$ waves:

$$\begin{pmatrix} E^r_p \\ E^r_s \end{pmatrix} = \begin{pmatrix} r_{pp} & r_{ps} \\ r_{sp} & r_{ss} \end{pmatrix} \cdot \begin{pmatrix} E^i_p \\ E^i_s \end{pmatrix}.$$

With these notations, we may say that the direct detection scheme measures r_{ps} or r_{sp}. It also can measure pure variation of reflectivity, which is not associated with any depolarization. This option may be useful in measuring the transversal Kerr effect, when the coefficient r_{pp} depends on magnetization. The homodyne technique measures $\mathrm{Re}(r_{sp}\, r_{pp})$ or $\mathrm{Re}(r_{ps}\, r_{ss})$. The heterodyne technique measures $\lvert r_{sp}\, r_{ss}\rvert$ or $\lvert r_{ps}\, r_{pp}\rvert$. This is summarized in Table 8.2.

Analytical expressions for the reflection coefficients r_{sp}, r_{ps}, r_{pp}, and r_{ss} are given in Table 8.3.

Comprehensive summary of the formulas related to the Kerr effects can be found in the article by Z. J. Yang and M. R. Scheinfein, "Combined three-axis surface magneto-optical Kerr effects in the study of the surface and ultrathin-film magnetism", Journal of Applied Physics, volume 74, number 11, pp. 6810–6823 (1993). Regretfully, there is a mistake in formula (12) for transverse Kerr coefficients. Although corrected in the same article in the formula (16), it initiated numerous mistakes in subsequent publications of others as a result of purely parasitic compilation. Detailed derivation of the formulas can be found in the

Table 8.3 Magneto-optical reflection coefficients

Coefficient	Polar	Longitudinal	Transversal
r_{pp}	$\frac{n\cos\theta_i - \cos\theta_t}{n\cos\theta_i + \cos\theta_t}$	$\frac{n\cos\theta_i - \cos\theta_t}{n\cos\theta_i + \cos\theta_t}$	$\frac{n\cos\theta_i - \cos\theta_t}{n\cos\theta_i + \cos\theta_t} + iQ\frac{\sin 2\theta_i}{(n\cos\theta_i + \cos\theta_t)^2}$
r_{ss}	$\frac{\cos\theta_i - n\cos\theta_t}{\cos\theta_i + n\cos\theta_t}$	$\frac{\cos\theta_i - n\cos\theta_t}{\cos\theta_i + n\cos\theta_t}$	$\frac{\cos\theta_i - n\cos\theta_t}{\cos\theta_i + n\cos\theta_t}$
r_{sp}	$\frac{-iQn\cos\theta_i}{(n\cos\theta_i + \cos\theta_t)(\cos\theta_i + n\cos\theta_t)}$	$\frac{iQn\cos\theta_i\tan\theta_t}{(n\cos\theta_i + \cos\theta_t)(\cos\theta_i + n\cos\theta_t)}$	0
r_{ps}	r_{sp}	$-r_{sp}$	0

n—complex refraction index of the magnetic material
θ_i—real angle of incidence; θ_t—complex angle of transmission: $\sin\theta_i = n\sin\theta_t$
Q—complex Voigt parameter, proportional to magnetization in the polar, longitudinal, or transversal direction. When $Q = 0$ there is no magnetization, although magnetic field may exist

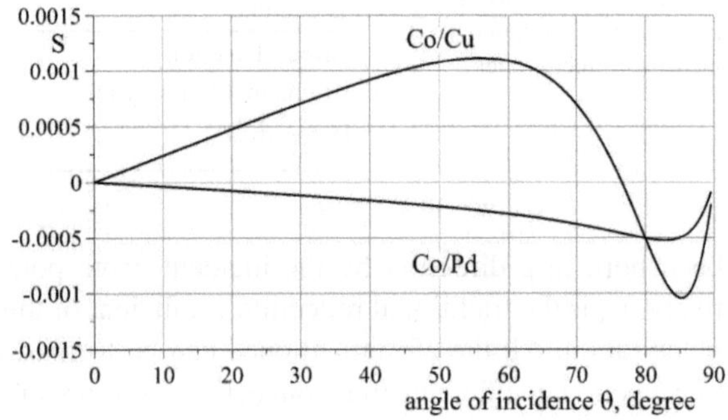

Fig. 8.5 Transversal Kerr effect. Differential signal S in the direct detection technique

books by W. Voigt, by G. S. Krinchik, and by K. Zvezdin and V. A. Kotov. The book by W. Voigt does not elaborate on the transversal effect, however.

Designing magneto-optical system, it is important to understand what angles of incidence deliver the maximum signal. For that, it is necessary to compute angular dependence of coefficients summarized in the Table 8.2, using exact analytical expressions of the Table 8.3 and numerical values for n and Q. For demonstration purpose, we shall consider two ferromagnetic systems, widely used in magnetic memory: Co/Cu with $n = 1.58 + i3.58$ and $Q = 0.0177 - i0.0063$, and Co/Pd with $n = 2.04 + i4.06$ and $Q = 0.00038 - i0.00314$ at the wavelength 633 nm. For the beginning, consider transversal Kerr effect and direct detection technique, reacting to r_{pp}. Actually, photodetector responds to intensity, therefore we should compute $|r_{pp}(n, Q)|^2$. To characterize the difference between the magnetized and non-magnetized surfaces, Fig. 8.5 shows $S = |r_{pp}(n, Q)|^2 - |r_{pp}(n, 0)|^2$. Similarly, it is possible to compute angular dependence for other cases. Figure 8.6 presents the results. One important practical conclusion follows immediately from these non-trivial results: magneto-optical systems that are supposed to measure the polar Kerr effect should be designed for zero angle of incidence, whereas those for measuring longitudinal or transversal effects—for highly oblique incidence with $\theta \approx 50$–$60°$.

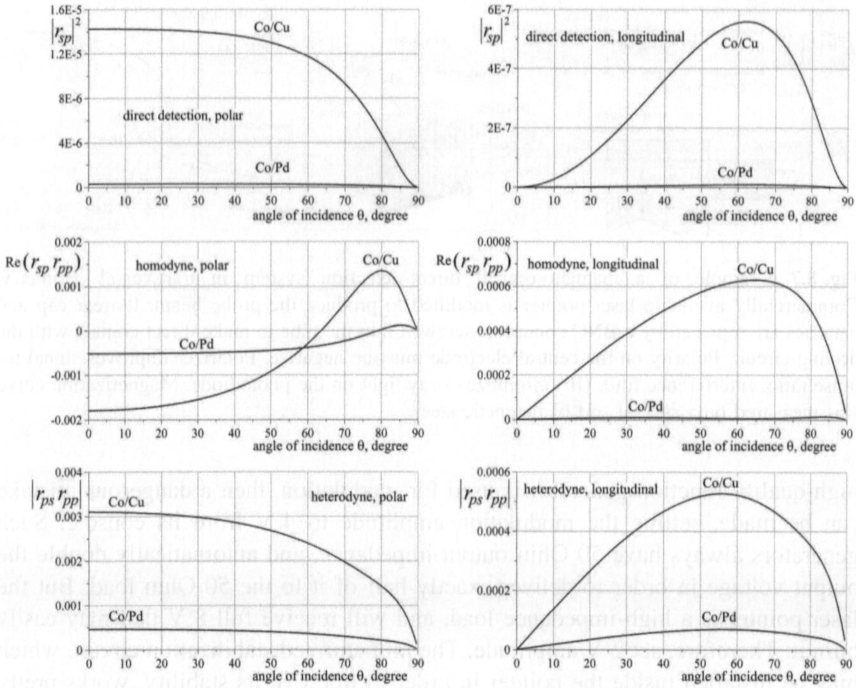

Fig. 8.6 Angular dependence of signals in three magneto-optical techniques shown in Fig. 8.3

Direct detection scheme requires minimum optical components, which is the reason why it is used most often for monitoring magnetization processes. Unlike the homodyne or heterodyne schemes, it does not provide intrinsic amplification, so that the photodetector always works in extremely low optical fluxes. Therefore, photomultipliers must be used (see Chap. 3), and the system must be strongly protected from stray light. Nevertheless, even with these precautions, filtering (synchronous detection, for instance, see Chap. 4) is necessary to obtain reasonable signal-to-noise ratio. The only exception, when photomultipliers cannot be used with the direct detection scheme, is the transversal Kerr effect. An example of such a system is shown in Fig. 8.7. There is no depolarization in the transversal Kerr effect ($r_{sp} = r_{ps} = 0$, Table 8.3), but instead of it, intensity of the reflected beam depends on magnetization (r_{pp} is proportional to Q). Thus, the detector must receive the entire power of the laser beam, and the photomultiplier would be completely blinded. As such, the photodiode must be used in this case. An ordinary laser pointer, working in the red domain (650 nm), is a perfect choice because it easily accepts electrical modification for pulsed operation (see Chap. 2) that is needed to implement synchronous detection. The simplest solution is to apply square-wave modulation signal directly to the power contact instead of the three 1.4 V batteries, commonly powering the laser. Thus, the amplitude of modulation must be around 4 V. Standard TTL output suffices. However, if a

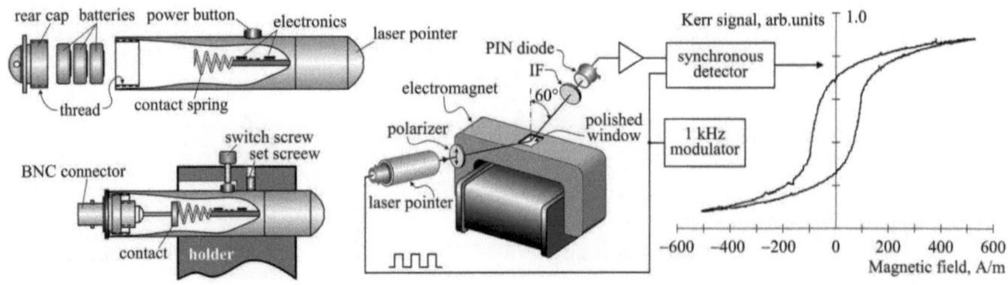

Fig. 8.7 Example of a magneto-optical direct-detection system in transversal geometry. Commercially available laser pointer is modified to produce the probe beam. Its rear cap and batteries are replaced by a BNC connector screwed into the tube to make direct contact with the driving circuit. Polarity on the central electrode must be negative. Polarizer improves signal-to-noise ratio. Interference filter (IF) minimizes stray light on the photodiode. Magnetization curve was measured on a special sort of magnetic steel

high-quality function generator is used for modulation, then a dangerous mistake can be made, setting the modulation amplitude to 4 V from its console. Such generators always have 50 Ohm output impedance, and automatically double the output voltage in order to deliver exactly half of it to the 50 Ohm load. But the laser pointer is a high-impedance load, and will receive full 8 V that may easily burn it. Therefore, set 2 V amplitude. The miniaturized stabilization circuit, which may be installed inside the pointer in order to improve its stability, works pretty well with modulation frequencies up to 1–5 kHz. The cheaper laser pointers, with no stabilization circuitry, provide much high modulation frequencies in the megahertz range. Figure 8.7 explains all other details.

Theoretically, for transversal geometry, the probe beam may even be unpolarized because r_{ss} does not depend on magnetization, meaning that *s*-polarization does not contribute to the signal. However, it does contribute to photodetector noise. Therefore, *p*-polarized probe beam delivers maximum signal-to-noise ratio. Although laser diode generates already polarized light, low polarization ratio $\sim 1/20$ slightly decreases signal-to-noise ratio. Therefore, an ordinary dichroic polarizer is always a plus.

As an example of the alternative design, Fig. 8.8 explains a heterodyne magneto-optical system for longitudinal Kerr effect. Heterodyning implies mixing of a strong (amplifying) and a weak (signal) optical components (Fig. 8.4). The strong component is actually a direct laser beam, which would not only saturate but even damage a photomultiplier if it were used as the photodetector. Therefore, unlike the direct detection scheme, photodiodes are the only choice in heterodyne and homodyne systems. In this design, for easy manufacturing, the angle of incidence is chosen to be 45°, which is not far from the optimum for the heterodyne technique (Fig. 8.6). Expecting micro-scale magnetic structures to be measured, the laser beam is focused onto the sample by a lens. The second identical lens is used to recollimate the beam before it enters the polarizer in order to achieve maximum possible extinction ratio. Due to geometrical limitations, these two lenses cannot be placed close to each other, dictating relatively long focal lengths of 40 mm.

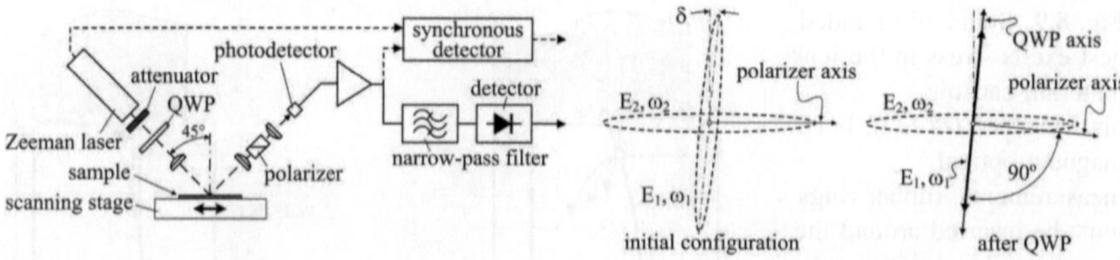

Fig. 8.8 With heterodyne technique, two options are available: narrow-pass filtering with direct detection, or synchronous detection with reference signal from the Zeeman laser (see Chap. 4). Both give practically the same results. Quarter-wave plate (QWP) serves to compensate for initial ellipticity of one of the Zeeman laser partial waves (see Chaps. 2 and 5)

With the wavelength 633 nm and beam diameter 5 mm, such a long focal length gives only diffractional resolution about 5 micron, if not to count spherical aberration.

With extremely small polarization rotation in magneto-optical measurements, optical elements must be mounted very gently. For instance, a standard threaded mounting shown in Fig. 8.9 is inapplicable because it exerts stress in the lens material, causing noticeable birefringence and uncontrollable variations of the signal that are not associated with magnetization.

Initial adjustment of the scheme shown in Fig. 8.8 is performed on a non-magnetized sample in such a way that to minimize the output signal. The lower limit of the signal is zero, which can be achieved only when one of the partial waves makes zero projection onto the polarizer axis, i.e. when it is linearly polarized. However, partial waves of the Zeeman laser output are always elliptically polarized, with ellipticity around 10^{-2}–10^{-3}. Therefore, regardless the orientation of the polarizer, partial waves are always coupled to the extent of the same value (Fig. 8.8), which makes it practically impossible to detect the Kerr depolarization that is of the order of 10^{-4}–10^{-5}. In order to minimize ellipticity of at least one partial wave, the quarter wave plate is used. When the axis of the quarter-wave plate stands along the polarization ellipse, the ellipticity transforms to linear polarization. The second partial wave remains elliptical or even becomes more elliptical than before, but since it is used only for amplification of the depolarized component, this transformation does not play a big role.

Finally, the role of the attenuator must be explained. A focusing lens with the sample in its focus is nearly perfect retroreflector, returning scattered radiation exactly back into the laser cavity, destabilizing it. Thus, optical isolation is needed. However, with two cross-polarized components of the Zeeman laser, traditional isolators of the Faraday type (Chap. 5) are inapplicable in this case. The simplest solution is just an attenuator. With the laser power, far exceeding the minimum necessary for operation of the phorodetector, it can be attenuated without significant drop of the signal-to-noise ratio. Then the parasitic wave, returning backwards, is attenuated to the second degree, substantially improving stability of the laser. Additional trick, useful to disrupt unwanted feedback to the laser, is to tilt

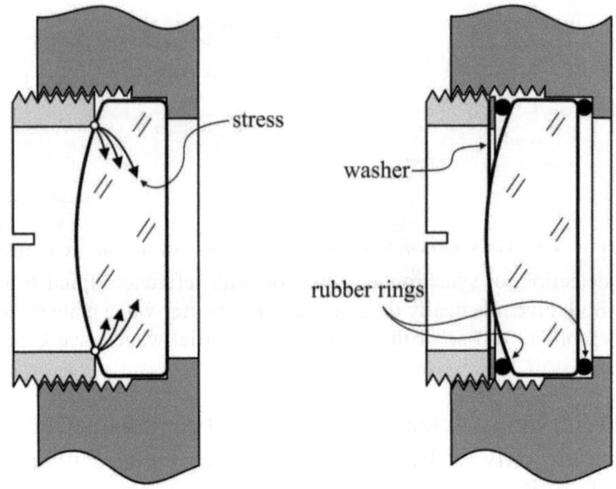

Fig. 8.9 Standard threaded nest exerts stress in the lens material, causing birefringence (*at left*). For magneto-optical measurements, rubber rings must be inserted around the lens (*at right*). Friction-relief washer facilitates easy rotation of the threaded ring

slightly the polarizer and the photodetector (actually, its front window), thus decoupling them from the laser cavity.

Among the most convenient test-objects for magneto-optical measurements are the plates of a hard disk drive. Every magnetic disk has the so-called service magnetic tracks that are written almost radially with magnetization being directed tangentially (Fig. 8.10). This regular and evenly spaced structure is very convenient in a sense that a narrow region of magnetization can be easily found by any type of scanning: either linear or angular. Typically, the width of the track is tens of microns—quite within spatial resolution of a long-focus optical system that is used. Figure 8.11 explains detailed structure of a particular magnetic track and presents the oscilloscope trace of a magneto-optical signal as it is recorded during scanning.

8.3 Magneto-Optical Imaging

Magneto-optical imaging makes it possible to study microscopic structure of magnetic domains. It is actually a magneto-optical microscopy, and there are two types of such systems: instant imagers and scanners. Instant imagers are based on traditional polarizing microscopes equipped with imaging cameras or visual ports (eyepieces), with only minor modifications of the illuminating system needed to maximize relevant Kerr effect (Fig. 8.12). When magnetization in the plane of observation is important, i.e. m_x or m_y components, then illumination system must maximize the longitudinal Kerr effect. Transversal Kerr effect is inappropriate in this situation because the image contrast would be too small on the background of blinding direct optical flux. On the contrary, the longitudinal Kerr contrast is formed through crossed polarizers, which is usually called the dark-field contrast, and may be potentially high. When the normal component of magnetization is of

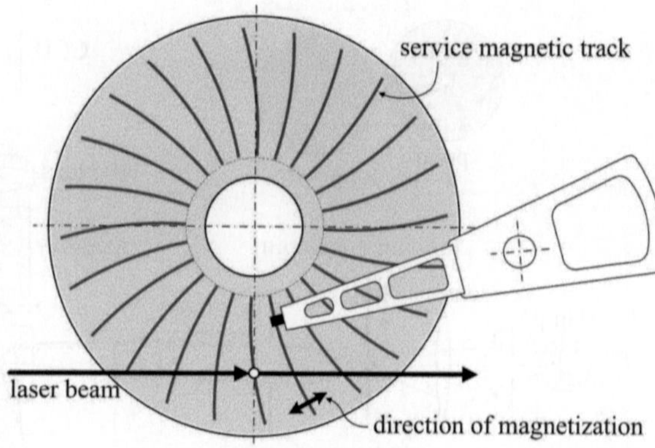

Fig. 8.10 Geometry of service magnetic tracks on a hard disk plate. When the probe beam strikes the disk perpendicularly to its diameter, magnetization is practically in longitudinal configuration

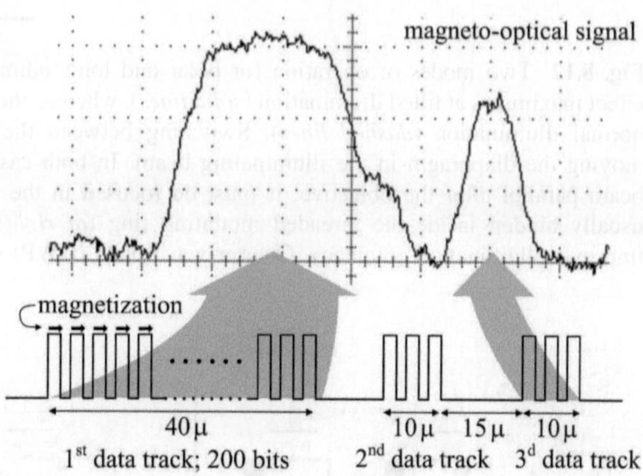

Fig. 8.11 Spatially resolved longitudinal Kerr signal from magnetic tracks. A single magnetic bit cannot be resolved by the long-focus optical system with the focal spot bigger than 5 micron, therefore the entire bit sequence (track) produces consolidated signal

importance, i.e. m_z, then illumination system must be readjusted to produce normal illumination of the sample, which maximizes the polar Kerr effect.

Formation of the contrast in magneto-optical microscopy is a very important factor, which may be not very easy to understand. For the beginning, consider the simplest case shown in the left panel of Fig. 8.13: polarizer and analyzer are fixed perpendicular to each other. Then analyzer blocks completely illumination wave E and transmits only the depolarized component v, regardless its sign. Therefore, in longitudinal illumination, longitudinal magnetization components both $+m_x$ and $-m_x$ will be seen equally bright. Transversal components $\pm m_y$ would not be seen in longitudinal illumination, unless they are combined with polar magnetization $\pm m_z$, which produces some depolarization (Fig. 8.6, polar configuration, angle of incidence less than 90°). In polar illumination, situation is simpler: both $+m_z$ and $-m_z$ magnetization produce the same signal $|v|$, whereas any orthogonal components m_x or m_y do not (Fig. 8.6, polar configuration, angle of incidence 90°). Thus, information about direction of magnetization is lost in this configuration.

Information about direction of magnetization may be retrieved with more complicated adjustment explained in the right panel of Fig. 8.13. In it, the quarter-

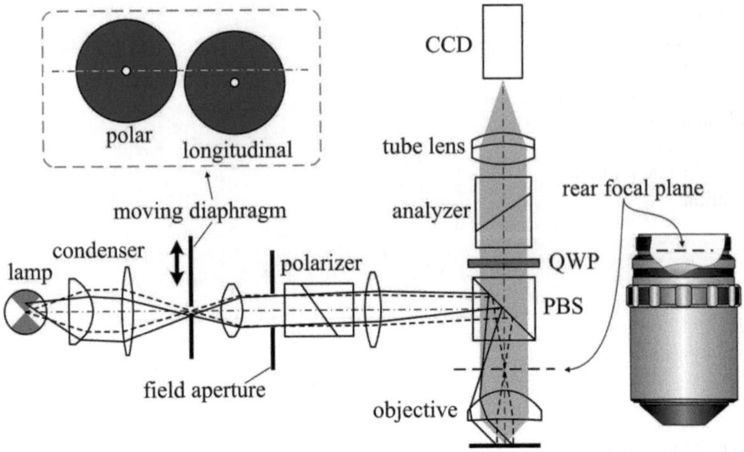

Fig. 8.12 Two modes of operation for polar and longitudinal Kerr effects. Longitudinal Kerr effect maximizes at tilted illumination (*solid lines*), whereas the polar Kerr effect is a maximum at normal illumination (*dashed lines*). Switching between the two modes is accomplished by moving the diaphragm in the illuminating beam. In both cases, in order to make illuminating beam parallel after the objective, it must be focused in the rear focal plane of the objective, usually hidden inside the threaded mounting ring (*at right*). Polarizing beam splitter (PBS) improves illumination efficiency. Quarter-wave plate (QWP) maximizes the contrast

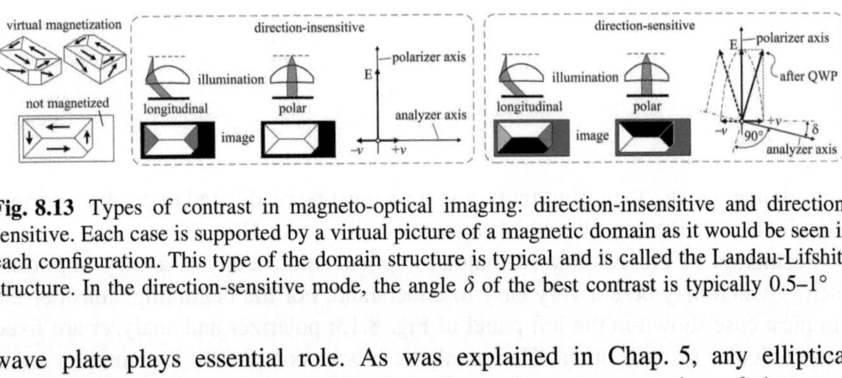

Fig. 8.13 Types of contrast in magneto-optical imaging: direction-insensitive and direction-sensitive. Each case is supported by a virtual picture of a magnetic domain as it would be seen in each configuration. This type of the domain structure is typical and is called the Landau-Lifshitz structure. In the direction-sensitive mode, the angle δ of the best contrast is typically 0.5–1°

wave plate plays essential role. As was explained in Chap. 5, any elliptical polarization can be transformed back to linear by proper rotation of the quarter-wave plate. New direction of linear polarization makes some angle δ with the initial direction. Now, adjusting the angles of both the polarizer and the quarter-wave plate, it is possible to find position where some magnetized areas become completely dark. Namely, these areas correspond to maximal magnetization in the direction of illumination (depolarized component v is maximal), and the analyzer is directed 90° relative to polarization from these areas. Then the areas with the opposite magnetization direction, including even the non-magnetized areas, will produce bigger projections onto the analyzer (dashed-line vector in Fig. 8.13), and will look brighter. This is how direction of magnetization can be rendered in magneto-optical imaging.

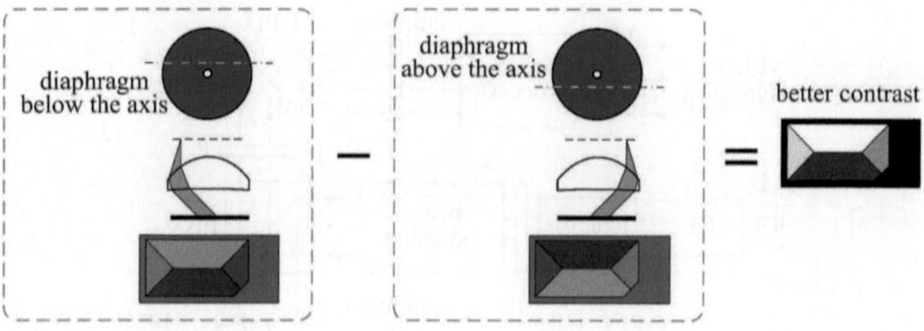

Fig. 8.14 Contrast-enhancement technique uses linear dependence of contrast on direction of illumination in longitudinal Kerr effect (tangent function in r_{sp} in Table 8.3). Subtraction of two images with opposite illumination enhances the contrast of longitudinal Kerr effect and cancels contributions from the polar Kerr effect

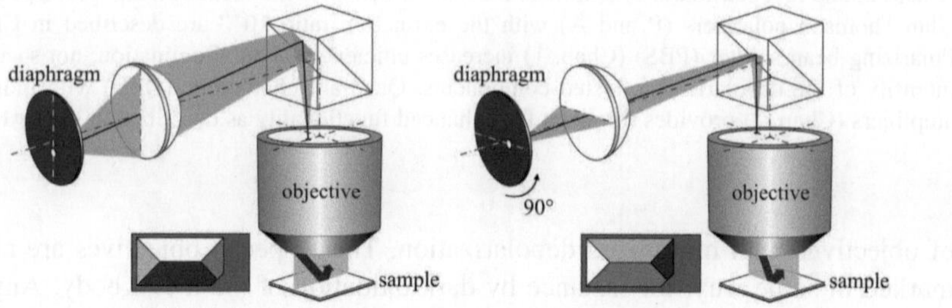

Fig. 8.15 In transverse illumination, the same longitudinal Kerr effect illumination scheme is used, but the diaphragm may be turned 90°, making the incident beam coming to the sample from perpendicular direction. This enhances the contrast on orthogonally magnetized parts of the sample, leaving the sample in place. Neither sample rotation nor image subtraction are needed. Other necessary elements like polarizer, analyzer, etc. are not shown for clarity

But not only this: altering direction of illumination and subtracting the images, it is possible to improve the contrast. Figure 8.14 is self-explanatory. Yet, the side elements (triangles) in the virtual domain shown in Figs. 8.13 and 8.14 can hardly be differentiated in contrast. The situation changes with the so-called transverse illumination shown in Fig. 8.15. The term «transverse illumination» may sound somewhat confusing, making parallel with the transverse Kerr effect. Therefore, it must be stressed again, that transversal Kerr effect is irrelevant to this application because strong direct probe beam would blind the image sensor.

Longitudinal Kerr effect commonly requires big angles of incidence about 60°, corresponding to numerical aperture of the objective NA = sin60° ≈ 0.86—a value close to a maximum for non-immersed objectives. This is why magneto-optical microscopes always have very high NA: it is not a matter of spatial resolution but the requirement for high magneto-optical signal. Such objectives always represent a multi-element combination, which results in high depolarization of the output beam and, consequently, low contrast of the image. Manufacturers know about it and for polarization-sensitive applications offer special types

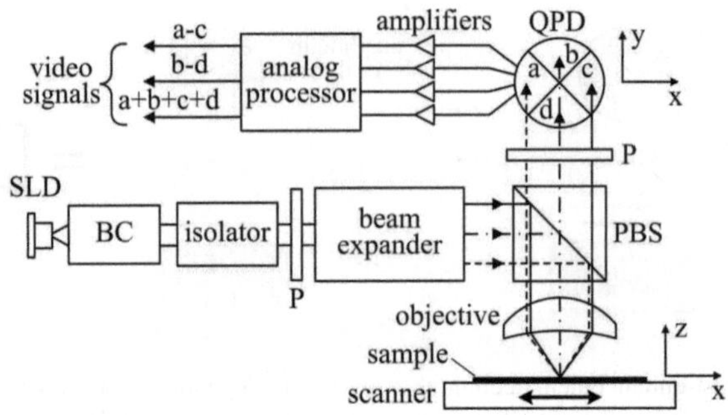

Fig. 8.16 Functional scheme of the scanning magneto-optical microscope. Superluminescent diode (SLD) (Chap. 2) is chosen as the light source because its relatively wider spectrum decreases the level of speckles. The beam conditioner (BC) includes anamorhic prismsAnamorhic prisms (Chap. 2) and first collimator to collimate the narrow beam for the Faraday isolator(Chap. 5). The Glan-Thomson polarizers (P and A) with the extinction ratio 10^{-6} are described in Chap. 5. Polarizing bean splitter (PBS) (Chap. 1) increases efficiency of the illumination, not sacrificing intensity of the depolarized reflected components. Quadrant photodiode (QPD) with individual amplifiers (Chap. 3) provides the basis for enhanced functionality as described in the text

of objectives with minimized depolarization. These special objectives are always marked in some way, for instance by dark oxidation of the metal body. Anyway, low contrast is the biggest problem of magneto-optical imaging, so that it can be practically useful only in combination with digital contrast-enhancement techniques.

In scanning magneto-optical microscopes, image is formed by scanning the sample with a focused probe beam. Typical applications include analysis of magnetic domains and inspection of magnetic tracks on hard-disk plates. Obvious disadvantage of scanning systems—low speed—is compensated for by some additional features that are unavailable in instant imagers: much wider image area limited only by scanning stage, separate and simultaneous imaging of both in-plane magnetization components m_x and m_y, spot magnetometry, phase sensitivity. The basic scheme of a scanning magneto-optical microscope with the feature of simultaneous imaging of m_x and m_y is shown in Fig. 8.16. All the opto-electronic elements that it is combined of are described in detail in other chapters, therefore only general functional description is needed. The quadrant photodetector intercepts the entire beam reflected from the sample. Signals from individual sub-sensors a, b, c, and d are proportional to spatial integrals over their sensitive areas, and may be considered as intensities of the rays coming to their geometrical centers. If magnetization projections are m_x, m_y, and m_z in the system of coordinates shown in Fig. 8.16, then, considering the analyzer A being crossed with the polarizer P,

$$\begin{cases} a = I + \alpha m_x + \beta m_z \\ b = I + \alpha m_y + \beta m_z \\ c = I - \alpha m_x + \beta m_z \\ d = I - \alpha m_y + \beta m_z \end{cases},$$

where I is the background common for all the four areas, α is the coefficient proportional to longitudinal Kerr effect, and β is the coefficient proportional to polar Kerr effect. Thus, the three magnetization components can be evaluated relative to one another by combining signals as follows:

$$m_x \to a - c$$
$$m_y \to b - d \quad .$$
$$m_z \to a + b + c + d$$

Since these three signals may be combined simultaneously, using analog processing with operational amplifiers, the three independent pictures can be obtained simultaneously in one scan. Typically, one scan for 100 × 100 pixels image may take 5–15 min, depending on the signal-to-noise ratio. Spatial resolution depends on the numerical aperture NA of the objective and is not far worse than the theoretical limit λ/NA for the particular wavelength λ. For the NA = 0.8, for instance, it is about 1 μ in visible domain.

Industrial application of magneto-optics is mostly limited to inspection of magnetic data tracks on the plates of hard disk drives. It was already explained in Sect. 8.2 that this application totally relies on the longitudinal Kerr effect. The cross-polarized direct detection scheme is commonly used to obtain images by means of two-dimensional angular-radial scanning (Fig. 8.17). In production line or in any other manufacturing facility, there is no time for precise angular adjustment of the polarizer and analyzer. Therefore, they are permanently set orthogonally to each other without option of detecting direction of magnetization. Signal-to-noise ratio is always small because the focal spot on the disk covers many magnetic bits, averaging the signal, and the speed of rotation is high, widening necessary bandwidth of the electronic system (Chap. 4).

Magneto-optical techniques with intrinsic amplification, like homodyne or heterodyne, provide better signal-to-noise ratio and better image quality when used in scanning microscopes. For example, the same scanning configuration like in Fig. 8.17 may work with heterodyne technique as it was already explained in Fig. 8.8. Sample images of magnetic tracks are presented in Fig. 8.18 for comparison with the image in Fig. 8.17. Not only the quality is better here, but also direction of magnetization may be rendered by proper alignment of the polarizer relative to the quarter-wave plate.

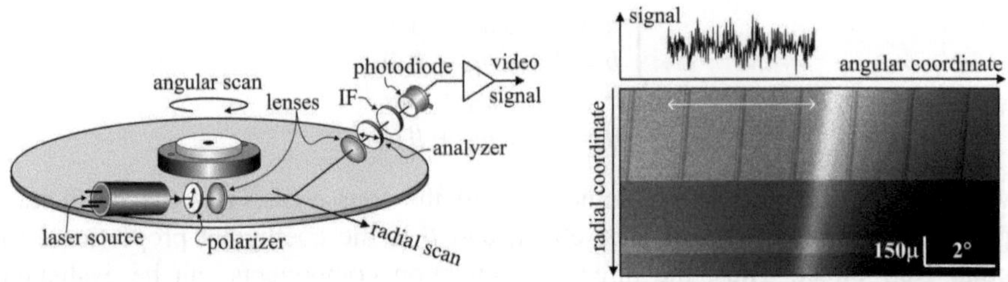

Fig. 8.17 In a magneto-optical imager for inspection of magnetic disks, the disk plate rotates permanently with high speed, while the opto-electronic head slowly drifts radially. Angular and radial encoders (not shown) generate coordinates of the focused spot on the disk, thus creating two-dimensional map of the video-signal—the image. Signal-to-noise ratio is so small that magnetic signal can hardly be visible in a single line scan, as shown in the trace above the image. The white arrow in the image identifies where the trace is taken from. However, when combined in a picture, human eye definitely recognizes magnetic pattern (*vertical lines*). The system detects not only magnetization but any depolarizing pattern on the disk that is coated with rather complicated multilayer structure. The white diagonal strip across the picture and black lower right corner exemplify this type of defects

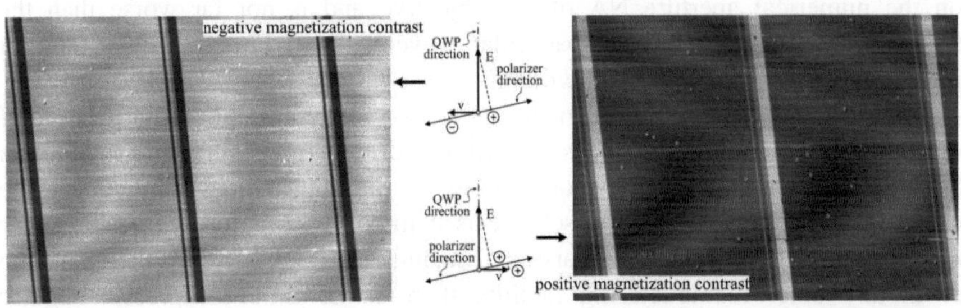

Fig. 8.18 Heterodyne technique (as well as the homodyne one) may be sensitive to magnetization direction. For that, polarizer must be slightly declined from the perpendicular to the direction of the quarter-wave plate (QWP). In heterodyne technique, QWP is needed to correct ellipticity of one of the partial waves (Fig. 8.8)

8.4 Magneto-Optical Liquids

All the modalities discussed above were developed for non-intrusive or non-contact magneto-optical imaging. If, however, the contact with the sample is not prohibited then there is one much simpler and handy tool to observe magnetic structures on microscopic scale of spatial resolution: magnetic fluids. Such fluids contain nano-particles of magnetic materials, like mono-grain iron oxide (Fe_xO_y), dispersed in a quickly drying solvent like isoparaffin or slowly evaporating carrier liquids like water. Size of the particle, typically 10 nm, is so small that they are in thermal equilibrium with the molecules of the surrounding liquid, and do not precipitate. Spatial resolution on magnetic structures is estimated as 20 μ. Vendors offer this product in two types of packages, optimized for two types of applications

Fig. 8.19 Magneto-optical fluids can be purchased either in a manicure-type bottle or as a felt-tipped marker

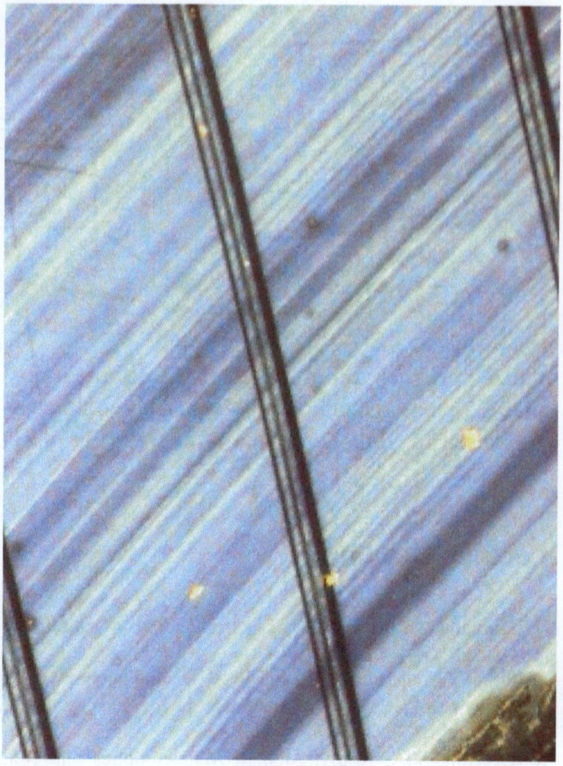

Fig. 8.20 Image of the same magnetic tracks as in Fig. 8.18 developed by magnetic fluid. Spatial resolution around 20μ can be easily obtained. White-light illumination

(Fig. 8.19). The marker contains quickly drying fluid that may be used for non-volatile magnetic structures. Open the cap, mark the magnetized surface evenly, and let it dry for several seconds. Then thin layer of wmagnetically-patterned and optically transparent film develops on the sample. Magnetic field distributes nano-particles in such a way that optical properties of the film change, and magnetic structure becomes visible, directly or through a microscope. If necessary, this film can be easily removed with a proper solvent and a soft tissue. Figure 8.20 helps to

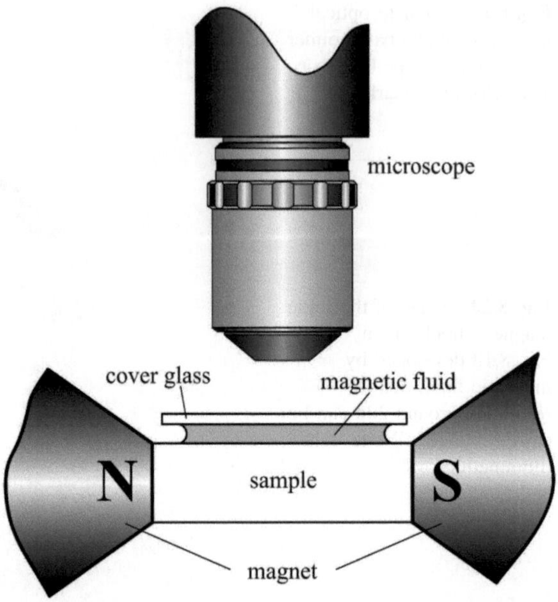

Fig. 8.21 In time-dependent phenomena, nano-particles, sustained in water, restructure to track changes of magnetization. This can be seen through the microscope. Be sure that the objective is designed for cover glass, otherwise the image will be blurred (Chap. 1)

realize how such a film makes magnetic structures visible through a conventional microscope. Larger amounts of water-based fluids available in bottles are supposed to be used for time-dependent processes (Fig. 8.21).

Further Reading

L. D. Landau, E. M. Lifshitz, Electrodynamics of Continuous Media, 2nd ed., Pergamon Press, Oxford, 1984.

W. Voigt, Magneto- und Electrooptik, B. G. Teubner Verlag, Leipzig, 1908.

G. S. Krinchik, Physics of magnetic phenomena, 2nd ed., Moscow State University Press, Moscow, 1985.

K. Zvezdin, V.A. Kotov, Modern Magnetooptics and Magnetooptical Materials, Institute of Physics Publishing, Bristol, UK, 1997.

V. Protopopov, Laser heterodyning, Springer, Heidelberg, 2009.

U. Ebels, Scanning Kerr Microscopy of Magnetic Domains in Epitaxial Thin Film Systems, PhD thesis, University of Cambridge, 1995.

Chapter 9
Spectrometers and Monochromators

With all the variety of spectrometers on the market, it is better to know what to request from the vendor in order not to have problems later in the laboratory. And, do not be afraid to make your own device or modify the existing one if necessary.

Abstract The first section of this chapter introduces three main options for compact fiber-optic spectrometers available on the market: the Czerny-Turner, the Wadsworth, and the Rowland schemes. This is followed by detailed derivation of the basic diffraction grating equation in blazed configuration and analysis of this formula and the formulas for angular and linear dispersions. Detailed figures with graphical representations are helpful to understand basic phenomenology and geometry of the spectrometers. Focusing optical systems provide imaging capabilities, however, aberrations are imminent. The most important practical problem that may be encountered is the order-sorting filter. The purpose and design principle of this element is comprehensively explained. The Wadsworth and the Rawland schemes are analyzed in full detail, emphasizing the curvature of the spectrum, which is particularly important in compact spectrometers that use flat linear detector arrays. Newly purchased spectrometer frequently requires calibration, which can be made with the help of an argon-mercury calibration module. Most frequently used argon and mercury calibration lines are summarized and listed in the table for quick reference. But it is not enough: general understanding of how the entire spectrum looks like is needed to correctly identify the lines. For reference purpose, a typical spectrum is graphically presented and calibration procedure is explained. Some particular doublets of the mercury spectrum may serve as a mean for estimating spectral resolution of a particular spectrometer, and the nomogram for that is provided. Some ways of increasing energy efficiency of fiber-optic spectrometers and typical mistakes are explained, like the use of fiber bundles with random organization of fibers. The second section is devoted to imaging spectrometers—devices that not only measure spectrum but also link it to spatial position of the source in one dimension. Interpretation of such a map may be tricky without initial experience, therefore the experimental pictures in two and three dimensions are presented with comprehensive explanation. The option of using imaging spectrometers for spectral interferometry (Chap. 6) is a very efficient way to measure one-dimensional profiles of microscopically patterned surfaces. The optical schemes and impressive examples are presented and explained in detail. The next section explains the concept, design, and typical features of gated intensified spectrometers. Starting from the principle, design, and parameters

V. Protopopov, *Practical Opto-Electronics*, Springer Series in Optical Sciences 184,
DOI: 10.1007/978-3-319-04513-9_9,

of microchannel image intensifiers, the reader is guided through mathematics, explaining advantages of optical amplification, to design features of a gated spectrometer that can be built in the laboratory. Experimental results, presented here, show that, with optical amplification around 4000, the intensified spectrometer offers significantly better signal-to-noise ratio than ordinary spectrometers. This section ends with experimental results, showing temporal resolution of spectral measurements equal to about one meter of optical fiber. The fourth section explains principles, features, and practical know-how of Fourier-transform spectrometers in visible domain. Although the idea of the Fourier spectroscopy is widely known, its implementation in visible domain is commonly considered as unreliable. The facts presented in this section oppose this stereotype. Explanation begins with explicit mathematical formulas that directly connect the Fourier transform array to the spectrum. This mathematical part is followed by the example of experimental spectrometer, assembled from standard optical elements available from ordinary vendors. Peculiarity of Fourier spectroscopy is that the resultant spectrum appears self-calibrated if the scanning stroke is known. Compensated and non-compensated beamsplitters, spectral resolution, importance of scanning linearity, phase measurements are explained. Typical experimental modulation curves and restored spectra are presented, including high-frequency spectra with phase-sensitive detection. The next section introduces the Fabry-Perot interferometer—actually a spectrometer with exceptionally high spectral resolution, commonly used to analyze mode structure of lasers. The section begins with the simplified theory of the interferometer with flat mirrors, introducing the Airy formula, followed by explanation of its practical use. Interferometers with flat mirrors are extremely unstable, therefore in practice they were superseded by confocal versions. The nature of unique stability of confocal interferometers is explained in full detail with comprehensive graphical representation. This theoretical part is followed by description of the design of commercially available interferometers, practical recommendations, and experimental oscilloscope traces. The last section of this chapter compares the two most frequently used types of monochromators: the grating monochromators and interference filters. Discussion of monochromators covers the following topics: formula of the monochromatic condition, linearity of tuning, sine-bar mechanism, higher-order ghosts in broad-band spectrum, experimental results. Their rivals—interference filters—are actually interferometers Fabry-Perot but on much smaller scale of separation. As such, they are also governed by the Airy formula, predicting multiple transmission peaks. To suppress them, practical devices are combined with colour glass filters and additional multilayer cavities. Typical experimental transmission curves finalize this chapter.

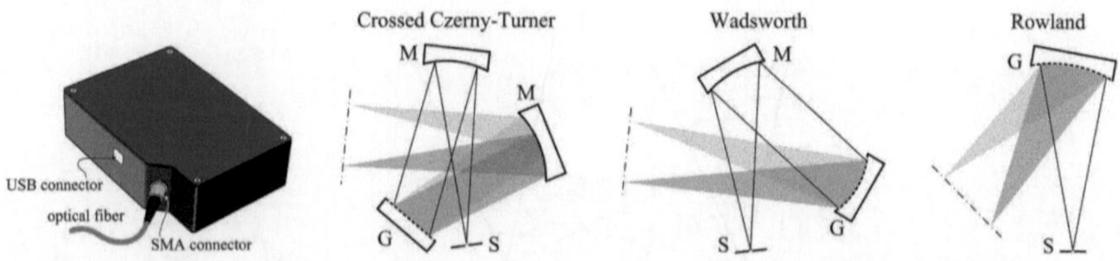

Fig. 9.1 Compact fiber-optic grating spectrometers are ready for use without any additional adjustment. Commonly, they arrive already calibrated at manufacturing site and do not need calibration. Calibration parameters are stored in operating program. Some versions allow changes of these parameters. Inside, there may be three different schemes explained in the text. «G» stands for the grating (flat or concave), «M»—for the mirror; «S»—for the slit. Spectrum plane is shown in *dashed-dotted line*

9.1 Compact Grating Spectrometers

Most of spectroscopic needs in general optical applications can be met by compact grating spectrometers. Sealed in small brick-size boxes, these devices require nothing more than an optical fiber with SMA connector (Chap. 7) and universal serial bus (USB) cable to the computer (Fig. 9.1).

Typically, compact grating spectrometers cover spectral range 200–1000 nm with spectral resolution 1 nm and exposure time varying from 5 ms to 5 s. Exposure time—the integration time of the photodetector—is the only parameter that can be altered in order to adjust necessary sensitivity. Sensitivity greatly depends on the width of the slit. Basically, the optical scheme of the spectrometer projects the image of the slit into the spectral plane that coincides with the photodetector. Therefore, the narrower the slit the better spectral resolution. Thus, sensitivity and spectral resolution are the competing factors. Practically, to achieve spectral resolution of about 0.5 nm, the slit must be around 5 μ, and this proportion is roughly a constant for compact wide-range spectrometers. For instance, if your spectrometer shows spectral resolution 1.5 nm, then its slit is about 20 μ.

Blazed diffraction grating—flat or curved—is the key part of the spectrometer. The term blazed means that each and everyone groove of the grating is tilted by a certain angle α, which concentrates the diffracted light selectively at a certain angle of reflection (Fig. 9.2). Since, in the spectrometer, each angle of reflection corresponds to a certain wavelength, this feature makes it possible to concentrate energy at a certain part of the spectrum, thus compensating for possible deficiencies of spectral response of the photodetector or reflectivity of the mirrors. Nowadays, blazed gratings are manufactured by deep lithography and subsequent slanted ion etching.

Let $F(\psi, \varphi, \varepsilon)$ be angular distribution of the wave with the wavelength λ diffracted on the slanted (blazed) part s of the grating period t. Then the complex amplitude E of the wave diffracted on the entire grating is the sum of $F(\psi, \varphi, \varepsilon)$ over all the N grooves taken with respective phases:

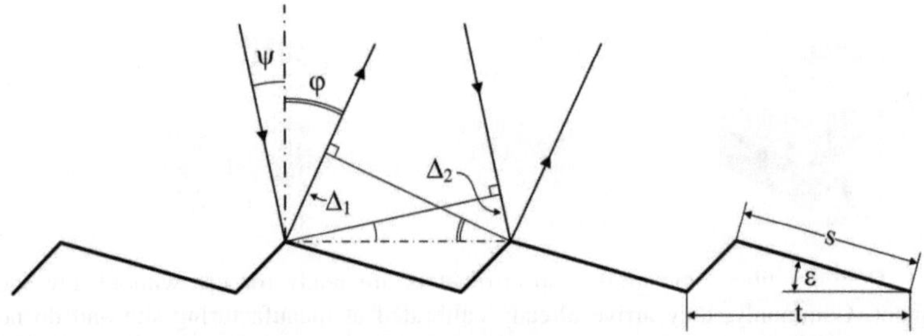

Fig. 9.2 Two processes determine angular profile of the reflected beam: diffraction on the blazed section s and interference of waves spaced periodically by t

$$E = F(\psi, \varphi, \varepsilon) \sum_{n=0}^{N-1} e^{-in\delta},$$

with $i = \sqrt{-1}$ and

$$\delta = \frac{2\pi}{\lambda}(\Delta_1 - \Delta_2) = \frac{2\pi}{\lambda}(t \sin\varphi - t \sin\psi).$$

We assume that short connecting parts of the profile between adjacent blazed surfaces do not contribute to the wave. Geometrical series sums to

$$1 + q + q^2 + \cdots + q^m = \frac{1 - q^{m+1}}{1 - q},$$

giving

$$E = F(\psi, \varphi, \varepsilon) \frac{1 - e^{-iN\delta}}{1 - e^{-i\delta}}$$

and intensity $I = |E|^2$

$$I = |F(\psi, \varphi, \varepsilon)|^2 \frac{\sin^2 N\nu}{\sin^2 \nu}, \qquad \nu = \frac{\pi}{\lambda}\, t\, (\sin\varphi - \sin\psi).$$

The function

$$\frac{\sin^2 N\nu}{\sin^2 \nu}$$

is periodic with narrow peaks of the width $2\pi/N$ as shown in Fig. 9.3. Each maximum corresponds to

$$\nu_m = \mp m\, \pi, \, m = 1, 2, 3, \ldots.$$

The values of m are called the orders of diffraction. Thus, the angles φ_m, at which the intensity is concentrated, are wavelength-dependent, and this is how diffraction

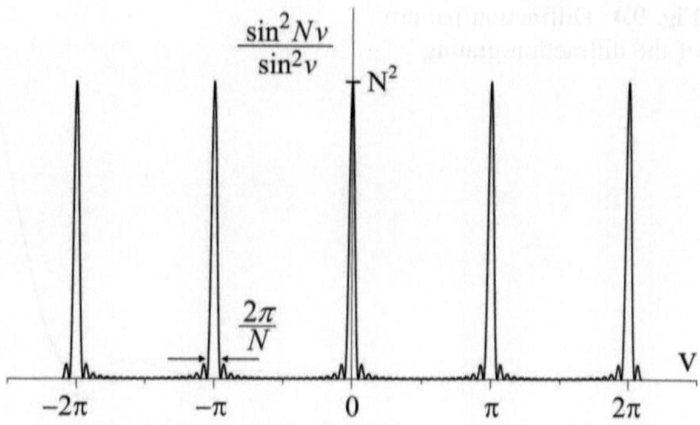

Fig. 9.3 Interference pattern of the diffraction grating

grating decomposes light into its spectral components. Note, that upon the convention, orders with $\varphi < \psi$ are called the positive orders, whereas the negative ones are those with $\varphi > \psi$. Positive orders are commonly used.

Angular distribution $F(\psi, \varphi, \varepsilon)$ is proportional to the wave-propagation integral over s, which reduces to one dimension because the grating is a one-dimensional object:

$$\int_0^s E(x)\, e^{-i\frac{2\pi}{\lambda}r(x)}\, dx,$$

where x is the coordinate along s and r is the distance to the point of observation. In Fraunhofer diffraction, when only plane waves are considered, the observation point is placed at infinity, making $r(x) = r(0) - x \sin\varphi$. Then, disregarding constant phase introduced by $r(0)$, the integral transforms to

$$\int_0^s e^{i\frac{2\pi}{\lambda}x[\sin(\varphi-\varepsilon)-\sin(\psi+\varepsilon)]}\, dx = se^{iu}\frac{\sin u}{u},\ u = \frac{\pi s}{\lambda}[\sin(\varphi-\varepsilon) - \sin(\psi+\varepsilon)],$$

and

$$|F(\psi, \varphi, \varepsilon)|^2 = s^2\frac{\sin^2 u}{u^2}.$$

This function maximizes at $u = 0$ and is shown in Fig. 9.4. With obvious trigonometric manipulations,

$$u = \frac{\pi s}{\lambda} 2 \sin\frac{\varphi - \psi - 2\varepsilon}{2} \cdot \cos\frac{\varphi + \psi}{2}.$$

To better feel the physical meaning of this result, consider the particular case of $\varphi \approx \psi$ and $\varepsilon = 0$—reflection on a narrow stripe of the width s:

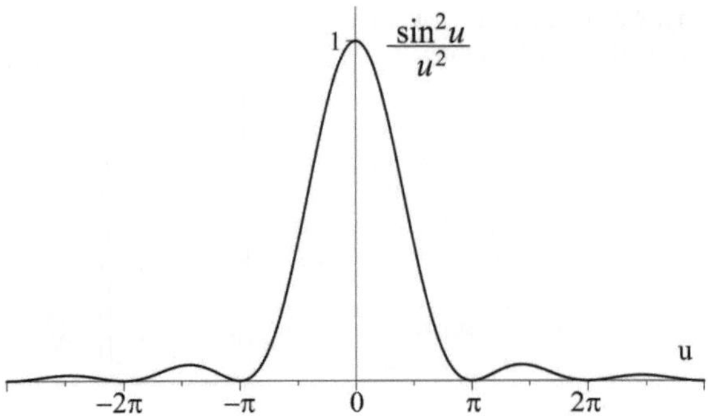

Fig. 9.4 Diffraction pattern of the diffraction grating

$$u \approx \frac{\pi s \cdot \cos \psi}{\lambda} (\varphi - \psi).$$

Then the function $|F|^2$ has the central maximum around $\varphi \approx \psi$ (meaning that the angle of incidence is equal to the angle of reflection) with the angular width at half-maximum

$$\approx \frac{\lambda}{s \cdot \cos \psi}.$$

It is the same as in diffraction on a slit of the width $s \cdot \cos \psi$—projection of the stripe width s on the wavefront of the incident wave.

Returning to the intensity reflected from the blazed grating, we see that it is proportional to the product of the two functions

$$I(\varphi) = \frac{\sin^2 u}{u^2} \cdot \frac{\sin^2 N v}{\sin^2 v}, \qquad v = \frac{\pi}{\lambda} t (\sin \varphi - \sin \psi),$$

$$u = \frac{\pi s}{\lambda} 2 \sin \frac{\varphi - \psi - 2\varepsilon}{2} \cdot \cos \frac{\varphi + \psi}{2}.$$

The overlapping pattern is shown schematically in Fig. 9.5. Thus, choosing the blaze angle ε, it is possible to direct optical power not only to a particular diffraction order of interest, but also to location of the particular spectral component in this order. For that, it necessary to satisfy the condition $u = 0$ at the specified diffraction angle ψ, which means $\varphi - \psi = 2\varepsilon$.

The most important parameter of a flat diffraction grating is its angular dispersion D_φ

$$D_\varphi = \frac{d\varphi}{d\lambda}.$$

It determines the angle by which the grating separates close wavelengths. Differentiating the equation $v_m = \pm\, m\pi$ over λ and understanding that $\psi = const$, obtain

$$D_\varphi = \frac{m}{t \cos \varphi}.$$

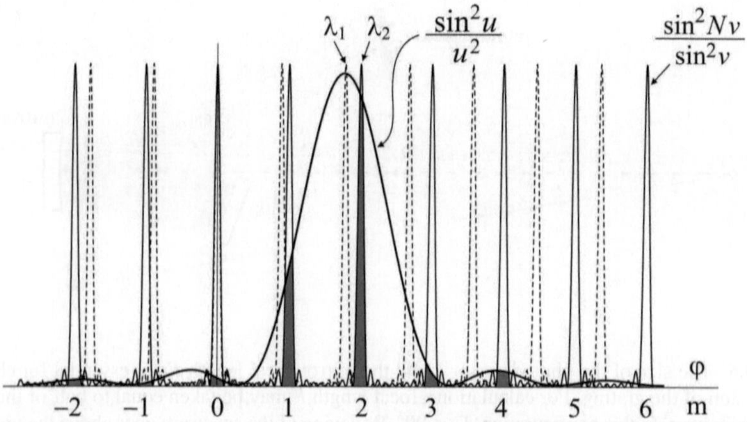

Fig. 9.5 Blazed grating may be designed to support any particular spectral component, for example wavelength λ_1 in the second order (*dashed line*). Then other components will be attenuated, like λ_2 in the same order (*solid line*), or strongly suppressed like other orders

Thus, the smaller grating spacing t the bigger angular dispersion. In catalogs and specifications on gratings, instead of direct values of t another equivalent parameter is used: number of grooves per millimeter, which ranges typically from 100 to 1200 grooves per millimeter. With it, angular dispersion can be easily calculated. Another important conclusion that follows from this formula is that angular dispersion increases with the diffraction order m. Therefore, when very high spectral resolution is needed, researches work in high diffraction orders. However, for general purpose spectrometers even more important parameter is the wavelength range $\Delta\lambda$. This wavelength span, when focused onto photodetector, must fit the length L of its sensitive area that is typically around 30 mm (Fig. 9.6):

$$D_\varphi \Delta\lambda \cdot F \approx L \sin\Gamma, \text{ or } L \approx \frac{D_\varphi F}{\sin\Gamma} \Delta\lambda.$$

The dimensionless quantity

$$D_l \equiv \frac{L}{\Delta\lambda} = \frac{D_\varphi F}{\sin\Gamma}$$

is named linear dispersion, and is usually measured in [mm/nm]. Sometimes, the reciprocal value is used, measured in [nm/mm]. Usually, the acute angle Γ is close to 90°, and the sine in denominator may be considered as unity.

With all the aforementioned, it is easy to choose particular grating needed for particular application. For example, we need to cover wavelength range from 200 to 800 nm, i.e. $\Delta\lambda = 600$ nm. Dimensions of the spectrometer give the first estimate for F: typically it is about 80 mm. Photodetector length L is fixed even more precisely: 30 mm is almost a standard. So, linear dispersion of the grating should be 30 mm/600 nm = 0.05 mm/nm, and angular dispersion 0.05/80 = 620 mm^{-1}.

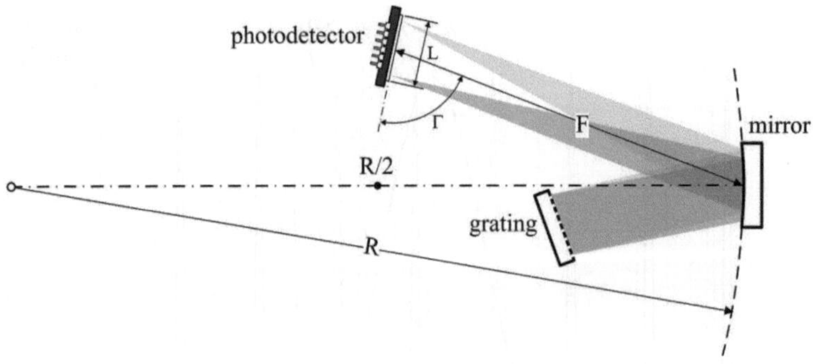

Fig. 9.6 The size of the photodetector L and the mirror focal length F are essential for choosing dispersion of the grating. For calculations, focal length F may be taken equal to half of the radius R of the mirror. In this configuration, $\Gamma < 90°$. Red wing of the spectrum goes above the violet one

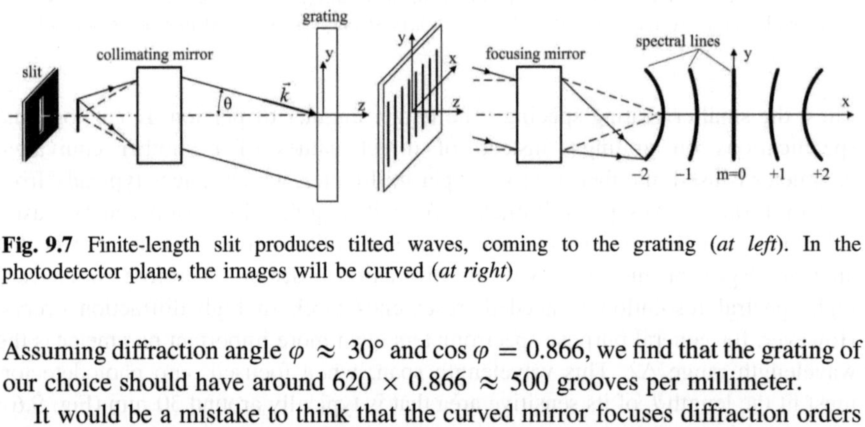

Fig. 9.7 Finite-length slit produces tilted waves, coming to the grating (*at left*). In the photodetector plane, the images will be curved (*at right*)

Assuming diffraction angle $\varphi \approx 30°$ and $\cos\varphi = 0.866$, we find that the grating of our choice should have around $620 \times 0.866 \approx 500$ grooves per millimeter.

It would be a mistake to think that the curved mirror focuses diffraction orders exactly into point-like spots located along a straight line—the photodetector plane. Not at all: curved mirror is an aberrated imaging element, especially at off-axial illumination as shown in Fig. 9.6, which may be called geometrical distortion, and diffraction grating produces additional distortion of purely spectral nature. The nature of the latter can be better understood with the scheme in Fig. 9.7. The ends of the slit produce plane waves, coming at finite angle θ to the grating. It means that normal component of the wave vector $\vec{k}$ decreases to

$$k_z = \frac{2\pi}{\lambda}\cos\theta = \frac{2\pi}{\lambda'}$$

with $\lambda' > \lambda$. As such, these waves will diffract at larger angles φ, and will be focused farer from the zero-order point. This leads to curviness of spectral lines in the plane of the photodetector as shown in Fig. 9.7. The longer the slit the more it affects spectral resolution. However, for compact fiber-optic spectrometers with the height of one sensitive element of the photodetector of only 200 μ, this phenomenon is not critical.

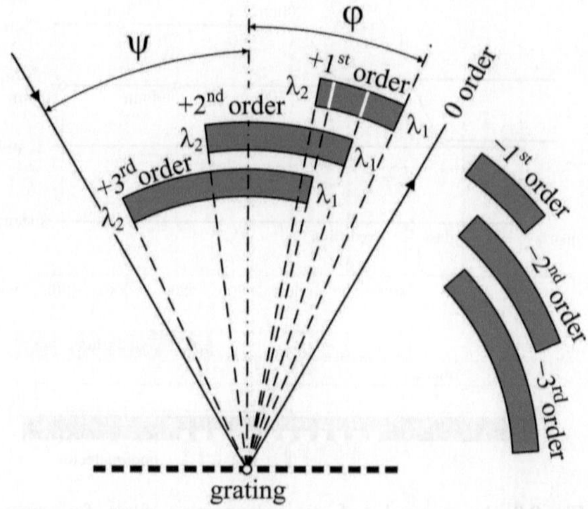

Fig. 9.8 In wide-range spectrometers, multiple diffraction orders may angularly overlap. The photodetector size L, angular dispersion of the grating D_{φ}, and focal length of the mirror F (Fig. 9.6) are designed to intercept the 1st order in between the wavelengths λ_1 and λ_2. Multiple orders may also be focused here, as indicated by *white* marks on the 1st order section

What is really critical and causes numerous mistakes, is the overlapping of diffraction orders. The basic diffraction grating equation

$$\sin\varphi - \sin\psi = \mp m\frac{\lambda}{t}$$

may be presented schematically in the form shown in Fig. 9.8. For small $\varphi < 30°$, $\sin\varphi \approx \varphi$, and approximately for positive orders

$$\varphi \approx \psi - m\frac{\lambda}{t}.$$

Thus, when $2\lambda_1 \leq \lambda_2$, the 2nd order meddles into the interval of the 1st order diffraction. If $3\lambda_1 \leq \lambda_2$ then the 3rd order appears in this interval, and so on. For instance, the spectrometer designed to work between 200 and 650 nm would show the strongest 253 nm mercury line not only at 253 nm but also at 506 nm, making it impossible to identify real spectrum. Or, in case of a wide-range source with strong ultra-violet (UV) emission, like xenon lamp (Chap. 2), the entire UV wing around 300 nm will relocate to visible part of the spectrum around 600 nm, confusing everything. The majority of compact spectrometers, being in use in laboratories, cover very wide spectral interval between 200 and 800 nm or even 1100 nm. Then what do they do to eradicate higher-orders intrusion? The solution is the so-called order-sorting filter.

There are two types of order-sorting filters, being used in compact fiber-optic spectrometers: the detachable glass plates and thin-film coating of the photodetector input window. Principle of functionality is the same, but the first type is more versatile. It consists of two, maximum three, glass plates, one on top of another, geometrically and spectrally designed to block multiple orders from reaching the photodetector (Fig. 9.9).

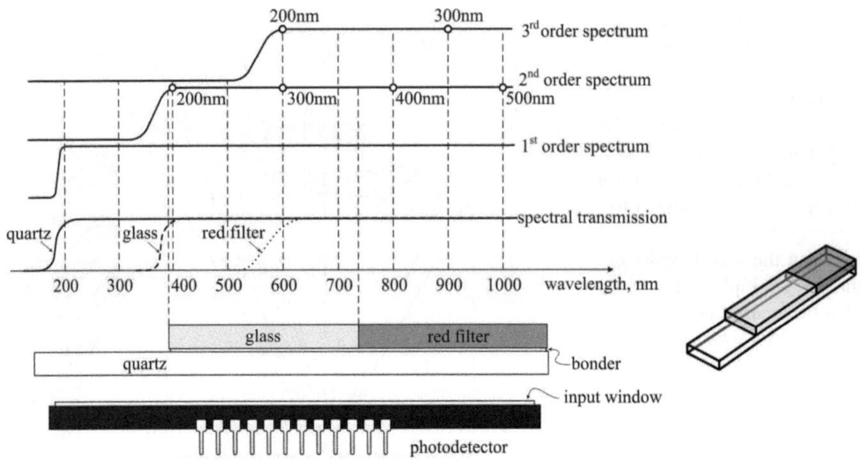

Fig. 9.9 An example of an order-sorting filter for spectral interval 200–1000 nm. Two materials—glass and red filter—sequentially block the second and third orders from reaching the photodetector. Glass blocks all the spectral components below 400 nm, and the red filter does the same below 600 nm. The glass section of a proper length absorbs UV radiation in the interval 200–350 nm that might be able to reach specific area of the photodetector to which all the visible components from 400 to 700 nm are focused. The red section shields the photodetector area assigned for 700–1000 nm, blocking both the UV radiation, infiltrating with the third order, and the visible light, coming through the second order. Quartz plate does not block anything but serves as a support for the two other sections. The input spectrum never contains components below 200 nm as they are fully attenuated by a feeding optical fiber (Chap. 7)

Order-sorting filter is a very important point of negotiations when ordering a new spectrometer. Manufacturers offer great variety of diffraction gratings for their spectrometers, listing all the necessary details in their catalogs, but never mention the necessity of the order-sorting filters. This commonly leads to a very pitiful situation when the user calculates parameters of the grating, satisfying his needs in spectral range and blazed wavelength, and then just orders the spectrometer with this grating without even mentioning the order-sorting filter. The aftermaths are dramatic: total confusion of spectral data and the vendor does not accept any responsibility because the filter was not formally ordered.

We discussed the flat type of diffraction gratings—the most easily manufactured one, and the crossed Czerny-Turner scheme shown in Fig. 9.1 is commonly used with them in compact spectrometers. This scheme is named «crossed» because another Czerny-Turner scheme exists, in which optical beams do not intersect (Sect. 9.6). However, the uncrossed Czerny-Turner scheme requires more space, and therefore is not used in compact spectrometers. Requirements of small space and least possible number of optical components—the most essential for mass production—dictate usage of curved diffraction gratings that perform two functions simultaneously: dispersion and focusing. It is even more important because reduction of the number of optical components decreases scattered light and

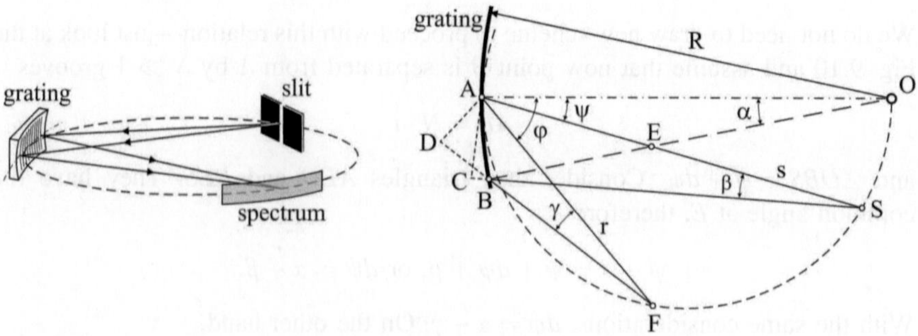

Fig. 9.10 In the Rowland scheme, the entrance slit and the spectrum are positioned on the circle (*dashed line*), touching the diffraction grating (*at left*). This *circle* is called the Rowland *circle*. Diameter of it is equal to the radius R of the grating. On this *circle*, every point S is imaged into another point F—dispersed or not dispersed (*at right*)

aberrations. The basic scheme of that kind is known as the Rowland circle (Fig. 9.10). Consider a source of monochromatic rays with wavelength λ at the point S. Let the A and B be the two adjacent grooves on the grating and O—the center of the grating curvature with the radius R. The rays SA and SB make the angles ψ and $\psi + \Delta\psi$ with the local normals to the grating (lines OA and OB). Diffracted rays AF and BF come to the point of focusing F at the angles φ and $\varphi + \Delta\varphi$ to the local normals OA and OB. Constructive interference of these rays takes place when

$$(SA + AF) - (SB + BF) = m\,\lambda; \quad m = 0, \pm 1, \pm 2, \ldots$$

Continue the sections SB and BF to the points C and D to make their lengths equal to SA and AF. Then

$$(SA + AF) - (SB + BF) = BC + BD.$$

Since the single groove width $t \ll R$, the angles $\beta \ll 1$ and $\gamma \ll 1$, which makes $\angle ACB \approx \pi/2$, $\angle ADB \approx \pi/2$ and $\angle CAB \approx \psi$, $\angle DAB \approx \varphi$. As such, $BC \approx t \cdot \sin\psi$ and $BD \approx t \cdot \sin\varphi$. Then the basic grating equation transforms to

$$t\,(\sin\varphi + \sin\psi) = m\,\lambda; \;\; m = 0, \;\; \pm 1, \;\; \pm 2, \ldots$$

It is the same equation as for the flat grating, taking into consideration that, according to the previous convention, the angle φ is negative (compare to Figs. 9.2 and 9.8).

Next, we are going to show that the Rowland scheme possesses the quality of focusing, i.e. all the rays from S with the same wavelength λ and slightly different angles of incidence $\psi + d\psi$ come to the same point P within the same diffraction order m. If the angle of incidence ψ changes then the angle of diffraction φ also changes to $\varphi + d\varphi$, according to the basic grating equation. To find $d\varphi$ as a function of $d\psi$, we have to take differential of the above equation:

$$\cos\varphi \cdot d\varphi + \cos\psi \cdot d\psi = 0.$$

We do not need to draw new scheme to proceed with this relation—just look at the Fig. 9.10 and assume that now point B is separated from A by $N \gg 1$ grooves

$$AB = N \cdot t$$

and $\angle OBS = \psi + d\psi$. Consider then triangles AEO and BES. They have the common angle at E, therefore

$$\psi + \alpha = \psi + d\psi + \beta, \text{ or } d\psi = \alpha - \beta.$$

With the same considerations, $d\varphi = \alpha - \gamma$. On the other hand,

$$\alpha = \frac{Nt}{R}, \quad AD = Nt \cdot \cos\varphi; \quad AC = Nt \cdot \cos\psi; \quad \beta = \frac{AC}{s}; \quad \gamma = \frac{AD}{r}.$$

Altogether gives the differential

$$\cos\varphi \cdot \left(\frac{1}{R} - \frac{1}{r}\cos\varphi\right) + \cos\psi \cdot \left(\frac{1}{R} - \frac{1}{s}\cos\psi\right) = 0.$$

For this equation to be satisfied for an arbitrary ψ and φ, the following must hold true:

$$r = R\,\cos\varphi \quad \text{and} \quad s = R\,\cos\psi.$$

This is the equation of a circle with the diameter R, which proves focusing properties of the Rowland circle. It may be useful to understand that not only spectrally dispersed components with $m = \pm 1, \pm 2, \ldots$ are focused on this circle but also zero-order wave $m = 0$, which means focusing property of an ordinary spherical mirror, not necessarily the diffraction grating.

The analysis above considers only meridional rays, i.e. the rays in the plane perpendicular to both grating and grooves. The skew rays or sagittal rays, i.e. the rays that propagate at some angle to this plane, do not focus in the same point as the meridional rays. This produces the aberration called astigmatism. Astigmatism of the Rowland scheme is stronger the bigger the numerical aperture of the grating. The Wadsworth scheme (Figs. 9.1 and 9.11) produces much smaller astigmatism. In it, the grating is illuminated by a parallel beam of rays. As before, constructive interference of these rays takes place when

$$(CD + r) - (AB + \rho) = m\lambda; \qquad m = 0, \pm 1, \pm 2, \ldots.$$

Applying standard trigonometric identities, we have

$$CD - AB = t \cdot \sin(\alpha + \beta).$$

The next goal is to determine r and ρ. For that, solve the triangle ODF to find

$$y = \sqrt{R^2 - 2Rr\,\cos(\theta - \beta) + r^2}, \quad \sin\varepsilon = \frac{r\,\sin(\theta - \beta)}{\sqrt{R^2 - 2Rr\,\cos(\theta - \beta) + r^2}}.$$

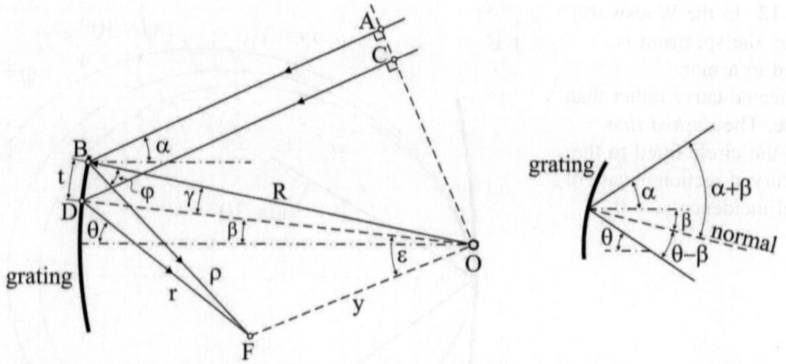

Fig. 9.11 In the Wadsworth scheme, curved diffraction grating is illuminated by a parallel beam of rays, represented by *AB* and *CD*. Points *B* and *D* on the grating are spaced by one groove step *t*. Then the dispersed rays are focused in *F*. Point *O* is the center of grating curvature with the radius *R*

Similarly, solve the triangle *OBF* to find

$$\rho = \sqrt{R^2 - 2Ry\cos(\varepsilon + \gamma) + y^2}.$$

Since $t \ll R$,

$$\cos(\varepsilon + \gamma) \approx \cos\varepsilon - \frac{t}{R}\sin\varepsilon.$$

The next sequence of calculations includes the following:

$$\rho - r \approx \frac{yt}{\sqrt{p}}\sin\varepsilon \text{ and } p = R^2 + y^2 - 2Ry\cos\varepsilon \text{ with the same condition } t \ll R;$$
$$p = r^2.$$

With this, the initial equation transforms to

$$\sin(\alpha + \beta) - \sin(\theta - \beta) = \frac{m\lambda}{t},$$

which is again the basic diffraction grating equation as it follows from the right scheme in Fig. 9.11. The angle θ, to where diffracted light is focused, is the function of the wavelength λ.

Now, we are going to find the radius of focusing r and, applying the same formalism as before, assume that the point D is separated from B by a macroscopic space of $N \gg 1$ grooves: $N \cdot t$. The differential of the above equation over the two independent variables $\theta + d\theta$ and $\beta + d\beta$ with

$$d\beta = \frac{Nt}{R}, \quad d\theta = \frac{Nt}{r}\cos(\theta - \beta)$$

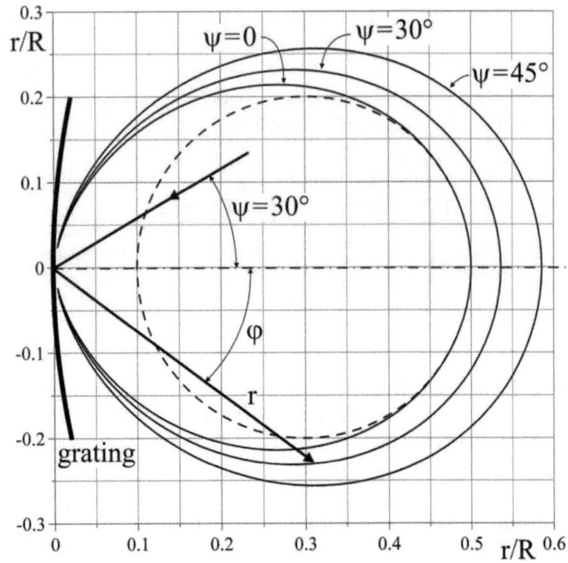

Fig. 9.12 In the Wadsworth scheme, the spectrum is focused to a more complicated curve rather than a circle. The *dashed line* shows the circle fitted to the most curved section in case of normal incidence $\psi = 0$

gives

$$r = \frac{R\ \cos^2(\theta - \beta)}{\cos(\alpha + \beta)\ + \cos(\theta - \beta)}.$$

Since the radius of focusing r does not depend on rotation of the entire picture around the center of curvature O, we may rotate it until $\beta = 0$. In this configuration, the line OD coincides with horizontal axis, and the angle α becomes the angle of incidence ψ that we introduced in Fig. 9.2. Similarly, the angle θ becomes the angle of diffraction φ, and finally we obtain:

$$r = \frac{R\ \cos^2 \varphi}{\cos \psi\ + \cos \varphi}.$$

This is not a circle like in the Rowland case, and the line determined by this formula is shown in Fig. 9.12.

For practical reasons, it is interesting to realize the scale of curviness that the Wadsworth scheme provides. From Fig. 9.12 follows that its maximum corresponds to $\psi = 0$ and $\varphi = 0$. In polar coordinates $r(\varphi)$, radius of curvature ν is determined by the following formula:

$$\nu = \frac{r^2(0)}{r(0) - r''(0)} = \frac{R}{5}.$$

Thus, the minimum radius of field curvature in the Wadsworth scheme is 2.5 times less than in the Rowland scheme, which is not a very good news. However, recently developed technology of varied line-space (VLS) gratings minimizes field

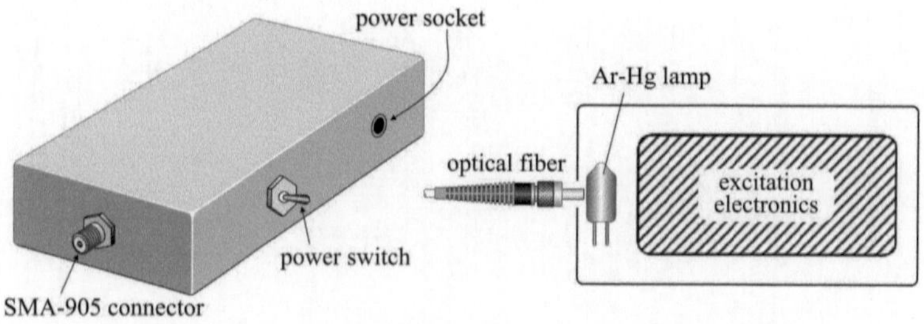

Fig. 9.13 The argon-mercury calibration module contains a miniature lamp and high-voltage excitation circuit powered by any low-voltage source of 12–24 V. Designed to work with the fiber-optic spectrometers, the output is coupled to a fiber by a standard SMA-905 connector. Normally, no special optical interface system is needed to couple the fiber to the lamp

curvature in the Wadsworth scheme. Well-defined variation of spacing between straight parallel grooves from one side of the grating to another flattens the surface to which the dispersed rays are focused. In our mathematical calculations above, it means that t varies smoothly over the grating. The zero-order focus has no relation to t, and therefore the curved mirror would focus on the curved surface as described above. But diffracted (dispersed) rays focus in nearly a plane.

Although fundamental for large-scale spectrometers, the Rowland scheme could not be used in compact versions for a long time because high numerical aperture of the grating produced big astigmatism and field curvature. The very compact Rowland scheme in Fig. 9.1 is the product of recent technological developments like independent surface profiling in meridional and sagittal planes and VLS: the first one minimizes astigmatism and the second flattens field of view. For such gratings, only linear dispersion makes sense and it is always carefully specified by manufacturer together with relative positions of slit and focal plane.

Raw spectral information is retrieved from the spectrometer as a one-dimensional array s_j of electrical signals detected in the j-th pixel of a line detector array, commonly the charge-coupled device (CCD) photodetector. The number of pixels is typically 1024, 2048, or 3648, and a single-pixel geometry is a rectangle $a \times 200\ \mu^2$. With the length of such an array being always around 30 ± 1 mm, the width a can be easily calculated. Next, the pixel number j must be associated with the wavelength λ, and for that the calibration procedure should be applied. As a rule, spectrometers are shipped already factory-calibrated, and no additional calibration is needed. However, with time or by any other reason, calibration may be disturbed, and then it should be restored. The best tool for that is an argon-mercury lamp, readily available from the same vendor that supplied the spectrometer itself (Fig. 9.13). In it, mercury covers the UV and visible parts of spectrum, whereas argon fills the near-infrared interval (Fig. 9.14). Relative intensities of argon and mercury lines may drift significantly with time as chemical reactions develop inside the bulb. Nevertheless, this process does not affect stability of specific spectral lines of the both gases. Argon and mercury characteristic lines are exceptionally well calibrated and readily available in the literature.

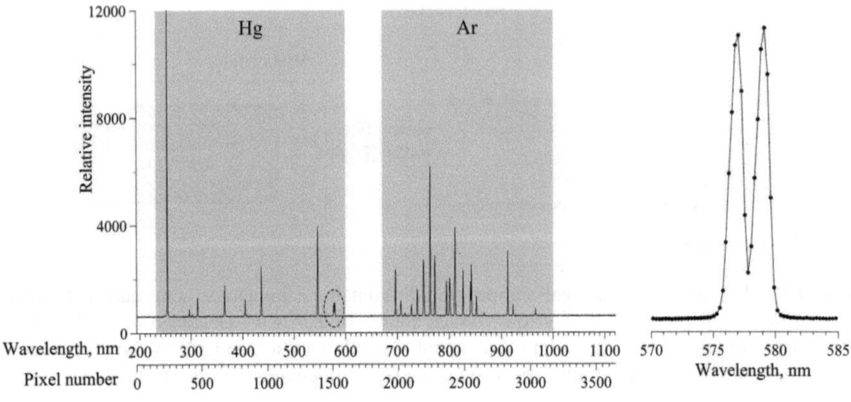

Fig. 9.14 Typical calibration spectrum of the Ar-Hg lamp. The doublet 576.960 and 579.066 nm confined within the *dashed ellipse* is magnified separately at right. Table 9.1 presents wavelengths of clearly seen *spectral lines*

Table 9.1 Most frequently used argon and mercury spectral lines

Hg	253.652, *296.728*, 313.155, *365.015*, 404.656, *435.833*, 546.074, *576.960*, 579.066
Ar	696.543, *706.722*, 727.294, *738.398*, 750.387, *763.511*, 772.376, *794.818*, 800.616, *811.531*, 826.452, *842.144*, 912.297, *922.450*

However, do not think that the knowledge of wavelengths alone is enough: the real practical problem, especially for the beginners, is to identify the designated spectral line in the spectrum, especially among the argon lines. These lines are narrowly spaced, and mixing the one for its neighbor is easy if not to take into consideration their relative amplitudes. Therefore, the entire picture of relative intensities like the one presented in Fig. 9.14, is very helpful. Because of possibly different quantum efficiencies of CCDs, blazing efficiencies of gratings, design of order-sorting filters, and other particular details each spectrometer has its own relative intensity distribution. That is why the picture shown in Fig. 9.14 should be considered only as a typical one, with significant variations quite possible.

After calibration spectrum is recorded and characteristic lines identified, the function $\lambda(j)$—the wavelength λ in nanometers as the function of the pixel number j—should be defined. The question is how to know the pixel number, corresponding to specific wavelength. For that, the operating program supplied with the spectrometer always indicates pixel numbers along with the wavelength. It is not necessary to identify all the lines that are visible in the spectrum: 4–5 most distantly located points will suffice. The farer the two end points are located from one another, the better the precision of calibration will be. Therefore, commonly strong 253.652 nm mercury and 763.511 nm argon lines terminate the set of points. The resultant two-column table is used for calibration, which is commonly performed automatically by the operating program, using the calibration option.

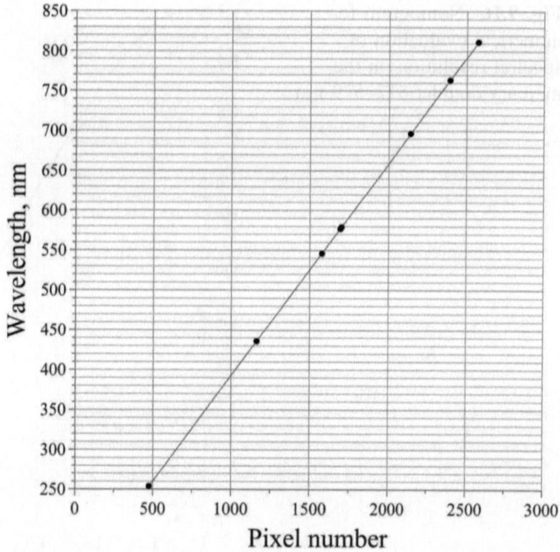

Fig. 9.15 Calibration curve for 100 mm focal length grating and 3648 pixels CCD photodetector. *Dots* mark the points of the table, the line is the *least-square* linear fitting. The mercury doublet 577–579 nm is also included in calibration table—these two *dots* almost coincide

What this option actually does is the least-square fitting of a polynomial into the set of points of the calibration table.

With all the explanation above about focusing of the spectrum, the function $\lambda(j)$ should be expected non-linear. And in general it is, although in relatively narrow spectral intervals and for long-focused spectrometers non-linearity may be hardly noticeable. For example, Fig. 9.15 shows calibration points measured with the diffraction grating, having 100 mm focus. Simple linear function fits exceptionally well, and if only it were known beforehand then only two points would suffice. However, linearity is never guaranteed, therefore at least three points are needed to fit the second-order polynomial.

Calibration source with argon-mercury lamp is also useful for testing spectral resolution of the spectrometer. Some regretful cases are known when even new spectrometers failed to show specified spectral resolution. The test is simple: turn on the lamp and record the spectrum around the mercury doublet 577–579 nm. Spacing between these lines is almost exactly 2 nm. According to the Rayleigh criterion, two spectral lines are considered resolved if they produce the gap in between of them more than 20 % of the values in maxima. For example, the gap of the doublet shown in Fig. 9.14 is 80 %, meaning that spectral resolution of this spectrometer is significantly better than 2 nm. This simple assessment technique may be converted from the domain of guessing to numerical evaluation if we recall that theoretical shape of a single spectral line with the wavelength λ_0 is defined by diffraction on the entrance slit:

$$\left[\frac{\sin a(\lambda - \lambda_0)}{a(\lambda - \lambda_0)}\right]^2,$$

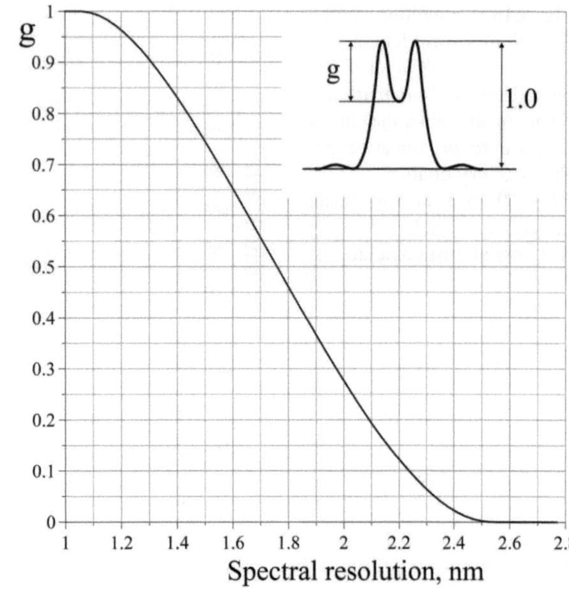

Fig. 9.16 Nomogram for numerical evaluation of spectral resolution on the mercury doublet 577–579 nm

where λ is the wavelength and a is the unknown scaling parameter. Then, computing the sum of two of these functions nested at λ_0 and $\lambda_0 + \Delta\lambda$ and evaluating the relative gap g in the middle as a function of $\Delta\lambda$, it is easy to find that the 20 % Rayleigh criterion satisfies when

$$a = \frac{\pi}{\Delta\lambda}.$$

It means that the spectrometer with spectral resolution $\Delta\lambda$ forms the shape of a single spectral line λ_0 as

$$\left[\frac{\sin\frac{\pi}{\Delta\lambda}(\lambda - \lambda_0)}{\frac{\pi}{\Delta\lambda}(\lambda - \lambda_0)}\right]^2.$$

The next step is to take $\lambda_0 = 576.960$ nm for one line and $\lambda_0 = 579.066$ nm for another and compute g as a function of $\Delta\lambda$. The result in the form of a nomogram is presented in Fig. 9.16. It can be used to quantitatively estimate spectral resolution of spectrometers. For instance, spectral resolution of the spectrometer that was used to obtain the data in Fig. 9.14 may be estimated as 1.4 nm—a very good value for a wide-range compact spectrometer.

Spectral resolution and sensitivity of a spectrometer are two competing parameters. All the three typical optical schemes shown in Fig. 9.1 can be presented in a generalized form as in Fig. 9.17. Since the space is limited and the total length $L + F$ may be considered constant, equal maxima of both L and F, necessary to minimize numerical apertures and minimize aberrations, are reached

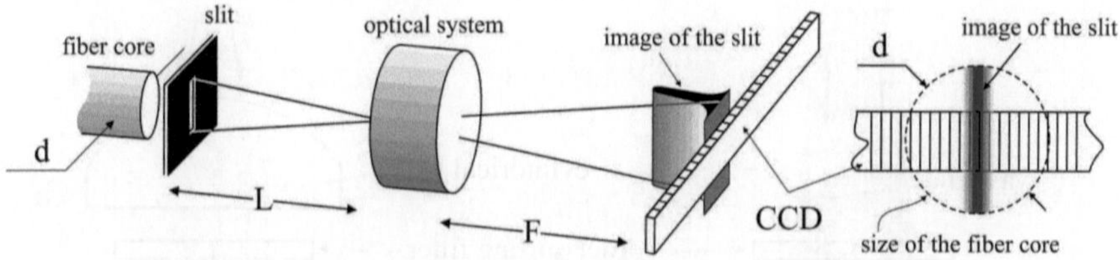

Fig. 9.17 In a spectrometer, photodetector receives the image of the slit magnified by the factor F/L

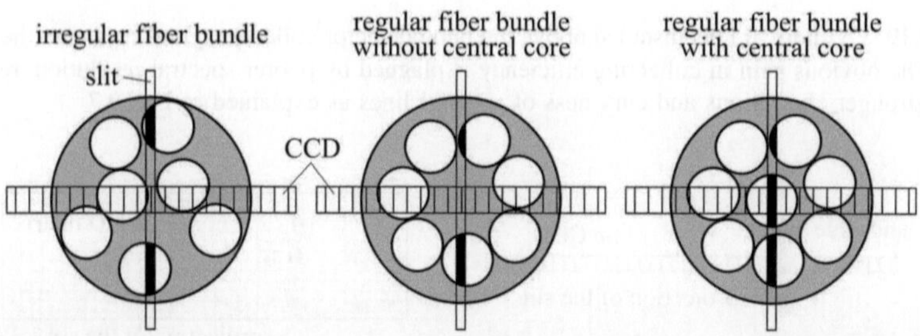

Fig. 9.18 Standard SMA-905 connectors may terminate optical fiber bundles with multiple cores. Only those of them that have a central core deliver light to CCD (charge-coupled device)

when $L = F = (L + F)/2$. This corresponds to unity magnification, meaning that the slit is projected onto CCD one-to-one. From the point of view of sensitivity, the more optical flux comes through the slit the better, and thus the wider the slit the more sensitive the spectrometer. On the other hand, slit width limits spectral resolution since it is projected one-to-one. For example, for a wide-range spectrometer designed to work from 200 to 800 nm with spectral resolution 1 nm on 30 mm long CCD, the slit width must be less than 30 mm/(600/1 nm) = 50 μ. The priority is always spectral resolution, therefore, in wide-range compact spectrometers slit width never exceeds 30 μ. But even this infinitesimal amount of energy is not used completely because diameter of the fiber core d exceeds transversal dimension of the CCD (Fig. 9.17). In order to ensure axial alignment of the optical fiber to the slit within manufacturing tolerances, typical core diameter must be $d > 400$ μ, whereas the height of the CCD pixel is only 200 μ, wasting half of the usable optical flux. Therefore, it is an illusion to think that thicker optical fiber, like 1 mm core, may improve sensitivity of your spectrometer. However, the thick-core optical fiber is not a mistake, whereas the use of a fiber bundle may be real mistake. This is explained in Fig. 9.18. Only fiber bundles with central core may work reliably.

Understanding deficiency of sensitivity, manufacturers offer an option that increases efficiency of transferring light to photodetector at the expense of somewhat poorer spectral resolution (Fig. 9.19). Not everyone manufacturer offers this option, and preliminary enquiries should be made prior to ordering.

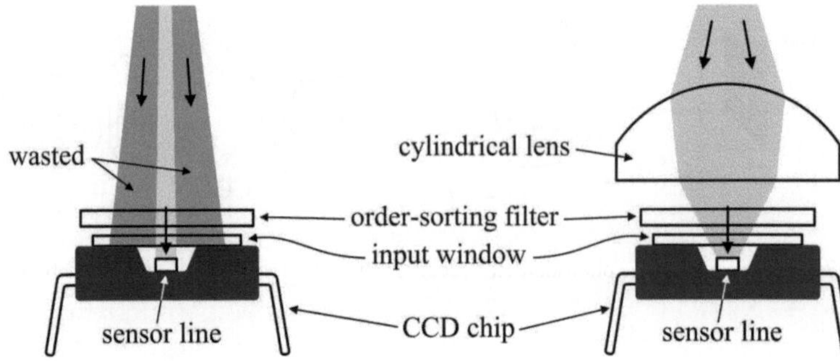

Fig. 9.19 Cylindrical lens installed above the photodetector collects sagittal rays onto the sensor line. The obvious gain in collecting efficiency is plagued by poorer spectral resolution, resulting from stronger aberrations and curviness of spectral lines as explained in Fig. 9.7

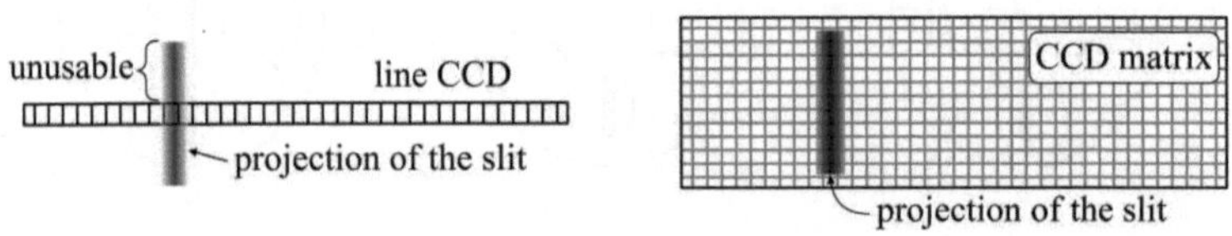

Fig. 9.20 With the CCD matrix, the entire optical flux may be intercepted. The read-out protocol combines only vertical elements of the matrix, which is often called «full vertical binning». (FVB). The noise amplitude in the resultant signal also increases as it sums individual noise signals in each vertical pixel. However, they are statistically uncorrelated, producing the increase in signal-to-noise ratio proportionate to the square root of the number of vertical pixels

Another, more efficient but at the same time more expensive, way to increase sensitivity is to use CCD matrix with special read-out technique instead of the line array (Fig. 9.20). Substantially more expensive, this option is used only in research-grade spectrometers. But the real breakthrough in sensitivity can be achieved only with image intensifiers that are used in gated spectrometers—rather expensive devices described below in Sect. 9.3. Except for the cost, these spectrometers offer numerous other advantages. For instance, in compact spectrometers described above, the only way to change (adjust) amplitude of the signal is to change exposure time. This may present significant practical discomfort when temporal evolution of the process is essential. The only way to perform time-resolved measurements with adjustable sensitivity is to use image intensifiers.

9.2 Imaging Spectrometers

From the previous section we know that optical system of the spectrometer projects image of the slit in the plane of the photodetector. Then, if the CCD matrix is installed in this plane as shown in Fig. 9.20 and a standard imaging mode is

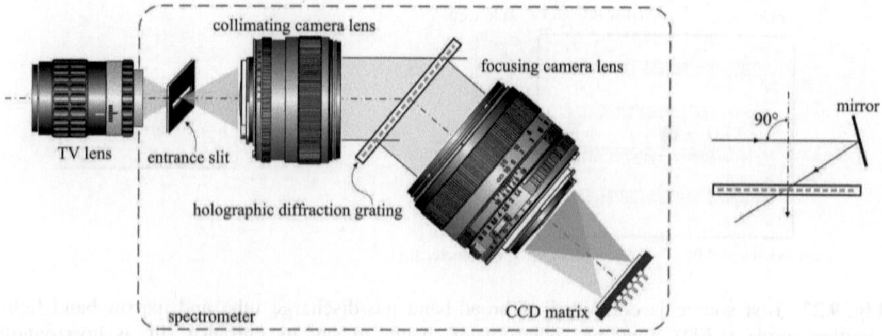

Fig. 9.21 Imaging spectrometer, working in visible domain, may be greatly simplified in design by using standard high-quality camera lenses and transparent holographic diffraction grating. When design considerations prevail, a folding mirror may be installed to make 90° angle between the input and output axes. The focused image is most symmetrical at all the wavelengths when the grating is oriented parallel to the input plane of the focusing lens. It also makes the shortest design. Anamorphism of the grating, when the diffracted beam is wider than the input one, requires bigger focusing lens with respect to the collimating. TV lens projects the image of a radiating object into the plane of the entrance slit

activated (not the FVB—full vertical binning), the one-dimensional image of the slit will be obtained along vertical rows at each horizontal pixel of the matrix. As to the horizontal rows, they will be showing spectral intensity at each vertical pixel of the matrix. Thus, the mixed spatial-spectral picture will be the result of the measurement. Does anybody need it? Definitely yes. There are numerous applications where spectral radiance of the source is not spatially uniform and this spatial non-uniformity is the purpose of the research. Imaging spectrometers cover special niche of the market, being much more expensive than ordinary compact spectrometers primarily due to two factors: cheap line CCD photodetector is substituted for high-quality CCD matrix and aberrations of optics are reduced to a minimum. It is well known that aberrations of reflecting optical elements—mirrors—are always stronger than those of the refracting ones—lenses—because reflection does not leave much room for optimization except for only the profile of reflecting surface. In lens design, not only refractive index is an optimization parameter but also a combination of multiple optical components that form a lens. Therefore, if high spatial resolution is needed then the priority should be given to lenses. But typical glass does not work below 400 nm and the red wing of visible domain ends at 800 nm. It means that spectral orders do not overlap and the order-sorting filter is not necessary. As such, lens-based imaging spectrometers may be simplified without loss of quality to a typical scheme shown in Fig. 9.21. Photographic camera lenses are always highly corrected for any type of aberrations, chromaticity, and field curvature. Being designed for standard 36 mm photographic films, their corrected field of view in the focal plane far exceeds dimensions of any CCD matrix, thus leaving no doubts about image quality. The new element here is the transmission-type holographic diffraction grating, working in

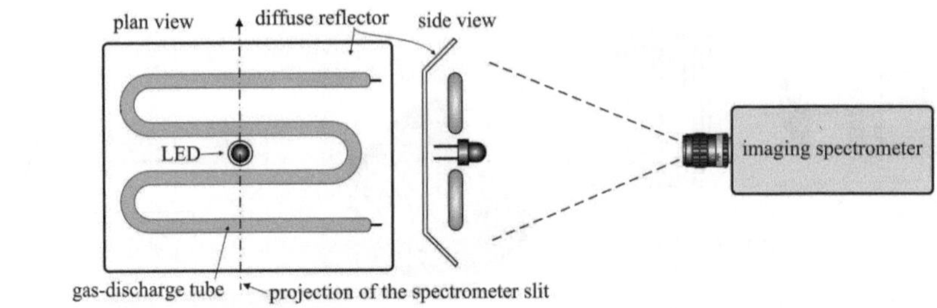

Fig. 9.22 Test source is combined of broad-band gas-discharge tube and narrow-band light-emitting diode (LED). Imaging spectrometer, better to say its entrance slit, is horizontally precisely aimed at the center of the LED

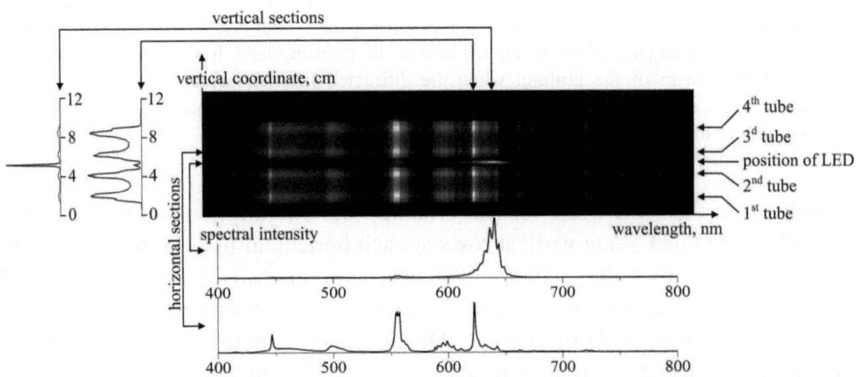

Fig. 9.23 Spatial-spectral picture of the test source shown in Fig. 9.22. Its 1024 × 256 matrix is composed of 1024 columns, each representing single wavelength, and 256 rows, representing vertical spatial coordinates. Each horizontal section gives a spectrum radiated by a single point on the source. Each vertical section gives vertical distribution of spectral intensity at a single wavelength. Two horizontal sections are made along the LED position and along the axis of third horizontal part of the gas-discharge tube. Two vertical sections are made along 640 nm—maximum spectral intensity of the LED and along 622 nm—red spectral maximum of the gas-discharge tube. In the picture, diffuse component returned from the reflector smoothes sharp physical edges of the tubes

parallel beams. Functionally, it is full analogue of the flat reflection-type diffraction grating already discussed in the previous section.

Peculiarity of pictures that imaging spectrometers provide requires some accommodation. As a simplified example, consider a test source shown in Fig. 9.22. Its picture recorded by an imaging spectrometer is shown in Fig. 9.23. The best impression gives the three-dimensional picture (Fig. 9.24).

Imaging spectrometers are often used in combination with spectral interferometry (Chap. 6) to measure thickness profile of thin films and, when equipped with interference objectives (Chap. 6),—vertical profile of micro-patterned surfaces. The advantage is speed: entire profile is measured in one frame of the CCD,

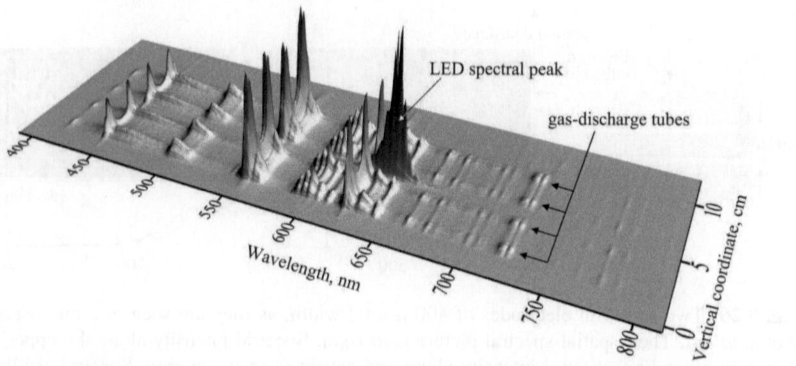

Fig. 9.24 Spatial-spectral map recorded by an imaging spectrometer can be better analyzed in three-dimensional presentation

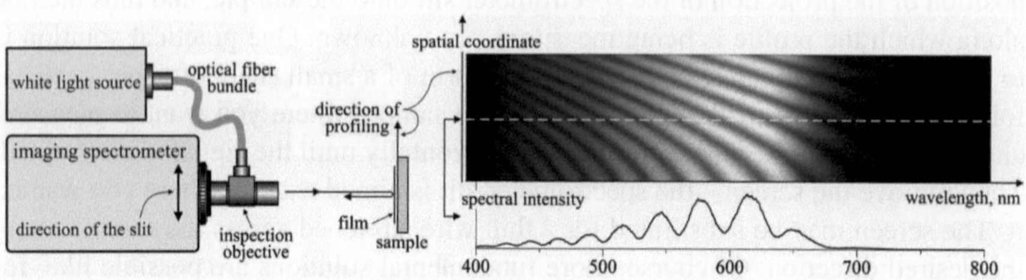

Fig. 9.25 In spectral interferometry mode, imaging spectrometer displays oscillating but not periodical (Chap. 6) spectra in every point of the slit along *vertical axis*. One such spectrum along the *white dashed line* is portrayed below the picture. Oscillations can be converted to the film thickness in this particular point, using algorithm described in Chap. 6. The spectrum of an ordinary tungsten-halogen lamp is not uniform, with lower spectral emission in the short-wavelength part of the spectrum (Chap. 2). This explains black area in the left part of the picture. Maximum emission falls into the infrared domain, which is blocked by multilayer coating of the lamp reflector and additional filters in order to protect outer optical elements from over-heating. This explains *black area* in the *right part* of the picture. It is clearly seen that the source is optimized for visible domain

without mechanical point-by-point scanning. To quickly make spectral interferometer from an imaging spectrometer, TV lens should be replaced by an inspection objective (Chap. 1) with C-mount flange and illumination port as shown in Fig. 9.25. When there is no pattern on the film, the resultant picture looks very much like an ordinary interference pattern, but it is not: one axis of it represents wavelength.

A clear composition of the picture can be seen on regularly patterned films, like the one presented in Fig. 9.26. The two strips are the thin-film leads on the glass of a mobile phone display. Information about film thickness is encoded in spectral oscillations, and the theory of extracting it is explained in Chap. 6.

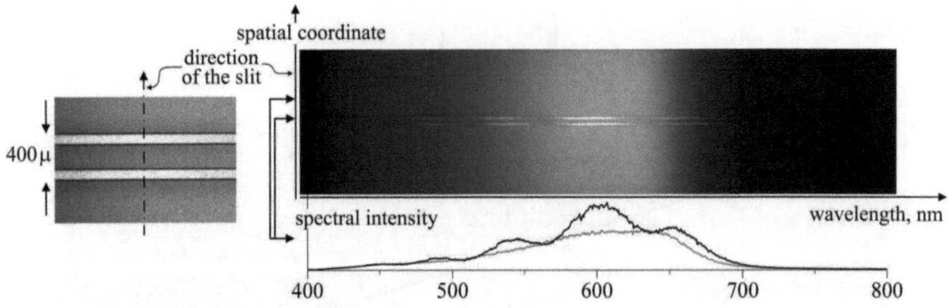

Fig. 9.26 Two thin-film electrodes of 400 μ total width, as they are seen in a microscope, are shown *at left*. Their spatial-spectral picture is *at right*. Spectral intensity along the upper strip is shown in *black line*, spectral intensity along non-patterned area—in grey. Spectral oscillation is clearly visible

The most confusing thing with the device described above is aiming: exact position of the projection of the spectrometer slit onto the sample, and thus the line along which the profile is being measured, are unknown. One practical solution is to insert temporarily an opaque screen in a form of a small sheet of paper or metal foil with straight vertical edge in front of the sample where you want to measure, and move the sample with such a screen horizontally until the signal is interrupted. Then remove the screen—the spectrometer slit is aimed exactly where you wanted it. The screen may be substituted for a thin wire stretched across the sample along the desired direction. Of course, more fundamental solutions are possible like, for instance, permanent TV monitor coupled through a beamsplitter.

Being connected to a microscope with interference objective of the Mireau or Michelson type, imaging spectrometers can measure not only thickness of thin films but also the profile of bare vertically patterned surfaces (Chap. 6). Also, this technique can be effectively extended even to vibrating surfaces, like in industrial environment, if fast gating is used. If the measurement is performed in much shorter time than the period of vibrations, say in one microsecond, then vibration does not produce any noticeable effect. Gated spectrometers are considered in the next section.

9.3 Gated Intensified Spectrometers

Charge coupled devices (CCDs) are unrivaled photodetectors for spectrometers. However, minimum exposure time that can be achieved with CCDs is only about 5 ms—too long to resolve nanosecond-scale processes in biological or chemical kinetics. And even if it were a nanosecond, it would be completely useless because only few photons could be recorded in a single pixel during such short a time. The solution for studying fast kinetics, which was successfully commercialized and is available on the market, is to use gated image intensifiers: fast optically switching devices with high optical gain. Such spectrometers are an order of magnitude more

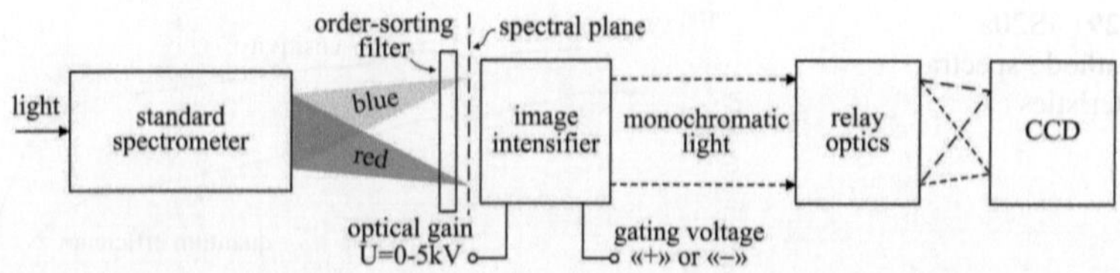

Fig. 9.27 The gated spectrometer is a combination of an ordinary spectrometer with image intensifier accompanied by relay optics that conveys the image in monochromatic light to photodetector, usually charge-coupled device (CCD)

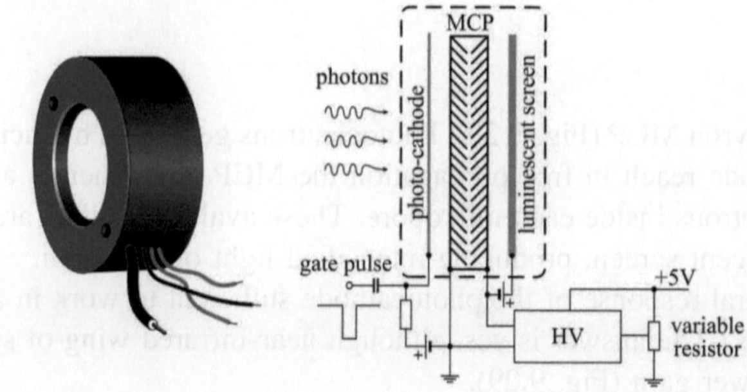

Fig. 9.28 All the components of an image intensifier are assembled in a vacuumized enclosure and sealed in an isolating compound with four leads for the gate pulse (usually a coaxial cable), high voltage, and ground. High voltage power supply (HV) sets variable voltage from 0 to 5 kV on the micro-channel plate (MCP), which performs amplification of the photoelectron flux from the photocathode. Close proximity between the photocathode and MCP ensures small divergence of photoelectrons needed for high image quality. Luminescent screen is covered in phosphor that radiates almost monochromatic light, usually green around 500 nm, when excited by secondary electrons emerging from the MCP. Diameters of the input aperture range from 18 to 100 mm

expensive than compact fiber-optic spectrometers described in Sect. 9.1, but their performance is worth it.

To begin with, consider generalized concept of a gated spectrometer (Fig. 9.27). Its key component that brings numerous new features is the image intensifier (Fig. 9.28). The concept of image intensification is based on spatially-resolved amplification of photoelectrons in a micro-channel plate (MCP). The MCP is a high-technology product. It is a thin, usually about 0.5 mm, plate of a dielectric material with numerous densely packed tiny holes—micropores, piercing it from one side to another. Diameter of each micropore is typically 10 μ and their spacing, determining spatial resolution, is about 20 μ. Ratio of the length to diameter—the aspect ratio—ranges from 40 to 60. Inner surface of each micropore is covered in emissive material, producing multiple secondary electrons on each incoming photoelectron. In order to avoid direct flight of photoelectrons through micropores, the latter are slightly tilted to about 8°. For higher gain, two such plates may be stacked together in a sandwich with opposite tilts, making what is

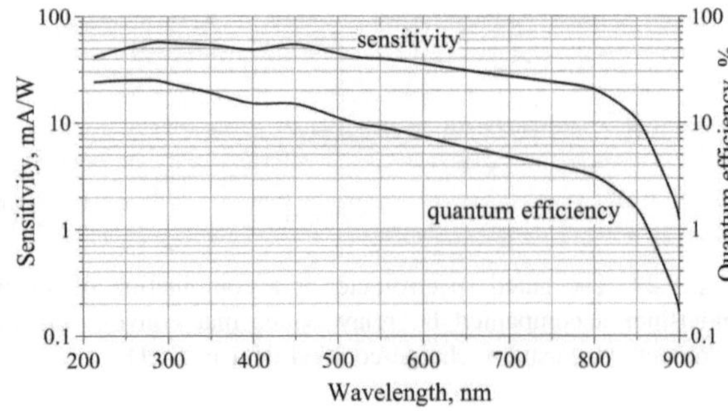

Fig. 9.29 «S20» photocathode spectral characteristics

called the Chevron MCP (Fig. 9.28). Photoelectrons generated by incident light on the photocathode reach in free propagation the MCP and generate avalanches of secondary electrons inside each micropore. These avalanches then are accelerated to the luminescent screen, producing intensified light on collision.

Is the spectral response of the photocathode sufficient to work in a wide range of wavelengths? The answer is yes, although near-infrared wing of spectrum will suffer from lower gain (Fig. 9.29).

Gating function is performed by applying permissive electrical pulses to photocathode. Polarity of these pulses depends on the polarity of the permanent voltage applied to photocathode. Negative permanent voltage across the photocathode-MCP gap accelerates photoelectrons, therefore to stop them the gate pulse must be positive and of a higher potential than the permanent voltage. In this mode, short gating pulses interrupt optical flux. In another option—the most frequently used—much lower positive permanent voltage ~50 V locks photoelectrons near the photocathode. To unlock them and accelerate, the negative gate pulse of about 300 V must be applied. In this mode, gating pulses open spectrometer for input light. Making 300 V amplifier for handling nanosecond-scale pulses is not a simple job that can be done easily. Therefore, manufacturers of image intensifiers always supply proprietary gating modules, specifying minimum pulse width that can be achieved with them. But even with this well-designed electronics, no one guarantees exact open time of the image intensifier at the lower limit of the pulse width. Calibration curve is always necessary, like the one shown in Fig. 9.30.

Another control wire of the image intensifier is the high-voltage accelerating potential on the MCP. It controls the optical gain: the higher the accelerating potential the bigger the number of secondary electrons and the brighter luminosity of the phosphor screen. Optical gain is measured in units Watt/Watt—the ratio of power of the input monochromatic light at some specific wavelength to power of the phosphor screen luminosity at its characteristic wavelength. It is a highly nonlinear function of the control voltage (Fig. 9.31).

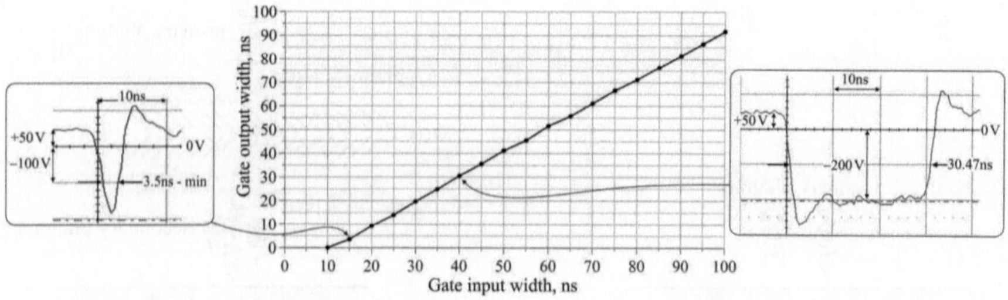

Fig. 9.30 A gating module specified for 3 ns minimum pulse width does make this voltage, but is the voltage level at 3 ns pulse width sufficient to open optical gate? Only optical measurements can give the answer. Nonetheless, there is no doubt for longer pulses

Fig. 9.31 Typical gain *curve* of a one-stage MCP at 400 nm monochromatic input flux

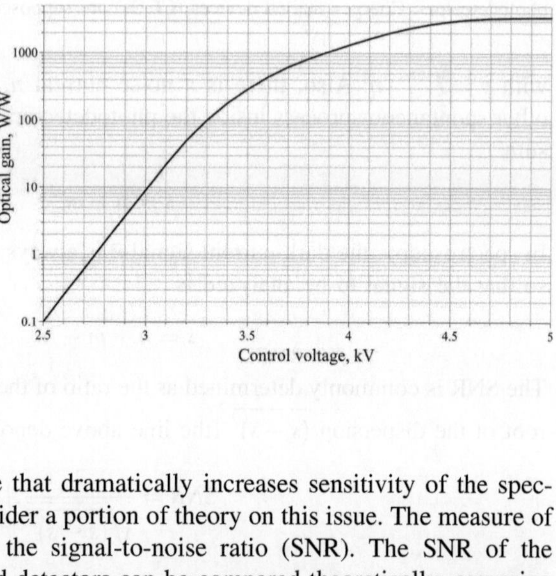

Optical gain is the feature that dramatically increases sensitivity of the spectrometer. To begin with, consider a portion of theory on this issue. The measure of the quality of the signal is the signal-to-noise ratio (SNR). The SNR of the intensified and non-intensified detectors can be compared theoretically, assuming Poisson statistics of photo-electrons. Figure 9.32 presents schematically the detection process in the non-intensified and intensified photodetectors. Consider the case when the flux I (photons per second) of the primary photons is constant during the exposure time T. Without image intensifier, each photon creates a carrier inside a pixel with quantum efficiency η. This process is random, and the total amount m of carriers generated during the exposure time is described by the Poisson probability function:

$$P(m) = \frac{\nu^m e^{-\nu}}{m!}$$

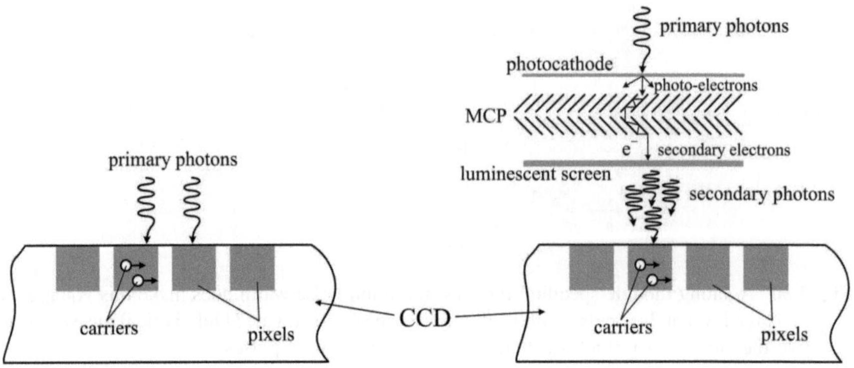

Fig. 9.32 Generalized schemes of the non-intensified (*at left*) and intensified (*at right*) photodetectors. Charge-coupled devices (CCDs) are supposed to be used in both cases

with $v = I \cdot T \cdot \eta$. Also, there is a noise current n, generated by thermal or any other spontaneous process inside the photodetector. Then the detector signal s is a sum

$$s = n + m.$$

In spectroscopy, the dark-current signal $\bar{n}$ is always subtracted from the spectrum, so that the signal to be analyzed is

$$s = n + m - \bar{n}.$$

The SNR is commonly determined as the ratio of the average signal $\bar{s}$ to the square root of the dispersion $\overline{(s - \bar{s})^2}$ (the line above denotes averaging):

$$SNR = \frac{\bar{s}}{\sqrt{\overline{(s - \bar{s})^2}}}.$$

Having in mind that for the Poisson statistics

$$\overline{m} = v; \overline{m^2} = v^2 + v,$$

it is easy to obtain for the non-intensified detector

$$SNR = \frac{v}{\sqrt{\sigma_n^2 + v}} = \frac{N\eta}{\sqrt{\sigma_n^2 + N\eta}},$$

where $\sigma_n^2 = \overline{n^2} - \bar{n}^2$—the detector noise and $N = I \cdot T$—the number of photons.

For the intensified detector not only the carrier generation process is random, but also the process of intensification, which brings additional noise. However, this additional noise may be less important than the detector noise σ_n^2, and then it is possible to expect a better SNR. Consider this in more detail. Again, let the input photon flux I be constant. Then the number M of the photons, reaching the detector, is random with the Poisson statistics

$$P(M) = \frac{\varepsilon^M e^{-\varepsilon}}{M!}$$

with $\varepsilon = I \cdot T \cdot \mu G$. Here G is the intensifier gain and μ—quantum efficiency of the photocathode. Thus, unlike the first case, the number of photons at the detector is random, therefore the probability of generating m carriers is now

$$P(m) = \sum_{M=0}^{\infty} \frac{(M\eta)^m e^{-M\eta}}{m!} \cdot \frac{\varepsilon^M e^{-\varepsilon}}{M!}.$$

With some straightforward manipulations, one can obtain the following relations:

$$\overline{m} \equiv \sum_{m=0}^{\infty} m P(m) = \eta\varepsilon;$$

$$\overline{m^2} \equiv \sum_{m=0}^{\infty} m^2 P(m) = \eta^2 (\varepsilon^2 + \varepsilon) + \eta\varepsilon;$$

$$\overline{m^2} - \overline{m}^2 = (1+\eta)\,\eta\varepsilon.$$

The SNR is then

$$SNR = \frac{N\eta\,(\mu G)}{\sqrt{\sigma_n^2 + (1+\eta)(\mu G) N\eta}}.$$

Physically, the product μG represents the optical gain in units Watt/Watt. It shows how many photons are generated at the output of the intensifier per one photon at its input.

Figure 9.33 compares the SNR as the function of the number of input photon N. Obviously, with the high enough μG, the intensified detector is always better than the non-intensified one.

Consider two limiting cases: very low input flux such as $N\eta\,(\mu G) \ll \sigma_n^2$, and very strong flux $N \to \infty$. At low flux, the non-intensified detector gives

$$SNR = \frac{N\eta}{\sigma_n},$$

whereas the intensified detector will have

$$SNR = \frac{N\eta}{\sigma_n}(\mu G),$$

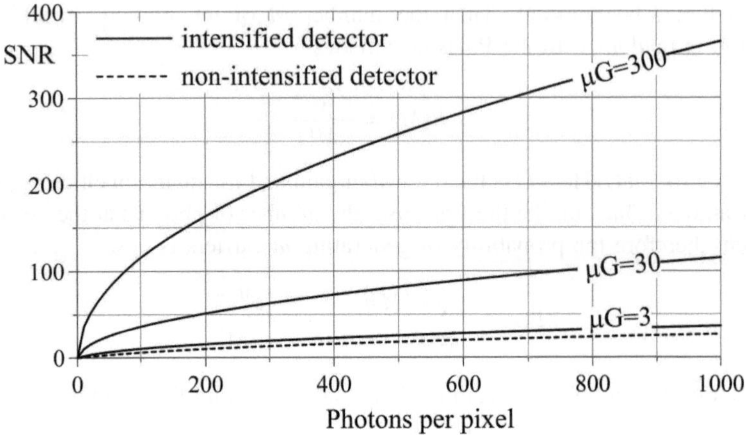

Fig. 9.33 SNR of the intensified photodetector. Quantum efficiency is taken $\eta = 0.8$ and the noise level $\sigma_n^2 = 100$. The non-intensified photodetector is shown in the *dashed line*

i.e. μG times bigger SNR. At very strong fluxes, this superiority is roughly $\sqrt{\mu G}$ times smaller:

$$SNR = \sqrt{N\eta}$$

for the non-intensified and

$$SNR = \frac{1}{\sqrt{1+\eta}} \sqrt{N\eta\,(\mu G)}$$

for the intensified detector. At first glance, it may look strange that we are considering the case of very strong flux with image intensifier: if the flux is strong then why do we need intensifier? However, there is one very important practical consideration why intensifier is needed even with strong input fluxes: stabilization. We shall consider this option of the intensified spectrometer below.

Commercially available gated spectrometers are always a combination of two separately sold parts: the spectrometer itself and the intensified CCD camera (ICCD) with proprietary software. The design of a spectrometer was already explained in Sect. 3.1, now it is time to shed some light on what is inside the ICCD (Fig. 9.34). First, the photodetector is always a CCD matrix. Cost of the ICCD is so high that manufacturers have no motivation to make it a bit cheaper by replacing the matrix with a linear array, like in compact spectrometers. And this is right decision because ICCDs separately find numerous applications in visualization of fast phenomena. Next, in order to minimize noise, the CCD matrix is always thermo-electrically cooled. This requires additional electrical circuit and fans. Finally, the image intensifier and the CCD matrix are permanently connected together in a solid block, being interfaced through imaging fiber-optic plate. This

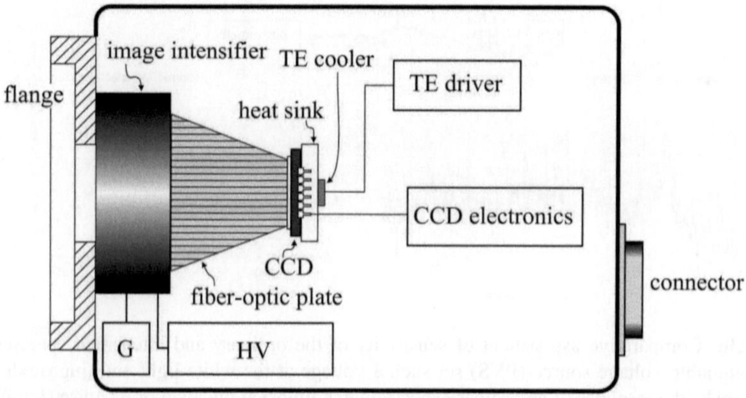

Fig. 9.34 ICCD contains an assembly of image intensifier, fiber-optic plate, and CCD matrix. The CCD is cooled by thermo-electric element (TE). Inside are also high-voltage power supply (HV) and gating module (G) for image intensifier

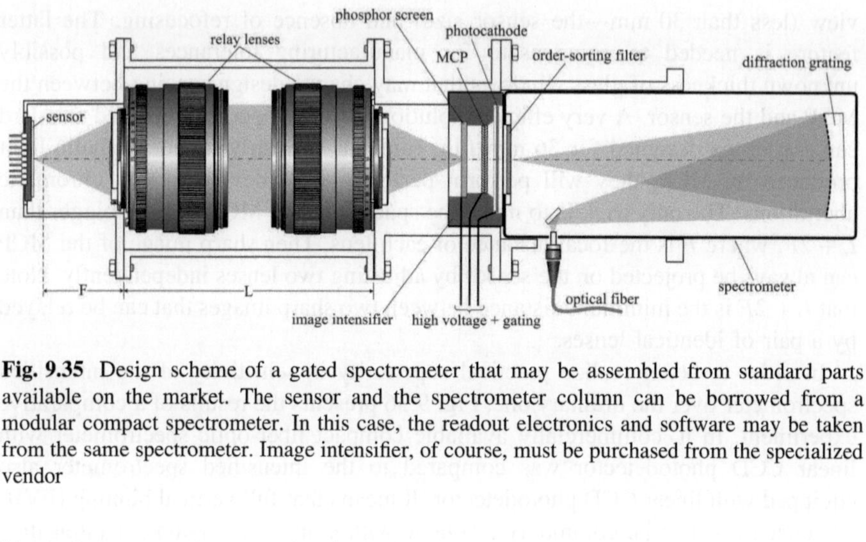

Fig. 9.35 Design scheme of a gated spectrometer that may be assembled from standard parts available on the market. The sensor and the spectrometer column can be borrowed from a modular compact spectrometer. In this case, the readout electronics and software may be taken from the same spectrometer. Image intensifier, of course, must be purchased from the specialized vendor

block, being proprietary technological achievement of a manufacturer, is the most expensive part of an ICCD, requiring high-grade facilities for assembling. Since the light at the output of the fiber-optic plate diverges from each single fiber, the input window of the CCD matrix must be removed and the sensor glued directly to the plate, with only thin layer of an optical bonder between them.

When size is not an issue, another, technologically simpler optical scheme can be used in combination with any standard photodetector without removing its input window (Fig. 9.35). This scheme uses relay lenses to project image of the MCP rear window onto the sensor. Commercially available relay lenses described in Chap. 1 are inappropriate in this case for two main reasons: insufficient field of

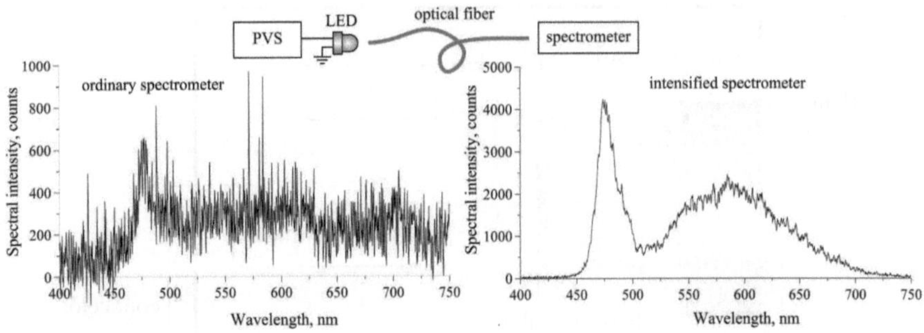

Fig. 9.36 Comparative assessment of sensitivity of the ordinary and intensified spectrometers. Programmable voltage source (PVS) set such a voltage at the white-light emitting diode (LED) that to make the response of an ordinary spectrometer almost at the level of its noise (*left picture*). Then optical fiber was reconnected to the intensified spectrometer with optical gain set to 10^3 (*right picture*). Sensitivity of both photodetectors is roughly the same: 200 V/(lx · s) for the ordinary and 160 V/(lx · s) for the intensified spectrometers. In both cases, exposure time was set to 300 ms and dark current subtracted

view (less than 30 mm—the sensor size) and absence of refocusing. The latter feature is needed to compensate for manufacturing tolerances and possibly unknown thickness of glass windows that may change design spacing between the MCP and the sensor. A very efficient solution is to use oppositely placed standard camera lenses designed for 36 mm film cameras. In nearly monochromatic light produced by MCP, they will perform perfectly well, devoid of any chromatic aberrations. The only trick is to make the space sensor—MCP slightly bigger than $L+2F$, where F is the focal distance of each lens. Then sharp image of the MCP can always be projected on the sensor by adjusting two lenses independently. Note that $L+2F$ is the minimum distance between two sharp images that can be relayed by a pair of identical lenses.

In order to better realize practical superiority in sensitivity of the intensified spectrometer over the ordinary one, Fig. 9.36 presents the results of a comparative experiment. In it, commercially available compact fiber-optic spectrometer with linear CCD photodetector was compared to the intensified spectrometer also equipped with linear CCD photodetector. It means that full vertical binning (FVB) advantage (Sect. 9.1), commonly available with enhanced versions of intensified spectrometers equipped with CCD matrices, could not be used.

Most of the gated spectrometers on the market are of 3 ns class, although some very expensive models claim sub-nanosecond gating. But even 3 ns temporal resolution opens unique possibilities in many laboratory applications, like, for instance, spectral dynamics of white-light LED emission (Chap. 4). Optical gate width cannot be measured directly, therefore, indirect experiments should be used to prove that gated spectrometer is capable of nanosecond optical sampling. Particularly, the technique of calibrated delays is very straightforward and convincing. In it, a short light pulse travels through optical fibers of different lengths, and the gating device must resolve the delay difference introduced by the

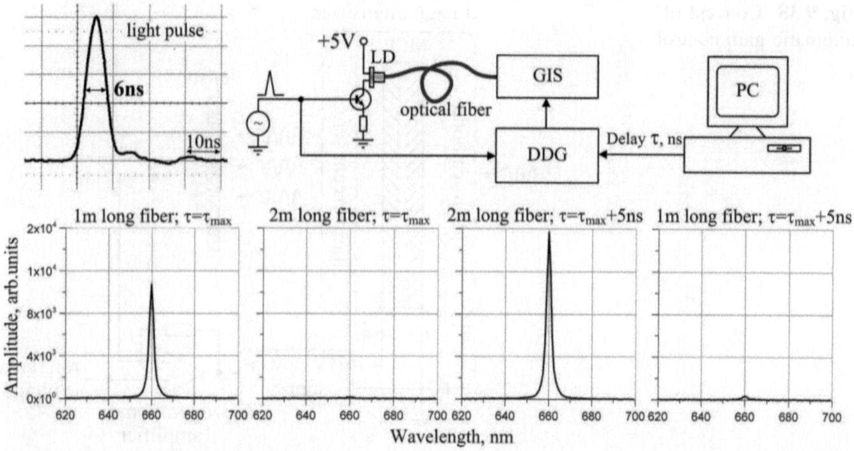

Fig. 9.37 Optical sampling scheme (*above*) and optically gated spectra (*below*). Electrical excitation pulse of the laser diode (LD) was set triangular with 5 ns edges, making full width at half-maximum (FWHM) 5 ns. Fast photomultiplier response (*in the upper left corner*) shows 6 ns optical width. Gated intensified spectrometer (GIS) was being sampled by 5 ns gate pulses from digital delay generator (DDG), controlled from the computer (PC). Sections of 1 m and 2 m optical fibers created delays that were clearly resolved by the spectrometer

difference in optical fiber length. Simple experiment explained in Fig. 9.37 gives an additional insight into gating capabilities of these spectrometers.

In the experiment, a short light pulse was generated by a high-speed laser diode (Chap. 3) powered by a nanosecond function generator. The FWHM of such a pulse should be 5 ns. The 1 GHz oscilloscope trace recorded by a high-speed photomultiplier with 80 MHz amplifier shows the FWHM about 6 ns. Initially, the laser diode was connected to the spectrometer through a 1 m long optical fiber. The delay between the function generator pulse and the triggering pulse was adjusted to make the spectral amplitude a maximum, which meant temporal coincidence of the laser pulse and the spectrometer gate. This spectrum is shown in the first picture. Next, this fiber was substituted for the 2 m long one. The amplitude dropped 75 times (second picture), but after increasing the delay by only 5 ns it became even bigger than with the shorter fiber (third picture). This increase in spectral amplitude can be attributed to discrete variation of the delay time, limited to 5 ns per step. Therefore, initial positioning of the trigger relative to the laser pulse might not exactly coincide with its maximum. Finally, with the delay unchanged and the original configuration restored, i.e. 1 m long fiber, the amplitude of the spectrum became much smaller than it was in the very beginning (the last picture). This result clearly shows that the spectrometer readily resolves 1 m long section of optical fiber, i.e. 3 ns delay.

An intensified spectrometer is not necessarily supposed to work in a pulsed mode: continuous-wave mode has its own advantages. For instance, a very useful practical feature that has nothing to do with gating is gain control. An ordinary

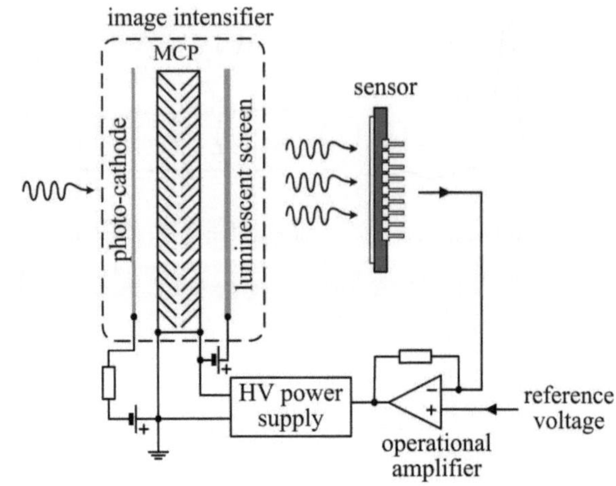

Fig. 9.38 Concept of automatic gain control

spectrometer offers only one option to change amplitude of the signal: exposure time. But it may be unacceptable on volatile processes. Moreover, the input signal may be so intense that even minimum exposure time does not protect from saturation—a common thing with argon-mercury calibration sources where the very strong mercury line at 253 nm always saturates. With intensified spectrometers, there are no such problems: merely set the intensifier gain as high or as low as to make signal amplitude acceptable. Even better option is automatic gain control, which is schematically explained in Fig. 9.38. The signal from the detector, unfolded into a spectrum, is digitally analyzed in the computer in order to find its maximum. This maximum amplitude is then compared to a reference value preset by the user in order to control optical gain of the image intensifier in such a way that to maintain the maximum amplitude of the spectrum at the comfortable level. With this option, a researcher does not need to spent time on manual adjustment of the gain, or even worse—the exposure time. He simply turns on the spectrometer with the exposure time he wants, and makes his measurements whatever the input flux is.

9.4 Fourier-Transform Spectrometers

Fourier-transform (FT) spectroscopy is an exceptionally well-established technology in infrared domain with wavelengths above one micron. However, in the domain of shorter wavelengths, FT spectrometers cannot compete with grating spectrometers for many reasons like cost, speed, sensitivity, ruggedness, etc. With one exception: modulation sensitive spectroscopy. There are applications like plasma-etching machines in semiconductor industry, aircraft jet engines, high-voltage coronas on transmission electrical lines, fluorescence spectroscopy, where optical emission is modulated by frequencies ranging from 50 Hz to several megahertz. How to separate weak modulated spectrum from strong non-modulated

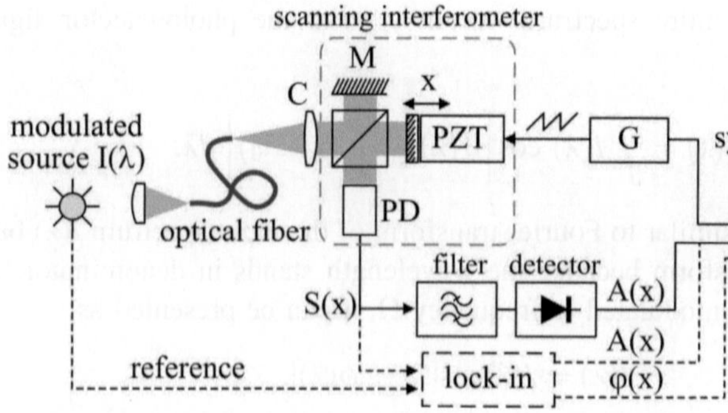

Fig. 9.39 Modulation-sensitive Fourier spectrometer in basic (*solid line*) and phase-resolved (*dashed line*) configurations. Scanning interferometer is combined of the fixed mirror (M), beam-splittingcube, and the moving mirror installed on the piezo-transducer (PZT). Collimator (C) makes parallel beam at the input of the interferometer. Saw-tooth voltage from generator (G) drives the scanning mirror linearly in one direction of x axis and returns it quickly back. Electrical signal $S(x)$ from the photodetector (PD) passes through the narrow-pass filter adjusted to modulation frequency and is rectified by a detector (Chap. 4) to produce amplitude of modulation $A(x)$. Analog signals are digitized by data acquisition board (DAQ) installed in a computer. DAQ starts acquisition when synchronization impulse comes from the generator, marking initial position of the moving mirror. If reference signal is available, synchronous detection can be implemented to extract additionally phase $\phi(x)$ of modulation (Chap. 4)

background? Traditional spectrometers cannot do that. The solution is modulation-sensitive Fourier spectroscopy. There are no commercial devices of that type on the market so far, but it is quite possible to make a laboratory version, and we are going to show how to do that, using readily available blocks and modules described in other chapters of this book. The concept of the FT spectroscopy is sketched in Fig. 9.39.

Traditional spectrometers cannot track high-frequency modulation of optical power because their photodetectors, usually charge-coupled devices (CCDs), are slow. But even if they were fast, it would be quite impractical to put narrow-pass electrical filter with a rectifier on every one of thousands of spectral channels. As to the FT spectrometer, it has only one channel, therefore all the filtering capabilities of radio-electronics are in our disposal to select weak modulated signal on strong background of non-modulated optical flux. This is the idea, but first we need to recall theoretical backgrounds of FT spectroscopy. Consider spectral intensity I of an input optical flux as a function of wavelength λ. In the output plane of the interferometer with 50 % beamsplitter, variable part of the intensity in a quasi-monochromatic (narrow) spectral interval $d\lambda$ is (Chap. 6)

$$I(\lambda)\cos\Phi \cdot d\lambda,$$

where Φ is the total phase difference between the waves in two shoulders of the interferometer. Splitting the total phase in two components, accounting for spectral dispersion of optics $\psi(\lambda)$ and non-zero initial position of the moving mirror x_0, and summing over the entire spectrum, we may write the photodetector signal as proportional to

$$S(x) = \int I(\lambda) \cos\left[\psi(\lambda) + \frac{4\pi}{\lambda}(x - x_0)\right] d\lambda.$$

This formula looks similar to Fourier transform of the input spectrum $I(\lambda)$ but it is not the Fourier transform because the wavelength stands in denominator. When input optical flux is modulated at frequency Ω, it can be presented as

$$I(\lambda) = I(\lambda)\cos[\Omega t + \varphi(\lambda)],$$

where $\varphi(\lambda)$ is the phase shift of a particular spectral component at the wavelength λ. Thus, spectral intensity $I(\lambda)$ and the phase shift $\varphi(\lambda)$ are mixed together under the integral. If phase is not a subject of a particular application then the best way to simplify the problem is merely to ignore the phase:

$$S(x, t) = \cos \Omega t \cdot \int I(\lambda) \cos\left[\psi(\lambda) + \frac{4\pi}{\lambda}(x - x_0)\right] d\lambda,$$

and only amplitude of modulation

$$A(x) = \int I(\lambda) \cos\left[\psi(\lambda) + \frac{4\pi}{\lambda}(x - x_0)\right] d\lambda$$

should be measured.

Next, we need a practical algorithm to extract $I(\lambda)$ from this measurement. Denoting

$$u = \frac{2}{\lambda} \quad \text{and} \quad \delta = x - x_0,$$

the relation between the measured signal $S(\delta)$ and spectrum $I(\lambda)$ formalizes to

$$S(\delta) = \int F(u) \cos(2\pi \delta u)\, du, \qquad F(u) = -\frac{2}{u^2} I\left(\frac{2}{u}\right).$$

This already is the Fourier transform, and $F(u)$ can be recovered by inverse fast Fourier transform (FFT). The FFT is defined on a discrete mesh of M points $l = 1, 2, 3, \ldots, M$ uniformly separated on the scanner stroke Δx:

$$\delta_l = \frac{l-1}{M} \Delta x,$$

making the recorded array

$$S_l = \int_{-\infty}^{+\infty} F(u)\, e^{i 2\pi u \delta_l}\, du, \qquad i = \sqrt{-1}.$$

The result of the FFT is the array V_j of exactly the same size as S_l:

$$V_j = \sum_{l=1}^{M} S_l \cdot e^{-2\pi i (j-1)(l-1)/M}.$$

Substituting S_l as the integral, bringing summation under the integral, and realizing that only those terms will remain, for which the argument of the complex exponent is zero

$$u\,\Delta x = j - 1,$$

we obtain:

$$V_j = M \cdot F\left(\frac{j-1}{\Delta x}\right).$$

The function $F(u)$ has already been defined, so that it is only a matter of arithmetical manipulations to obtain the final result:

$$I\left(\frac{2\,\Delta x}{j-1}\right) = \frac{(j-1)^2}{2\,M\,\Delta x^2}\,V_j.$$

There are two fundamental results in one formula. The first establishes absolute calibration of the final spectrum:

$$\lambda_j = \frac{2\,\Delta x}{j-1}.$$

It says that if you know exactly the displacement of the mirror Δx (in micrometers, millimeters, or any other units) then the wavelength associated with the number j in the final array V_j depends on nothing but only Δx and the number of samples M that you have recorded. This dependence is non-linear: wavelengths are ascribed to j-th element of the array inversely proportional to j—1. From this formula, spectral resolution $\Delta\lambda = \lambda_j - \lambda_{j+1}$ of the FT spectrometer easily follows:

$$\Delta\lambda = \frac{2\,\Delta x}{j(j-1)} \approx \frac{2\,\Delta x}{(j-1)^2} = \frac{\lambda_j^2}{2\,\Delta x}.$$

Thus, the bigger mirror travel the better spectral resolution. The mirror travel, or the scanner stroke as it is sometimes called, is the first parameter that determines spectral resolution.

The second result says how to correctly compute intensities: they should be multiplied by the factor $(j-1)^2$ for every one element of the final array V_j. Otherwise, the scale of intensities will be incorrect.

These deeply theoretical considerations are of great practical value because without them the final result—the spectrum—cannot be obtained. Now it is time to address the design of the FT spectrometer (Fig. 9.40). Standard TV lens (Chap. 1)

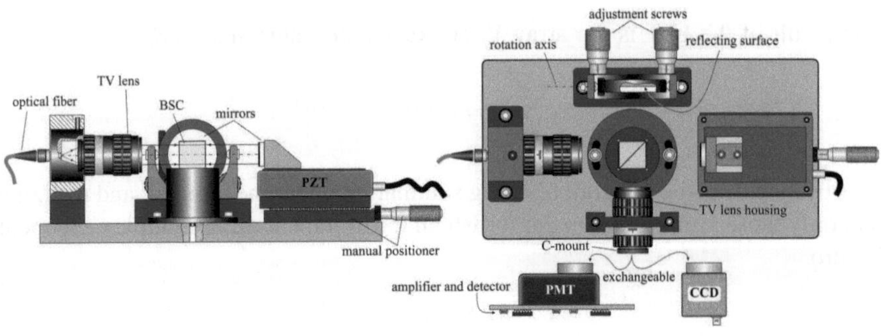

Fig. 9.40 FT spectrometer that works. Standard *TV lens* coupled to an optical fiber forms a collimator. *TV lens* always has focus adjustment ring, which makes collimation easy. Also, the iris ring on the *TV lens* is very useful to truncate beam diameter (*see text*). Beam-splitting cube (BSC) is mounted on the rotation pedestal, enabling coarse adjustment. Final adjustment is done by the fixed mirror. According to what has been told in Chap. 6, reflecting surface of this mirror coincides with the rotation axes: the vertical and horizontal. Fine micrometer screws are needed for precise alignment. Pizeo-transducer (PZT) carries the second (non-adjustable) mirror. Its zero position must be adjusted manually by a micrometer screw. Output beams go through *TV lens* housing with the second iris diaphragm in it (*see text*). Its C-mount flange makes it easy to exchange the CCD camera with the photomultiplier (PMT), as explained in the text. The compact PMT module (Chap. 3) was used in this design

with an optical fiber in its focal plane is a very comfortable tool to create collimated beam. Angular divergence α of the beam is the second parameter that determines spectral resolution. Even without any additional drawings, it is clear that the ray, coming at the angle α, traverses $\cos^{-1}\alpha$ longer path. As such, the optical path difference increases by approximately $\Delta x \cdot \alpha^2$, after expanding the cosign of a small angle. For the constructive interference still taking place for these rays, this additional path difference must be smaller than half of the wavelength:

$$\alpha^2 < \frac{\lambda}{2\Delta x}.$$

Combining this with the last formula for spectral resolution, obtain

$$\alpha^2 < \frac{\Delta\lambda}{\lambda}.$$

Practically, it means that for $\Delta\lambda \approx 2$ nm in visible domain divergence of the beam must be smaller than 60 mrad or 3°. For comparison, an optical fiber ∅0.6 mm in the focal plane of a TV lens with focal distance 17 mm (C-mount standard) produces three times smaller divergence. Another useful feature of a TV lens is the iris diaphragm: with it, fitting beam diameter to the photodetector size, thus reducing amount of scattered light, is easy.

The second TV lens is used at the photodetector side. In it, the lens system is completely removed, leaving only the iris with its adjustable ring and the C-mount flange. This iris serves to confine sensitive area of the photodetector, increasing

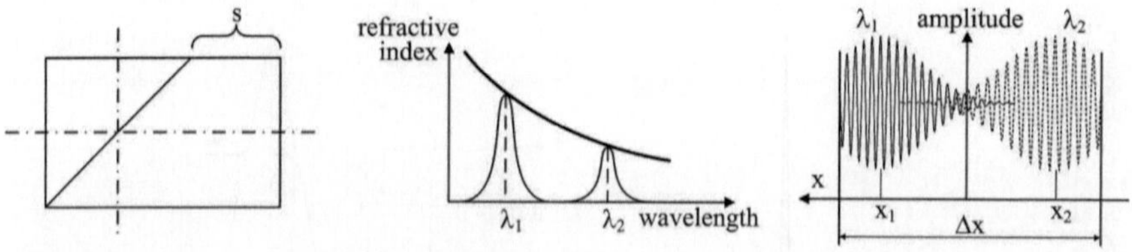

Fig. 9.41 In the beamsplitter, geometrical difference *s* in two orthogonal directions must be a minimum (*at left*). Refractive index increases for shorter wavelengths (*in the middle*). As mirror moves, this causes relative displacement of fringe curves for different wavelengths (*at right*)

contrast of fringes. The C-mount is used to quickly interchange the photodetector with the CCD camera needed for initial alignment of the interferometer (Chap. 6).

Beam-splitting cube must be of non-polarizing type and meet stringent requirements on polarization. Any priorities in polarization contribute to lower contrast of fringes. This issue is discussed in Chap. 6 and shall not be repeated here. Another requirement, which is never specified by manufacturers of beamsplitters, is equality of optical paths. Consider what happens when vertical thickness of the cube is different from the horizontal one (Fig. 9.41). Since refractive index $n(\lambda)$ is a function of wavelength, optical path difference $s \cdot n(\lambda)$ will be different for short and long wavelengths. Consequently, zero optical path difference will correspond to different displacements x_1 and x_2 of the mirror. If s becomes too big, say 100 μ, then $x_1 - x_2 = s[n(\lambda_1) - n(\lambda_2)]$ may become comparable to the mirror travel range Δx, which causes truncation of useful parts of the interference pattern as shown in Fig. 9.41. It means that useful spectral information is lost. To prevent this from happening, high-quality beamsplitters for FT spectrometers are made of two glass plates cut from one precisely parallel substrate and bonded together with semi-transparent coating deposited on one of them. Such beamsplitters are called compensated. Nonetheless, even standard beam-splitting cubes work satisfactory, although some non-compensation is always visible (Fig. 9.44).

The most expensive part of the spectrometer is the piezo-transducer with controller. Their functionality is explained in Chap. 10. It is always a trade-off between the maximum displacement of the mirror and motion stability. If spectral resolution about 2–3 nm is tolerable then better to choose very stiff piezo-stage with not very big stroke, less than 100 μ. For example, Fig. 9.42 shows the spectrum obtained with 30 μ piezo-stage on 2.6 MHz modulated output of the Zeeman He-Ne laser (Chap. 4). Since spectrum of the laser source itself is megahertz-wide, all the spectral spread may be attributed to final spectral resolution of the FT spectrometer.

Linearity of scanning is the first priority in Fourier-transform (FT) spectrometers. Practically, piezo-stages are the only suitable option, even though they are expensive and limited in maximum displacement. In them, linearity is guaranteed by a feedback from a capacitive sensor. Some would suggest a simpler less

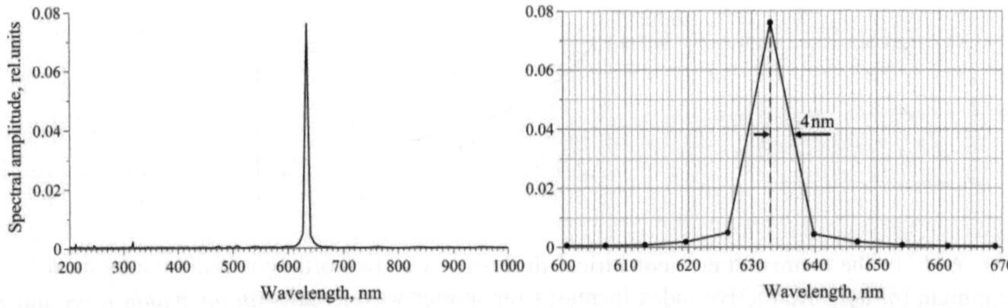

Fig. 9.42 2.6 MHz spectrum of the He-Ne Zeeman laser at 632.8 nm processed in wide range from 200 to 1000 nm. Scan time 2 s

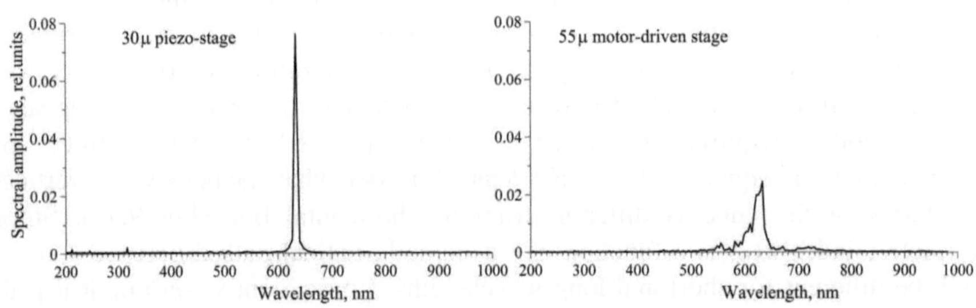

Fig. 9.43 2.6 MHz spectra of the He-Ne Zeeman laser obtained with a piezo-stage and 30 μ stroke (*at left*) and ball-bearing scanning stage driven by a stepping-motor with 55 μ stroke (*at right*). Even though the motor-driven stage has almost two times longer scan, spectral resolution is much worse: the entire spectrum is torn apart by irregularities of scanning. The area under the curves is the same due to energy conservation principle

expensive solution like, for example, precision motor-driven scanning stage. Alas! Fig. 9.43 clearly shows the difference.

Narrow-band sources, like He-Ne laser, produce well-defined sinusoidal modulation $A(x)$ during scanning. Wide-band sources, white-light LED for instance, have coherence length of about micrometers, and modulation curve $A(x)$ is clearly seen only near $x = x_0$ position of the scanning mirror (zero optical path difference). Modulation curves of such sources only distantly resemble a sinusoid (Fig. 9.44). In order to precisely position this rather narrow (several micrometers) and most important part of the modulation curve within the scanning range of the mirror, the manual micrometer positioner is used as shown in Fig. 9.40.

Traditional spectroscopy deals with the spectral intensity $I(\lambda)$ only. Modulation-sensitive Fourier spectroscopy offers additional possibility of measuring phase shift $\varphi(\lambda)$, which is of great importance for biology and plasma diagnostics. In biology particularly, phase shift determines the lifetime τ of an optical transition:

$$\tau = \frac{\tan\varphi}{\Omega},$$

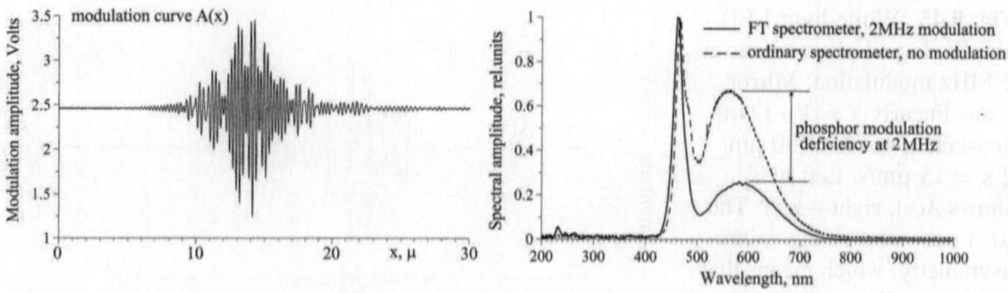

Fig. 9.44 Modulation curve $A(x)$ of a white-light LED at 2 MHz pumping (*at left*). Signal is concentrated in the middle of the scan, where x_0 is located. Note that the curve is not symmetrical around its center, which is a clear sign of not entirely compensated beamsplitter. The right picture shows the spectrum $I(\lambda)$ restored from the modulation curve. In the same scale, the spectrum of not modulated LED measured by an ordinary spectrometer is shown in the *dotted line*. The difference between them in the *red* part of spectrum around 600 nm is caused by slower mechanism of fluorescence: *blue* laser diode excites at 460 nm faster than subsequent spontaneous fluorescence responds in red

and it can be measured, using synchronous detector. Some mathematics below shows how $I(\lambda)$ and $\varphi(\lambda)$ can be separated in the measurements.

Substitute $I(\lambda)$ into formula for $S(x)$, decompose cosine of the sum into the sum of products, group the result into sine and cosine components to obtain the following:

$$S(x) = A(x) \cdot \cos[\Omega t + \alpha(x)];$$

$$A(x) = \sqrt{\left[\int I(\lambda)\,\cos\varphi(\lambda)\cdot\cos\Phi(x,\lambda)\,d\lambda\right]^2 + \left[\int I(\lambda)\,\sin\varphi(\lambda)\cdot\cos\Phi(x,\lambda)\,d\lambda\right]^2};$$

$$\cos\alpha(x) = \frac{\int I(\lambda)\,\cos\varphi(\lambda)\cdot\cos\Phi(x,\lambda)\,d\lambda}{A(x)};$$

$$\sin\alpha(x) = \frac{\int I(\lambda)\,\sin\varphi(\lambda)\cdot\cos\Phi(x,\lambda)\,d\lambda}{A(x)};$$

$$\Phi(x,\lambda) = \psi(\lambda) + \frac{4\pi}{\lambda}(x - x_0).$$

Here $A(x)$ and $\alpha(x)$ are the amplitude and phase of high-frequency modulation. Form the products:

$$A(x)\cdot\cos\alpha(x) = \int I(\lambda)\,\cos\varphi(\lambda)\cdot\cos\left[\frac{4\pi}{\lambda}x + \theta(\lambda)\right]d\lambda;$$

$$A(x)\cdot\sin\alpha(x) = \int I(\lambda)\,\sin\varphi(\lambda)\cdot\cos\left[\frac{4\pi}{\lambda}x + \theta(\lambda)\right]d\lambda,$$

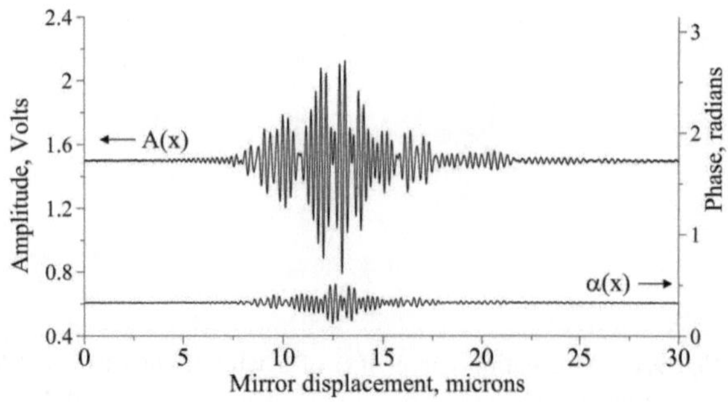

Fig. 9.45 White-light LED amplitude and phase traces at 2 MHz modulation. Mirror scans linearly $x = V \cdot t$ with constant speed $V = 30$ μm/2 s = 15 μm/s. Left axis shows $A(x)$, right—$\alpha(x)$. The $A(x)$ trace also shows minor asymmetry, which means that the beamsplitter is not completely compensated

where $\theta(\lambda) = \psi(\lambda) - \frac{4\pi}{\lambda}x_0$. Apply complex Fourier transforms to these products, implying that physical reality requires $I(-\lambda) = 0$:

$$F_c(\omega) \equiv \int A(x) \cdot \cos\alpha(x) \cdot e^{-i\omega x}\,dx = \pi I\left(\frac{4\pi}{\omega}\right)\cos\varphi\left(\frac{4\pi}{\omega}\right)e^{i\theta(4\pi/\omega)};$$

$$F_s(\omega) \equiv \int A(x) \cdot \sin\alpha(x) \cdot e^{-i\omega x}\,dx = \pi I\left(\frac{4\pi}{\omega}\right)\sin\varphi\left(\frac{4\pi}{\omega}\right)e^{i\theta(4\pi/\omega)}.$$

Denote $\frac{4\pi}{\omega} \equiv \lambda$ to obtain

$$F_c\left(\frac{4\pi}{\lambda}\right) = \pi I(\lambda)\cos\varphi(\lambda)\,e^{i\theta(\lambda)}; \qquad F_s\left(\frac{4\pi}{\lambda}\right) = \pi I(\lambda)\sin\varphi(\lambda)\,e^{i\theta(\lambda)}.$$

Although F_c and F_s are complex variables, their ratio is not:

$$\tan\varphi(\lambda) = \frac{F_s\left(\frac{4\pi}{\lambda}\right)}{F_c\left(\frac{4\pi}{\lambda}\right)}.$$

From here, the final result follows:

$$I(\lambda) = \frac{1}{\pi}\sqrt{|F_c|^2 + |F_s|^2}, \qquad \varphi(\lambda) = \arctan\frac{F_s\left(\frac{4\pi}{\lambda}\right)}{F_c\left(\frac{4\pi}{\lambda}\right)}.$$

Now consider how this theory works in practice. From spectra presented in Fig. 9.44 it follows that red fluorescence of a white-light LED lags behind its blue excitation at 460 nm. The idea of the experiment is to observe the phase difference between blue and red spectral wings. Modulation frequency was chosen to be equal to 2 MHz—not very high to preserve good modulation contrast, and not too low to obtain reasonable values of the phase. Figure 9.45 shows the amplitude $A(x)$ and phase $\alpha(x)$ signals at the two outputs of the synchronous detector (Fig. 9.39). Computed phase delay $\varphi(\lambda)$ along with the restored spectral amplitude $I(\lambda)$ are shown in Fig. 9.46.

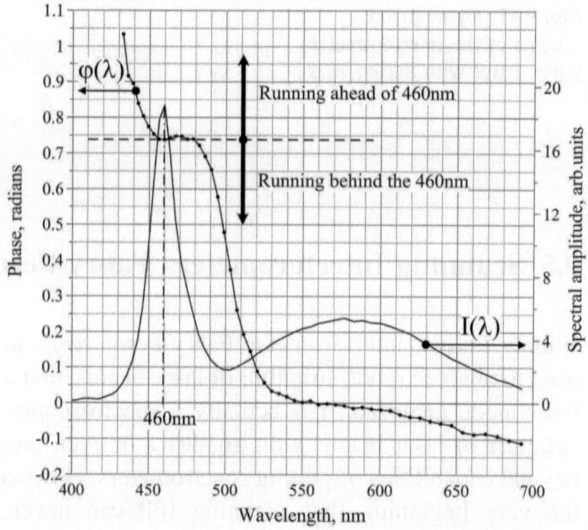

Fig. 9.46 Restored phase delay $\varphi(\lambda)$ is shown in *solid line* with dots, representing discrete Fourier transform nods; spectral amplitude $I(\lambda)$ is shown in a *solid line*

The absolute value of the phase is of no avail for us because it is measured relative to reference electrical signal that is composed of many uncertain delays in electronic circuits and cables. We are interested in the phase relative to some fixed spectral component, for example, the maximum of blue luminescence at 460 nm because it may be considered as the pumping optical source for red fluorescence. Figure 9.46 shows that phases of all the red spectral components are smaller relative to the phase of the 460 nm excitation, meaning that red fluorescence of the phosphor falls behind the excitation. Time delay can be computed according as

$$\tau(\lambda) \approx \frac{\varphi_\lambda - \varphi_{460\text{nm}}}{2\pi \cdot 2 \cdot 10^6} \text{ [s]}.$$

Negative sign means lagging, positive—running ahead of the 460 nm pumping component. For example, for $\lambda = 500$ nm $\tau = -24$ ns, which means that this spectral component runs 24 ns after excitation. At $\lambda = 500$ nm, blue excitation wing overlaps with red fluorescence, which physically means that this spectral component does not represent a single damped oscillator and the decay constant phenomenology is inapplicable here. Alternatively, all the spectral components of red fluorescence beyond 550 nm are free from blue excitation contribution, which means that they may be considered as damped oscillators with definite decay constants. Figure 9.46 shows roughly the same phase delay for all of them and the decay constant may be computed as $\tau \sim 60$ ns. Violet wing phase below 460 nm excites even before the central line at 460 nm, reacting faster to electrical signal.

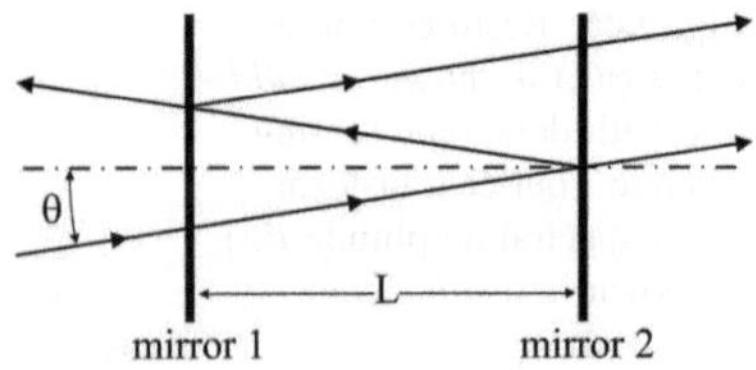

Fig. 9.47 Generalized scheme of the interferometer Fabry-Perot with flat *mirrors*

9.5 Scanning Interferometers Fabry-Perot

Widely known interferometer Fabry-Perot (IFP) in its scanning modification is used to analyze mode structure of laser beams, and as such is frequently called the laser mode analyzer. It is actually a spectroscopic tool designed to resolve fine structure of laser beams with gigahertz or even megahertz spectral resolution far beyond capabilities of grating spectrometers. However, it must be understood from the very beginning that scanning IFP can never measure absolute values of wavelengths rather than by comparison with another laser source already calibrated with required precision. Therefore, it is better to say that scanning IFP is an indicator of spectral structure of lasers.

Commercially available scanning IFPs, being very simple and reliable in design, are at the same time far more theoretically complicated devices than traditional IFP known from university courses. Therefore, theoretical overview of basic multi-beam interference phenomena is necessary to realize how they work. To simplify mathematics, consider for the beginning a simplified model of a flat-mirror IFP (Fig. 9.47), which is also relevant to interference filters explained in Sect. 9.6. Let the amplitude transmission and reflection coefficients at the mirrors be t and r. Then the plane monochromatic wave with unity amplitude and the wavelength λ, coming at the angle θ, will reflect many times inside the IFP cavity, making the outgoing wave an infinite sum of partial components with phase differences δ:

$$t^2 + t^2r^2e^{-i\delta} + t^2r^4e^{-i2\delta} + \cdots + t^2r^{2(j-1)}e^{-i(j-1)\delta} + \cdots, \qquad \delta = \frac{4\pi}{\lambda}L\cos\theta,$$

$i = \sqrt{-1}$, $j = 1, 2, 3, \ldots$. In the limit of $j \to \infty$, this geometrical progression converges to

$$a\sum_{j=0}^{\infty} q^j = \frac{a}{1-q},$$

giving the resultant amplitude of the transmitted wave

$$\frac{t^2}{1 - r^2e^{-i\delta}}.$$

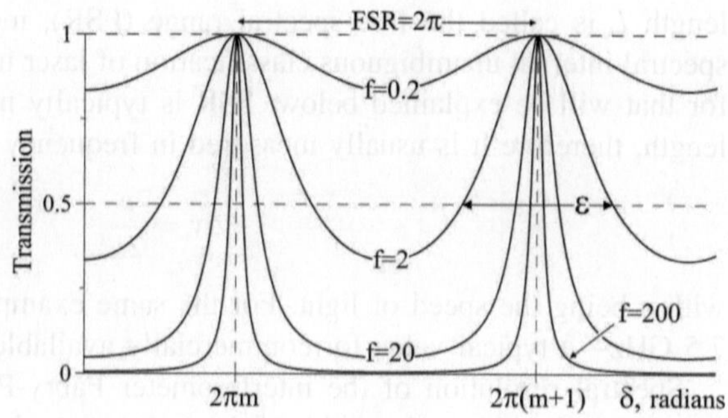

Fig. 9.48 The Airy formula gives periodical function for transmission of the interferometer Fabry-Perot (IFP). With reflectivity of the mirrors $R \to 1$ the shape factor $f \to \infty$, making the peaks narrower

In order to minimize unnecessary details, we allowed some inexactness in this derivation, assuming that transmission coefficients on the entry into the IFP and on the exit of it are the same t. Nevertheless, if this is followed by another minor inexactness in the form of writing energy transmission and reflection coefficients as $T = t^2$ and $R = r^2$, then the final result is exact:

$$\frac{T}{1 - Re^{-i\delta}}.$$

Transmission coefficient of the entire IFP is the square modulus of it, which is better be written in the following form:

$$\frac{T^2}{(1-R)^2 + 4R\sin^2\frac{\delta}{2}}.$$

This formula is known as the Airy formula, which must not be mistaken with the famous Airy function that relates to diffraction on a round hole. In the ideal case of the absence of absorbance, i.e. $T = 1 - R$, the formula transforms to

$$\frac{1}{1 + f\sin^2\frac{\delta}{2}}, \quad f = \frac{4R}{(1-R)^2}.$$

The basic result that follows from here is that transmission of IFP is periodical function of δ with transmission peaks the narrower the higher the reflection coefficient R (Fig. 9.48). The shape reproduces itself every $\delta = 2\pi m$, $m = 0, \pm 1, \pm 2, \ldots$.
Recalling that

$$\delta = \frac{4\pi}{\lambda} L\cos\theta,$$

it means that interferometer is transparent for wavelengths λ_m that satisfy condition

$$\lambda_m = \frac{2L}{m}$$

at $\theta = 0$. Integer m is a very big number. For example, for $L = 20$ mm and visible domain $\lambda \sim 500$ nm $m \sim 4{\cdot}10^4$. Spectral separation $\lambda_m - \lambda_{m\text{-}1}$ for the particular length L is called the free spectral range (FSR), meaning that only within this spectral interval unambiguous classification of laser modes is possible. The reason for that will be explained below. FSR is typically much smaller than the wavelength, therefore it is usually measured in frequency units:

$$\Delta\nu = \frac{c}{2L}$$

with c being the speed of light. For the same example with $L = 20$ mm, FSR is 7.5 GHz—a typical value for commercially available IFPs.

Spectral resolution of the interferometer Fabry-Perot (IFP) is determined by instrumental width —the width of the resonant peak in the Airy curve at half the maximum. The number of spectral peaks that can be resolved within the FSR characterizes the quality of an IFP and is called finesse:

$$F = \frac{\text{FSR}}{\varepsilon}.$$

In the domain of phase differences FSR $= 2\pi$, and ε can be derived by applying Taylor expansion for $\varepsilon \ll 1$ to equation

$$\frac{1}{1 + f\,\sin^2\frac{\varepsilon}{4}} = \frac{1}{2}.$$

Then

$$\varepsilon = \frac{4}{\sqrt{f}} \quad \text{and } F = \frac{\pi}{2}\sqrt{f} = \frac{\pi\sqrt{R}}{1-R} \approx \frac{\pi}{1-R}.$$

This very important formula shows that spectral resolution of the flat-mirror IFP increases as reflection coefficient of the mirrors approaches unity. For the finesse to be in the range 100–200, the reflection coefficient of flat mirrors must be as high as 0.98, and it really is in the specified spectral interval. Finesse and free spectral range (FSR)—are the two basic parameter always included in specifications of scanning IFPs.

Now we are ready to understand how the scanning IFP works (Fig. 9.49). Initially, highly reflective mirrors totally block the laser beam from reaching the photodetector. Probability of incidental resonant matching $2L = m\lambda$ between the cavity spacing L and laser wavelength λ is negligible. Fine changes of L are needed to reach full transparency of the IFP. This is done by scanning one of the mirrors. From Fig. 9.48 it is clear that when L changes more then by $\lambda_m/2$ the same wavelength λ_m will be marked for the second time, and may overlap with the wavelength of an adjacent order $\lambda_{m\pm1}$. This happens when spectral width of the laser mode structure is wider than FSR of the IFP. Since longitudinal modes of a laser are spaced in frequency domain by $c/2Z$ where Z is the length of the laser cavity, the full mode structure of a particular laser may be covered by IFP with $L < Z$. However, it does not mean that small spacing L is always better: when superfine mode structure is to be analyzed, then longer spacing makes better

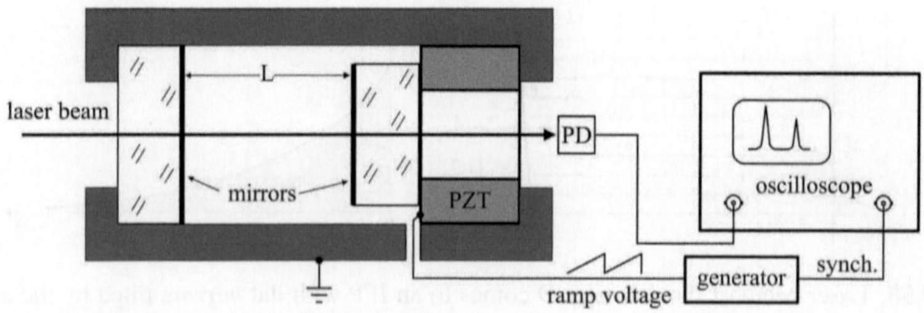

Fig. 9.49 Piezo-transducer (PZT) changes cavity spacing L according to ramp voltage of about 100 V from electrical generator. At the moments when the cavity becomes transparent for a specific wavelength, photodetector (PD) excites short bursts at the screen of the oscilloscope synchronized to the generator. Stable picture of laser modes displays on the screen

magnification of the frequency interval of interest. Spectral resolution determined by finesse F does not depend on spacing in our approximation.

Now it is time to make a shocking statement: scanning IFP with plane mirrors never works in laser mode analysis. There are two reasons for that. First, high finesse can be achieved only at normal incidence ($\theta = 0$), and this alignment creates strong reflected wave directed back into the laser cavity, disturbing its operation. Second, tolerance on parallelism of the mirrors is so high that makes volume manufacturing impractical even on the scale of small scientific production. While the first reason is obvious, the second one requires explanation. Figure 9.50 helps to grasp the idea.

With the tilt α, average dephasing within the beam diameter D is

$$\phi \sim 2\frac{2\pi}{\lambda}\xi = \frac{2\pi D\alpha}{\lambda}.$$

Each ray undergoes on the average $\bar{n}$ reflections until total extinction, and accumulates total dephasing $\bar{n}\,\phi$. This total dephasing must not exceed π:

$$\bar{n}\,\phi < \pi.$$

Average number of reflections $\bar{n}$ is closely related to finesse F and may be estimated as the integral average (Fig. 9.50):

$$\bar{n} \cdot I_0 = \sum_{n=0}^{\infty} I_n = \frac{I_0}{1-R},$$

giving

$$\bar{n} = \frac{1}{1-R} = \frac{1}{\pi}F.$$

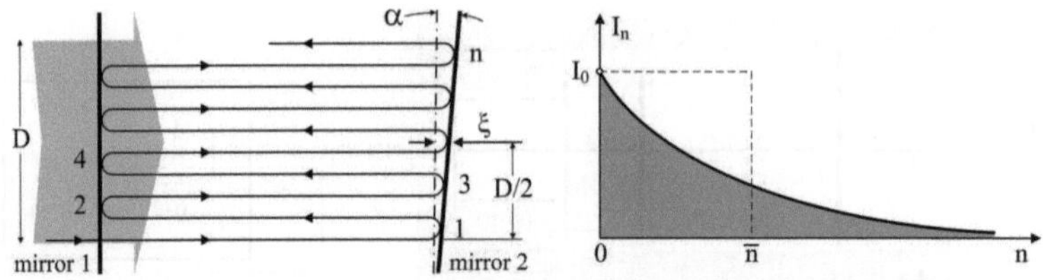

Fig. 9.50 Laser beam of the diameter D comes to an IFP with flat *mirrors* tilted by the angle α. Average variation of optical paths is 2ξ. Each ray exhibits multiple reflections 1, 2, 3,... n, each time loosing intensity by the factor R. Intensity of the ray after n-th reflection decreases as $I_n = I_0 \cdot R^n$

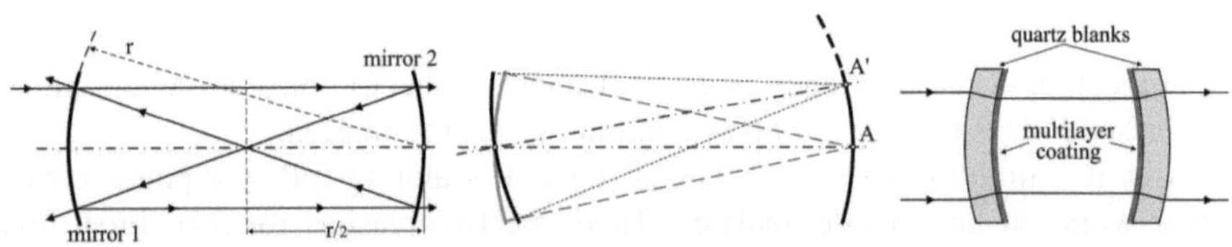

Fig. 9.51 Basic scheme of reflections in confocal IFP. Before recombining in wavefronts, each ray passes four times through the cavity, exiting also four times but never in the initial direction (*at left*). Tilting of mirrors does not change properties of the cavity: the point A only slips to a new position A′ on the same mirror, reproducing initial configuration (*in the middle*). Highly reflecting multilayer coatings are deposited on thin mirror blanks of zero optical power, i.e. concentric surfaces, thus preserving collimation at the output of the IFP (*at right*)

This gives the estimate for the maximum tilt angle

$$\alpha < \frac{\pi}{2} \cdot \frac{\lambda}{DF} \sim \frac{\lambda}{DF}.$$

Thus, the tilt angle must be F times smaller than the diffraction divergence of the laser beam—a very stringent requirement that cannot be reliably maintained in practice even if the beam is focused inside the cavity to a micrometer-scale size D.

The solution is the so-called confocal or spherical interferometer Fabry-Perot formed by two spherical mirrors placed at doubled focal length from one another (Fig. 9.51). The greatest advantage of this scheme is insensitivity to tilts of the mirrors. It makes fine angular adjustment redundant, dramatically simplifying manufacturing. The only requirement left is longitudinal adjustment of mirrors to ensure confocality, which can be easily accomplished by fine screws. The second problem of the flat-mirror IFP is also solved: reflected rays do not propagate backwards, making no disturbance to the laser.

Theoretical results obtained for the flat-mirror IFP hold true for the confocal IFP with only very simple modification: since rays reflect four times instead of two, R in the formulas above must be replaced with R^2. Thus, finesse

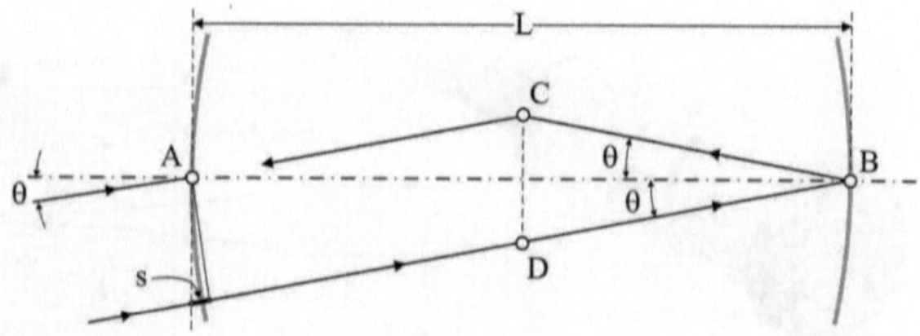

Fig. 9.52 In a confocal IFP, all the rays coming at an angle θ focus first time in the focal point C, then in the focal point D, and then in the focal point C again

$$F = \frac{\pi R}{1 - R^2} \approx \frac{\pi}{2\,(1 - R)};$$

Airy formula

$$\frac{T^2}{\left(1 - R^2\right)^2 + 4R^2\,\sin^2\frac{\delta}{2}};$$

shape factor

$$f = \frac{4R^2}{\left(1 - R^2\right)^2}.$$

Note that now finesse F is two times smaller than in the case of a flat-mirror IFP, and it brings complications: reflection coefficient of the mirrors must be higher—around 0.993 for $F = 200$.

Another important feature of the confocal IFP is that the phase difference δ does not depend on the angle of incidence θ like in the flat-mirror IFP, dramatically increasing stability of measurements. This is explained in Fig. 9.52. In order to determine phase difference δ after the round-trip circulation in the cavity, consider its first half from the point of entrance A on the left mirror to the right mirror, then to first focal point C, and then to A again. Instead of exploring cumbersome ray traces from point A to the right mirror and then to point A back, we can use the Descartes principle, saying that optical paths to the focus are the same for all the rays starting at one wavefront surface. As such, the path from A to C is the same as that along the ray passing through D-B-C:

$$\frac{L}{\cos\theta} - s + \frac{L}{2\cos\theta}; \qquad s = L\tan\theta \cdot \sin\theta.$$

From C the diverging cone of rays passes to the left mirror with the section C-A being

$$\frac{L}{2\cos\theta}.$$

Adding, obtain the first half of the optical path p

$$\frac{1}{2}p = L\cos\theta + \frac{L}{\cos\theta}$$

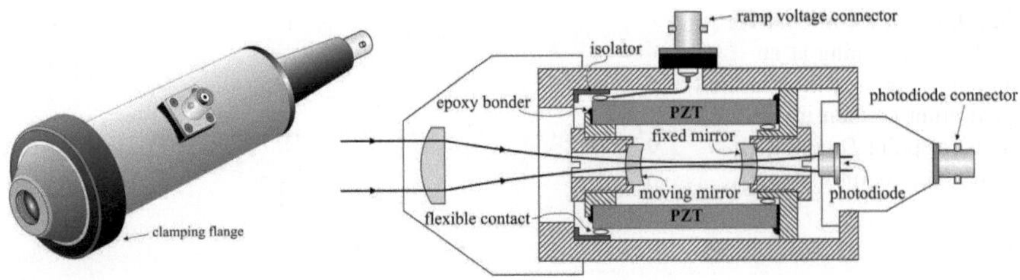

Fig. 9.53 Typical design of a scanning interferometer Fabry-Perot (IFP). Inner and outer cylindrical surfaces of the piezo-transducer (PZT) are metallized and connected to the ramp voltage through flexible contacts. The photodiode tail is usually removable in order to observe interference fringes when the device is illuminated from the *right side*. Then the input lens serves as a microscope. Mirrors are fixed in threaded nuts for adjusting the length of the cavity

and the full optical path

$$p = 2L\left(\cos\theta + \frac{1}{\cos\theta}\right).$$

For paraxial rays with $\theta \ll 1$ use Taylor expansion of cosine to obtain to the accuracy of the fourth order in θ

$$P = 4L \, \text{and} \, \delta = \frac{8\pi}{\lambda}L.$$

Thus, the phase difference in confocal IFP does not depend on the angle of incidence and is two times bigger than in the flat-mirror IFP. From another point of view, paraxial approximation $\theta \ll 1$ guarantees the absence of spherical aberrations on spherical mirrors. If, however, beam diameter D increases, spherical aberration becomes noticeable, adding corrections to the above formula for the path difference δ. This effect displays itself in a form of circular interference fringes that are formed in the middle plane of the cavity where rays are focused into the image of the source: wide central circle with ever tighter concentric circles around it. In spectroscopic applications, which are the subject of this section, fringes are only a hindrance because they decrease spectral resolution, and as such they should be eliminated. Therefore, the narrower the beam inside the cavity the better. The best solution is to focus the beam in the middle of the cavity. Then rays will go exactly the same way as in the case of a collimated beam (Fig. 9.51) but in a reverse order.

All the commercially available scanning IFPs are of the confocal type (Fig. 9.53).

Narrow focused laser beam can miss the photodetector if not properly aligned. For that, the IFP must be clamped in an adjustable mount with the laser beam approximately in the middle of the entrance lens (Fig. 9.54). Then adjust angularly the IFP to bring the reflected beam cone coaxially with the laser beam. Clamping flange is always designed to be roughly in the plane of the lens in order to avoid

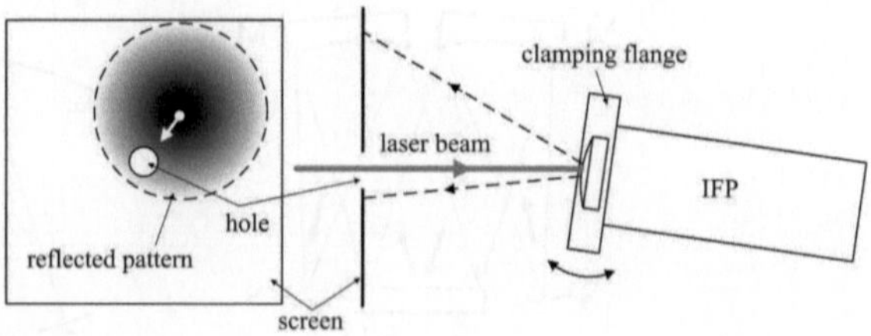

Fig. 9.54 Adjust IFP angularly so that the reflected pattern is roughly centered around the laser beam

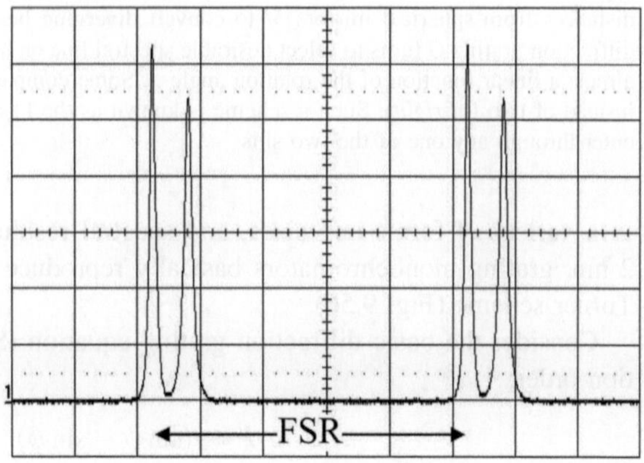

Fig. 9.55 Two laser modes displayed twice during scan over approximately 2 *FSR* range. The finesse is about 50

lateral displacements during angular alignment. The IFP is ready for use. If the reflected beam is hardly visible, then another option may be used: remove the photodetector tail and entrance lens; pass the laser beam through the IFP onto the screen; reinstall the lens and adjust the IFP angularly to superimpose the centers of the two beams. Connect the entire system as in Fig. 9.49 and observe the result like the one shown in Fig. 9.55.

9.6 Monochromators: Diffraction Gratings or Filters?

Monochromators are optical devices, selecting narrow spectral components from white light. It can be done either with the help of dispersive optical elements like diffraction gratings or spectrally-blocking devices like filters. Grating monochromators can arbitrarily select any spectral line from their working spectral interval whereas filters are designed for a fixed spectral line. Spectrometers (Sect. 9.1) do actually the same as grating monochromators but terminate spectrally decomposed beam by a relatively slow photodetector. In some applications, fast response is required at one specific spectral line, and then CCD detectors must be substituted for fast photomultipliers connected to the exit slit of a monochromator. Available

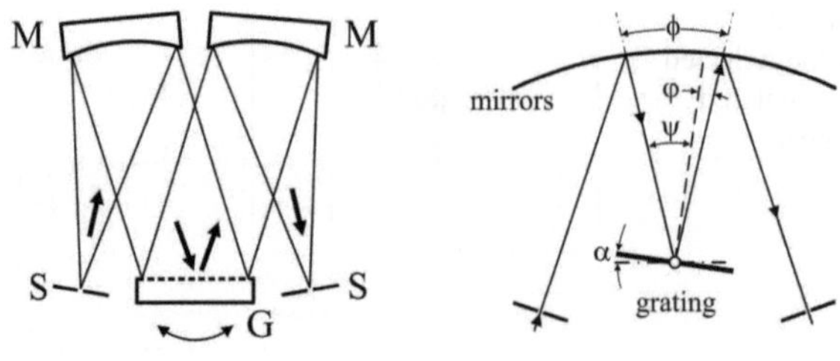

Fig. 9.56 Uncrossed Czerny-Turner monochromator (*at left*). Slits *S* are positioned at focal distances from spherical mirrors *M* to convert diverging beam to parallel and in reverse. Flat diffraction grating *G* turns to select desirable spectral line on the exit slit. Wavelength selection is almost a linear function of the rotation angle α. Some compact designs use a single long mirror instead of two (*at right*). Such a scheme is known as the Fastie-Ebert configuration. Light may enter through any one of the two slits

in a variety of forms and sizes, and spectral resolution varying from 0.1 nm to 2 nm, grating monochromators basically reproduce the same uncrossed Czerny-Turner scheme (Fig. 9.56).

Consider the basic diffraction grating equation (Sect. 9.1) for the +1st diffraction order:

$$\lambda = t(\sin\psi - \sin\varphi)$$

with t being the period of the grating. Since all the optical elements except grating are fixed, the rays coming to the grating and exiting through the exit slit are also fixed, making the angle

$$\phi = \psi + \varphi$$

constant. Practically, $\phi \sim 30$. In horizontal position of the grating $\psi = \phi/2$, and being rotated by the angle α grating comes to a new angle of incidence $\psi = \phi/2 + \alpha$, selecting the wavelength

$$\lambda = t\,[\sin\psi - \sin(\phi - \psi)] = 2\,t\,\cos\frac{\phi}{2}\cdot\sin\alpha$$

This function is shown in Fig. 9.57 along with the so-called sine-bar mechanism, providing linear wavelength tuning.

Grating monochromators do not have order-sorting filters like spectrometers do (Sect. 9.1) because the selected wavelength is an arbitrary choice of the user. Therefore, working with broad-band sources, always be prepared for higher-order ghosts, emerging from the exit slit along with the selected spectral component. Figure 9.58 exemplifies some typical situations.

In recent years, following the wave of compact fiber-optic spectrometers, the market received compact grating monochromators—handheld devices with manual selection of wavelengths and fiber-optic input-output, which can be a good solution for limited space on optical table (Fig. 9.59).

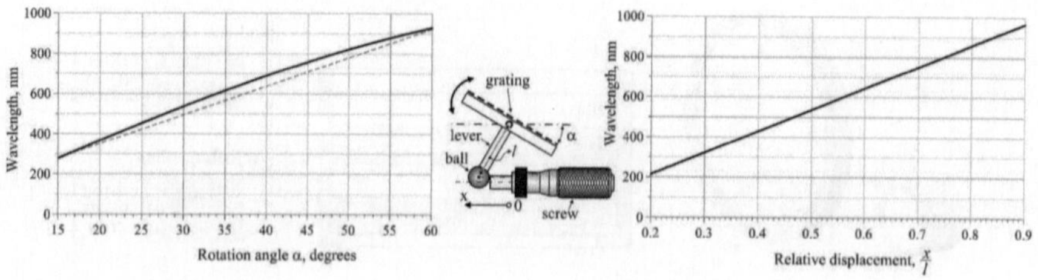

Fig. 9.57 Tuning curve $\lambda(\alpha)$ computed in the range 200–900 nm for the grating with 1800 grooves per millimeter and $\varphi = 30°$ (*at left*). Straight dashed line shows non-linearity of angular tuning. The sine-bar mechanism (*in the middle*) provides linear tuning as a function of the screw relative displacement (*at right*). With the ball at the end, the lever keeps constant hypotenuse *l*, making $\alpha = \arcsin(x/l)$

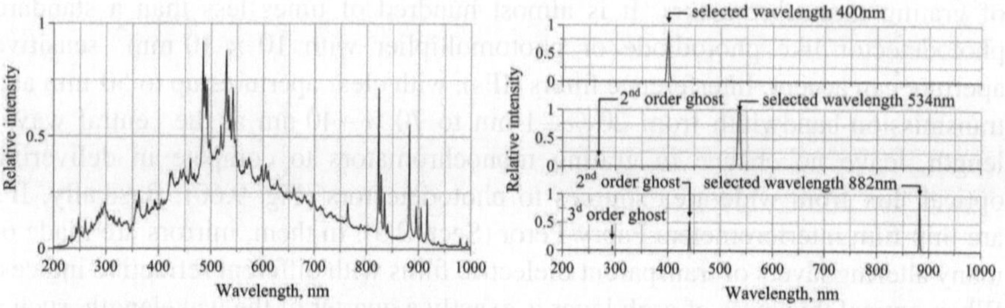

Fig. 9.58 Broad-band spectrum of Xe flash lamp (*at left*) creates multiple-order ghosts at the output of a monochromator without additional filters (*at right*). The lamp spectrum does not contain components below 220 nm, therefore when monochromator is set for 400 nm, no ghosts are visible (*upper spectrum at right*). Selection of the 534 nm line drags 534/2 = 267 nm ghost from the xenon spectrum (*middle section at right*). Even third-order ghost may accompany longer wavelengths (*bottom at right*)

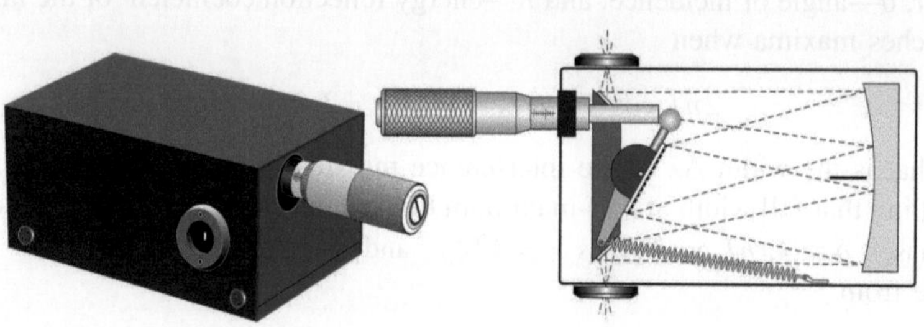

Fig. 9.59 Compact grating monochromators are commonly built on the Fastie-Ebert scheme. Belonging in the class of compact devices, they are not supposed to show too high spectral resolution, which is typically about 3 nm with 300 μ slits

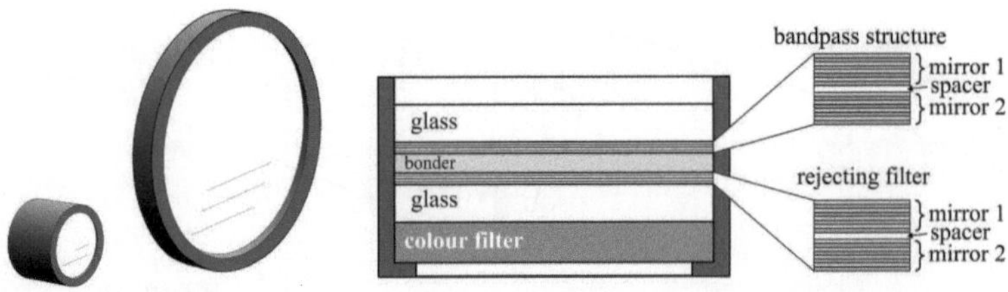

Fig. 9.60 Interference filters (IFs) are essentially closely spaced interferometers Fabry-Perot with multilayer mirrors

The indispensable feature of grating monochromators is the ability to arbitrarily select the wavelength with potentially high spectral resolution. On the other hand, small area of a slit, not exceeding $5 \times 0.3\ \text{mm}^2$, sets the limit to energy efficiency of grating monochromators. It is almost hundred of times less than a standard photodetector like photodiode or photomultiplier with $10 \times 10\ \text{mm}^2$ sensitive aperture can accept. Interference filters (IFs), with clear apertures up to 50 mm and transmission-bandwidth from 20 %–1 nm to 70 %–40 nm at the central wavelength, leave no chance to grating monochromators to compete in delivering optical flux from wide-area sources to photodetectors (Fig. 9.60). Basically, IFs are thin-film interferometers Fabry-Perot (Sect. 9.5). In them, mirrors are made of many altering layers of transparent dielectric films with different refractive indices. When optical thickness of each layer is exactly a quarter of the wavelength, such a structure creates constructive interference and behaves as a mirror. According to the Airy formula, transparency of the Fabry-Perot structure is proportional to

$$\frac{1}{1+f\,\sin^2\left(\pi\frac{2nL}{\lambda}\cos\theta\right)},\quad f=\frac{4R}{(1-R)^2}$$

with λ being the wavelength, n and L—refractive index and thickness of the spacer, θ—angle of incidence, and R—energy reflection coefficient of the mirrors. It reaches maxima when

$$2nL\cos\theta = m\lambda,\ m=\pm1,\pm2,\pm3,\ldots.$$

What is the width $\Delta\lambda$ of the interference maxima in the wavelength domain? Recalling that full width at half-maximum of each interference peak in the domain of phases $\delta \equiv 4\pi nL\cos\theta/\lambda$ is $\varepsilon = 4/\sqrt{f}$, and assuming $\theta = 0$, we can easily derive from

$$\varepsilon = \left|\frac{d\delta}{d\lambda}\right|\Delta\lambda$$

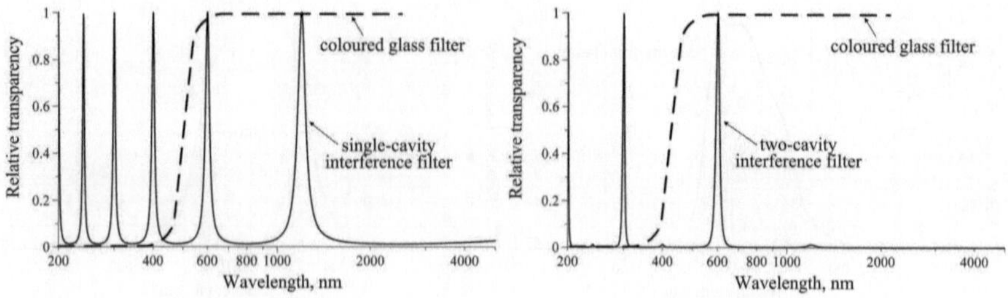

Fig. 9.61 Theoretical spectral transparency of a single Fabry-Perot cavity with $R = 0.8$ and $2nL = 1200$ nm (*at left*). Spectral peaks at 400 nm and below can be stopped by a coloured glass filter as shown in the dashed line. The second IF with $2nL = 600$ nm not only suppresses parasitic maximum at 1200 nm but also sharpens the design peak at 600 nm (*at right*)

that

$$\frac{\Delta\lambda}{\lambda} = \frac{\lambda}{\sqrt{f}\pi nL} = \frac{1}{mF},$$

where

$$F \approx \frac{\pi}{1 - R}$$

is the finesse (Sect. 9.5). Therefore, if we want narrow spectral transmission then both the interference order m and the finesse F must be high. Big finesse means that reflectivity of mirrors must be close to unity, which is quite achievable with dielectric multilayer coatings. As to the interference order m, it cannot be made high because then consecutive orders become inseparable. Therefore, in IFs, $m = 1 - 3$. Suppose, for example, we want to make an IF for 600 nm at normal incidence with reflecting stack with $R = 0.8$. This reflectivity gives finesse only 15. In order to make relative spectral width narrower we have to choose bigger interference order, say $m = 2$. For these parameters, theoretical transmission curve for a single cavity given by the Airy formula is shown in Fig. 9.61.

Now, the problem is how to block spectral maxima other than 600 nm. The shorter wavelengths can be blocked by a red filter, but the longer wavelengths cannot. This is usually done by inserting the second interference filter with $m = 1$. Transmission of such a sandwich is the product of the two individual curves, which gives already acceptable result (Fig. 9.61). Not only transmission at 1200 nm is blocked but also the peak at 600 nm is narrower.

An example of real transmission curve of a most popular IF with 10 nm bandwidth is shown in Fig. 9.62.

Some attention should be paid to angular properties of interference filters (IFs). From the aforementioned, it follows that the resonant wavelength decreases with

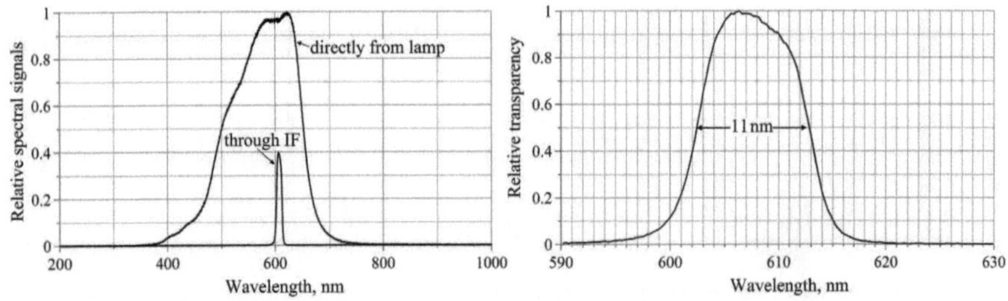

Fig. 9.62 Spectral transmission of an IF designed for 610 nm. Design width 10 nm. Broad angular illumination. Comparison with direct flux of tungsten-halogen lamp shows transmission efficiency better than 40 % (*at left*). Spectral width is approximately 11 nm at half maximum (*at right*)

the angle of incidence as cosθ. Therefore, if very narrow spectral line is needed then it is better to place the IF in a parallel beam. In this configuration, some spectral tuning is possible by tilting the IF relative to the beam. On the contrary, if the beam is diverging then some very minor widening of spectral transmission may be expected.

List of Common Mistakes

- connecting spectrometer to a source through a fiber bundle without central light guide;
- fiber bundle with improper spectral transmission;
- wide-range spectrometer without order-sorting filter;
- erroneous calibration of a spectrometer on closely positioned argon lines.

Further Reading

M. Born, E. Wolf, Principles of Optics, Cambridge University Press, 7th ed., 1999.
E.G. Loewen, E. Popov, Diffraction Gratings and Applications, CRC Press, 1997.
W.H. Steel, Interferometry, Cambridge University Press, 2nd ed., 2009.
H.A. Macleod, Thin-film Optical Filters, Institute of Physics Publishing, 3nd ed., 2003.

Chapter 10
Beam Alignment and Positioning Techniques

When it comes to adjustment, some simple practical advises may save much time.

Abstract In four sections, this chapter guides the reader through variety of mechanisms for controlling the beam. Manual stages for angular alignment are considered in the first section: three-contact gimbals, flexible hinges, rotating wedges. Displacement curves and deflection nomogram with analytical formula are presented for the reference purposes. Lateral alignment with the pairs of flats and mirrors is explained in the next section, also supported by the nomogram and formula The third section summarizes most common beam steering elements: the widely used galvano-mirrors and much more rare Risley prism. The rules of coupling galvano-mirrors to light sources through relay optics (Chap. 1) are explained and graphically illustrated. The reader will find design details, performance, and practical realization of two-dimensional scanners with the example of a simple laser marking arrangement. Angular precision of this type of scanners is determined by optical feedback, using quadrant photodiodes (Chap. 3). Experimental results presented in the form of oscilloscope traces give good impression of the capabilities of these devices. The Risley prism is a very efficient but much more expensive solution for beam steering. Beam steering devices often require F-theta lens for focusing. This special type of lenses, which was not mentioned in Chap. 1, is explained in detail in this section: the principle, design considerations, and mathematical explanation of its rather peculiar shape. The forth section is a source of concentrated information about manual and motorized translation stages that may be found in the laboratory. Some simple recommendations and practical know-how may significantly improve performance of translational stages. High-precision differential stages and micrometer screws are explained. Detailed discussion of stepping motors—connection diagrams, drivers, microstepping modes, electromagnetic noise, vibrations—illustrated with experimental oscillograms is supposed to clarify many practical questions. Scrutinized analysis of geared motorized stages unveils some rarely known facts about their precision. The section and the chapter end with detailed explanation of design and characteristics of piezo-stages. Although the basic principle of piezo-electric deformation is widely known, particular design and technology of contemporary multilayer piezo-actuators with 0.

V. Protopopov, *Practical Opto-Electronics*, Springer Series in Optical Sciences 184,
DOI: 10.1007/978-3-319-04513-9_10,

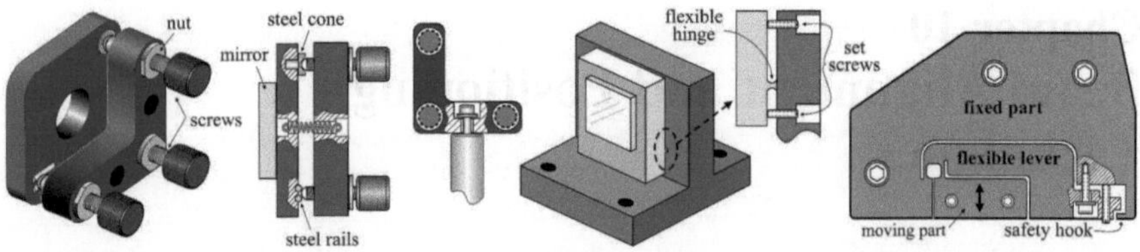

Fig. 10.1 General-purpose three-contact gimbal (*at left*). The only fixing option for it is single-bolt clamping (*in the middle*). Flexible hinges provide utmost stiffness but not very large adjustment range (*next to the right*). Flexible lever may be used to improve resolution (*at right*)

1 % excursion are not in the public domain. The effects of hysteresis, poling, and repolarization are explained in order to avoid possible mistakes in applications. Finally, the chapter presents impressive experimental comparison of a geared motorized stage with the piezo-stage.

10.1 Angular Alignment of Beams

Directing a beam of light is the most common necessity in practice. An obvious solution is the adjustable mirror. Traditional three-contact gimbal with spiral springs is a good choice if only some design requirements are satisfied (Fig. 10.1). The first of those is steel inserts to which the pushing screws contact. The frames are always made of aluminium in order to reduce the weight, and in some cheap versions hard steel screws press directly on soft aluminium surface of the frame, making small pits in it. This makes fine adjustment unreliable. High-quality gimbals always have steel hardened inserts under the screws. Besides, it is important that steel screws move inside bronze or brass nuts and not directly in threaded holes in the aluminium frame. Thread pitch is also an important quality of the gimbal. Inexpensive versions may have coarse pitch like 0.5 mm or even 0.75 mm—standard for M3 and M4 threads respectively. If we assume that 5° is the best that fingers can reliably reproduce on rotation, then with 20 mm lever such a screw gives not more than 2 arc minutes accuracy. High-quality screws usually have 0.35 mm or even 0.25 mm pitch, providing several times better precision.

In the three-contact gimbals, the mirror-carrying frame is fixed by nothing but a spring, and therefore may be vulnerable to incidental displacements. But the weakest point is their installation and bolting. As a rule, they are designed to be bolted to posts on a perforated optical table, with the only hole for M6 or M4 bolts in the frame. Much more stable design uses flexible hinges (Fig. 10.1). Rather complicated long and narrow (less than 1 mm) slashes in metal can be made by spark erosion (electro-discharge machines). The force needed to flex the joint is high, requiring screws not finer than M3. As such, precision of angular adjustment is supposed to be not very high for a flexible hinge alone. In order to overcome this limitation, various systems of flexible levers may be used.

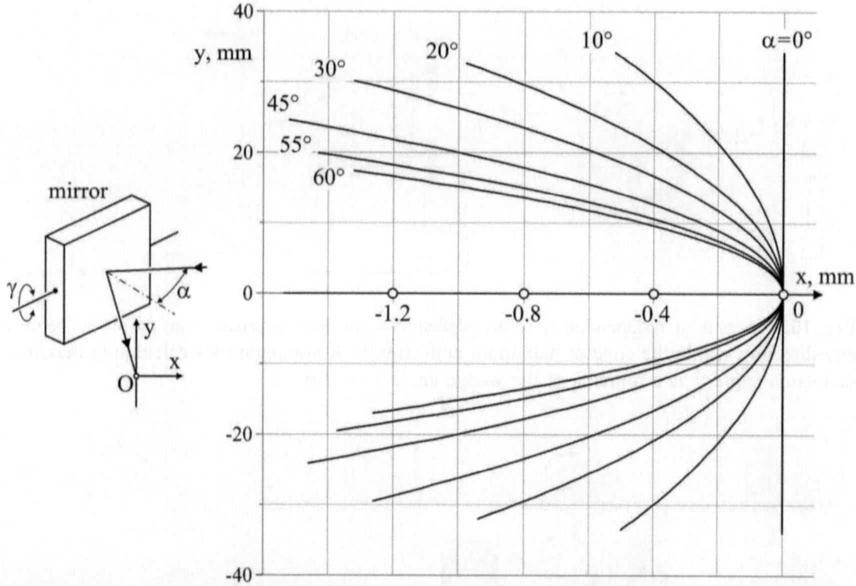

Fig. 10.2 Displacement curves of a beam at 200 mm away from the mirror. Each curve corresponds to vertical rotation from $\gamma = -5°$ to $\gamma = +5°$

The beam reflected from the mirror reacts differently to horizontal (in the plane of incidence) and vertical tilts. Horizontal tilts of the mirror always result in only horizontal deflection of the beam equal to double angle of mirror rotation. Vertical rotation γ, however, results in both vertical and horizontal deflection (Fig. 10.2). Although horizontal coupling is relatively small comparing to vertical deflection, it must be taken into account.

Angular position of the beam reflected from the mirror repeats all the vibrations it exerts. Alternatively, transmission optics is insensitive to vibrations. Therefore, whenever possible, use transmission optical elements for angular alignment of the beam. Not only it decreases the influence of vibrations but also makes the entire beam path shorter. Standard angular deflection element, known as the Risley prism, consists of a pair of independently rotating glass wedges (Fig. 10.3). From the Snell law, it is easy to obtain the formula for maximum deflection angle θ that can be obtained with the Risley prism of the refractive index n:

$$\theta = \arcsin\left(n \sin\left(\alpha + \arcsin\left(\frac{1}{n}\sin(\arcsin(n \sin\alpha) - 2\alpha)\right)\right)\right).$$

Cylindrical cavities in which the wedges are installed are always made on a lathe, with the nesting rim perpendicular to the axis of rotation. As such, front optical surface is always perpendicular to optical axis. It does not mean, however, that the reflection from these surfaces returns directly back to the laser: mechanical uncertainties of installation always diverge the reflected beam from penetrating the laser.

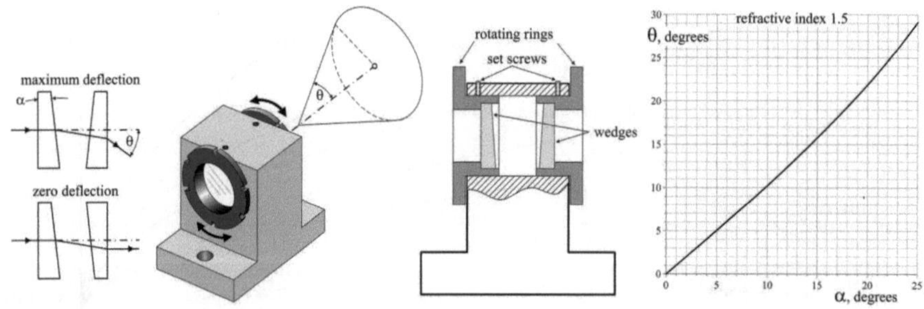

Fig. 10.3 A pair of independently rotating wedges—the Risley prism—can point the beam to any direction within the cone of maximum deflection θ. A nomogram for calculating maximum deflection angle θ as a function of the wedge angle α is *at right*

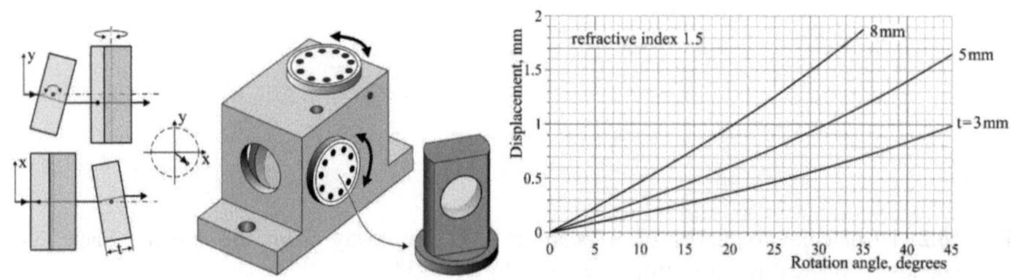

Fig. 10.4 Independent rotation of each optical flat produces parallel translation of the beam in two directions (*at left*). Nomogram at right may be used to estimate displacement of the beam after passing a single optical flat of thickness t (*at right*)

10.2 Lateral Alignment of Beams

When the beam must be shifted without changing direction, the scheme shown in Fig. 10.4 solves the problem. Simple formula gives the value of the displacement s in an optical flat of thickness t and refractive index n rotated by the angle α:

$$s = t\sin\alpha\left(1 - \frac{\cos\alpha}{n\,\cos\left(\arcsin\left(\frac{\sin\alpha}{n}\right)\right)}\right).$$

Sometimes, much bigger parallel displacements are required, on the scale of centimeters. This is not an issue of adjustment but better to say leveling of optical axes of different devices. For example, optical axis of a big laser may be higher than other components assembled on an optical table. There is no sense in mounting the entire optical system on a separate platform above the optical table only because the laser alone is too high. The solution is to bring the laser beam down, preserving its direction, with the simple arrangement shown in Fig. 10.5. Manufacturing technology is very important for preserving direction of the beam, and it should be explained to manufacturer beforehand in order to avoid mistakes.

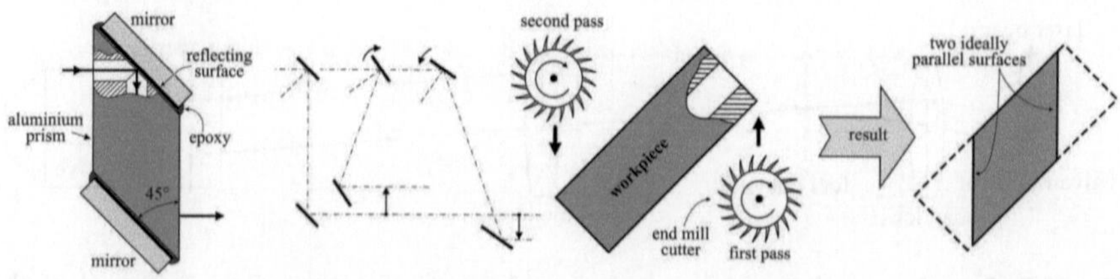

Fig. 10.5 A 45° prism with two mirrors bonded to parallel surfaces lowers (or rises) the beam, preserving initial direction (*at left*). Slightly tilting the assembly in vertical plane, it is possible to adjust vertical position of the beam without loosing initial direction. Two parallel approximately 45° slanted surfaces can be reliably manufactured on a vertical milling machine (*at right*). The workpiece must be clamped horizontally on the table at 45° to the direction of processing. At first pass, one triangular part of the workpiece is cut, at the second pass without reclamping—another one

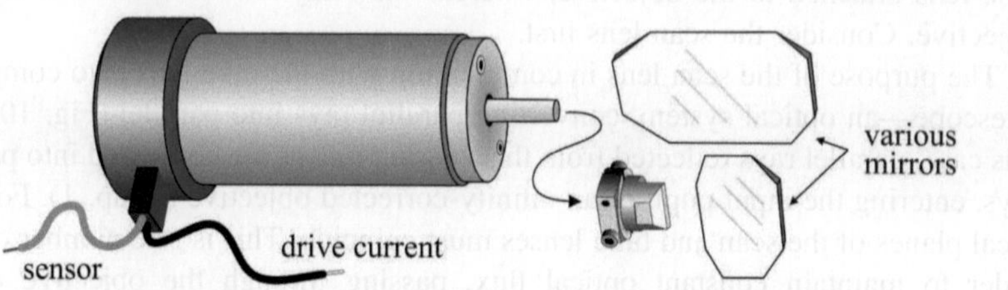

Fig. 10.6 Inside the motor, there are the stator coil and permanent magnet rotor coaxial with rotary encoder. Various geometry of mirrors is available, depending on application

10.3 Beam Steering

Widely accepted and well-established technique for beam steering is galvano-mirrors (Fig. 10.6). Usually, galvano-mirror works in the so-called closed loop mode when the driver compares the command signal with the signal of the encoder, making the difference close to zero. With maximum mirror deflection typically ±5°, the majority of the single-axis scanners follow bar-step command signals up to 100 Hz, and sinusoidal signals 2–3 times quicker. In small angle deflection, within ±1°, they are significantly faster, up to 1 kHz range.

With the manufacturer-supplied driver accurately setting angular positions of the laser beam, the user does not have to think much about precision of scanning, but he does have to think about accurate delivery of the beam. There is no great variety of applications where galvano-scanners are used: typically it is a kind of scanning microscopy or its equivalent, with laser beam being focused onto the sample through an objective lens. How to couple the scanner to an objective? There are two standard solutions: the so-called scan lens and the relay lens (Chap. 1). The difference between them is that the scan lens is used on an assembled microscope with

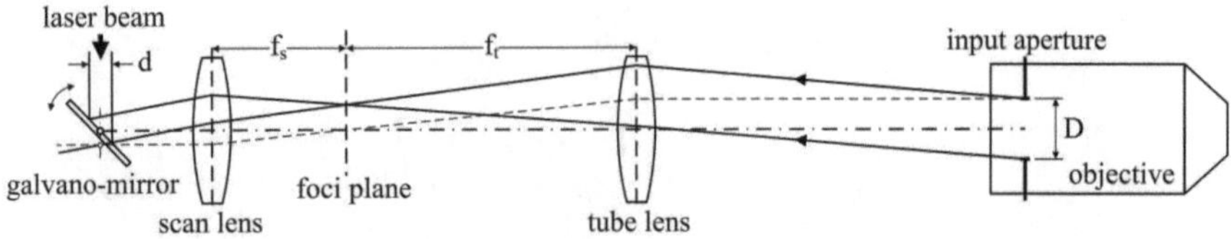

Fig. 10.7 Scan lens is supposed to couple laser beam to the microscope, composed of the objective itself and the tube lens. For the best performance, galvano-mirror must be placed in the input pupil of the optical system, and the input aperture of the objective—in its output pupil. The output and the input pupils of the system are the conjugated images of one another. Since the microscope is a given device, start with its input aperture *D*, tracing it back to galvano-mirror as shown in the figure. The image of the upper point of the input aperture of the objective is formed by intersection of two rays: one skewed and one parallel to the optical axis. This point defines position of the input pupil, where rotation axis of the galvano-mirror must be placed

tube lens attached to the objective, whereas the relay lens is used on a separate objective. Consider the scan lens first.

The purpose of the scan lens in combination with the tube lens is to compose a telescope—an optical system, converting parallel rays into parallel (Fig. 10.7). In this case, parallel rays reflected from the steering mirror are converted into parallel rays, entering the input pupil of an infinity-corrected objective (Chap. 1). For that, focal planes of the scan and tube lenses must coincide. This is rule number one. In order to maintain constant optical flux, passing through the objective during scanning, laser beam must permanently fill the entire input aperture of the objective. This property also guarantees maximum possible spatial resolution as the diffractional divergence is minimized. In the optical language, it means that galvano-mirror must be placed in the input and the objective in the output pupils of the optical system. This is rule number two. Diameter of the input aperture of the objective D and diameter of the laser beam d are two given parameters, and the focal distance of the scan lens f_s must be chosen so that to satisfy obvious relation:

$$\frac{f_t}{f_s} = \frac{D}{d}.$$

Depending on the distance between the tube lens and the objective in a particular microscope, galvano-mirror must be place closer or farer to the scan lens, according to considerations explained in the caption to Fig. 10.7. Since the input and output pupils of an optical system are actually its conjugated images, from general optical phenomenology it follows that bringing galvano-mirror closer to the scan lens results in moving the objective farer from the tube lens, and *visa versa*.

Pre-installed tube lens, designed for specific microscopic applications, is not optimized for scanning. Using special relay lens (Chap. 1), it is possible to reduce space between galvano-mirror and objective (Fig. 10.8).
For that, focal length of each of the pair of lenses should be short. Again, the system must be telescopic, converting parallel beam to parallel. Therefore, focal

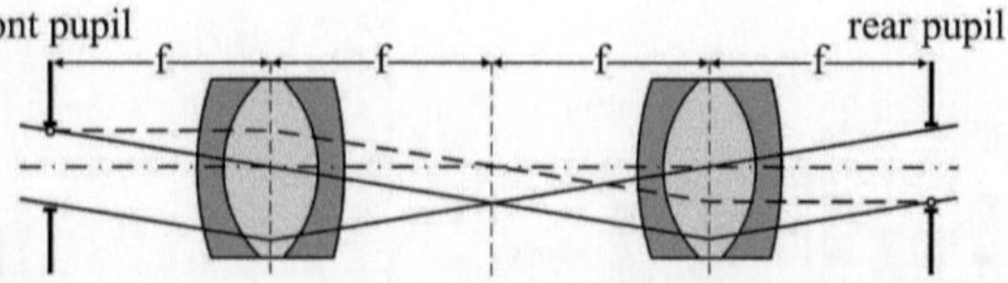

Fig. 10.8 Aberration-corrected achromatic triplets (Chap. 1) are commonly used in relay systems. Achromatism is not needed for particular laser line but it makes the product equally suitable for the various lasers. The system is symmetrical, with equal diameters of the input and output beams

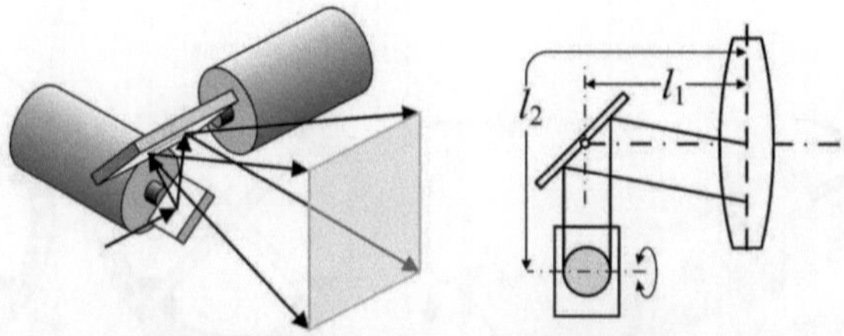

Fig. 10.9 The simplest solution for 2D scanning. With two separate mirrors, deflecting in perpendicular directions, it is impossible to place the both in the input pupil of the relay system: distances l_1 and l_2 are always different

planes of the two lenses coincide and, since the lenses are identical, separation between the front and rear pupils is minimized to $4f$.

All the above is good for one-dimensional (1D) scanning. But what about two-dimensional (2D) scanning? An obvious but not the best solution is simply to add another galvano-mirror, scanning in the perpendicular direction (Fig. 10.9). However, this configuration suffers from inadequate positions of the mirrors relative to lenses, which results in wandering of the laser beam over the input aperture of the objective during scanning. A better solution is the combined 2D steering mirror (Fig. 10.10). In it, one reflecting surface executes commands in two directions simultaneously. Since now there is no need in combining two identical deflecting mechanisms one close to another, space limitations are not as strong as they are in axial 1D design shown in Fig. 10.6. Consequently, the lever, applying torque to the mirror, may be longer, making the momentum stronger and speed higher. Such 2D deflectors may work reasonably well above 100 Hz on bar-step commands and up to 300 Hz on sinusoidal commands. However, long lever makes deflection angle smaller: about ±1.5° mechanical (±3° optical). To rise the speed, the 2D gimbal must be as light as possible. Therefore, ball bearings, routinely used in 1D axial motors, are replaced with flexible hinges—a very peculiar piece of high technology shown in Fig. 10.11.

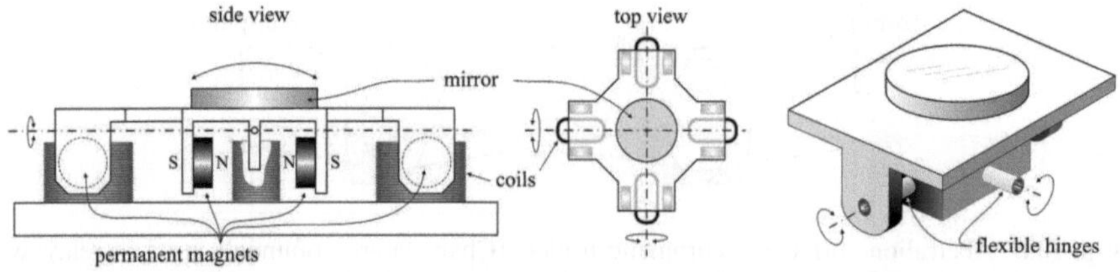

Fig. 10.10 2D galvano-mirror. Four torque motors rock the frame with the mirror mounted on the Cardan joint with flexible hinges: soft on rotations but hard on vertical and horizontal displacements

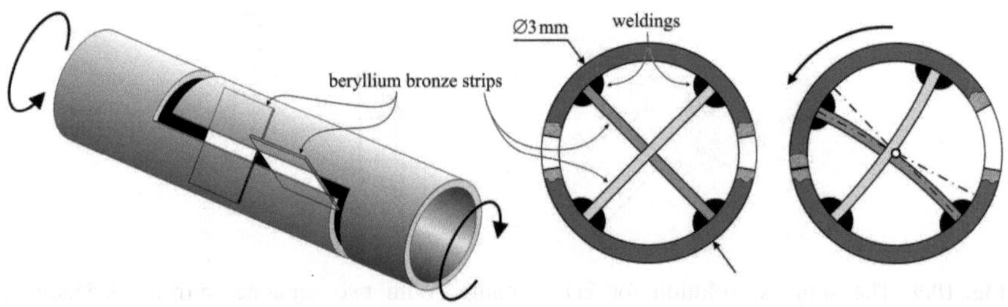

Fig. 10.11 Two sectioned pipes welded coaxially together on flexible strips of beryllium bronze form a flexible rotation hinge. The outer diameter is only 3 mm—ready to fit into small holes drilled in the Cardan joint

The next thing that should be understood about 2D galvano-mirrors is the angle control. While in the 1D versions standard rotary encoders are used to provide the feedback signal, in 2D systems another opto-electronic technique is implemented as shown in Fig. 10.12.

Consider now how galvano-mirrors work in practice. Typical application is high-power laser engraving (Fig. 10.13). An aluminium plate with black oxidation is a perfect sample for visual demonstration of the result: ablation of the black oxide layer opens white aluminium surface of the plate, clearly visible in microscope. The controller receives two independent signal in X and Y directions on the mirror. Suppose we want to make a circle on the sample. Then the X and Y signals must be $a \sin \omega t$ and $b \cos \omega t$. For the beginning, let $a = b$. Of course, it is possible to use only one generator and make quadrature signals (Chap. 4) by differentiation or integration. However, we want to demonstrate something more than a mere circle, therefore two function generators are needed. They must be synchronized, otherwise frequencies will drift away and the desired figure will be destroyed. Synchronization is not a simple feature, and it is available only on high-quality generators. Such generators have at least two connectors for synchronization signals, usually on the back panel: one for input and another for output. Connect the output on the first generator to the input on the second one. The one that sends a signal to another is a master generator, and the phase between

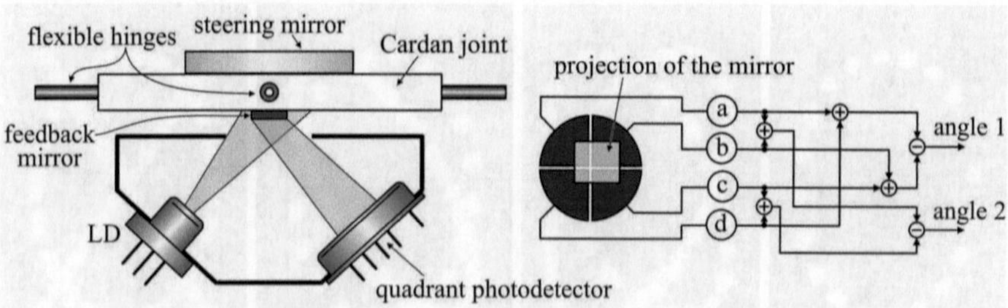

Fig. 10.12 The combination of a laser diode (LD) and quadrant photodiode (Chap. 1) coupled through the feedback mirror bonded to the bottom side of the Cardan joint produces electrical signal proportional to deflection angles 1 and 2 around two orthogonal axes. In order to maintain linearity between the angle of rotation and electrical signals, the mirror is made rectangular, smaller than the LD beam. It makes the reflected beam onto the quadrant photodiode almost a square, thus making electrical signals proportional to deflection

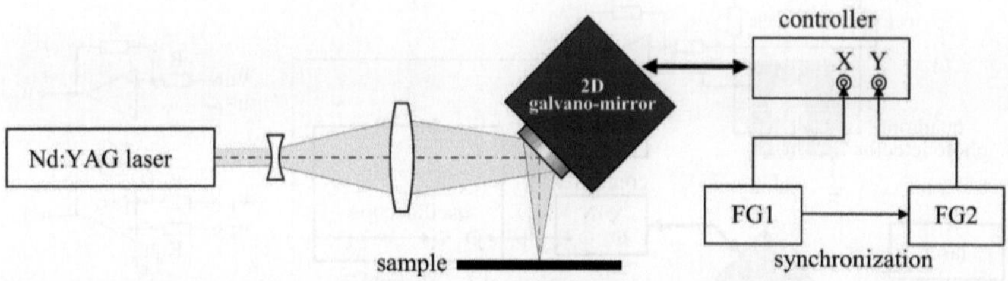

Fig. 10.13 In the laser engraving, the beam of a neodymium laser (Chap. 2) is focused onto the sample, reflecting from two-dimensional (2D) galvano-mirror. Two separate function generators (FG) drive two independent channels of the mirror. Synchronization of frequencies in necessary

them can be adjusted. The results are presented in Fig. 10.14. The first conclusion that follows from the pictures is that with equal amplitudes $a = b$ the circle we wanted becomes an ellipse. The reason for this follows from Fig. 10.2: at 45° angle of incidence, approximately 40° bigger angular deflections γ are needed to reach the same linear displacement y as for 0° angle of incidence. The second conclusion is that 2D galvano-mirror is quite a tame and reliable device, reacting adequately to control signals. But what about speed?

Commonly, the speed of a scanner is judged by the phase lag between the optical deflection and command signal for sinusoidal signals or by visual deterioration of the shape of optical deflection for bar-step signals. In order to make such measurements in laboratory, some hands-on job must be done above mere connection of cables. Figure 10.15 explains the simplest scheme that should be made. Relatively large quadrant or, even better, two-element photodiode will serve as the displacement sensor (Chap. 3). Cross-sectional shape of the beam is not essential since we are interested only in phase shift for sinusoidal signals. For bar-step signals, usually only the time between the front of the command impulse and the

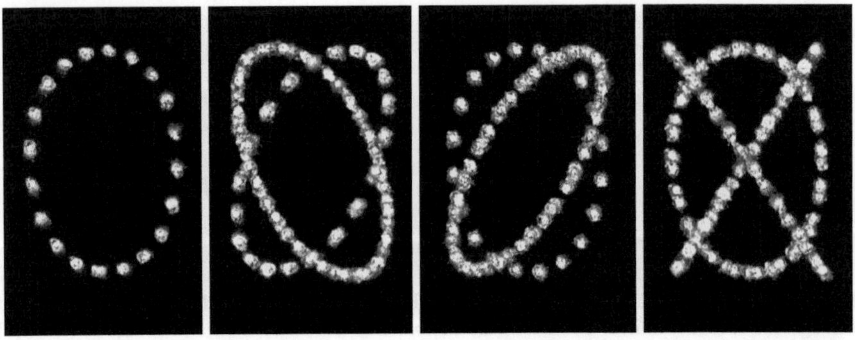

Fig. 10.14 Microscope images of the patterns generated by sine and cosine signals with different phases: 0° (*first from the left*); +30° and −30° (*the second*); 0° and 45° (*the third*); 0° and ±90° (*at right*). 20 Hz laser fire rate, 11 Hz scanning frequency. 45° angle of incidence on the mirror

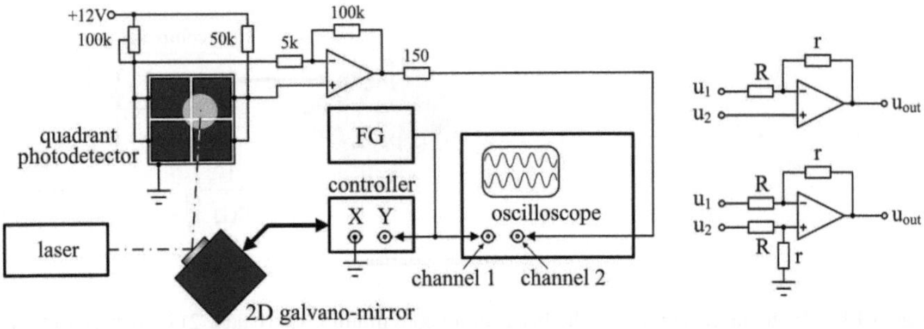

Fig. 10.15 Multi-element photodetector and differential amplifier are needed to measure phase delay on scanning. Function generator (FG) may be set either to sinusoidal wave or to bar-step output. Variable resistor in one biasing shoulder of the photodetector is needed to set zero point of the differential amplifier. The circuit with only a pair of resistors *R* and *r* is simple but performs only approximate subtraction. Additional *R–r* divider in non-inverting input of operational amplifier makes subtraction almost perfect

moment of stabilization is important, which again does not depend on the shape of the beam. Therefore, circular geometry of the beam does fine. Then, the differential amplifier subtracts the signals u_1 and u_2 from both sides of the photodetector: $u_{\text{out}} \sim u_2 - u_1$. This voltage is approximately proportional to the angle of deflection because the latter is small. However, the simplest scheme with directly connected non-inverting input of the operational amplifier shown in Fig. 10.15 performs subtraction only approximately, and this subject should be explained in order to avoid mistakes. Making standard assumption that no current flows into operational amplifier (input resistance about 10 MOhms) and voltage difference at its inputs is zero (open-loop gain about 10^6), balance of voltages gives the following equation:

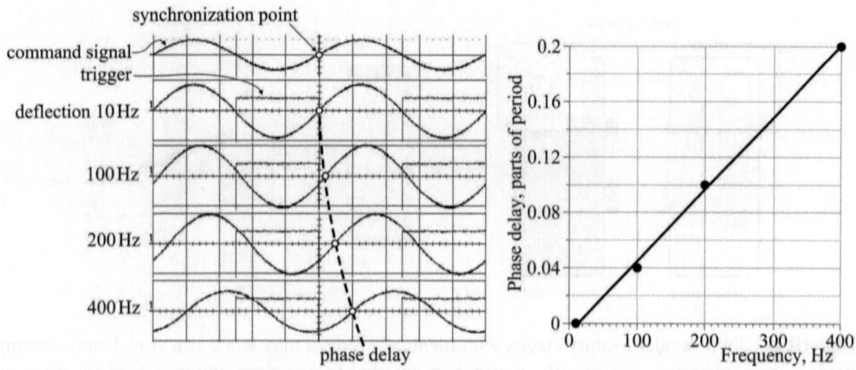

Fig. 10.16 On sinusoidal command, deflection remains sinusoidal but phase delay increases approximately linearly with frequency

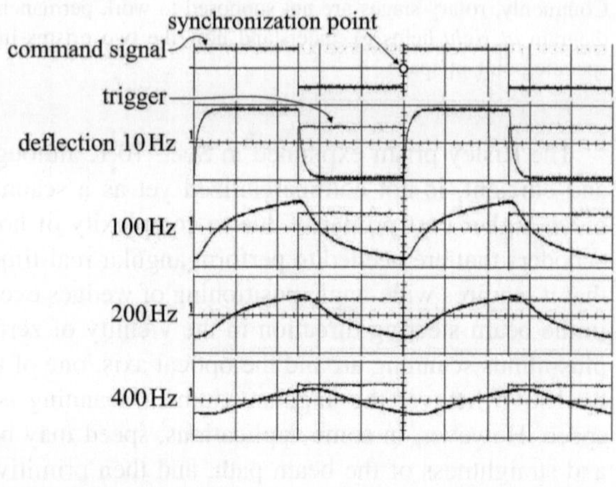

Fig. 10.17 On the bar-step command, even 10 Hz frequency is too high to follow the front and end edges reasonably well

$$u_1 - \frac{u_1 - u_{out}}{R + r} R = u_2.$$

From here

$$u_{out} = (u_2 - u_1)\left(1 + \frac{r}{R}\right) + u_1.$$

Only for high gain $G \approx \frac{r}{R} \gg 1$

$$u_{out} \approx (u_2 - u_1) \cdot G.$$

In our case $G = 20$, and this approximation holds true. If better precision is necessary then additional voltage divider must be added in the non-inverting input (Fig. 10.15).

Figure 10.16 shows the result for sinusoidal and Fig. 10.17—for the bar-step commands.

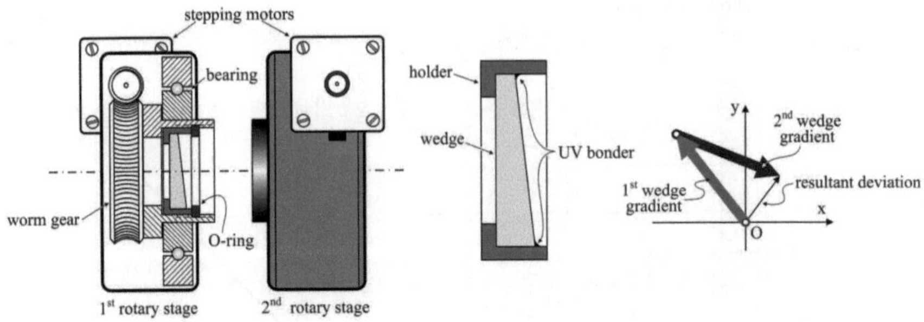

Fig. 10.18 Two standard rotary stages with identical wedges may work as a slow beam-steering device. In inexpensive versions, the worm gear coupled to a stepper motor is used for rotation, which is slow: about 10° per second. Important thing is that a wedge can never be reliably clamped in a cylindrical housing by a threaded O-ring. Therefore, the wedge should be primarily bonded in a separate cylindrical holder, and only after that clamped in the rotary stage. Also note that standard brass-steel worm gear wears off quickly after periodical motion, becoming loose. Commonly, rotary stages are not supposed to work permanently in periodic motion. Mnemonic diagram *at right* helps to understand how the two prisms must rotate to direct the beam to a specific point in space

The Risley prism explained in Sect. 10.1, although theoretically very compact and efficient, is not commercialized yet as a scanning device. The reason is its much higher cost primarily due to complexity of hollow-shaft torque motors and encoders that are needed to perform angular real-time scanning. Another reason is that it requires wide-angle positioning of wedges even for relatively small changes in the beam steering direction in the vicinity of zero beam deviation. Indeed, for plus-minus scanning around the optical axis, one of the prisms needs to be rotated by 180° whatever the angular width of scanning is. The result is low scanning speed. However, in some applications, speed may be sacrificed in favor of space and straightness of the beam path, and then primitive in-line combination of two standard rotary stages may work fine (Fig. 10.18). Assembling your own Risley scanner, it is necessary to understand that the two wedges must be almost identical, otherwise the so-called nadir problem occurs: inability to access the vicinity of zero beam deviation.

Fast versions of Risley scanners, available from specialty companies, use torque motors and rotary encoders (Fig. 10.19).

When galvano-mirrors or any other mirror-type scanning devices are supposed to steer focused laser beams, like in laser-marking machines for instance, the design of the entire system may be simplified, using the so-called F-θ (F-theta) lenses. All the rotating mirrors are controlled in terms of rotation angle θ. However, an ideal lens focuses parallel beam into a spot whose lateral position is defined as $F \tan\theta$ (Fig. 10.20). This functionality is unacceptable in laser printers, where a polygon mirror rotates with constant angular speed. It means that the focused laser spot moves with different speed across the paper, making exposure time different at the sides and in the middle of the paper sheet. The problem was

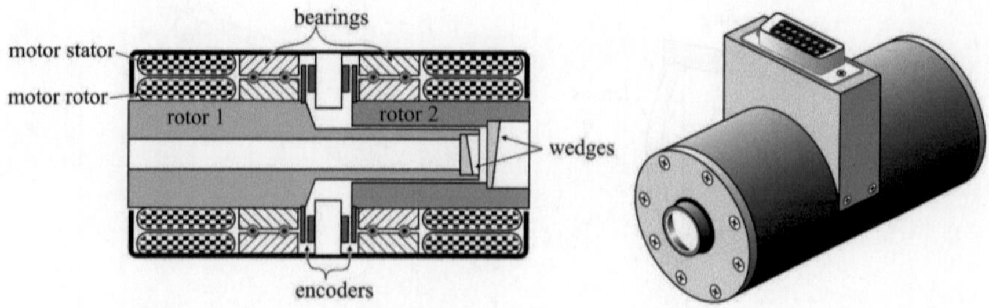

Fig. 10.19 Hollow-shaft torque motors with rotary encoders are the basic parts of fast Risley prism deflectors. Manufacturers claim around 0.1 s response time for 180° full move, and 1 mrad pointing accuracy with ±60° steering ability

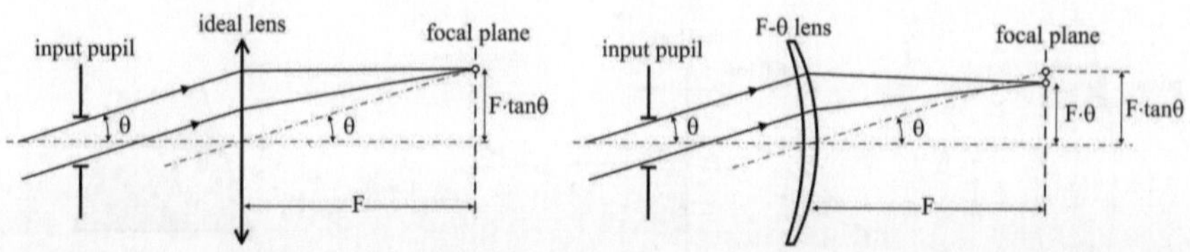

Fig. 10.20 In an ideal lens, all parallel rays come to a single focus point located in the focal plane. As such, this point may be identified as intersection of the principal ray, coming through the lens center, with the focal plane. Vertical coordinate of this point is F tan θ (*at left*). In the F-θ lenses, displacement is proportional to the angle of scanning θ and not to the tangent (*at right*)

solved by developing the F-θ lenses that deflect the focal spot not as F tan θ but as $F \cdot \theta$. Even if we are not making a laser printer, the F-θ lenses can simplify a beam-steering system by simplifying the control algorithm since the command signals may now be computed proportional to the angle of rotation θ. F-θ lenses are readily available on the market, with some representatives shown in Fig. 10.21. Very peculiar thin and deeply bent shapes of F-θ lenses spawn an almost irresistible question of everyone who sees them for the first time: what is the physics of this shape? Although lens design is not the subject of our discussion, the absence of physical interpretation even in special literature dictates to shed some light on this topic.

Consider a thin symmetrical lens with the radii of curvature R (Fig. 10.22). From school, we know that the focal length of a lens.

$$F = \frac{R}{2\,(n-1)},$$

where n is the refractive index. We are going to divide the spherical profile of the lens into many vertically narrow horizontal intervals, containing prisms instead of spherical sectors, and to tilt each and everyone of them so as to direct refracted

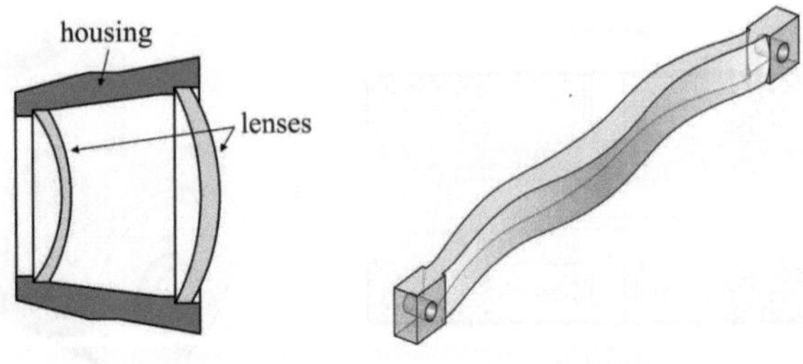

Fig. 10.21 A family of F-θ lenses: for visible domain made from glass (*at left*); plastic lens for laser printer (*at right*)

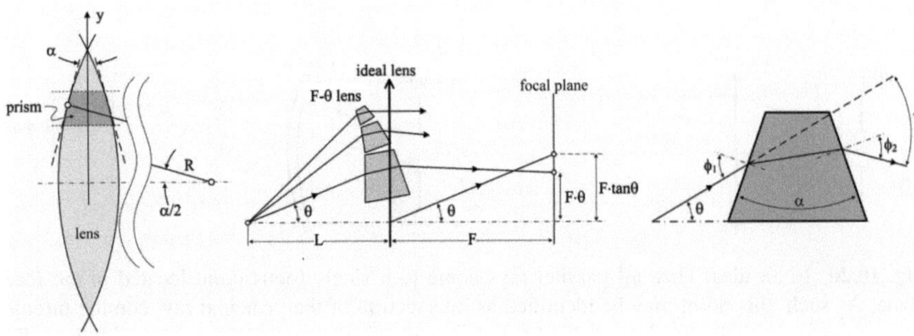

Fig. 10.22 Design of an F-θ lens as a combination of elementary prisms tilted so as to make vertical displacement of the ray in the focal plane equal to $F \cdot \theta$. Apex angle of each elementary prism is that of a section of an ideal lens. Therefore, without tilting, each ray intersects the focal plane at the point $F \cdot \tan\theta$

rays to the point $F \cdot \theta$ instead of $F \cdot \tan\theta$. Then the middle line of all the prisms will give the section of the F-θ lens.

For each elementary prism located at y,

$$\frac{\alpha}{2} = \arcsin\frac{y}{R},$$

or, expressing R in terms of the focal length F,

$$\alpha = 2\arcsin\left[\frac{y}{2F(n-1)}\right].$$

Next, consider angular deviation γ of a ray entering the prism at the angle of incidence φ_1. From the Snell law it is easy to derive:

$$\gamma = \varphi_1 + \arcsin\left[n\sin\left(\alpha - \arcsin\frac{\sin\varphi_1}{n}\right)\right] - \alpha.$$

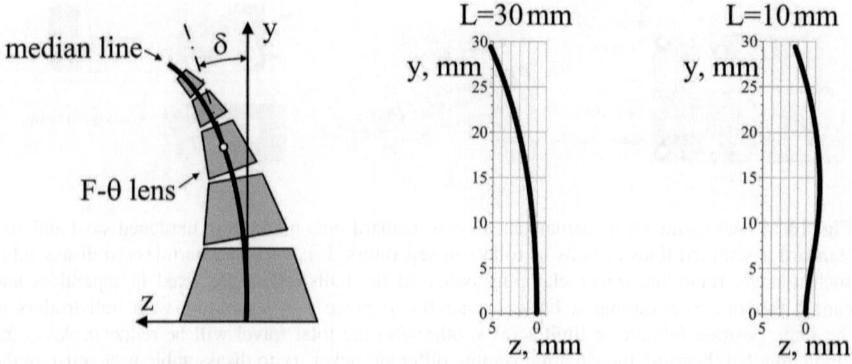

Fig. 10.23 Median line through the centers of all the elementary prisms defines the shape of the F-θ lens (*at left*). Computed shapes for two typical values of L (*at right*). $F = 100$ mm; $n = 1.5$. The vertical and horizontal scales of the curves are the same to facilitate comparison with Fig. 10.21

All the variables here are functions of y. If the prism is tilted by δ then the angle of incidence on the prism changes to

$$\varphi_1 = \theta + \frac{\alpha}{2} - \delta.$$

The essential parameter of any F-θ lens is the distance from the input pupil L. Since

$$\theta = \arctan\frac{y}{L},$$

the input angle θ becomes connected to y.

The sense of our basic equation is maintaining displacement equal to $F \cdot \theta$:

$$y + F\tan(\theta - \gamma) = F \cdot \theta.$$

With all the parameters defined above, it is an equation that determines the function $\delta(y)$. Of course, it is impossible to solve it analytically, but using standard mathematical routines for solving equations, available in such languages as Fortran, we can easily solve it numerically.

What we are going to do next is to find the shape of a median line, going through the centers of all the prisms. Physically, it is a median line, going through the cross section of the F-θ lens, i.e. the lens shape (Fig. 10.23). Since δ is the angle,

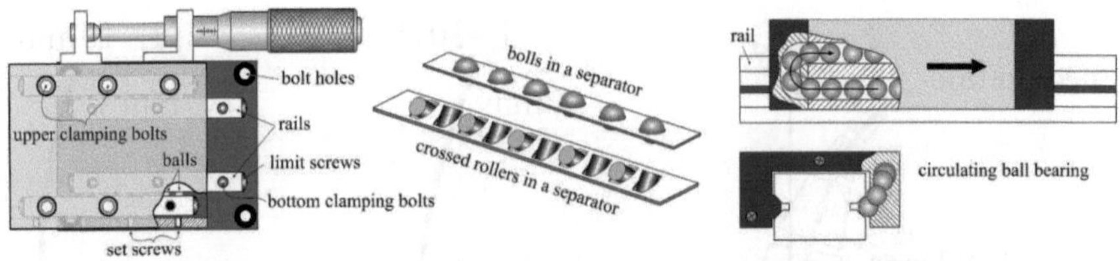

Fig. 10.24 All manual translational stages use standard rails made from hardened steel and also standard assembled lines of balls or rollers in separators. It is easy and harmless to disassemble such a stage, removing *upper* clamping bolts: all the balls/rollers are fixed in separators and cannot fall out. Assembling it back, be careful to place two separators with bolls/rollers at the same position relative to limit screws, otherwise the total travel will be reduced. As to the circulating ball bearing, the situation is quite different: never try to disassemble it or separate the moving part from the rail, otherwise the balls will fall out and scatter

$$\frac{dz}{dy} = \tan\delta(y),$$

and the shape is

$$z(y) = \int_0^y \tan\delta(y)\,dy.$$

It must be emphasized that all the above is not a strict computation of the F-θ lens but only qualitative explanation of how its shape is formed. Some relevant examples are shown in Fig. 10.23. Clearly, smaller ratio *L*/*F* requires more bending with even inflection points, which is a characteristic of a F-θ lens of the laser printer (Fig. 10.21). Note that paraxial rays, coming near $y = 0$, does not require any significant correction to the median line of the lens: here $\tan\theta \approx \theta$ and the F-θ condition is fulfilled even on an ideal lens.

10.4 Translation Stages

Precision manual translation stages are driven by a micrometer screw and use either linear ball bearings, crossed roller bearings, or circulating ball bearings (Fig. 10.24). A very simple thing should be known about manual translation stages: those that use bolls or crossed rollers in separators are adjustable, those that use circulating balls—are not. It is recommended to test a new translation stage before using it. Although the bearings and micrometer screws are the components that are precisely manufactured and assembled with high accuracy at specialized companies and, therefore, are highly reliable, the entire assembly may be done inaccurately. Test the sliding motion of the upper platform with your fingers: if you feel irregularity and interruptions in motion then the stage needs adjustment.

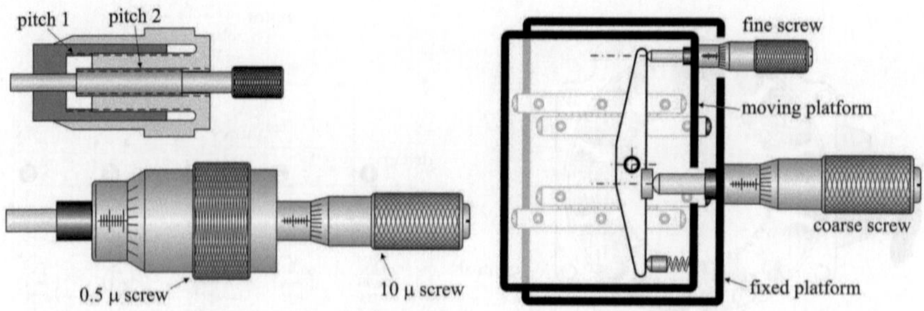

Fig. 10.25 In a differential screw, pitch of two connected threads is slightly different: when the nut goes right the central screw goes left, resulting in a smaller differential protrusion (*at left*). Differential lever system uses two separate micrometer screws connected through a lever with the pivoting axis on the moving platform (*at right*)

To do that, find the side of the stage with set screws and release slightly upper clamping bolts on this part of the assembly only. Do not unscrew them completely. Then release all the set screws—the rail will be freed and the entire bearing becomes loose. Very gently return the set screws, each one individually, to the initial position and test the motion. Repeat this until you are satisfied with the smoothness of motion and finally fix the upper row of bolts.

Circulating ball bearings are used when long travel is needed. This type of stages does not require any adjustment.

Precision of positioning is determined by the micrometer screw. Single-thread micrometer screws typically have resolution 10 μ per division, and it is the real lower limit that fingers can set on the knurled knob. Interferometry requires substantially better precision, which can be satisfied either by differential screws or differential translation stages (Fig. 10.25). On a differential screw, fine division is typically marked as 0.5 μ, and practically it is the real accuracy of such devices. Differential lever systems show the same scale of precision.

Motorized translation stages are supposed to produce long displacements and, therefore, always use linear circulating ball bearings and ball lead screws. For opto-electronic applications, where fine and well-controlled positioning is a necessity, the most appropriate driving system is stepper motor with a controller (driver). Specifically, five-phase stepper motors produce most versatile control over rotation. For opto-electronic engineer, it is only necessary to understand how to connect them to a driver (controller) and how to choose proper settings on it. It is explained in Fig. 10.26. Typical driver has two contacts for direct current 24 V—its power source, and five contacts for phases marked as 1, 2, 3, 4, 5 or «A», «B», «C», «D», and «E». The motor leads are supposed to be connected in a certain order, and for that, they are coloured. One widely used colour convention is explained in Fig. 10.26. If the motor you have does not match this convention then it is easy to sort the wires out, using an ohmmeter. Choose arbitrarily any one wire and mark it as «A». Find two other wires that have minimal resistance relative to

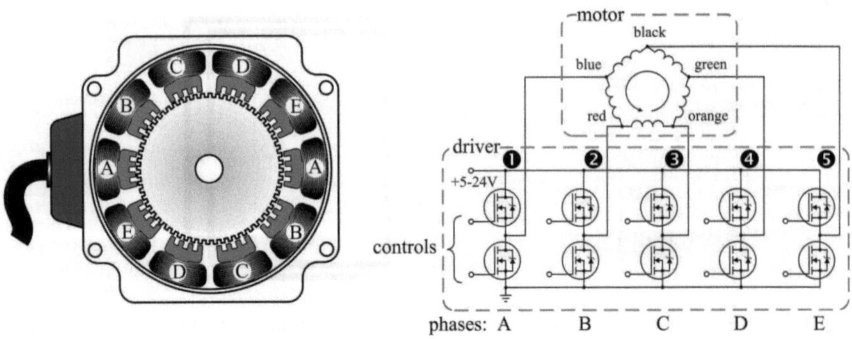

Fig. 10.26 For the motor to rotate, the wires must be connected to the driver in a proper sequence: either clockwise or counterclockwise. Choose arbitrarily the first phase, say A, and connect it to the driver's contact number 1. Then connect B to 2, C to 3, etc. In order to facilitate connection, most of the manufacturers use certain convention of coloured wires as shown in the figure. For instance, if you take the *blue* wire first and want to go counterclockwise, then B will be *red*, C—*orange*, and so on. The same algorithm for the clockwise direction. If other colours are used, for example *brown* or *yellow*, then coils resistance gives the answer (see the text)

«A», and mark then arbitrarily «B» and «E». Let the resistance of each winding be r. Then there are only two options: either the measurement shows the value.

$$r||4r = \frac{r \cdot 4r}{r + 4r} = \frac{4}{5}r$$

or

$$2r||3r = \frac{2r \cdot 3r}{2r + 3r} = \frac{6}{5}r,$$

which is 1.5 times bigger than the first one. Typical minimum value is 16 Ohm. Then another measurement can only be 24 Ohm. Next, find the minimal resistance between «B» and others. There will also be two wires, one of which is «A». Name the second one «C», and the last unnamed is obviously «D». The problem is solved.

When the power is applied, the motor will be still until the pulsing sequence—the series of electrical pulses of TTL level that determine the rate of rotation—is applied. Note that this input does not require power current: this signal is only a command. In the simplest algorithm, one pulse turns the rotor by the angle between adjacent teeth divided by the number of poles. This is typically either 0.72° or 0.36°. The angular step is always marked on the label on the motor, if it exists. Commercial drivers usually divide one step rotation by an integer number called the resolution. Resolution can be set from the front panel of the driver between 1/1 and 1/500. This value shows how much the speed of rotation will be decreased relative to the input frequency of the pulsing sequence. This option called microstepping is introduced to improve smoothness of motion (Fig. 10.27).

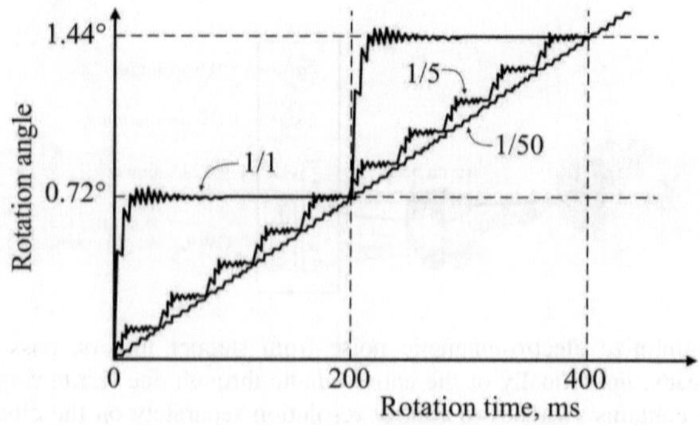

Fig. 10.27 Actual rotation as a function of time for different resolution. The 1/1, 1/5, and 1/50 curves are shown on the same time scale, which requires progressively higher frequency of pulsing sequence: 5 Hz, 25 Hz, and 250 Hz respectively. Maximum command frequency cannot exceed the frequency of the input pulsing sequence applied to the driver

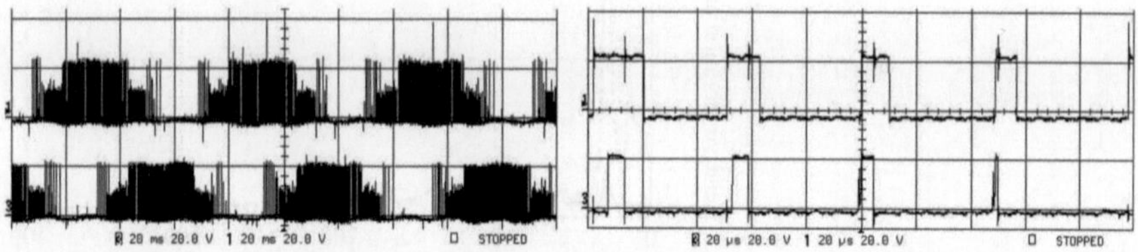

Fig. 10.28 Voltage across two different poles of a stepper motor in the resolution mode 1/500. The *right* picture shows the fragment of the *left* picture on the 10^3 finer time scale (numbers below the traces: the values are per division)

The algorithm of angular step division uses simultaneous excitation of multiple phases with short bursts of voltage at frequency of the input pulsing sequence. The current through the motor winding evolves like an integral of voltage, overlapping in time on adjacent poles and creating net rotation momentum that evolves slower than the frequency of the input sequence of pulses. Typical voltages that develop on two poles of a stepper motor are shown in Fig. 10.28. Among other details, these pictures show a lot of high-frequency noise that stepper motor creates. With 20 V pulse amplitude and megahertz frequency, stepper motors act like small radio transmitters, especially with long cables. It may be a real and insurmountable obstacle when working along with sensitive photoreceivers (Chap. 3). The worst mistake that can be done is to power the motor and the photoreceiver from the same power supply, say +24 V. Practical advice is to ground heavily the driver and the measurement equipment and to pass the motor wires through ferrite beads (Fig. 10.29).

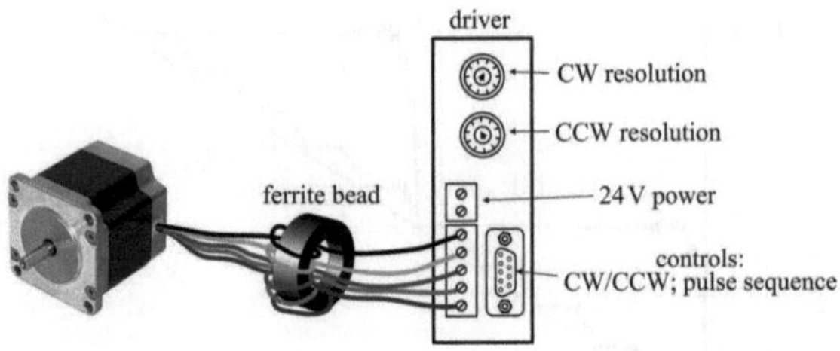

Fig. 10.29 To minimize electro-magnetic noise from stepper motors, pass the motor wires through ferrite beads: individually or the entire bundle through one ferrite ring. The drive front panel commonly contains switches to control resolution separately on the clockwise (CW) and counterclockwise (CCW) rotations. It is a convenient option when you want slow direct motion and fast return

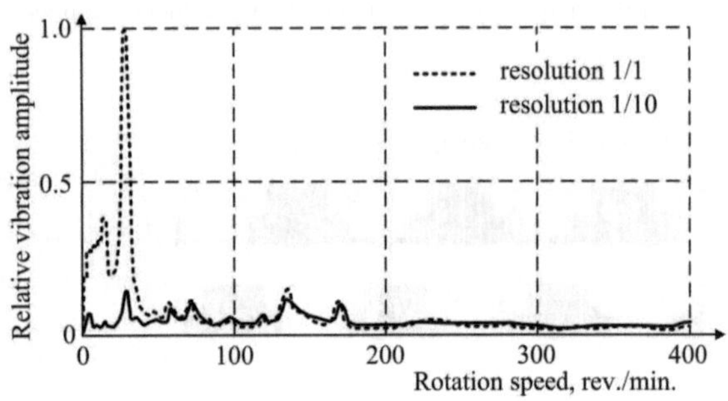

Fig. 10.30 Typical vibrations of a stepper motor with 0.72° step

Microstepping, i.e. the ability of stepper motors to work in a high-resolution mode with a step angle much less than 0.36° or 0.72°, often creates fundamental misunderstanding of the precision that may be expected from translational stages driven by stepper motors. Suppose the driver supports 1/100 resolution on a 0.36° motor installed on a translation stage with 1 mm pitch lead screw. Theoretical angular step is then 0.0036°, which corresponds to linear step as small as 1 mm/ 360° × 0.0036° = $1 \cdot 10^{-5}$ mm = 100 Å. However, this arithmetics does not account for rotor backlash, static friction, elasticity of couplers that connect the motor shaft to the lead screw, grease in the bearings, etc. What we really have with high resolution option is lower motor speed and smaller vibrations. As to the positioning precision, its practical limit is about 0.5 μ whatever the driver resolution is. Nevertheless, smaller vibrations is already a big deal, especially in optical applications (Fig. 10.30). Level of vibrations strongly depends on the type of a coupler that connects the motor to the lead screw. As the motor shaft and the lead screw never can be exactly coaxial, a kind of the Cardan joint must be between them. A variety of couplers are available on the market, some of them shown in Fig. 10.31.

Smoothness of motion can be improved even without microstepping by using fine mechanical gears between the stepper motor and the lead screw (Fig. 10.32).

Fig. 10.31 The coupler with hard plastic insertion, shown *at left*, is the worst for accurate applications. High friction between the plastic and metal parts exerts strong momentum on the motor, making its rotation inaccurate. Other types are equally good. Unexpectedly, a short section of a polyurethane hose, of the inner diameter tightly fitting the shaft, is one of the best solutions when vibrations are a serious problem. Its flexibility, however, makes positioning less precise

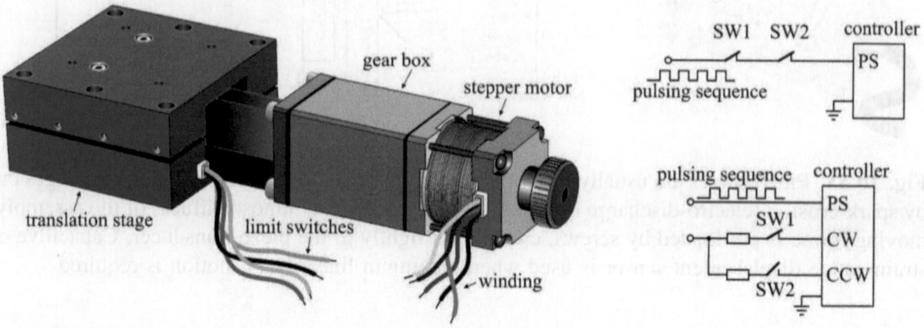

Fig. 10.32 Not a typical combination: stepper motor is connected to a translation stage through a fine gear (*at left*). For this option, limit switches are absolute necessity to prevent mechanical damage of the stage. Two functions of limit switches SW1 and SW2 may be realized: disconnect the pulsing sequence (PS) or change direction of rotation between clockwise (CW) and counter-clockwise (CCW) (*at right*). The latter option is commonly used for scanning

It is a common illusion (not a mistake) that a geared stage improves precision. Typical considerations sound like this: if the pitch of the screw in my stage is 1.5 mm, motor step is 0.72°, and gear ratio is 100 then positioning precision is $1.5 \times 0.72/360/100 = 0.03\ \mu$. The mistake is that this value, usually called linear resolution, is an arithmetically calculated number for reference purposes only, not a precision of positioning. Actual precision, as it was already mentioned, is limited by elasticity, friction, backlash, and other imperfections. Actual parameters that may be expected from a geared stage are listed in the Table 10.1.

A peculiar problem with geared stages is end (or limit) switching. Translational stage always has finite travel range established by limit screws installed to protect the bearing from disassembling (Fig. 10.24). When a stepper motor is connected to a stage only through a coupler, rotation momentum is insufficient to make any harm. The opposite situation is with the gear: if the stage crashes into limit screws, driving force of the gear is so strong that may easily damage the lead screw, the gear, or the coupler. Therefore, the most dangerous and costly mistake that can be made, working with the geared stages, is to disable or disconnect limit switches.

Table 10.1 Precision of a typical geared motorized stage

Parameter	Value
Maximum travel	30 mm
Positioning precision in one direction	5 μ
Repeatability	0.5 μ
Backlash	0.5 μ
Straightness of motion	3 μ
Linear resolution	0.02 μ

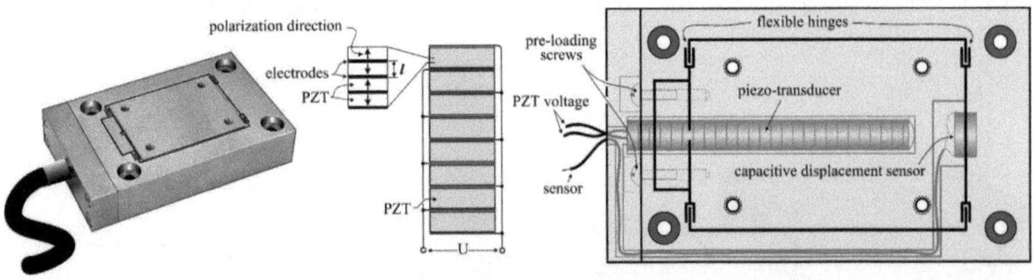

Fig. 10.33 Piezo-stages are usually made of aluminium or stainless steel with flexible hinges cut by spark-erosion (electro-discharge machine, EDM). To achieve utmost stiffness of the assembly, moving frame is pre-loaded by screws, clamping it tightly to the piezo transducer. Capacitive or strain-gauge displacement sensor is used when maximum linearity of motion is required

For nano-positioning, the only choice is piezo-electric stages, or merely piezo-stages (Fig. 10.33). In them, precisely controlled micrometer-scale displacements are produced by piezo-electric actuators or transducers. The lead zirconate titanate ($Pb[Zr_xTi_{1-x}]O_3$) ceramic—PZT in abbreviation—is commonly used as a material for piezo-electric transducers. Initially, it has randomly oriented polycrystalline structure and does not display piezo-electric activity as a macroscopic body. Nonetheless, each individual crystallite is piezoelectric, and it is possible to polarize the substance macroscopically by applying strong electrical field in a certain direction. This process is called poling. After the field is removed, the residual polarization remains in the entire specimen, which becomes macroscopically piezo-electric. Two things may disturb polarization: high temperature, exceeding the so-called Curie temperature, and strong electric or magnetic field.

Once the substrate is macroscopically polarized, external electrical field exerts strain σ, i.e. relative deformation of the material. A cube with dimensions $l \times l \times l$ along three Cartesian coordinates x, y, z is deformed to $l + \Delta x$, $l + \Delta y$, $l + \Delta z$. This defines the vector of strain.

$$\sigma = \begin{pmatrix} \Delta x/l \\ \Delta y/l \\ \Delta z/l \end{pmatrix} = \begin{pmatrix} d_{11} & d_{12} & d_{13} \\ d_{21} & d_{22} & d_{23} \\ d_{31} & d_{32} & d_{33} \end{pmatrix} \cdot \begin{pmatrix} E_x \\ E_y \\ E_z \end{pmatrix}.$$

The matrix d determines proportionality coefficients between deformation and electrical field inside the material. It is common to choose axis z along polarization. Then d_{33} determines deformation along polarization when electrical field is applied in the same direction: $E_x = 0$, $E_y = 0$, $E_z = E$. Assuming that the electrical field E created in the material by the voltage U is uniform over the ceramic thickness l.

$$E = \frac{U}{l},$$

absolute value of elongation Δz is proportional to U independently of l:

$$\Delta z = d_{33} U.$$

For PZT ceramic, $d_{33} \sim 200$–500 pm/V. Non-uniformity of the electrical field inside the ceramic decreases elongation. Therefore, a thin disk with the voltage applied to its parallel surfaces expands more than a long thin rod with the same voltage applied to its ends. N thin disks stacked together in a sandwich, with electrodes altering between them, increase total excursion to $N \cdot \Delta z$. Typical PZT transducers are designed to provide total displacement $\sim$0.1 % of the stack length at $U = 100$ V. For 10 cm long actuator, for instance, it means 100 μ displacement. Higher voltage is rarely used as it would require sophisticated amplifiers to drive the stage. How many thin disks must be sandwiched to produce this excursion? Let us take the maximum $d_{33} = 5 \times 10^{-10}$ m/V. With $U = 100$ V each layer produces $\Delta z = 5 \times 10^{-8}$ m $= 0.05$ μ. To obtain 100 μ excursion $N = 2000$ layers must be stacked one upon another. This can be done only with multilayer technology, when very thin, about 10 μ, PZT layers are sintered onto metal foil and then co-fired in a stack to produce 0.5–1.0 mm thick substrate with about 20 microscopic PZT layers inside. At this phase of the process, the substrate is poled by heating to the Curie temperature around 400 °C and applying polarizing voltage about 150 V. After that, the entire actuator is bonded of many multilayer substrates into a long prism with contacts on the opposite sides of it (Fig. 10.33).

Piezo-transducer responds to applied voltage non-linearly and with substantial hysteresis (Fig. 10.34). The residual displacement l_r is roughly proportional to maximum displacement l_{max} that has been applied to the actuator, and typically is less than 10 % of the maximum displacement specified for the particular actuator.

In order to avoid repolarization of the ceramic, it is always recommended to apply only positive voltage to the actuator, i.e. only in the direction of polarization. However, carefully controlled negative voltage can extend the overall displacement by 20 %. Manufacturers often use this option in their controllers, adding well-determined negative voltage to the signal supplied by the user. With this option, total displacement may extend to 0.2 % of the transducer length, which is an additional reason to use specially designed controllers (drivers). But the main reason for that is the option of closed-loop operation that removes non-linearity of

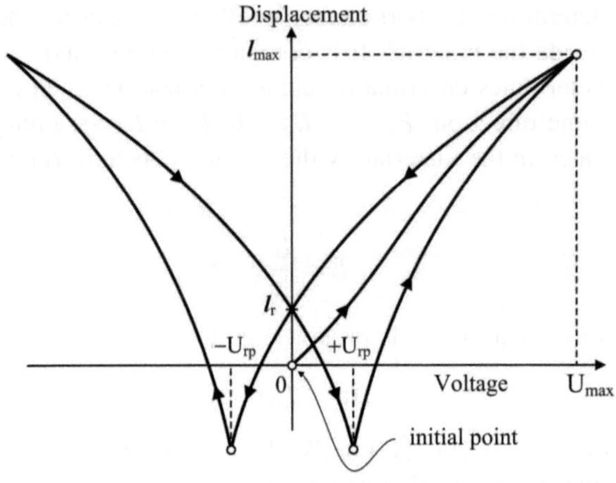

Fig. 10.34 Starting from the initial position, voltage applied along polarization drives the actuator to its maximum displacement limited by the breakdown. This maximum voltage must never be exceeded. Then, with the voltage coming back to zero, the actuator returns to a new state, slightly longer than the initial one. In order to make the actuator length equal to the initial one, negative voltage must be applied, i.e. the electrical field in the ceramic is directed against the polarization. Further increase of negative voltage shrinks the actuator, making the displacement negative relative to the initial. At some moment, electrical field exceeds the strength of permanent polarization, resulting in total repolarization of the medium in the opposite direction. This voltage is $-U_{rp}$. From this moment, the entire picture is mirrored, and further increase of negative voltage repeats the process in reverse order

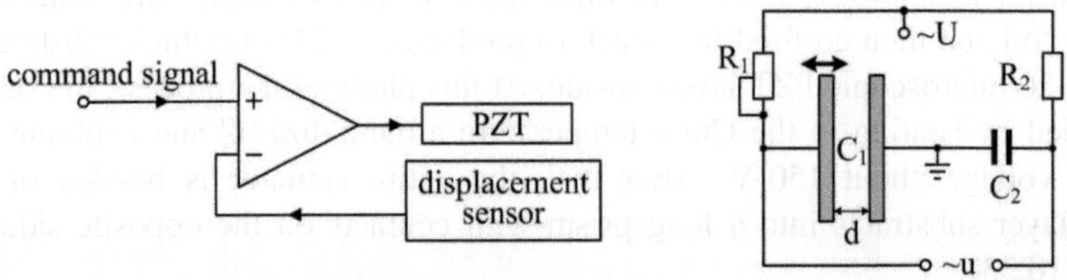

Fig. 10.35 In the closed-loop system, the command signal is a low-voltage potential between 0 V and 10 V, and the same must be the average signal of the displacement sensor (*at left*). The difference between them is amplified to high voltage necessary to drive the piezo-transducer (PZT). Displacement sensor may be implemented as a strain gauge or as a capacitive bridge (*at right*)

piezo-stage almost completely. This is achieved by using precise displacement sensor to control actual displacement of the stage (Fig. 10.35). A popular type of such a sensor is the capacitive bridge. In it, the sensing element is a capacitor with split electrodes: one fixed on the motionless part and another—on the moving part of the stage. Spacing between them only slightly exceeds the maximum displacement. It is powered by radio-frequency voltage *U*. When zero condition.

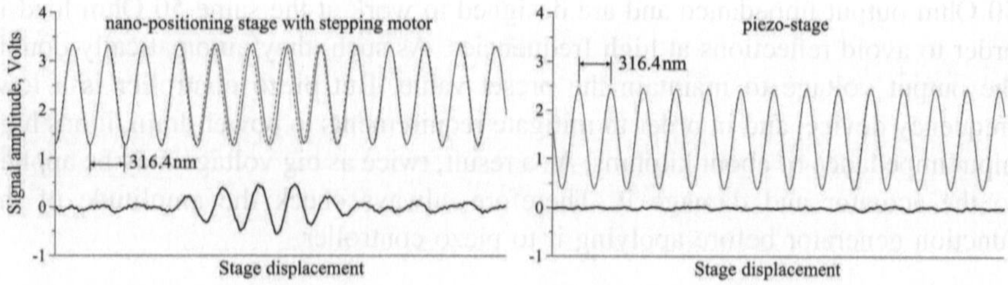

Fig. 10.36 Interference patterns recorded in the Michelson interferometer from a stabilized HeNe laser on a motorized geared nano-positioning stage as in Fig. 10.32 (*at left*) and a piezo-stage (*at right*). Theoretical modulation period is equal to half the wavelength, i.e. 316.4 nm. Any irregularity to displacement is seen as deviation of the recorded trace (*solid lines*) from a pure sinusoid, which is shown in each panel in a *dashed line*. In the *right panel*, this curve almost completely coincides with the recorded trace. Below each trace, the arithmetical difference is plotted in a *solid line*, oscillating around zero. These variations around zero show how precise the stage is

$$R_1 C_1 = R_2 C_2$$

is fulfilled, the output voltage is zero. Capacitance of the sensing element is inversely proportional to the gap d, i.e. to displacement. Initial zero position is adjusted by variable resistor R_1. Since the sinusoidal excitation is used, synchronous detector (Chap. 4) is needed to provide the output signal of variable polarity: positive when the bridge is shifted to one direction from zero position and negative otherwise.

Performance of the closed-loop piezo-stages is exceptionally high: manufacturers claim linearity to the command signal of about 0.1 % and resolution in sub-nanometer scale—a value that can hardly be verified in practice. All that can be achieved only by individual adjustment of the stage and controller. Therefore, if you want to change the piezo-stage for another one and expect the cost associated solely with the new product, then it is a mistake: you also have to ship the old controller to manufacturer for adjustment with the new stage.

Motorized stages, even with nanometer-scale resolution, cannot compete with the piezo-stages in linearity and precision. In laboratory, the only way to compare their performance is interferometry. Figure 10.36 presents such a comparison in a form of interference pattern recorded from stabilized helium–neon laser (632.8 nm) on the Michelson interferometer (Chap. 6).

Usually, application of piezo-stages is bound to linear scanning. In this case, the command that has to be applied to the controller is a ramp or saw-tooth signal. As we are striving for nano-precision, high linearity is needed, and it can be obtained only with high-quality function generators. In the simplest scheme, controller accepts the low-voltage analog signal from our function generator and amplifies it to hundreds of volts needed for the piezo-actuator. Therefore, it is crucial not to exceed the maximum voltage prescribed to the particular stage, and here a funny and dangerous mistake awaits us. High-quality function generators always have

50 Ohm output impedance and are designed to work at the same 50 Ohm load in order to avoid reflections at high frequencies. As such, they automatically double the output voltage to maintain the preset value. But piezo controller is a low-frequency device, and in order to mitigate requirements to power drain, it has high input impedance of about kilohms. As a result, twice as big voltage may be applied to the actuator and damage it. Therefore, always check the amplitude of the function generator before applying it to piezo controller.

List of Common Mistakes

- the use of geared motorized stages with disabled limit switched;
- powering the motor and sensitive electronics from the same power supply;
- overpowering a scanner or piezo stage by double voltage from 50 Ohm function generator.

Further Reading

Handbook of Optical and Laser Scanning, G.F. Marshall ed., Marcel Dekker, 2004.
J.H. Moore, C.C. Davis, M.A. Coplan, Building Scientific Apparatus, 4th ed., Cambridge University Press, 2009.

Chapter 11
Supporting Techniques

Imaging and data acquisition support opto-electronic research.
Some practical details of these technologies may be useful.

Abstract Video cameras and data acquisition techniques always support opto-electronic research. However, these areas are far from typical specialties of those who work in optics. Therefore, well filtered information on these subjects, compressed into four sections of this chapter, may be useful. The first section presents physical principles of charge coupled devices (CCDs) and complementary metal-oxide structures (CMOS)—the two types of video sensors that dominate contemporary market. They have absolutely the same principle of photon detection but many structural and operational differences, carefully explained in historical succession: first APS (active pixel structer) CMOS sensor, CCD transfer cycle and matrix, sequential and parallel nature of readout. Fill-factor, rolling and global shutter—the main principle differences between the CMOS and CCD. However, in practice, not these rarely known factors determine the choice of a sensor but simple geometrical considerations: the format. The table and a comparative figure with dimensions clarify the standard notations of the type «1/1.8″». The influence that different format may have on the image is clearly explained, presenting two pictures obtained with sensors of different format. Using colour cameras, it is important to realize the difference between the colour sensor and the so-called 3CCD cameras: the typical mistake of using the same type of lenses for them and polarization differences are explained. The second section discusses video cameras and how to connect them in a variety of schemes: power voltages, digital and analog outputs, advantages and disadvantages of USB and Ethernet connections. Awareness of the simplest practical tricks like extension rings may sometimes solve the problem. The next section introduces beam profilers: a video camera equipped with neutral filters and special software for numerical measurement of beam parameters. Numerical data from the image is a very important option, but it is not necessary to pay extra money for it, purchasing rather expensive beam profiler, if the user has the program that converts standard graphic file into mathematical array. The transcript of such a program written in Fortran is presented together with the result of its application. Data acquisition is often considered as something that an optical engineer must not do himself, totally relying on the help of others. However, with the National Instruments LabVIEW technology this view may be overturned, and the last section of this chapter presents

V. Protopopov, *Practical Opto-Electronics*, Springer Series in Optical Sciences 184,
DOI: 10.1007/978-3-319-04513-9_11, © Springer International Publishing Switzerland 2014

the concentrated guide of how to do it. Types of data acquisition boards (DAQs), connectors, installation into the computer, cables, terminal blocks—everything that may be needed to assemble the measurement system is explained succinctly in the beginning of this section. This is followed by practical instructions of how to create the simplest DAQ virtual instrument. The connectivity know-how—proper use of connection options—is very important to avoid mistakes, and this topic is explained next, including identification of grounded and isolated sources. The rest of the section guides the reader through testing of the first virtual instrument, explains the difference between the differential and pseudo-differential connection, and shows what may happen when connectivity rules are breached.

11.1 Video Sensors

Today, two types of video cameras are available for laboratory applications: the so-called CCD (charge coupled device) and CMOS (complementary metal–oxide–semiconductor) cameras. It is important to understand from the very beginning that both types have the same principle of converting light into electrical signal—generation and separation of carriers in a *pn*-junction (Chap. 3)—and the names CCD and CMOS only differentiate them on the read-out technique and topology. Brief historical overview of these two, nowadays competing, technologies is very instructive to better realize the difference between them.

Since the technology of single-element *pn*-photodiodes (Chap. 3) matured in the 1960s, it became clear that multi-element photodiode arrays, capable of detecting images, are feasible. However, the idea remained impractical because traditional photodiodes cannot accumulate light—a feature necessary for sufficiently high sensitivity. The solution was found in the form of a MOS capacitor (metal–oxide–semiconductor) shown in Fig. 11.1. The basic cycle, producing a signal proportional to the energy of light accumulated during exposure time, looks as follows.

1. Beginning of cycle: mechanical shutter is closed, blocking the light from reaching the sensitive structure, and the read switch is open, isolating the element from other circuits. The reset switch is closed, applying reverse bias to the *pn*-junction. At this moment, short burst of current charges both the C_{MOS} and C_{pn} capacitances to the same charge Q_0, making the equation

$$U = \frac{Q_0}{C_{\text{MOS}}} + \frac{Q_0}{C_{pn}}.$$

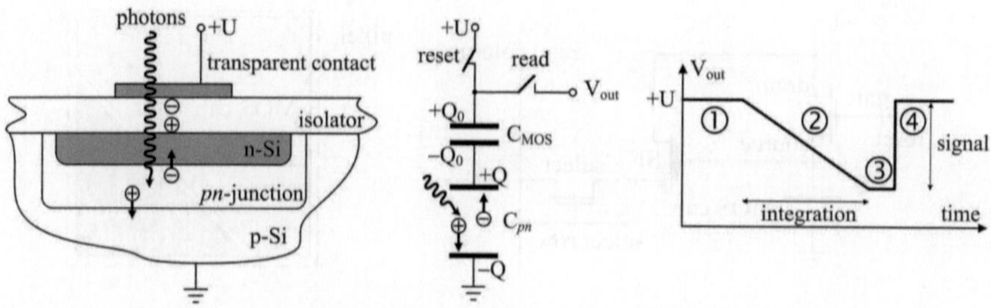

Fig. 11.1 MOS capacitor (*at left*) is formed by a transparent contact—heavily doped polycrystalline silicon—and silicon dioxide isolator deposited onto *p*-doped silicon. The *n*-doped channel is formed under the contact. Two capacitors—the one on the isolator layer with the capacitance C_{MOS} and the second on the *pn*-junction with the capacitance C_{pn}—are connected in series. Positive voltage is applied to the contact, producing reverse bias on the *pn*-junction. Two switches for reset and read are needed to form a single element signal (*in the middle*). Then integration is performed in a cycle of four steps (*at right*) as described in the text

2. Mechanical shutter and the reset switch open simultaneously. Photons reach the *pn*-junction, generating carriers in an amount

$$\eta\, n(t),$$

where η is quantum efficiency and $n(t)$ is the number of photons absorbed by the time t. Separated by the intrinsic electric field inside the *pn*-junction (Chap. 3), carriers drift to opposite electrodes of the C_{pn} capacitance, decreasing the charge to

$$Q(t) = Q_0 - e\,\eta\, n(t)$$

with e being the elementary charge. Voltage on the C_{pn} capacitance lowers and so does the voltage on the two capacitances:

$$V(t) = \frac{Q_0}{C_{\text{MOS}}} + \frac{Q_0}{C_{pn}} - \frac{e\,\eta\, n(t)}{C_{pn}}.$$

3. After the exposure time T elapses, both the mechanical shutter and read switch close. The output voltage

$$V_{\text{out}} = U - \frac{e\,\eta\, n(T)}{C_{pn}}$$

connects to the read-out circuitry that accomplishes measurement in some short time.

4. The read switch opens and the reset switch closes. The circuit returns to initial state.

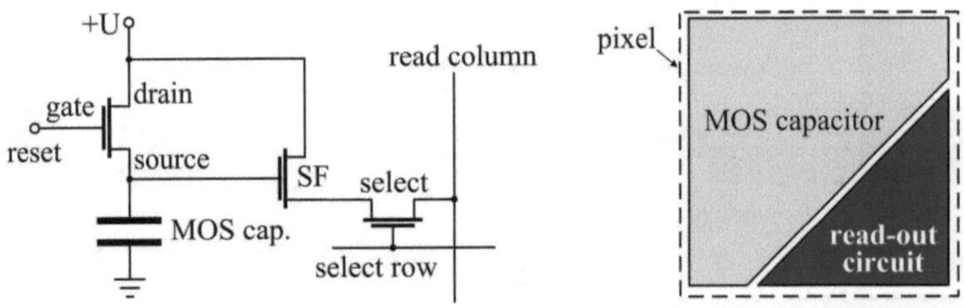

Fig. 11.2 T3-APS scheme. Source follower (SF) transfers the voltage from the MOS capacitor to the read column when the select switch is activated. Even that simple read-out circuit takes much space away from the sensitive area

Although mechanical shutter is quite a realistic option in some scientific applications and especially in photo-cameras, its function can be even better performed by applying positive voltage to the silicon substrate—the technique commonly used in commercial video cameras.

The above cycle was first implemented in the early CMOS sensors that preceded the invention of CCD sensors. The function of reset and read switches was performed by MOSFET (metal oxide semiconductor field effect transistor) transistors grouped around each sensitive element—the MOS capacitor. These elements are indispensable for CMOS sensors and remain until now in various topologies, called the APS—active pixel structure. The simplest APS is composed of at least three transistors and is called the T3-APS (Fig. 11.2). The T3-APS suffered from many drawbacks such as thermal noise and was superseded by more complicated circuits with the bigger number of elements among which are both *n*-MOSFET and *p*-MOSFET transistors—a very efficient switching pair of complementary types of transistors, which explains the name CMOS: complementary metal-oxide-semiconductor circuit. As a result, the fill factor of CMOS topology, i.e. the ratio of the sensitive area to the entire area of a pixel, is substantially less than unity, making sensitivity low.

Great breakthrough was made in the 1970s with the CCD technology. With the same MOS capacitors in each pixel, they differed from CMOS structures by the principle of transferring charge to the read-out amplifier. In CCDs, instead of erecting read-out electronics around each MOS capacitor, the charge propagated through adjacent capacitors, following traveling wave of potential wells (Fig. 11.3). Not only the higher fill factor and, consequently, better sensitivity were achieved, but the noise turned out to be smaller. Figure 11.4 compares the principles of CCD and CMOS readout.

Until now, we discussed only how the still image is transferred to the output. But what happens if the image is moving? Consider the CMOS sensor first. As the row selection wave runs downwards the matrix, static parts of the image will be transmitted correctly, but the moving parts may be recorded more than once by each pixel. Even if the exposure time is short compared to changes in the image, the readout time, i.e. the time in which the row selection wave rolls to the bottom

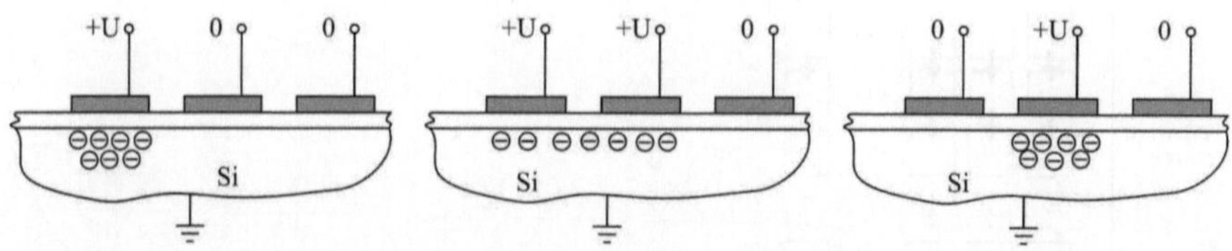

Fig. 11.3 Basic principle of charge transfer in CCDs. The wave runs from *left* to *right*, moving the accumulated charge from one pixel to another. In reality, more sophisticated algorithms may be used, even including avalanche multiplication during transfer from one element to another

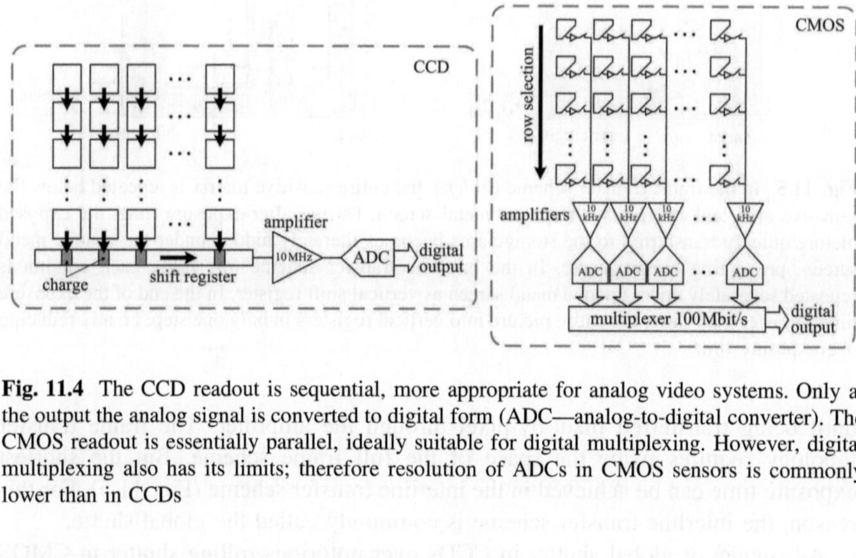

Fig. 11.4 The CCD readout is sequential, more appropriate for analog video systems. Only at the output the analog signal is converted to digital form (ADC—analog-to-digital converter). The CMOS readout is essentially parallel, ideally suitable for digital multiplexing. However, digital multiplexing also has its limits; therefore resolution of ADCs in CMOS sensors is commonly lower than in CCDs

of the matrix, may be longer. Then the recorded picture will have peculiar artifacts like skewed shapes of a moving car. The feature of the exposure wave running down the CMOS matrix is commonly referred to as the rolling shutter and is the topic of debates around consumer electronics with CMOS sensors.

CCDs have several different readout topologies, of which the one shown in Fig. 11.4 is called the full frame transfer. In it, the minimum exposure time is equal to the product of the total number of pixels by the inverse bandwidth of the amplifier. Since the bandwidth cannot be increased without loss of signal-to-noise ratio, minimum exposure time in this scheme is relatively long, blurring moving images. Nevertheless, full frame transfer topology offers the highest possible fill factor, which results either in highest possible number of pixels or largest pixel area, i.e. sensitivity. These unique features make the full frame transfer scheme the best choice for scientific applications with still images, like astronomy.

In order to reduce blurring of moving objects and overcome the bottleneck of the amplifier bandwidth, the so-called frame transfer topology was introduced (Fig. 11.5). In it, the entire exposed image is quickly row-by-row transferred one-to-one into optically shielded storage area, where it may wait indefinitely long

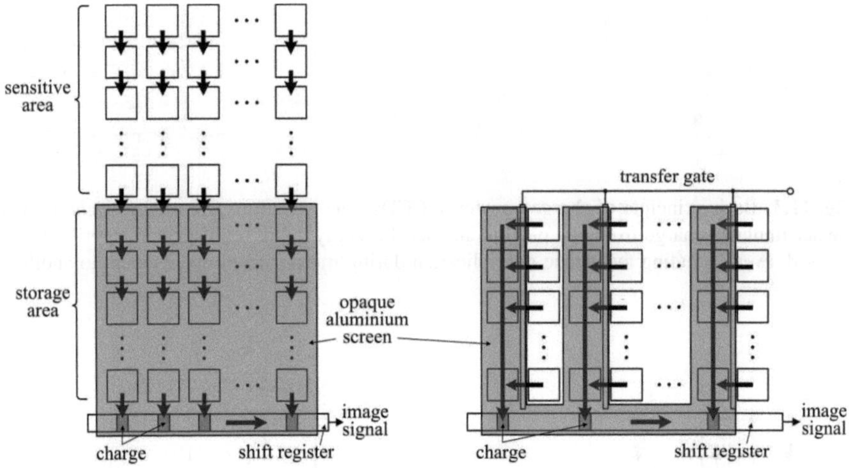

Fig. 11.5 In the frame transfer scheme (*at left*), the entire sensitive matrix is repeated below the sensitive area and covered with opaque metal screen. During after-exposure time, the exposed picture quickly transferred to the storage area becomes literately hidden under the opaque metal screen, preventing overexposure. In the interline transfer scheme (*at right*), each column is repeated separately under opaque metal screen as vertical shift register. In the end of the exposure time, transfer gate shifts the entire picture into vertical registers in only one step, greatly reducing overexposure time

until being transferred pixel-by-pixel through the amplifier. The frame transfer topology requires twice the space of the full frame scheme. But the shortest exposure time can be achieved in the interline transfer scheme (Fig. 11.5). For this reason, the interline transfer scheme is commonly called the global shutter.

Advantage of global shutter in CCDs over notorious rolling shutter in CMOS cameras inspired developers of CMOS sensors to accept the global shutter concept (Fig. 11.6). Nowadays, all the high-quality CMOS video cameras are built with global shutter option. When exposure time ends, accumulated charges in each and everyone pixel are immediately transferred to storage capacitors isolated from light by opaque metal layers. Here, they can wait undisturbed until common for CMOS sensors the rolling shutter process ends.

CMOS sensors consume an order of magnitude less power than CCDs that is considered to be their biggest advantage. However, in laboratory applications, power consumption is not a strong point. Other widely advertised advantages are speed and price. On the other hand, sensitivity and noise figure, especially the so-called fixed noise—permanent regular structure in the background of the image due to topology—are believed to be better in CCD cameras. Table 11.1 summarizes some parameters available from commercialized products. Recommendation may be straightforward: since power consumption is not a limiting factor in the laboratory, use CCD sensors.

All the parameters in this table have already been explained except for the sensor format, which becomes clear from Fig. 11.7. Pixel size determines optical

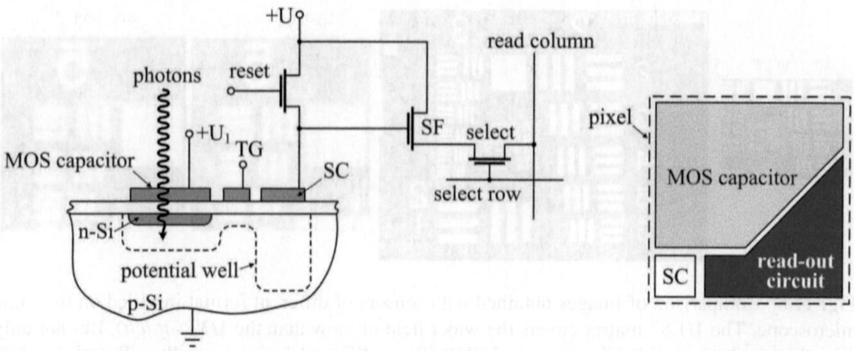

Fig. 11.6 In CMOS sensors, global shutter option is implemented by connecting the reset MOSFET not to the photosensitive MOS capacitor like in Fig. 11.2 but to a specially created storage capacitor (SC), which claims new portion of area from the photosensitive element. The charge from MOS capacitor is transferred to SC through the transfer gate (TG) that is normally kept at lower voltage than U_1 and $U > U_1$, erecting potential barrier for the electrons. When the TG voltage rises to the value between U and U_1, potential barrier dissolves, enabling transfer of the electrons to SC. Storage area is isolated from light by opaque metal layer like in CCDs (Fig. 11.5)

Table 11.1 Parameters of image sensors

Parameter	CCD			CMOS		
Resolution	640 × 480	1280 × 960	1600 × 1200	640 × 480	1280 × 960	1600 × 1200
Format	1/3″	1/3″	1/1.8″	1/3″	1/3″	1/1.8″
Pixel size (μ)	7.4	3.75	4.4	6	3.75	5.3
Frame rate (Hz)	100	30	25	120 300	40 120	60
Data format (bit)	14	14	14	12 8	12 8	10
Exposure time	10 μs – 10 s			10 μs – 1 s		
Shutter type	Global			Global/Rolling		

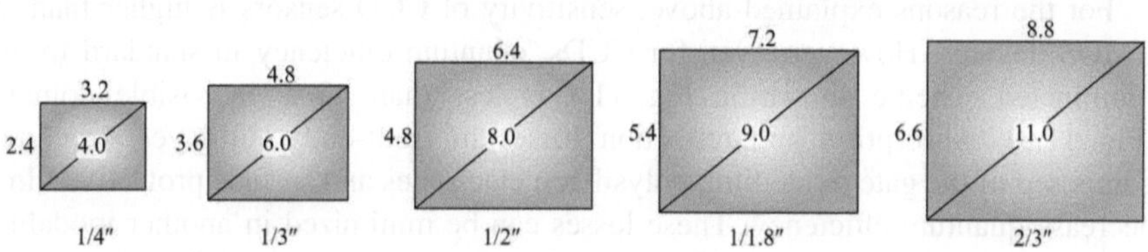

Fig. 11.7 Formats of image sensors. Aspect ratio 4:3. All the dimensions are in millimeters

resolution, and format—field of view. Therefore, substitution for the bigger format may result not only in wider field of view but also in better optical resolution, depending on the pixel size. This is exemplified in Fig. 11.8.

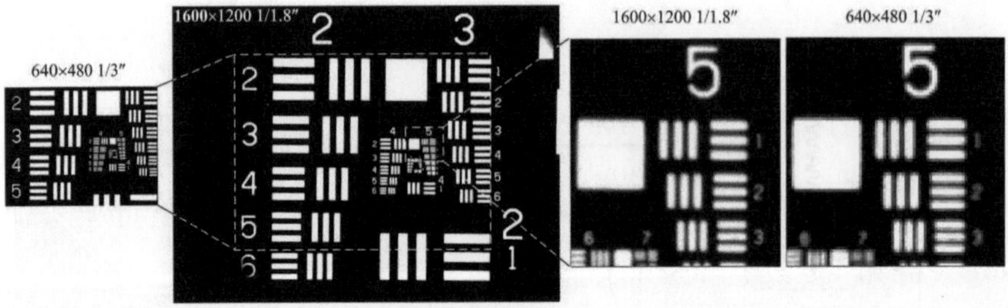

Fig. 11.8 Comparison of images obtained with sensors of different format installed on the same microscope. The 1/1.8″ matrix covers the wider field of view than the 1/3″ (*at left*). But not only this: the pixel of the 1/1.8″ matrix is 1600/640 × 1.8/3 = 1.5 times smaller. Therefore, with equal magnification, pixel structure on the 1/1.8″ matrix is less visible than on the 1/3″ (fragments *at right*)

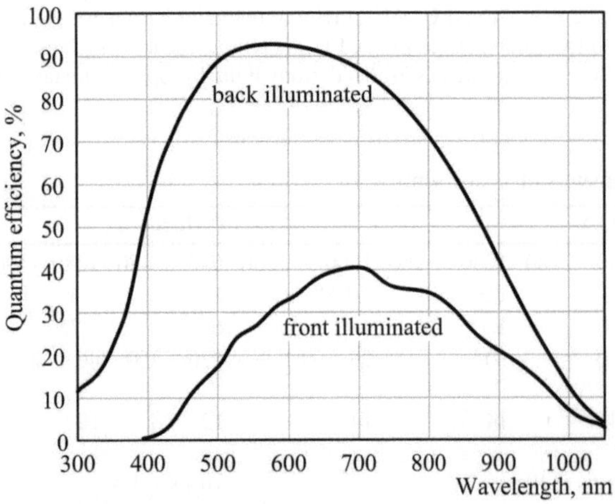

Fig. 11.9 Quantum efficiency of a CCD sensor

For the reasons explained above, sensitivity of CCD sensors is higher than in CMOS devices. However, even for CCDs, quantum efficiency in standard front-illuminated scheme shown in Fig. 11.1 is less than 40 % in visible domain (Fig. 11.9). Absorption and reflection losses in front-end multilayer structure, composed of the gate oxide film, polysilicon electrodes and surface protective film, decrease quantum efficiency. These losses can be minimized in another modality called the back-illumination CCD sensor. The idea is simple: since there is no need to deposit or form any films on the bottom side of a sensor, this area, exactly under the system of electrodes, may be thinned to about 20 μ and used for illumination, avoiding the aforementioned optical losses. Of course, manufacturing costs are higher, but the result is worth it: quantum efficiency more than doubles with respect to a front-illuminated sensor.

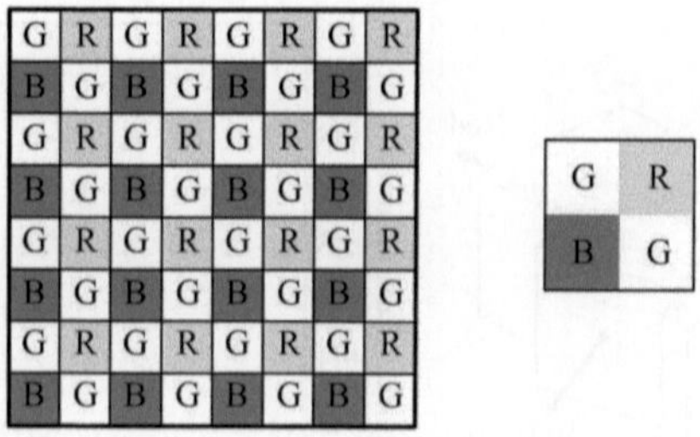

Fig. 11.10 The Bayer filter with the period composed of *red*, *blue*, and two *green* pixels is the most popular colour matrix. Two *green* pixels make colour sensitivity closer to human visual perception with its higher responsivity to *green* spectral component

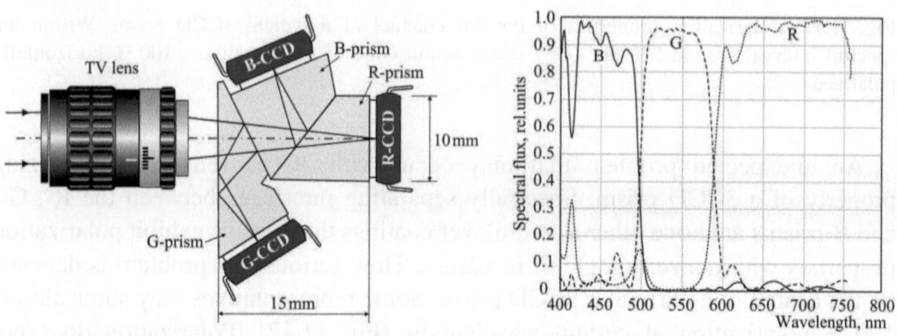

Fig. 11.11 The 3CCD prism is composed of three separate R-, G-, and B-prisms bonded together on the optical surfaces with multilayer coatings, separating red, green, and blue parts of the visible spectrum. Before bonding, R- and B-prisms must be aligned relative to the G-prism in order to equalize focal distances in three channels with better than 2 μ accuracy. This alignment is done by sliding the prisms along their contact surfaces. Therefore, a peculiar tooth is made on the G-prism to accommodate the R-prism during displacements. An ordinary TV lens (Chap. 1) cannot be used in this configuration because optically thick 3CCD prism would cause unacceptable aberrations. Typical non-polarized spectrum is shown at *right*

Colour images can be obtained in two ways: with colour filter matrix and with the so-called 3CCD prism. Colour filter arranged for each pixel in a form of a periodical red, green, and blue matrix is the simplest and earliest technique (Fig. 11.10). With colour filter matrix, actual resolution of a sensor is two times worse than with the same sensor in grey-scale mode because each colour pixel is now composed of 2 × 2 pixels of a CCD matrix. In order to overcome this disappointment, the 3CCD technology was introduced (Fig. 11.11). In it, light is decomposed into red (R), green (G), and blue (B) spectral components with the so-called 3CCD prism, and then each one of the three images is recorded by a separate sensor. A common mistake is to use an ordinary TV lens (Chap. 1) with a 3CCD sensor: 15 mm thick bulk glass of the prism introduces substantial aberrations, blurring the image. Only specially designed 3CCD lenses deliver sharp images. They are always marked as «3CCD» on the cylindrical side of the lens.

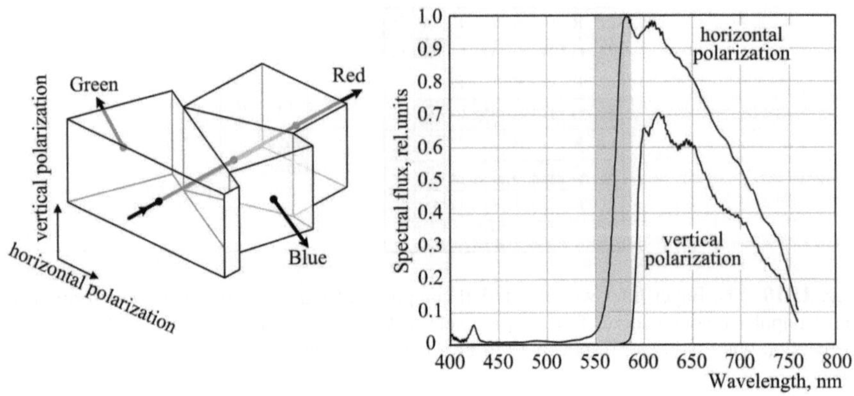

Fig. 11.12 Polarization properties of the red channel of a typical 3CCD prism. Within the spectral interval around 570 nm (*grey area*) output cone of rays is almost 100 % horizontally polarized

An unexpected problem that may occur with 3CCD sensors is polarizing property of a 3CCD prism. Spectrally separating interfaces between the R-, G-, and B-prisms are none other as multilayer coatings that always exhibit polarization properties with nonzero angle of incidence. How serious this problem is depends on the design of a particular 3CCD prism. Some representatives may show almost 100 % polarization at certain wavelengths (Fig. 11.12). Polarization does not mean much in photography, but if you are going to monitor interference patterns produced by polarized sources like lasers, for instance, then irreverence to the aforementioned phenomenon may make the result inexplicable.

11.2 Video Cameras

Video cameras are compact and reliable devices (Fig. 11.13). The only thing that the user needs to know is how to connect it to computer or display. Connection options are explained in Fig. 11.14. Analog cameras does not have USB or Ethernet connectors. In option 1, the camera module both supplies +12 V to the camera and receives analog video signal that is translated to the BNC connector. The camera itself also outputs analog video signal through a separate BNC connector. Either one of the two video outputs may be connected to an analog TV display. In option 2, the camera module is unavailable. Any standard +12 V power supply may be connected to the camera through its Hirose connector. In this case, only one analog output is available through BNC connector on the camera. In option 3, all the job is done by a PCI frame grabber card installed in the computer: it supplies +12 V to and receives analog video signal from the camera. Proprietary software for the frame grabber is needed. The image can be viewed both on the

Fig. 11.13 Typical CCD/CMOS camera. In applications, requiring highly coherent sources like lasers, protective glass window may be a source of parasitic interference fringes. In this case, it may be removed by unscrewing two screws inside the C-mount flange (Chap. 1) (*at left*). Not all the connectors shown in the *right* picture may be present. The BNC connector is used for analog output video signal. The Hirose connector is commonly used for +12 V power and analog output video signal. The USB connector is used for both the +5 V power and digital output video signal. The Ethernet connector is commonly used only for digital output video signal, but some options called POE (power on Ethernet) also apply power through it

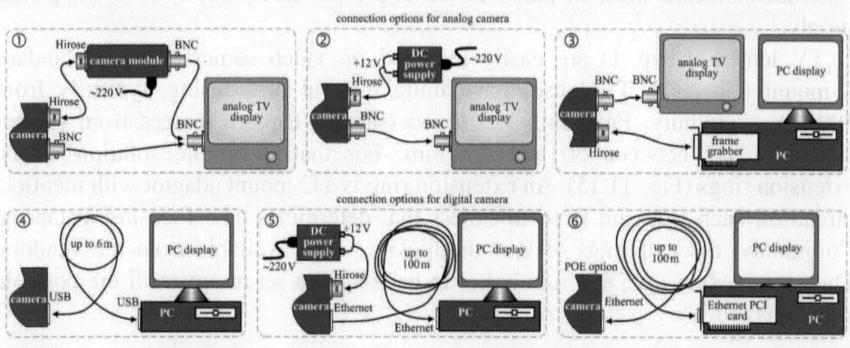

Fig. 11.14 Six basic options exist to operate the video camera: three analog and three digital (see the text)

analog TV display and on the computer (PC) display. If there is a digital connector on the camera, like USB or Ethernet, there is no BNC output on it: the camera is digital. In option 4, the simplest one, only USB cable is needed to connect the camera to a PC. Of course, proprietary software driver for the particular type of a camera is needed. The length of the USB cable is typically limited to 6 m—at a longer cables data transfer may be unreliable. In option 5, the Ethernet connector is used. Two things must be done to activate this option: software driver for the camera and the Ethernet port driver in the PC must be installed. Standard Ethernet protocol enables reliable data transfer up to 100 m, but it does not provide power. Therefore, in option 5, a separate direct current (DC) power supply is needed.

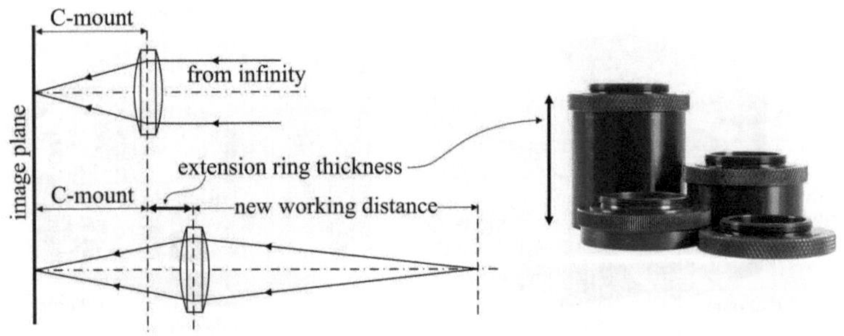

Fig. 11.15 A set of extension rings for C-mount adaptor is composed of several rings of discrete thicknesses. With continuously variable focal distance of a TV lens, such a set typically covers all the working distances from 50 mm to infinity

The camera may be purchased in the so-called POE option (power over Ethernet). In it, two contacts of the Ethernet connector are reserved for the DC 44–57 V and 350–400 mA current to power the camera. This is option 6. However, standard Ethernet connector on the PC never supplies current to a linked device: a special Ethernet PCI card must installed inside and connected directly to a PC power supply.

TV lenses (Chap. 1) are easily installed on video cameras, using standard C-mount (Chap. 1). TV lenses have finite interval of focusing, typically from 300 mm to infinity. Frequently, it is necessary to capture images from shorter distances, say between 50 and 200 mm. For that, a simple solution exists: extension rings (Fig. 11.15). An extension ring is a C-mount adaptor with identical thread on each side and fixed thickness that determines new focusing distances. Commonly, a set of rings of various thicknesses is available from the vendors. They are inexpensive, and it is better to have such a set to cover all the possible situations.

11.3 Beam Profilers

Beam profiler is a very useful part of opto-electronic equipment (Fig. 11.16). It is more expensive than a standard video camera because it incorporates several additional features: adjustable exposure time, ranging from 10^{-4} to 1 s; extensive software capable of measuring geometrical parameters of the beam, including fitting predetermined shapes like Gaussian or Lorentzian; and optical filters. Logically, there is no need in colour CCD matrix since we are interested in measuring monochromatic laser beams. But presentation of measurements in pseudo-colours is a very useful feature, facilitating visual analysis.

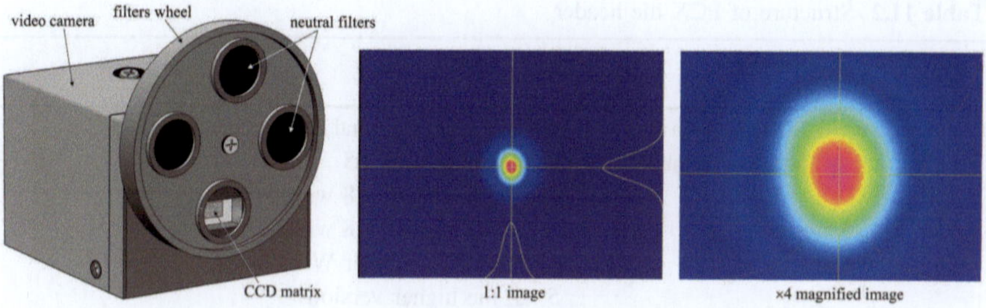

Fig. 11.16 Beam profiler is actually a CCD camera with a wheel of neutral density filters in front of it (*at left*). Filters of different optical density are needed to prevent saturation of the matrix on relatively strong laser beams. Typical software packages render the measurements in pseudo-colours, like the image of a single-mode 1 mm He–Ne laser beam (*at right*). *Yellow* curves show cross-sectional intensity distributions. 640 × 480 pixels; pixel size 9 μ

Coherent laser light is a very treacherous subject: even faintest contaminations or dust on the surface of neutral filters result in annoying diffraction rings in the image. Contamination of the cover glass of the CCD matrix is not as noticeable as the dirt on the filter: the filter is farer from the sensor, and propagation distance makes any interference effects stronger. Therefore, always keep the filters clean.

There is another very useful software option commonly supplied with beam profilers: access to the array of image intensities. When an image is recorded in a matrix of M × N pixels, intensity a forms a two-dimensional array $a[i,j]$ with the indices $1 \leq i \leq M$, $1 \leq j \leq N$. An option of storing this array as an ASCII (American standard code for information interchange) formatted file with three columns, easily read by Fortran or C mathematical routines or graphic engines like Golden Software or Origin, makes it possible to perform free mathematical computations on the image and plot impressive 3D (three-dimensional) views for presentations. There is no such an option in a standard video camera. Does it mean that the user has to pay extra money if he does not have beam profiler software? No, it does not. Even utterly simplified software drivers supplied with standard monochrome video cameras provide an option of storing the framed image as one of the standardized image files: BMP, PCX, TIFF or any other. Such a file is a binary file that can be read, decrypted, and transformed into the ASCII format. We are now going to show how it can be done in practice, using simple Fortran code.

The most widely used BMP format is a non-compressed bitmap file, thus occupying much space on disk. Therefore, it is better to consider the also widely used PCX format that employs simple RLE (run-length encoding) compression algorithm and, therefore, requires much less disk space. The RLE algorithm counts the number of consecutive pixels with equal intensities in one line of the image and writes the result in two adjacent bytes: first—the number of pixels, second—the intensity itself. In order to provide a mean for checking correctness of

Table 11.2 Structure of PCX file header

Shift	Size, bytes	Content	Description
0	1	Identification flag	Always «10» decimal («0A» hexadecimal)
1	1	Version number	0: PC Paintbrush 2.5 2: PC Paintbrush 2.8 with palette 3: PC Paintbrush 2.8 without palette 4: PC Paintbrush for Windows 5: all the higher versions
2	1	Coding	1: group coding
3	1	Digital resolution	Number of bits per pixel
4	8	Image size	Limits of the picture (in pixels): X_{min}, X_{max}, Y_{min}, Y_{max}
12	2	Horizontal print resolution	Dots per horizontal inch in printed version (irrelevant to our case)
14	2	Vertical print resolution	Dots per vertical inch in printed version (irrelevant to our case)
16	48	Palette	Always «0» for monochrome cameras
64	1	Reserve	Always «0»
65	1	Planes	Number of colour planes: always «0» for monochrome cameras
66	2	Bytes per line	Number of bytes required to store 1 line of the image
68	2	Palette code	1: colour or black-and-white image 2: grey-scale image
70	2	Horizontal screen size	Number of pixels in horizontal line on the display
72	2	Vertical screen size	Number of pixels in vertical line on the display
74	54	Blank	

encoding, the first byte (the flag) always has a definite structure: the first two bits are always «1» if the encoding is correct. Thus, only 6 lower bits of the flag give the count, i.e. the number of consecutively equal intensities. The code should check this property to confirm correctness of decoding. In the code below, this test is implemented in the subroutine *bina*.

The next property to be understood is the structure of the header—the first 128 bytes of the PCX file (Table 11.2). Below is the Fortran code that reads the PCX file «*PCX-file-name.pcx*», determines the number of bytes required to store one line of the image «*maxnumbyte*», number of pixels in one line of the image «*numpixelsinline*», number of rows in the image «*numrows*», and saves the three-column image table in the ASCII file «*picture.dat*». For reference, the code also prints out the header of the PCX file. Maximum size of the image is limited to $10^4 \times 10^4$ pixels as declared by the size of the array *a*. Instruction btest is the intrinsic Fortran logical command, testing bits of an integer argument.

```
c This program transforms PCX format to an array.

        program pcx
        byte b0(4), b5(58), b2(48), b3(2), flag, intensity     !b2-palette
        integer(1) count, bit(8)
        integer(2) b1(6), b4(2)            !b4(1)-maxnumbyte, b4(2)=2 grey scale
        integer(2) a(10000,10000)          !image map
        open(unit=1, err=103, file='PCX-file-name.pcx', readonly, form='binary',
status='old')
        read(1) b0, b1, b2, b3, b4, b5     !Binary reading of the file header
        maxnumbyte=b4(1)
        numpixelsinline=b1(3)-b1(1)+1
        numrows=b1(4)-b1(2)+1

c The header is printed out:
        print*, b0
        print*, b1
        print*, b2
        print*, b3
        print*, b4
        print*, b5
        print*, 'maxnumbyte=',maxnumbyte
        print*, 'number of pixels in line=',numpixelsinline
        print*, 'number of rows in the picture=',numrows

c From this point on, begin reading image data:
        irow=1
    1 line=0
        numbyte=0
        do while (numbyte.LT.maxnumbyte)
        read(1) flag                        !number of repeated intensities
        call bina(flag, bit, count)         !analysis of the flag bit
        if(count.NE.0) then
        read(1) intensity                   !the value of the repeated intensity
        do 2 i=1, count
        line=line+1
        numbyte=numbyte+1

c Integer and Byte variables are signed, i.e.,the 9-th bit gives sign.
c The next is the checking of bits to avoid negative values:
        bit=0
        if(btest(intensity,7)) bit(8)=1
        if(btest(intensity,6)) bit(7)=1
        if(btest(intensity,5)) bit(6)=1
        if(btest(intensity,4)) bit(5)=1
        if(btest(intensity,3)) bit(4)=1
        if(btest(intensity,2)) bit(3)=1
        if(btest(intensity,1)) bit(2)=1
        if(btest(intensity,0)) bit(1)=1

c The next is equal to: a(line,irow)=intensity
        a(line,irow)=bit(8)*128+bit(7)*64+bit(6)*32+bit(5)*16+bit(4)*8+bit(3)*4+bit(2)*
2+bit(1)
    2 continue
        else
        line=line+1
        numbyte=numbyte+1
        bit=0
        if(btest(flag,7)) bit(8)=1
        if(btest(flag,6)) bit(7)=1
        if(btest(flag,5)) bit(6)=1
        if(btest(flag,4)) bit(5)=1
        if(btest(flag,3)) bit(4)=1
        if(btest(flag,2)) bit(3)=1
        if(btest(flag,1)) bit(2)=1
        if(btest(flag,0)) bit(1)=1

c The next is equal to: a(line,row)=flag
        a(line,irow)=bit(8)*128+bit(7)*64+bit(6)*32+bit(5)*16+bit(4)*8+bit(3)*4+bit(2)*
2+bit(1)
        end if
        end do
        irow=irow+1
        if(irow.le.numrows) go to 1
        close(1)

c Writing the image map a(i,j) to a file:
        open(unit=2, err=103, file='picture.dat')
        do 3 i=1, numpixelsinline
        do 3 j=1, numrows
        write(2, 100) i, j, a(i,j)
    3 continue
        close(2)

  103 continue
  100 format(1x,3I10)
        end

        subroutine bina(flag, bit, count)      !Analyzes bit structure of flag
        integer(1) count, intensity, bit(8)
        byte flag
        bit=0
        if(btest(flag,7)) bit(8)=1
        if(btest(flag,6)) bit(7)=1
        if(btest(flag,5)) bit(6)=1
        if(btest(flag,4)) bit(5)=1
        if(btest(flag,3)) bit(4)=1
        if(btest(flag,2)) bit(3)=1
        if(btest(flag,1)) bit(2)=1
        if(btest(flag,0)) bit(1)=1
        if(bit(8).EQ.1.AND.bit(7).EQ.1) then       !see RLE for PCX files
c RLE -- run-length encoding algorithm, a simple lossless compression.
        count=bit(6)*32+bit(5)*16+bit(4)*8+bit(3)*4+bit(2)*2+bit(1)
        if(count.EQ.0) print*,'error: count=0'
        else
        count=0
        end if
        return
        end
```

With such a program, any picture recorded by a monochrome video camera can be converted into a two-dimensional array of numerical data and subjected to any desired mathematical analysis on a pixel level. As an example, Fig. 11.17 portrays three-dimensional (3D) view of the image shown in Fig. 11.8.

One comment should be added to the aforementioned. Some graphic formats, like BMP or PCX, do not alter measured intensities recorded by a camera in each pixel. Others, however, like JPG for instance, do it. Therefore, if the original file was saved as BMP but your code supports conversion from only PCX format, then it is quite possible to upload the BMP file to some graphic engine and re-save it in the PCX format without any loss of information. But if you have saved the original image as JPG format, then be sure that intensities are somewhat different from the reality, and the data file obtained after conversion may, and most probably does, differ from reality.

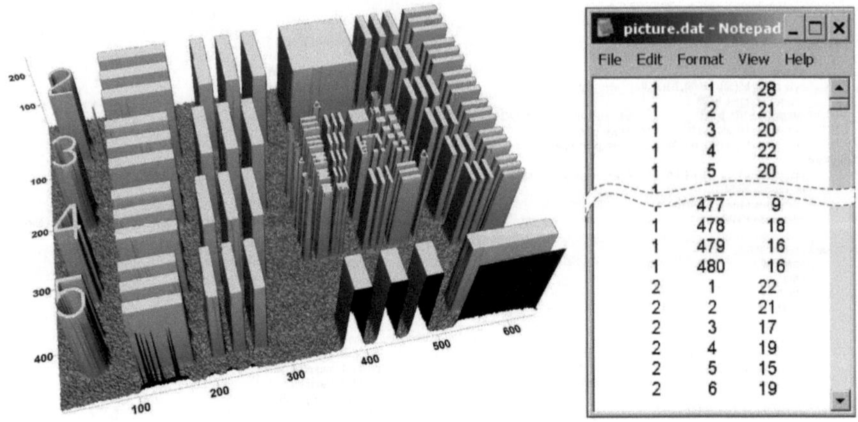

Fig. 11.17 3D presentation of the image recorded by a monochrome 640 × 480 1/3″ camera in Fig. 11.8. The level of noise is clearly visible in the bottom of the picture. The format of the output file «*picture.dat*» is clear from the snapshot at right. The 3D picture was generated by the «Surfer» program from the Golden Software graphic package

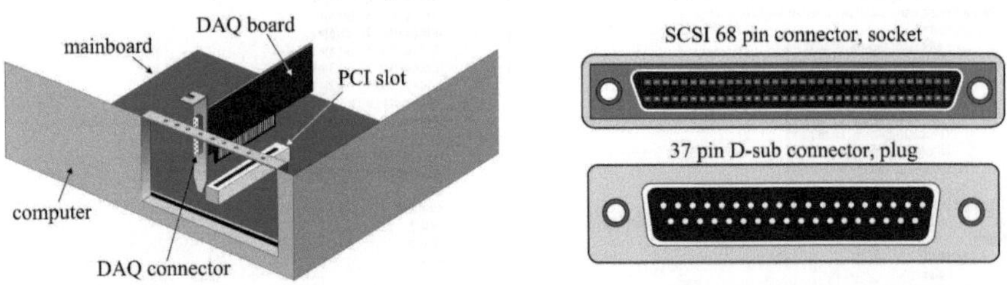

Fig. 11.18 PCI type DAQ board installs into PCI slot on the mainboard inside computer enclosure (*at left*). Two basic types of connectors determine peripheral options of the system (*at right*; see the text)

11.4 LabVIEW Technology

Strictly speaking, LabVIEW (laboratory virtual instrument engineering workbench) is only a visual programming language developed by National Instruments (NI) to support its family of data acquisition (DAQ) devices. However, integration with DAQ hardware is so complete, variety of devices supported by LabVIEW and simplicity of building measurement systems are so attractive that LabVIEW may be considered as a measurement technology. The purpose of this section is not to teach the LabVIEW language but to explain how to quickly build a simple measurement system with minimum mistakes.

To make a measurement system, three components are needed: personal computer with at least one PCI (peripheral component interconnect) slot, DAQ, and two compact disks (CDs)—one with the LabVIEW operating system and the

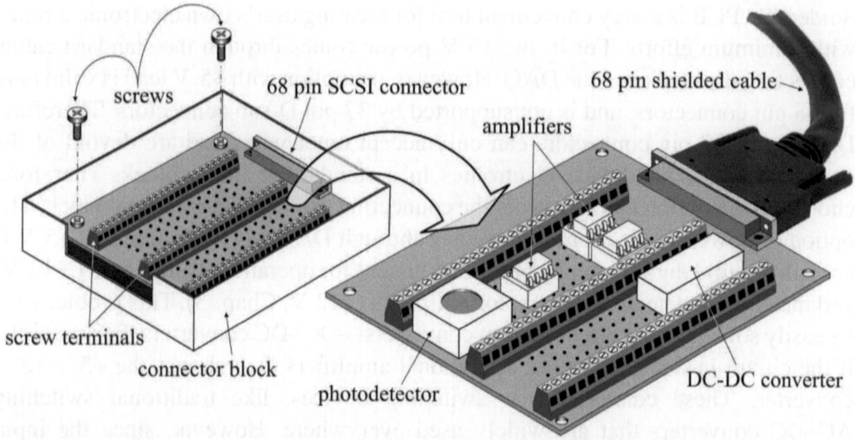

Fig. 11.19 Standard terminal block with 68 pin connector contains solderable PCB with perforated area between the terminals. The PCB is fixed with screws to the bottom of the block and through the connector to its side wall. Remove all the screws, release the connector, and take the PCB out. From now on, this PCB is your own opto-electronic module that may contain photodetectors, amplifiers, and other electronic systems. The DAQ board inside the computer receives the signals through the cable and digitizes them in real time

second one (or two) with device drivers. Make sure that the DAQ you have gotten is on the list of device drivers on the disk. The first question is what form-factor to choose for the DAQ? For simple laboratory applications with the project on a budget, the right choice is the PCI DAQ—the data acquisition board with PCI connector to be installed inside the computer on its mainboard (Fig. 11.18). Additional argument in favor of this choice is the minimized space that the entire system will occupy. The alternative to PCI is the more expensive so-called PXI geometry that requires its own frame with power supply, connectors, and all other attributes. Eventually, it also needs a computer. The only argument in favor of the PXI DAQ is that this way does not require from the user any hands-on skills of working with a screwdriver. But the present book is not for that kind of users.

The next practical question is what connector to choose on the DAQ? National Instruments (NI) support two basic types of connectors: 37 pin D-sub and 68 pin SCSI (small computer system interface) (or VHDCI—very high density cable interconnect). The VHDSI is merely a more compact and gentle version of the SCSI. There is no principle difference between DAQs with SCSI and VHDSI connectors, but from practical point of view, the SCSI is preferable owing to its durability: thick and rigid connecting cables can easily damage gentle VHDSI if someone incidentally tugs it. As to the 37 pin D-sub connector, it may be a real mistake, and the following explains why. An inexperienced user always wants to assemble the measurement system as easily and quickly as possible. Understanding it, NI offer standard terminal blocks: enclosures with perforated solderable PCB (printing circuit board) inside and rows of screw terminals on it for outer wires (Fig. 11.19).

Solderable PCB is a very convenient tool for creating user's own electronic circuits with minimum efforts. For it, the +5 V power comes through the standard cable, connecting the block and the DAQ. However, the option with +5 V lead is valid only for 68 pin connectors, and is not supported by 37 pin D-sub connectors. Therefore, DAQs with 37 pin connectors can only accept outer wires and are devoid of the option to create handmade electronics in a standard terminal block. Therefore, choose 68 pin option and also order the connecting cable and the terminal block. The option of powering electronic from +5 V through DAQ has its limitations: +5 V is enough to build digital circuits but is insufficient for operational amplifiers ($\pm$12 V) and many photodetectors like photomultipliers (+12 V, Chap. 3). This problem can be easily solved, using direct current converters—DC–DC converters. For example, if the circuit is supposed to use operational amplifiers then choose the +5–$\pm$12 V converter. These converters are switching devices, like traditional switching AC–DC converters that are widely used everywhere. However, since the input sampling voltage is only 5 V in contrast with 380 V in the AC–DC power supplies, the DC–DC converters are incomparably less noisy. Nonetheless, additional low--pass filter after it is always a plus.

The first step in making the system work is to establish communication between the DAQ and the computer. Install the LabVIEW operating system. This process requires additional disk(s) with device drivers. When it is finished, turn off the computer, disconnect the power cord, and install the DAQ board into its PCI slot. Reboot the computer and let the operating system find appropriate driver for your DAQ. Then launch the LabVIEW and from the drop-out menu «New» choose «Blank VI» (VI: virtual instrument). There are always two active panels: the «front panel» and the «block diagram». The diagram shows the wiring, connections, and the entire set of components. The front panel displays controls and indicators, like the graphic window where the real time trace of the input signal will appear. To create a measurement system, we need the block diagram. Right-click on the blank field and among the «Functions» menu point the cursor to the «Input» and choose the «DAQ Assistant»—this function will solve all our initial problems (Fig. 11.20). Simple consecutive clicking on the icons shown in Fig. 11.20 has led us to the DAQ Assistant menu. In order to qualify for editing it, we need to know more about physical channels and their connections. The wires, connecting our PCB (Fig. 11.19) with the DAQ board in the computer, end at the inputs of operational amplifiers, and there are two basic options of how to distribute them (Fig. 11.21). In the RSE mode, the user can make the maximum number of measurement channels equal to the overall number of the DAQ channels. This is the only advantage of the RSE mode over the differential one where only half of the number of channels may be used for measurements. However, with 16 channels that even the simplest DAQs provide, simultaneously all of them are never used. Therefore, this advantage is imaginary. But the disadvantage of the RSE mode is very real: noise that comes through distributed grounds may be overwhelming. This problem is almost completely solved in the differential mode (Fig. 11.21). Whether or not the user needs additional leakage resistors depends on the connection of the source (Fig. 11.22). The rule may be formulated as follows. If you

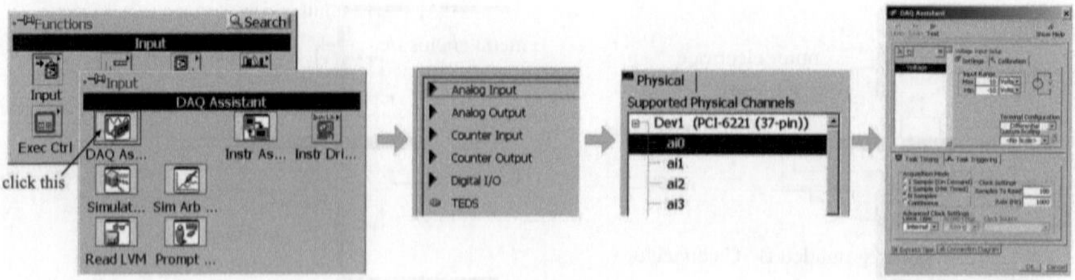

Fig. 11.20 From the «Input» icon, select «DAQ Assistant», and in the next menu choose «Analog Input». The following menu prompts to choose particular analog channel of the DAQ board: from $ai0$ (analog input 0) to aiN—as many as the particular DAQ board supports. The DAQ with 37 pin connector will always be automatically identified and marked in order to emphasize its disability, like in the figure: «PCI-6221 (37 pin)». At this point, we do not know much about the physical channels but assume that at least one of them must work anyway. Therefore, click $ai0$ for the beginning. This will lead us to the DAQ Assistant menu—the most important and very simple tool to configure the data acquisition

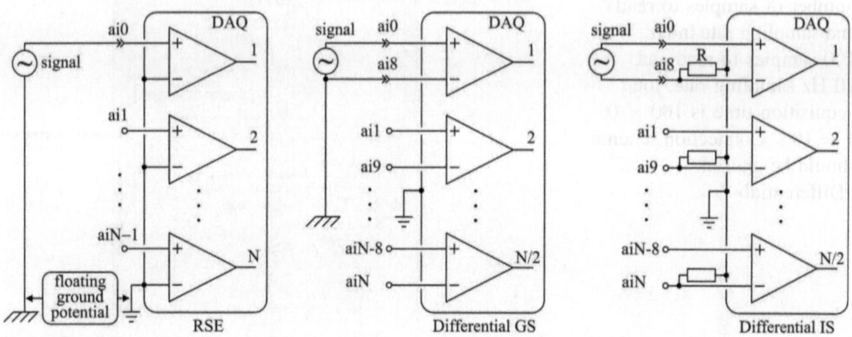

Fig. 11.21 In referenced single-ended (RSE) connection (*at left*), inverting inputs of operational amplifiers are grounded to the DAQ ground. When the power is off, DAQ ground is disconnected from the computer ground (chassis ground). With the power on, these two grounds become connected. However, the computer ground and the signal ground may be connected together through long grounding leads, exerting floating ground potential, i.e. noise. This noise voltage acts between the inputs of operational amplifiers, freely passing to the output exactly like the signal itself. In the differential mode with grounded source (GS) (*in the middle*), inverting inputs are disconnected from the ground. Therefore, only signal alone acts between the inputs of the amplifiers, rejecting completely the floating ground potential. The differential mode with isolated source (IS) (*at right*) does exactly the same, but requires leakage resistors *R* (see the text)

see grounded BNC connector on the source enclosure, use differential mode in case the number of measurement channels is less than N/2, and RSE in the opposite case. If the source connector is of the isolated type, then check with a multimeter whether its outer electrode is actually isolated from the source enclosure. If it is, then the only option is to use differential mode with leakage resistors. If the outer electrode is grounded on the source, then this is the case of the grounded BNC connector, and use the previous rule.

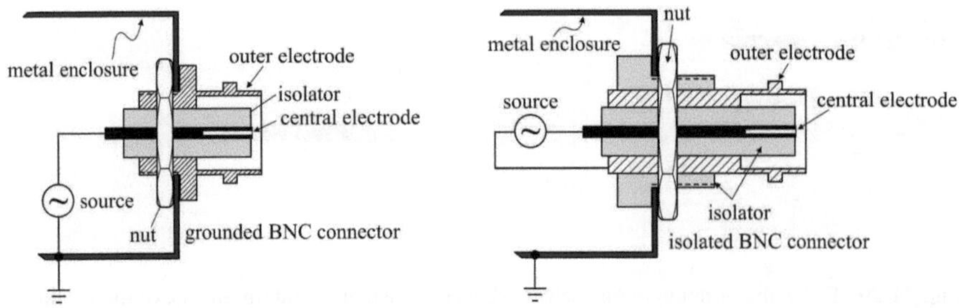

Fig. 11.22 Grounded BNC connector indicates grounded voltage source (*at left*). Isolated BNC connector is commonly used with isolated sources (*at right*)

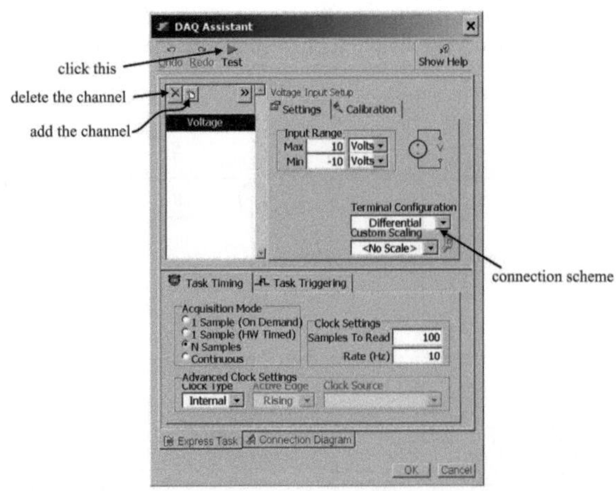

Fig. 11.23 The DAQ Assistant menu prompts to set input range of voltages, terminal configuration, number of samples to read and sampling rate in Hz. With 100 samples to read and 10 Hz sampling rate, total acquisition time is $100 \times 0.1 = 10$ s. Connection scheme should be chosen «Differential»

The case of the isolated source requires some more hands-on job because you need to add a resistor *R* in the input circuit. This can be usually done within the terminal block (Fig. 11.19). The nominal value of the resistor may be about 10–100 kOhm—just to discharge the input capacitance of the operational amplifiers. Input resistance of a typical MOSFET operational amplifier is as big as 10^{10}–10^{12} Ohm. Therefore, input capacitance may be quickly charged to substantial voltages by even infinitesimal unipolar currents. The output will be quickly saturated. When the source is grounded (differential mode), saturation by charging never happens because amplifier's inputs are grounded through the source (Fig. 11.21). On the contrary, with isolated sources, running differential mode without leakage resistors is a common mistake, leading to saturation of the amplifier in few seconds. We shall see this effect below.

Now, we are ready to return back to Fig. 11.20 and edit the DAQ Assistant menu (Fig. 11.23). With all the aforementioned, we can choose the connection scheme: it must be differential. At this moment, we have no preferences about input voltage range: let it be as is—±10 V. As to the sampling rate and number of

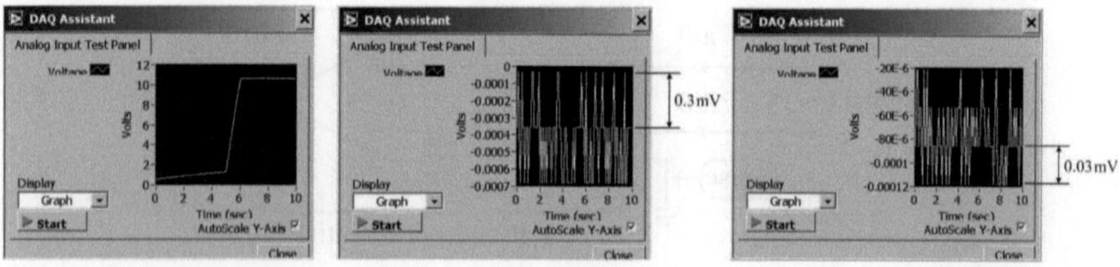

Fig. 11.24 In a differential mode, i.e. with isolated $ai0$ and $ai8$ pins, the DAQ saturates in several seconds (*at left*). This is a typical case of the differential mode, running on isolated source—nothing is connected to DAQ. The problem is solved by grounding the inverting input of the operational amplifier, i.e. $ai8$, through 100 kΩ resistor (*in the middle*). Now, the system works and measures its own noise. With the narrower voltage range 0–1 V, the discrete becomes an order of magnitude smaller (*at right*)

samples to read, we want an oscilloscope trace lasting for about 10 s with the number of points not more than 100.

From the point of view of hardware, our primary concern is whether the acquisition system is working, i.e. does the DAQ board digitize and store any signal? We have already seen that the operating system does sense the presence of the DAQ board and recognize it (Fig. 11.20). But can we measure something? To answer this question we do not need to connect any function generators to the DAQ: let it measure the noise at its input. For that, make short circuit between $ai0$ and $ai8$. To make necessary connections, we need to know the pin-out of the particular connector, which is always specified in the DAQ (data acquisition board) datasheet. Organization of pinning is always very efficient, both on the 37 and 68 pin connectors: the $ai0$ pin is placed near $ai8$, the $ai1$ near $ai9$, etc. Now, click the triangular button «Test» and see what happens (Fig. 11.24). The oscilloscope window opens, and after 10 s we see the output voltage soaring from 0 V to above the maximum value 10 V specified in the DAQ Assistant menu. The signal is saturated by charging the isolated inputs of the MOSFET amplifier. But now we know what to do: the inverting input must be grounded somewhere—either on the source side or on the DAQ side. In our primitive experiment, without actual source, grounding on the DAQ side is the only option. With pin $ai8$ grounded through 100 kOhm resistor, we get our first measurement: noisy trace of $\approx$0.3 mV amplitude (Fig. 11.24). It is easy to see that 0.3 mV is actually the value of one discrete of digitization—there are no measurement points between 0 V and 0.3 mV. Although this is not a very big voltage discrete even on the scale of fine measurements, we may think how to further decrease it, if necessary. Recall that we have set $\pm$10 V input voltage range in the DAQ Assistant menu, i.e. total 20 V range. With 16 bit analog-to-digital conversion, it gives 20 V/2^{16} = 0.305 mV—a very good agreement with our measurement. If we set the narrower voltage range, say 0–1 V, then the discrete will be smaller (Fig. 11.24). Therefore, if the voltage range of the source is limited by smaller values, it is always better to set the actual voltage range rather than the maximum $\pm$10 V.

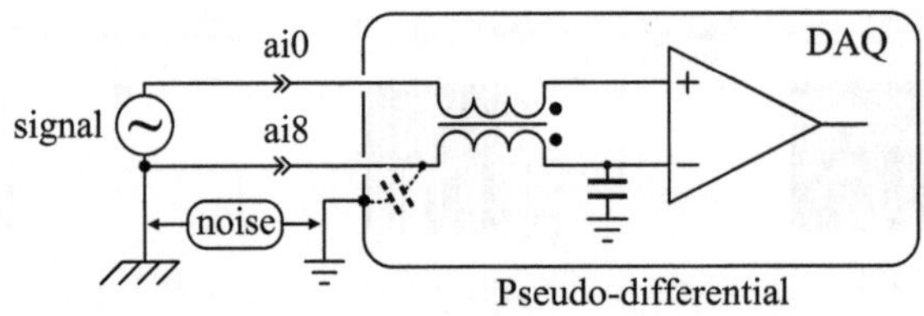

Fig. 11.25 Pseudo-differential connection. The idea is to suppress high-frequency noise, acting between the two grounds. The transformer is connected to generate compensating voltages of the same polarity in both the inverting and non-inverting inputs, thus subtracting them from the differential signal. These compensating voltages can be generated in the transformer only if there is any current in the coils. But the input impedance of the operational amplifier is almost infinity, making current nearly zero. To increase this current at high frequencies, the capacitor of about 10 nF shunts the alternating signal to ground

For the convenience of users, in LabVIEW, differential channels are shifted by 8: the $ai0$ must be paired with $ai8$, $ai1$ with $ai9$, etc. It means that if we want to add a new measurement channel, then in the DAQ Assistant menu press the «add the channel» button, and choose any physical channel from the list between $ai1$ and $ai7$: it will be automatically paired to an appropriate channel between $ai9$ and $ai15$ for the 16-channel DAQ.

In some rare cases, the DAQ Assistant menu (Fig. 11.23) offers the only Terminal Configuration option: «Pseudo-differential». What does it mean? It means that the DAQ board has been chosen without proper consideration and serious problems may await the user. There are types of DAQ boards permanently wired for the so-called pseudo-differential connection (Fig. 11.25), and the user has no opportunity to change anything. With the good intention of suppressing high-frequency noise, acting between the two grounds, this configuration makes the system vulnerable to low-frequency noise. The capacitor simply adds this noise to the inverting input, while the transformer does not work efficiently at low frequencies, making compensation incomplete. As a result, considerable amount of inter-ground noise appears at the output.

Although not very common, the pseudo-differential option may appear in expensive high-speed simultaneously sampling models. The datasheets always clearly state this property, so that if you still have purchased it, then there is no one to blame. Consider the simplest test shown in Fig. 11.26. In it, the function generator with grounded BNC connector and the computer with DAQ are powered from different lines. This is the typical grounded-source scheme. Two similar high-speed simultaneously sampling DAQs were tested: differential and pseudo-differential. The pseudo-differential option shows such a big noise that it may be considered unusable.

The last thing to be done is to learn how to save the data into a file on disk. Although this is the topic of programming, which goes beyond the scope of the present book, it is an essential practical question. Therefore, in order to facilitate quick practical reference for the reader, this issue will be briefly explained below.

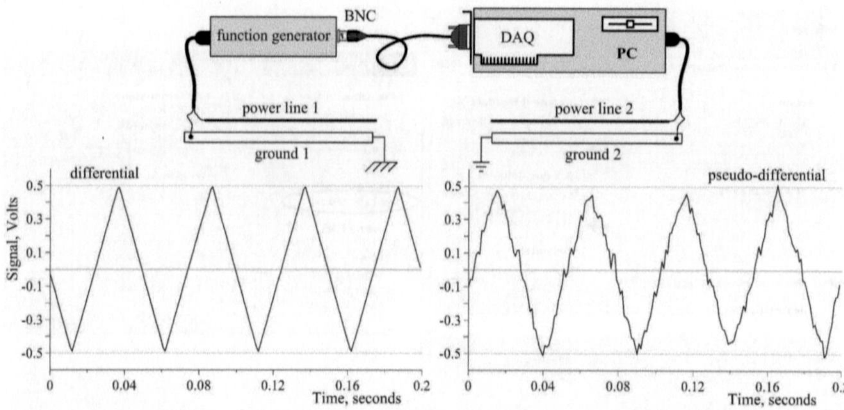

Fig. 11.26 An arbitrary function generator is programmed for triangular signal of 0.5 V amplitude and is connected through a BNC coaxial cable to the first channel ($ai0 - ai8$) of the differential DAQ. The power cord is plugged into one ~220 V power line, whereas the computer with the DAQ is plugged into another power line. The measured signal is shown in the left panel. After that, the differential DAQ was substituted for the adequate pseudo-differential one. The result is shown in the right panel: floating ground noise is clearly seen

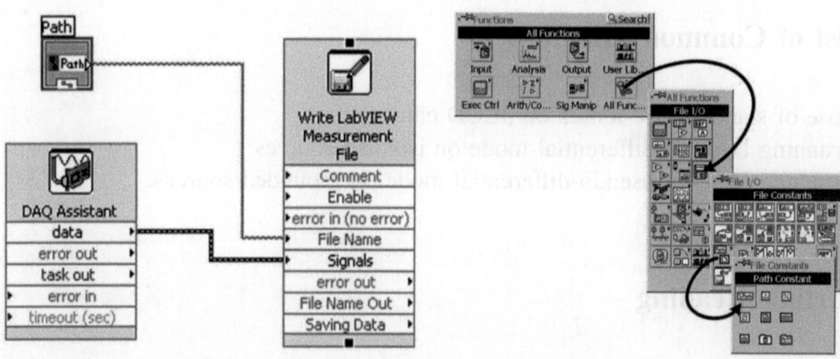

Fig. 11.27 The virtual wire, connecting «DAQ Assistant» with the «Write LVM», relays all the measurements in all the channels measured in the DAQ board. They will be recorded in the file on the hard disk under the name specified by the user in the «Path» control box on the front panel. The «Path» element can be found through the chain shown at *right*. Right-click the «Path» element on the block diagram to change it to «Control»

We have already chosen the «DAQ Assistant» on the block diagram (Fig. 11.20). Next, right-click the blank field and on the icon «Output» choose «Write LabVIEW Measurement File» (Fig. 11.27). There are many options how the data can be written into the file. Double-click the «Write LVM» element on the block diagram to open the menu (Fig. 11.28). Despite its simplicity, this basic configuration supports a wide variety of opto-electronic applications that may occur in the laboratory.

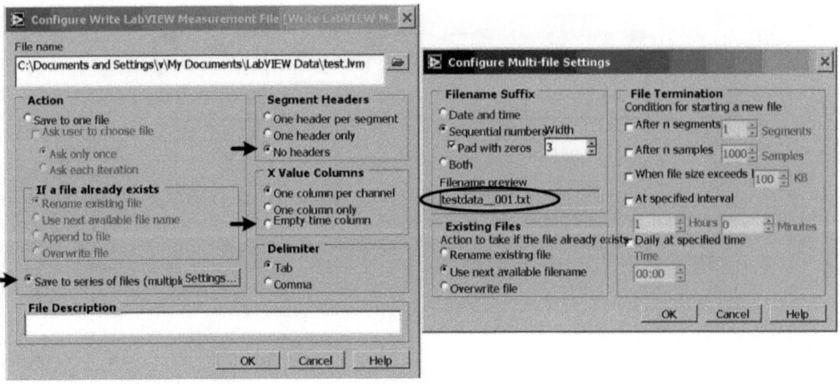

Fig. 11.28 A very convenient option is «Save to series of files» (*left panel*). It means that if the user has forgotten to change the file name to a new one before new measurement, then the program will automatically save the new trace into the file with new extension circled in the *right panel*. The «No header» option cancels all the annoying headers that contain irrelevant data and may be a hindrance to other programs, like graphic engines, reading the file. The first column written to the file will be the time when the measurement was made. It is the only thread, connecting the data with the real world. If the this circle is checked, then the data will be referenced only by their sequence number—the time scale will be lost

List of Common Mistakes

- use of standard TV lenses on 3CCD cameras;
- running DAQ in differential mode on isolated sources;
- running DAQ in pseudo-differential mode on grounded sources.

Further Reading

J.D. Murray, W. vanRyper, Encyclopedia of Graphics File Formats, O'Reilly Media, 2nd ed. (1996).

Compaq Fortran: Language Reference Manual, Compaq Computer Corporation, (1999).

Chapter 12
Beam Propagation

How to compute beam propagation and organize simple ray-tracing routines, using Fortran.

Abstract It is hardly possible to imagine a book on optics without mathematical laws of beam propagation. However, practical nature of all the previous chapters dictated to concentrate pure mathematics separately, confined in a single location that can be easily skipped if necessary. This chapter is this place. The first section presents the well-known ABCD technique—a simplified paraxial ray-tracing matrix formalism. The table in this section summarizes basic optical elements and their matrices. Although the ABCD formalism can be found in numerous other books, it is always better to have necessary tools at hand, so that this section is in harmony with other chapters. Particularly, the ABCD method is indispensable in calculating propagation of the Gaussian beams. The famous Kogelnik formula makes it possible to trace Gaussian beams easily, calculating their basic parameters - the waist and radius of curvature—without any numerical routines that are needed to trace the non-paraxial rays and that are presented in the subsequent sections of this chapter. Arbitrary three-dimensional ray tracing formalism of computing refractions is described in the second section. Mathematically, it is based on the vector presentation of the vector product—an operation that computes sine of the angle between two unity vectors. Transcript of the Fortran code of this routine is listed, so that more complicated simulating programs can be made with it. As the simplest example, spherical aberration of a plano-convex lens is demonstrated in the meridional plane, i.e. in two dimensions. To exemplify the true three-dimensional problem, the cross-sectional map of the coma aberration is computed next. Ray tracing with reflection is explained with application to the shear interferometer problem, discussed in Chap. 6. Reflection formalism is based on the vector presentation of the scalar product, and again, the transcript of the Fortran code is presented. However, the problem of the shear interferometer is so simple that can be solved even analytically, giving the result that is used in Chap. 6. Next, the computational routine is applied to trace the rays in the Czerny-Turner spectrometer, presenting impressive cross-sectional portraits around the focal plane. The last section of the chapter and of the book fulfils the obligation to

V. Protopopov, *Practical Opto-Electronics*, Springer Series in Optical Sciences 184,
DOI: 10.1007/978-3-319-04513-9_12,

compute refraction in birefringent media, given in Chap. 5. The equation devised in Sect. 5.1 is solved numerically here, using the ZREAL routine from the Fortran IMSL library. With it, deviation of rays in Nicol prism is computed and the transcript of the code is listed.

12.1 ABCD Technique

To begin with, consider the so-called ABCD technique—a simplified matrix formalism for propagation of paraxial rays. Although not accounting for aberrations and, therefore, inappropriate for detailed computations, it may be indispensable when computing propagation in long periodical structures, like laser cavities or microscopic guiding arrays.

In geometrical optics, paraxial approximation considers only rays, insignificantly inclined to optical axis (Fig. 12.1).
With z-axis along the optical axis, and $r_{1,2}$ being transversal displacements from the optical axis in planes $z = z_1$ and $z = z_2$, the tangent of the inclination angle

$$\tan\alpha = \frac{dr}{dz}.$$

For conciseness, it is convenient to denote the derivative as r'. Obviously,

$$r_2' = r_1' \; ; \; r_2 = r_1 + r_1' \cdot l.$$

Considering optical paths, we have to rewrite the second equality with the refractive index of the medium n:

$$r_2 = r_1 + r_1' \cdot ln.$$

These trivial geometrical identities may be written in a matrix form if we compose a two-dimensional ray vector (r, r'):

$$\begin{pmatrix} r_2 \\ r_2' \end{pmatrix} = \begin{pmatrix} 1 & ln \\ 0 & 1 \end{pmatrix} \begin{pmatrix} r_1 \\ r_1' \end{pmatrix}.$$

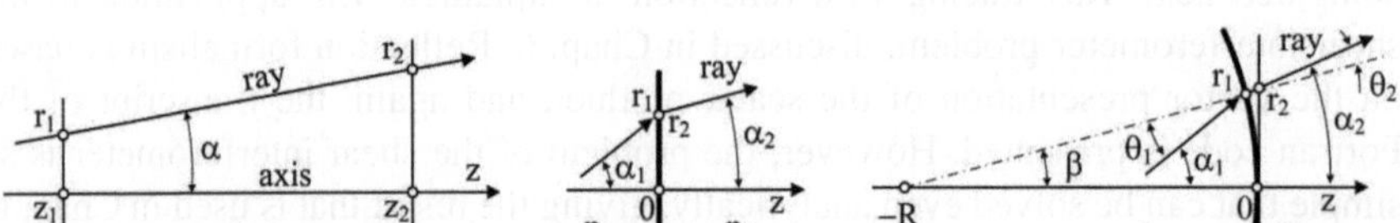

Fig. 12.1 Free propagation section (*at left*); flat interface between two media with refractive indices n_2 and n_1 (*in the middle*); curved interface (*at right*)

Thus, simple free-space propagation of a ray (not necessarily paraxial) may be represented by the so-called ABCD matrix:

$$\begin{pmatrix} A & B \\ C & D \end{pmatrix} = \begin{pmatrix} 1 & ln \\ 0 & 1 \end{pmatrix}.$$

Next, consider a ray, refracting at the interface between two media with refraction indices n_1 and n_2 (Fig. 12.1). The Snell law gives

$$n_1 \sin \alpha_1 = n_2 \sin \alpha_2.$$

At the interface $z = 0$, the ray vector transforms:

$$\begin{pmatrix} r_1 \\ r_1' \end{pmatrix} \to \begin{pmatrix} r_2 \\ r_2' \end{pmatrix}.$$

Obviously, $r_2 = r_1$, whereas r_2', being introduced above as the tangent, should be, strictly speaking, expressed, using trigonometric identities and the Snell law. Here, the paraxial approximation greatly simplifies analysis:

$$\sin \alpha \approx \alpha \approx \tan \alpha = r'.$$

Therefore,

$$\begin{pmatrix} r_2 \\ r_2' \end{pmatrix} = \begin{pmatrix} 1 & 0 \\ 0 & \frac{n_1}{n_2} \end{pmatrix} \begin{pmatrix} r_1 \\ r_1' \end{pmatrix}; \begin{pmatrix} A & B \\ C & D \end{pmatrix} = \begin{pmatrix} 1 & 0 \\ 0 & \frac{n_1}{n_2} \end{pmatrix}.$$

Curved interfaces are considered thin, with the reference plane going through the vertex (Fig. 12.1). Here, the sign convention is important: if the ray enters the concave surface, its radius R is considered positive; otherwise—negative. With yet another approximation

$$\beta \approx \frac{r}{R}$$

that holds true only for $r/R \ll 1$,

$$n_1\theta_1 \approx n_2\theta_2; \ \alpha_1 = \beta + \theta_1; \ \alpha_2 = \beta + \theta_2;$$

so that the matrix becomes

$$\begin{pmatrix} r_2 \\ r_2' \end{pmatrix} = \begin{pmatrix} 1 & 0 \\ \frac{n_2 - n_1}{n_2 R} & \frac{n_1}{n_2} \end{pmatrix} \begin{pmatrix} r_1 \\ r_1' \end{pmatrix}; \begin{pmatrix} A & B \\ C & D \end{pmatrix} = \begin{pmatrix} 1 & 0 \\ \frac{n_2 - n_1}{n_2 R} & \frac{n_1}{n_2} \end{pmatrix}.$$

Two consecutive concave-convex curved interfaces with refractive index n, surrounded by an air with unity refractive index, form the thin lens. Its matrix can be deduced by multiplying the two above matrices in a proper order and with correct signs of R:

$$\begin{pmatrix} 1 & 0 \\ \frac{1-n}{R} & n \end{pmatrix}\begin{pmatrix} 1 & 0 \\ \frac{n-1}{nR} & \frac{1}{n} \end{pmatrix} = \begin{pmatrix} 1 & 0 \\ -(n-1)\left(\frac{1}{R_1}+\frac{1}{R_2}\right) & 1 \end{pmatrix} = \begin{pmatrix} 1 & 0 \\ -\frac{1}{f} & 1 \end{pmatrix} = \begin{pmatrix} A & B \\ C & D \end{pmatrix}.$$

Here f is the focal length of a thin lens, obeying the well-known formula:

$$\frac{1}{f} = (n-1)\left(\frac{1}{R_1}+\frac{1}{R_2}\right).$$

ABCD matrices for basic optical elements are summarized in Table 12.1.

Note that the C element in the matrix of a thick lens is exactly the lens-maker equation given in Chap. 1.

Table 12.1 ABCD matrices

Element	Figure	ABCD matrix
Free space	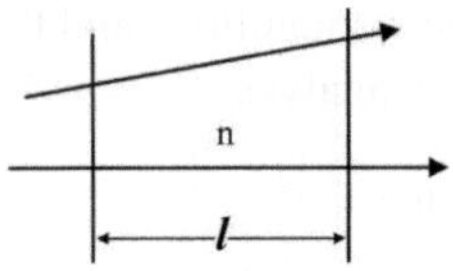	$\begin{pmatrix} 1 & ln \\ 0 & 1 \end{pmatrix}$
Flat interface	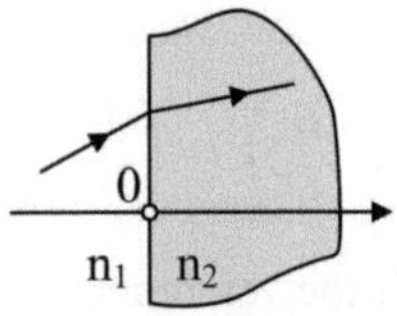	$\begin{pmatrix} 1 & 0 \\ 0 & \frac{n_1}{n_2} \end{pmatrix}$
Flat slab of refractive index n	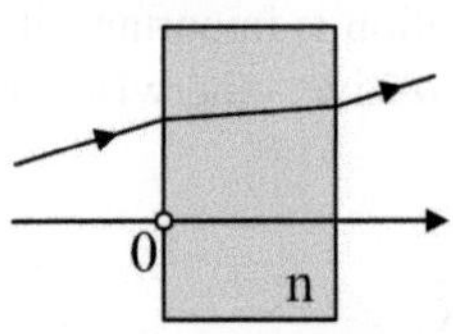	$\begin{pmatrix} 1 & \frac{l}{n} \\ 0 & 1 \end{pmatrix}$
Spherical interface	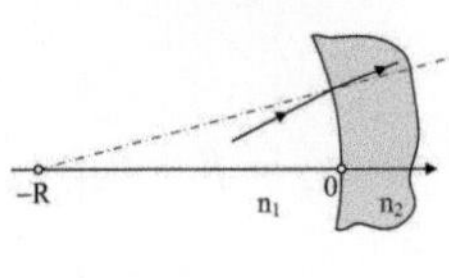	$\begin{pmatrix} 1 & 0 \\ \frac{n_2-n_1}{n_2 R} & \frac{n_1}{n_2} \end{pmatrix}$
Thin lens	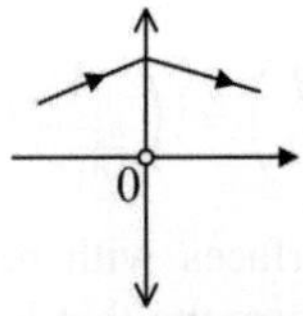	$\begin{pmatrix} 1 & 0 \\ -\frac{1}{f} & 1 \end{pmatrix}$
Thick lens	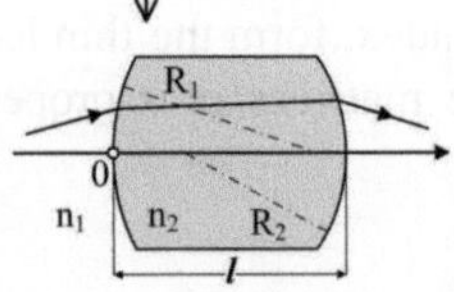	$\begin{pmatrix} 1-\frac{l(n_2-n_1)}{n_2R_1} & \frac{l n_1}{n_2} \\ \frac{(n_2-n_1)^2 l}{n_1 n_2 R_1 R_2} - \frac{(n_2-n_1)}{n_1}\left(\frac{1}{R_1}+\frac{1}{R_2}\right) & 1-\frac{l(n_2-n_1)}{n_2R_2} \end{pmatrix}$

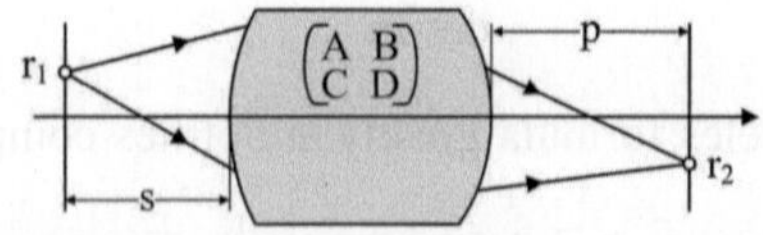

Fig. 12.2 All the rays coming from some point in the first of the two conjugated planes come to one point of the another. It means that displacement r does not depend on the angle r'

ABCD technique can help to find conjugated planes of a composite optical system, i.e. the planes that are the images of one another. Suppose the ABCD matrix of such a system is known, and the first and second conjugated planes are positioned at s and p from the vertices of the system (Fig. 12.2). Select any arbitrary point r_1 in the first plane and trace it to the second plane:

$$\begin{pmatrix} r_2 \\ r_2' \end{pmatrix} = \begin{pmatrix} 1 & p \\ 0 & 1 \end{pmatrix} \begin{pmatrix} A & B \\ C & D \end{pmatrix} \begin{pmatrix} 1 & s \\ 0 & 1 \end{pmatrix} \begin{pmatrix} r_1 \\ r_1' \end{pmatrix}.$$

Multiplying the matrices, obtain

$$r_2 = (A + pC)\, r_1 + (As + B + Csp + Dp)\, r_1'.$$

Conjugation means

$$As + B + Csp + Dp = 0.$$

This is the equation with two variables s and p, which can be solved for p with any arbitrary s. For example, the thin lens with the focal length f brings this equation to a form

$$s - \frac{1}{f} sp + p = 0,$$

which can be transformed by dividing it over s and p to a well-known lens formula:

$$\frac{1}{f} = \frac{1}{p} + \frac{1}{s}.$$

When multiple optical elements are stacked periodically in a large number N, it is always possible to identify the ray matrix of a single period $\mathbf{m}$ and present the ray matrix $\mathbf{M}$ of the entire optical system as the power:

$$\mathbf{M} = \mathbf{U} \cdot \mathbf{m}^N \cdot \mathbf{V}$$

with $\mathbf{U}$ and $\mathbf{V}$ being some simple ray matrices of the output and input respectively. As it follows from the Table 12.1, ABCD matrices are unimodular (determinant is unity) if they begin and end in the media with equal refractive indices. Therefore, it is always possible to organize one period so that

$$\det \mathbf{m} = 1.$$

Then the so-called Abeles formula greatly simplifies computations:

$$\begin{pmatrix} a_{11} & a_{12} \\ a_{21} & a_{22} \end{pmatrix}^N = \begin{pmatrix} a_{11}U_{N-1}(p) - U_{N-2}(p) & a_{12}U_{N-1}(p) \\ a_{21}U_{N-1}(p) & a_{22}U_{N-1}(p) - U_{N-2}(p) \end{pmatrix},$$

where U_m—Chebyshev polynomials of the second kind

$$U_N(x) = \frac{\sin[(N+1)\arccos x]}{\sqrt{1-x^2}},$$

$p = \frac{1}{2}(a_{11} + a_{22})$. Sometimes, the following property of Chebyshev polynomials may be useful:

$$U_{N+1}(x) = 2xU_N(x) - U_{N-1}(x).$$

Matrix formalism does not account for diffraction. However, propagation of one special type of self-reproducing beams—the so-called Gaussian beams—may be formulated in terms of matrices, accounting for diffraction. Gaussian beams are the product of an axially symmetrical laser cavity. The amplitude of a traditional TEM_{00} mode with the wavelength λ, propagating in the medium with refractive index n in z direction, is described in its cross section by complex Gaussian function

$$e^{-i\frac{kr^2}{2q(z)}},$$

where $i = \sqrt{-1}$, $k = \frac{2\pi}{\lambda}$ —the wave number, r—distance from the axis, and the complex parameter

$$\frac{1}{q(z)} = \frac{1}{R(z)} - i\frac{\lambda}{\pi n \rho^2(z)}$$

is composed of the radius $R(z)$ of wavefront curvature measured at the optical axis (the wavefront at large is parabolic) and the so-called waist $\rho(z)$, which characterizes the width of the beam. In the plane $z = z_0$, where the waist is a minimum $\rho(z_0) = \rho_0$, the wavefront is flat: $R(z_0) = \infty$. Hence, if we know $q(z)$, then separating the real and imaginary parts of its reciprocal value, it is possible to unambiguously determine the wavefront radius and the waist. Therefore, to completely determine the Gaussian beam, it is only needed to know how the $q(z)$ propagates.

At infinity, divergent Gaussian beams are simple cones with the angle determined by diffraction on the waist. As to the convergent beam, its minimum diameter is limited by diffraction, and it is always a practical problem to determine where the waist is positioned. The typical theoretical mistake, which can be heard in almost 100 % of cases, is to expect that the Gaussian beam with the waist on the lens (Fig. 12.3) will be focused to a minimal diameter in the focal plane. According to geometrical optics, this expectation is correct: since the wavefront in the plane of the lens is flat, the wave will be focused in the focal plane. However, this consideration

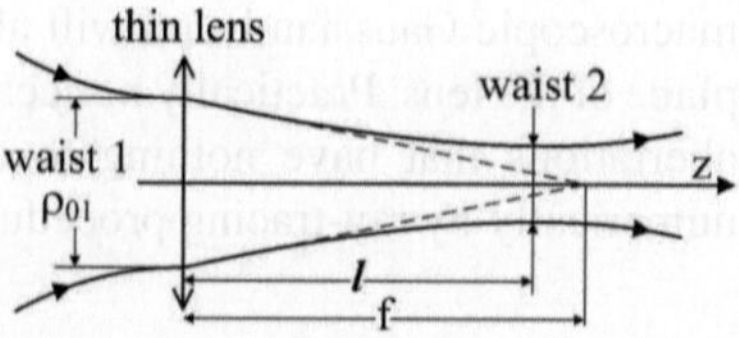

Fig. 12.3 With flat wavefront on the lens, Gaussian beam forms the waist in the plane before the focal plane

does not take into account diffraction. the correct answer can be obtained with the ABCD technique, which in the case of Gaussian beams takes the form of the Kogelnik rule: if the Gaussian beam propagates from the plane where $q = q_1$ to the plane where $q = q_2$, and the ABCD matrix between these planes is known, then

$$q_2 = \frac{Aq_1 + B}{Cq_1 + D}.$$

This holds true, of course, until the beam remains paraxial. The ABCD matrix of propagation from the lens to the plane at $z = l$ is

$$\begin{pmatrix} 1 & l \\ 0 & 1 \end{pmatrix} \begin{pmatrix} 1 & 0 \\ -\frac{1}{f} & 1 \end{pmatrix} = \begin{pmatrix} 1-\frac{l}{f} & l \\ -\frac{1}{f} & 1 \end{pmatrix} = \begin{pmatrix} A & B \\ C & D \end{pmatrix}.$$

Hence

$$q_2 = \frac{\left(1-\frac{l}{f}\right)q_1 + l}{-\frac{1}{f}q_1 + 1}.$$

We are interested in the waist, i.e. $R = \infty$. As such, q_2 must be imaginary as well as q_1. Therefore, denote $q_1 = ia$ and find the condition when q_2 is imaginary:

$$lf^2 + a^2 l = a^2 f.$$

From here

$$l = \frac{f}{1 + \left(\frac{f\lambda}{\pi n \rho_{01}^2}\right)^2} = \frac{f}{1 + (f/z_{01})^2} < f,$$

where

$$z_{01} = \frac{\pi n \rho_{01}^2}{\lambda}$$

is called the confocality parameter—distance from the waist to where the beam is increased by a factor of $\sqrt{2}$. So, indeed, theoretically, the waist is formed before the focal plane. But how far is l from f? For example, consider $f = 50$ mm; $\rho_{01} = 0.5$ mm (typical He–Ne laser); $\lambda = 633$ nm. Then, in air, $z_{01} = 1240$ mm and $l = 0.998f$—an infinitesimal difference. Thus, although incorrect theoretically, practically the macroscopic Gaussian beam will always be focused to a minimal spot in the focal plane of the lens. Practically noticeable deviations from this rule are associated with aberrations that have nothing to do with diffraction and can be analyzed only numerically by ray-tracing procedures. This is addressed in the next section.

12.2 Ray Tracing with Refraction

The Snell law of refraction at the flat interface with relative refractive index n is formulated as

$$\sin \alpha_1 = n \sin \alpha_2$$

where α_1 and α_2 are the angles that the rays make with the normal to the interface. The main problem of ray tracing in three dimensions (3D) is to associate α_1 and α_2 with Cartesian coordinates that describe complicated interfaces like spherical surfaces of lenses or tilted planes. We shall show how to do it in two steps: first, how to compute refraction on an arbitrarily inclined plane, and second, how to compute local normal to an arbitrary curved surface.

Let the axis z of the right triplet (x, y, z) be along the optical axis of the system, and unity directional vectors of the incident and refracted rays be respectively

$$\vec{v} = \begin{pmatrix} v_x \\ v_y \\ v_z \end{pmatrix} \text{ and } \vec{v}' = \begin{pmatrix} v'_x \\ v'_y \\ v'_z \end{pmatrix}.$$

It is important and necessary that these two vectors are unity vectors:

$$|\vec{v}|^2 = v_x^2 + v_y^2 + v_z^2 = 1.$$

Then, in the Cartesian system $\vec{r} = (x, y, z)$, the rays themselves are set parametrically as

$$\vec{r} = \vec{r}_0 + \vec{v}t, \quad \begin{pmatrix} x \\ y \\ z \end{pmatrix} = \begin{pmatrix} x_0 \\ y_0 \\ z_0 \end{pmatrix} + \begin{pmatrix} v_x \\ v_y \\ v_z \end{pmatrix} t.$$

Let the surface local normal be unity vector $\vec{s}$. Then the vector product

$$[\vec{v}, \vec{s}] = |\vec{v}| \cdot |\vec{s}| \cdot \sin \alpha = \sin \alpha.$$

This is how it works: the vector product of the surface normal and the directional vector of the ray give the sine of the incident angle, in any Cartesian system of coordinates. With it, the Snell law takes the form:

$$[\vec{v}, \vec{s}] = n\, [\vec{v}', \vec{s}].$$

In components, with the Cartesian triplet $\vec{e}_x$, $\vec{e}_y, \vec{e}_z$, the vector product may be computed as the determinant:

$$[\vec{v}, \vec{s}] = \begin{vmatrix} \vec{e}_x & \vec{e}_y & \vec{e}_z \\ v_x & v_y & v_z \\ s_x & s_y & s_z \end{vmatrix} = \vec{e}_x(v_y s_z - v_z s_y) - \vec{e}_y(v_x s_z - v_z s_x) + \vec{e}_z(v_x s_y - v_y s_x).$$

The Snell law suggests to solve the system of equations in order to find $\vec{v}'$:

$$\begin{vmatrix} \vec{e}_x & \vec{e}_y & \vec{e}_z \\ v'_x & v'_y & v'_z \\ s_x & s_y & s_z \end{vmatrix} = \frac{1}{n} \begin{vmatrix} \vec{e}_x & \vec{e}_y & \vec{e}_z \\ v_x & v_y & v_z \\ s_x & s_y & s_z \end{vmatrix}$$

or in components:

$$\begin{cases} v'_y s_z - v'_z s_y = \frac{1}{n}(v_y s_z - v_z s_y) \\ v'_x s_z - v'_z s_x = \frac{1}{n}(v_x s_z - v_z s_x) \\ v'_x s_y - v'_y s_x = \frac{1}{n}(v_x s_y - v_y s_x) \end{cases}.$$

It looks like we have three equations for three unknowns, so that each and every one of them can be determined exactly. Then, who guarantees that the result gives the unity vector, i.e. $v'^2_x + v'^2_y + v'^2_z = 1$? The answer is simple: determinant of this system of equations is zero, therefore only two variables can be determined as functions of the third one. As such, we need the fourth equation to unambiguously determine all the three unknowns, and this fourth equation is

$$v'^2_x + v'^2_y + v'^2_z = 1.$$

This last equation is of the second order, so that two solutions must be analyzed.

Everyone can solve this system, but the trick is to solve it in such a way as to avoid $s_{x,y}$ or $v_{x,y}$ in denominator, because otherwise horizontal rays or vertical interfaces will produce infinity in computations. The right solution in the order of computations is as follows:

❶: $v'_z = -G \pm \sqrt{G^2 - F}$; $\quad G = \frac{1}{n}\left[s_x(v_x s_z - v_z s_x) + s_y(v_y s_z - v_z s_y)\right]$;
$F = \frac{1}{n^2}\left[(v_x s_z - v_z s_x)^2 + (v_y s_z - v_z s_y)^2\right] - s_z^2$;
❷: or ❸: $v'_x = \frac{1}{s_z}\left[v'_z s_x + \frac{1}{n}(v_x s_z - v_z s_x)\right]$;
❸: or ❷: $v'_y = \frac{1}{s_z}\left[v'_z s_y + \frac{1}{n}(v_y s_z - v_z s_y)\right]$.

The next problem is what sign to choose in v'_z. It is not enough just to find it out by trial and error method what—plus or minus—to choose: it must be proved. For that, consider the case of n=1 and calculate analytically the term under the square root. With some patience and, taking into consideration the unity property of all the vectors, one obtains:

$$G^2 - F = (v_x s_x s_z + v_y s_y s_z + v_z s_z^2)^2.$$

Since $n = 1$, we have to have the identity $v_z' = v_z$ from ❶. This identity is true if only the sign is chosen «+».

The above set of formulas solves the refraction problem when the ray comes from air to glass with the refractive index n. On its way out of the glass, the same formalism works as well but with the substitution

$$n \to \frac{1}{n}.$$

And this finalizes the theory of the entire refraction routine.

The next problem is how to find the local normal $\vec{s}$ to a surface. It can be solved in two steps: first—find the local tangent plane, and second—calculate the normal to this plane. The local tangent plane to any surface can be found by linearization. Recall that the plane is given by the equation

$$Ax + By + Cz + D = 0,$$

and A, B, and C are the components of the normal vector $\vec{s}$:

$$\vec{s} = \begin{pmatrix} A \\ B \\ C \end{pmatrix}.$$

Now, if the surface of any complexity is given by the equation

$$W(x, y, z) = 0,$$

then make small perturbation (α, β, γ) to the point $\vec{r}_1 = (x_1, y_1, z_1)$ on the surface

$$\begin{pmatrix} x_1 + \alpha \\ y_1 + \beta \\ z_1 + \gamma \end{pmatrix}$$

and expand the W in the Taylor series over small (α, β, γ):

$$A\alpha + B\beta + C\gamma + D = 0.$$

You have found the normal vector $\vec{s}$.

Let us see how it works in practice with the simple sphere, which is frequently the front surface of a lens. The sphere of radius R with the axis z through its center and the vertex on the origin can be written as

$$(R - z)^2 + x^2 + y^2 - R^2 = 0.$$

Linearization in an arbitrary point $\vec{r}_1 = (x_1, y_1, z_1)$ gives

$$2x_1\alpha + 2y_1\beta - 2(R - z_1)\gamma + x_1^2 + y_1^2 + (R - z_1)^2 - R^2 = 0,$$

and the normal

$$\begin{pmatrix} 2x_1 \\ 2y_1 \\ -2(R - z_1) \end{pmatrix}.$$

This vector is the normal vector to the sphere in the point $\vec{r}_1$ but it is not the unity vector. It must be normalized:

$$\begin{pmatrix} x_1/N \\ y_1/N \\ -(R - z_1)/N \end{pmatrix}; \frac{1}{N}\sqrt{x_1^2 + y_1^2 + (R - z_1)^2} = 1; \ \frac{R}{N} = \pm 1; \ N = \pm R,$$

depending on the direction. Choose plus for certainty. Thus,

$$\vec{s} = \begin{pmatrix} x_1/R \\ y_1/R \\ -(R - z_1)/R \end{pmatrix}.$$

Next, consider a sample problem of tracing rays through a plano-convex lens with the axis z along its optical axis, and the point source, illuminating its spherical surface (Fig. 12.4).

Now, it is time for writing computer code, and an essential practical question is what particular computer language to use? The definite and clear answer is Fortran. Although other languages like C and its upgrades dominate the realm of applied software development, for mathematical computations Fortran is indispensable. Apart of airy arguments about superiority of speed of computations that may be only several percent over C, the strongest practical argument is the time required to make a working code. When two equally qualified programmers write the same code in C and Fortran, the Fortran code begins to work in two or three times shorter time. The reason is much better structured and economic coding language, which results in several times less number of syntax errors. The definite structure of Fortran syntax is also an argument to use it in this book in order to facilitate easy reading.

For the beginning, we are interested in meridional rays only, i.e. the rays lying in the plane of a drawing. The Fortran code for this program is below. All the notations in the transcript are consistent with the above theory and do need additional explanation.

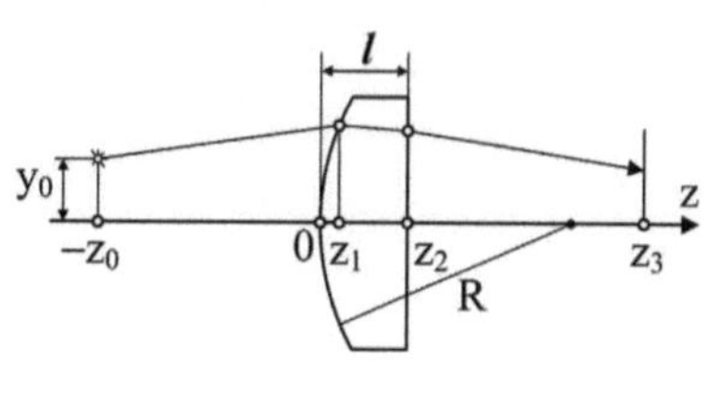

Fig. 12.4 Geometry of a sample problem

```
c This program traces 3D rays through plano-convex lens

      program threedrt
      real l, n
c Set lens parameters:
      R=20.                         !mm – front radius
      l=5.                          !mm – thickness on the vertex
      n=1.5                         !refractive index

c set initial point for the ray:
      x0=0
      y0=2.
      z0=-10.

c Set initial propagation vector:
      vx=0.
      vy=0.
      vz=sqrt(1.-vy*vy-vx*vx)

c Find intersection with the front spherical surface:
      U=-(R-z0)*vz+x0*vx+y0*vy
      H=(R-z0)**2+x0*x0+y0*y0-R*R
      t=-U-sqrt(U*U-H)
      x1=x0+vx*t
      y1=y0+vy*t
      z1=z0+vz*t

c Compute analytically normal to the surface:
      sx=x1/R
      sy=y1/R
      sz=-(R-z1)/R

c Compute refracted directional vector:
      gvalue=G(n,vx,vy,vz,sx,sy,sz)
      fvalue=F(n,vx,vy,vz,sx,sy,sz)
      vz1=-gvalue+sqrt(gvalue*gvalue-fvalue)
      vx1=dirx(n,vz1,vx,vy,vz,sx,sy,sz)
      vy1=diry(n,vz1,vx,vy,vz,sx,sy,sz)

c Find intersection with the second surface:
      t=(l-z1)/vz1
      x2=x1+vx1*t
      y2=y1+vy1*t
      z2=l

c Define normal vector at the second surface:
      sx=0.
      sy=0.
      sz=1.

c Compute refracted directional vector:
      n=1./n
      gvalue=G(n,vx1,vy1,vz1,sx,sy,sz)
      fvalue=F(n,vx1,vy1,vz1,sx,sy,sz)
      vz11=-gvalue+sqrt(gvalue*gvalue-fvalue)
      vx11=dirx(n,vz11,vx1,vy1,vz1,sx,sy,sz)
      vy11=diry(n,vz11,vx1,vy1,vz1,sx,sy,sz)

c Compute free propagation of this ray:
      t=40.
      x3=x2+vx11*t
      y3=y2+vy11*t
      z3=z2+vz11*t

c Ray drawing:
c Initial ray will be saved in the file "ray0.dat":
      OPEN(UNIT=1, FILE='3Dray0.dat')
      write(1,100)z0, y0
      write(1,100)z1, y1
      close(1)

c First refracted ray will be saved in the file "ray1.dat":
      OPEN(UNIT=2, FILE='3Dray1.dat')
      write(2,100)z1, y1
      write(2,100)z2, y2
      close(2)

c Third refracted ray will be saved in the file "ray2.dat":
      OPEN(UNIT=3, FILE='3Dray2.dat')
      write(3,100)z2, y2
      write(3,100)z3, y3
      close(3)

c Draw front surface:
      OPEN(UNIT=4, FILE='3Dfrontsurface.dat')
      do 1 i=1,100
      ro=0.1*(i-1)
      write(4,100) R-sqrt(R*R-ro*ro), ro
    1 continue
      close(4)

c Draw second surface:
      OPEN(UNIT=5, FILE='3Dsecondsurface.dat')
      do 2 i=1,100
      ro=0.1*(i-1)
      write(5,100) l, ro
    2 continue
      close(5)

  100 format(1x,2E12.3)
      end


      function G(n,vx,vy,vz,sx,sy,sz)
      real n
      G=(sx*(vx*sz-vz*sx)+sy*(vy*sz-vz*sy))/n
      return
      end

      function F(n,vx,vy,vz,sx,sy,sz)
      real n
      F=((vx*sz-vz*sx)**2+(vy*sz-vz*sy)**2)/(n*n)-sz*sz
      return
      end

      function diry(n,vz1,vx,vy,vz,sx,sy,sz)
      real n
      diry=(vz1*sy+(vy*sz-vz*sy)/n)/sz
      return
      end

      function dirx(n,vz1,vx,vy,vz,sx,sy,sz)
      real n
      dirx=(vz1*sx+(vx*sz-vz*sx)/n)/sz
      return
      end
```

In it, a ray parallel to the optical axis emerges from the point source at

$$\vec{r}_0 = \begin{pmatrix} 0 \\ 2 \\ -10 \end{pmatrix} \text{ mm.}$$

This ray intersects the front sphere of the radius $R = 20$ mm at z_1 and the flat surface necessarily at $z_2 = l = 5$ mm. Since this is a horizontal ray, it intersects the optical axis near the focal point for paraxial rays (Fig. 12.5). In the figure, this ray is composed of three sections: horizontal section before the lens (*ray0.dat* in the code), refracted section inside the lens (*ray1.dat*), and the final section on the way to focus (*ray3.dat*). It is instructive to compare the focal position around 40 mm with the formula of an ideal lens

$$\frac{1}{f} = (n-1)\left(\frac{1}{R_1} + \frac{1}{R_2}\right).$$

With $n = 1.5$ and $R = 20$ mm it gives $f = 40$ mm. This value might seem a good coincidence with ray-tracing computations, if it were not for the one thing: the formula describes the thin lens whereas the actual paraxial focus is 40 mm from the rear surface of the thick lens. Thus, uncertainty is about 5 mm.

Strictly speaking, 3D procedure is not necessary for tracing meridional rays: simple Snell law formula is easily applicable to this case. But for simulating cross sections of the beams, it is indispensable. The next Fortran transcript describes the code that computes focal plane distribution with off-axis illumination.

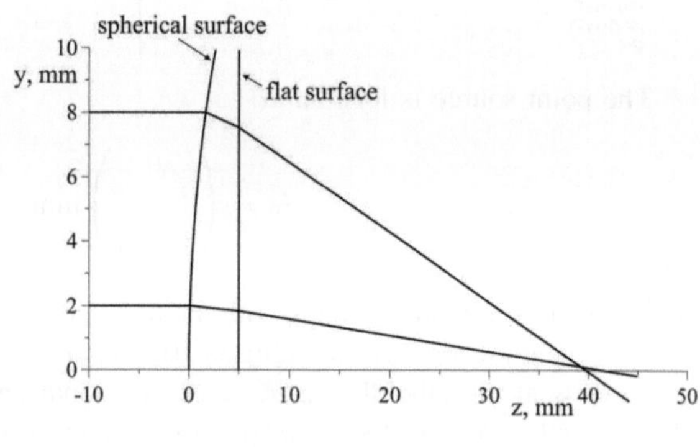

Fig. 12.5 Refraction in plano-convex lens. Two parallel rays intersect the optical axis at different points —this is the spherical aberration. Vertical and horizontal axes are not to scale in order to better show ray trajectories

```
c This program 3D traces rays through plano-convex lens and constructs
c cross-sections

      program sections
      parameter (M=200)
      double precision l,n,n1,x0,y0,z0,x1,y1,z1,vx,vy,vz,U,H,t,sx,sy,sz,
     &R,gvalue,fvalue,vx1,vy1,vz1,x2,y2,z2,x3,y3,z3,vx11,vy11,vz11

c Set lens parameters:
      R=20.                          !mm - front radius
      l=5.                           !mm - lens thickness on the vertex
      n=1.5                          !refractive index
      dalfa=0.05                     !angular discrete of simulation
      radius=10.                     !lens radius

c set initial point for the source in mm:
      x0=0
      y0=5.
      z0=-200.

      OPEN(UNIT=1, FILE='3Dray-section.dat')
c Set initial propagation vector:
      m2=M/2
      do 1 i=1,M
      do 1 j=1,M
      vx=dalfa*(-m2+i)/M
      vy=dalfa*(-m2+j)/M
      vz=dsqrt(1.-vy*vy-vx*vx)

c Rays, missing the lens, must be ignored:
      if ((vx*z0)**2+(vy*z0)**2.GE.radius) go to 1

c Find intersection with the front spherical surface:
      U=-(R-z0)*vz+x0*vx+y0*vy
      H=(R-z0)**2+x0*x0+y0*y0-R*R
      t=-U-dsqrt(U*U-H)
      x1=x0+vx*t
      y1=y0+vy*t
      z1=z0+vz*t

c Compute analytically normal to the surface:
      sx=x1/R
      sy=y1/R
      sz=-(R-z1)/R

c Compute refracted directional vector:
      gvalue=G(n,vx,vy,vz,sx,sy,sz)
      fvalue=F(n,vx,vy,vz,sx,sy,sz)
      vz1=-gvalue+dsqrt(gvalue*gvalue-fvalue)
      vx1=dirx(n,vz1,vx,vy,vz,sx,sy,sz)
      vy1=diry(n,vz1,vx,vy,vz,sx,sy,sz)

c Find intersection with the second surface:
      t=(l-z1)/vz1
      x2=x1+vx1*t
      y2=y1+vy1*t
      z2=l

c Define normal vector at the second surface:
      sx=0.
      sy=0.
      sz=1.

c Compute refracted directional vector:
      n1=1./n                        !here air and glass swap the places
      gvalue=G(n1,vx1,vy1,vz1,sx,sy,sz)
      fvalue=F(n1,vx1,vy1,vz1,sx,sy,sz)
      vz11=-gvalue+dsqrt(gvalue*gvalue-fvalue)
      vx11=dirx(n1,vz11,vx1,vy1,vz1,sx,sy,sz)
      vy11=diry(n1,vz11,vx1,vy1,vz1,sx,sy,sz)

c Compute free propagation of this ray:
      t=46.615             !adjusted to best focusing
      x3=x2+vx11*t
      y3=y2+vy11*t
      z3=z2+vz11*t

c Drawing cross section:
      write(1,100)x3, y3
    1 continue
      close(1)


  100 format(1x, 2E15.5)
      end

      double precision function G(n,vx,vy,vz,sx,sy,sz)
      double precision n,vx,vy,vz,sx,sy,sz
      G=(sx*(vx*sz-vz*sx)+sy*(vy*sz-vz*sy))/n
      return
      end

      double precision function F(n,vx,vy,vz,sx,sy,sz)
      double precision n,vx,vy,vz,sx,sy,sz
      F=((vx*sz-vz*sx)**2+(vy*sz-vz*sy)**2)/(n*n)-sz*sz
      return
      end

      double precision function diry(n,vz1,vx,vy,vz,sx,sy,sz)
      double precision n,vz1,vx,vy,vz,sx,sy,sz
      diry=(vz1*sy+(vy*sz-vz*sy)/n)/sz
      return
      end

      double precision function dirx(n,vz1,vx,vy,vz,sx,sy,sz)
      double precision n,vz1,vx,vy,vz,sx,sy,sz
      dirx=(vz1*sx+(vx*sz-vz*sx)/n)/sz
      return
      end
```

The point source is located at

$$\vec{r}_0 = \begin{pmatrix} 0 \\ 5 \\ -200 \end{pmatrix} \text{mm}.$$

and generates a bundle of rays with angular separation $dalfa = 0.05$ radians. The rays, missing the lens whose radius is 10 mm, are ignored as the wasted rays. This cycle ends at the label **1**, writing cross-sectional coordinates (x_3, y_3) in the file *3Dray-section.dat*. Unlike the previous code, accurate 3D simulation often requires the *double precision* capabilities of Fortran in order to avoid distortion of the finest features of the image. Longitudinal position of the analyzed section is adjusted to approximately the focal plane as much as the spherical aberration allows. Coma aberration—inevitable companion of the off-axis focusing in non-corrected systems—is clearly visible in Fig. 12.6.

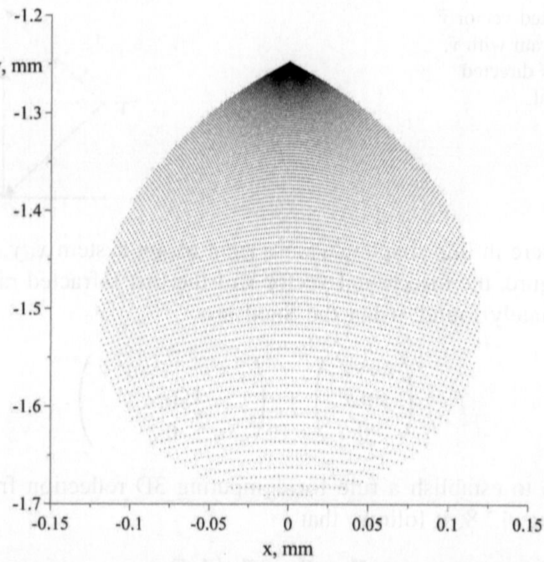

Fig. 12.6 Focal cross section in off-axis illumination. The same plano-convex lens as in Fig. 12.5. Note that the axes are not to scale for better visual resolution of *dots*, representing each ray

12.3 Ray Tracing with Reflection

In Chap. 6, we promised detailed derivation of formulas, describing shear-plate interferometer. This is the good example of ray tracing with both refraction and reflection. It is necessary to remind that the ray-tracing problem was to find displacement S (capital letter, must not be mixed up with the normal vector $\vec{s}$ written in lowercase) and inclination angle θ of the ray, passing inside the glass block. For the convenience of readers, Fig. 12.7 reproduces Fig. 6.20 in new Cartesian system of coordinates relevant to this problem.

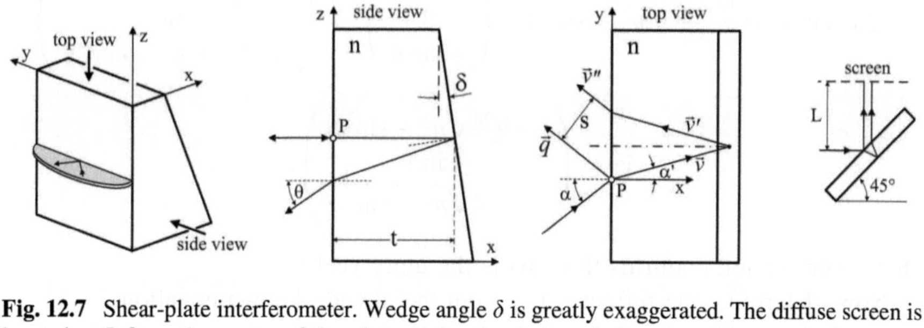

Fig. 12.7 Shear-plate interferometer. Wedge angle δ is greatly exaggerated. The diffuse screen is located at L from the center of the plate. Refractive index of glass is n. For practical reasons, angle of incidence α is always 45°

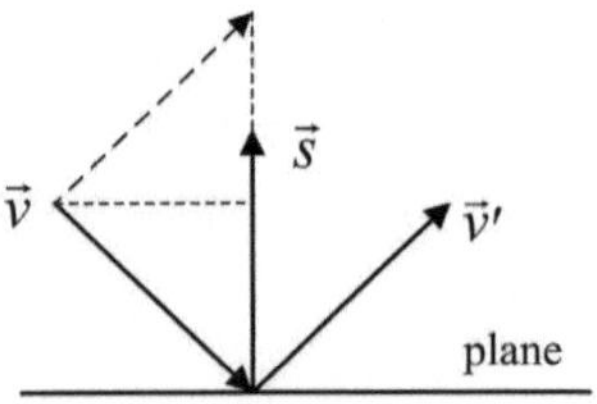

Fig. 12.8 Reflected vector $\vec{v}'$ makes parallelogram with $\vec{v}$, with the normal $\vec{s}$ directed along the diagonal

As everywhere in this chapter, we use right triplet system x, y, z. From the top view in this figure, the directional vector $\vec{v}$ of the first refracted ray at the point P can be immediately found, using the Snell law:

$$\vec{v} = \begin{pmatrix} \cos\alpha' \\ \sin\alpha' \\ 0 \end{pmatrix} = \begin{pmatrix} \frac{1}{n}\sqrt{n^2 - \sin^2\alpha} \\ \frac{1}{n}\sin\alpha \\ 0 \end{pmatrix}.$$

Now, we have to establish a rule for computing 3D reflection from an arbitrary plane. From Fig. 12.8, it follows that

$$\vec{v}' = \vec{v} - 2\vec{s}\cdot(\vec{v}, \vec{s}),$$

where the round brackets denote the scalar product. The negative sign stands because the angle between $\vec{v}$ and $\vec{s}$ is obtuse. It is easy to find that this transform produces unity vector $\vec{v}'$ if only $\vec{v}$ and $\vec{s}$ are unity vectors.

The normal to the back surface of the wedge is

$$\vec{s} = \begin{pmatrix} -\cos\delta \\ 0 \\ -\sin\delta \end{pmatrix},$$

and the directional vector of the reflected ray can be readily calculated:

$$(\vec{v}, \vec{s}) = \left(\frac{1}{n}\sqrt{n^2 - \sin^2\alpha},\ \frac{1}{n}\sin\alpha,\ 0\right)\begin{pmatrix} -\cos\delta \\ 0 \\ -\sin\delta \end{pmatrix} = -\frac{\cos\delta}{n}\sqrt{n^2 - \sin^2\alpha};$$

$$-2\vec{s}\cdot(\vec{v}, \vec{s}) = \frac{\cos\delta}{n}\sqrt{n^2 - \sin^2\alpha}\cdot\begin{pmatrix} -\cos\delta \\ 0 \\ -\sin\delta \end{pmatrix} = \begin{pmatrix} -\frac{2\cos^2\delta}{n}\sqrt{n^2 - \sin^2\alpha} \\ 0 \\ -\frac{\sin 2\delta}{n}\sqrt{n^2 - \sin^2\alpha} \end{pmatrix};$$

$$\vec{v}' = \begin{pmatrix} -\frac{\cos 2\delta}{n}\sqrt{n^2 - \sin^2\alpha} \\ \frac{1}{n}\sin\alpha \\ -\frac{\sin 2\delta}{n}\sqrt{n^2 - \sin^2\alpha} \end{pmatrix}.$$

Direct verification confirms that this is the unity vector.

Now, determine the refracted ray from the already known equation

$$[\vec{v}'', \vec{s}'] = n\,[\vec{v}', \vec{s}'],$$

where

$$\vec{s}' = \begin{pmatrix} -1 \\ 0 \\ 0 \end{pmatrix}$$

is the normal to the front surface. Simplicity of this equation is encouraging:

$$\begin{pmatrix} 0 \\ -v_z'' \\ v_y'' \end{pmatrix} = n \begin{pmatrix} 0 \\ -v_z' \\ v_y' \end{pmatrix}.$$

Remembering that we have to preserve unity module for $\vec{v}''$, the solution is

$$\vec{v}'' = \begin{pmatrix} -\sqrt{1 - n^2(v_z'^2 + v_y'^2)} \\ n\, v_y' \\ n\, v_z' \end{pmatrix}.$$

This is the complete solution for the problem, and the only thing that we have to do is to substitute $\vec{v}'$ and expand over the small wedge angle δ. The result is below:

$$\vec{v}'' \approx \begin{pmatrix} -\cos\alpha \\ \sin\alpha \\ -2\,\delta\sqrt{n^2 - \sin^2\alpha} \end{pmatrix}.$$

The ray reflected from the front surface has directional vector simply

$$\vec{q} = \begin{pmatrix} -\cos\alpha \\ \sin\alpha \\ 0 \end{pmatrix}.$$

Comparing $\vec{v}''$ and $\vec{q}$, we see that the refracted ray is tilted downwards by the small angle

$$\theta \approx -2\,\delta\sqrt{n^2 - \sin^2\alpha}.$$

Thus, one part of the initial problem is solved.

Next, we have to calculate displacement S. Unlike the ray reflected from the front surface (directional vector $\vec{q}$), the refracted ray with directional vector $\vec{v}''$ does not lie in the xy plane (Fig. 12.7). However, the inclination angle θ is small, of the order of the wedge angle δ, therefore, S may be approximately computed in the xy plane. As such, this has become a trivial trigonometric problem with the result

$$S = 2t\tan\alpha' \cdot \cos\alpha = t\frac{\sin 2\alpha}{\sqrt{n^2 - \sin^2\alpha}}.$$

The last parameter that we have to calculate in order to simulate interference pattern on shear interferometer, is the optical path difference Δ between the two rays. From Fig. 12.7, the approximate solution in the xy plane follows as

$$\Delta = n\frac{2\,t}{\cos\alpha'} - S\,\tan\alpha = 2\,t\,\sqrt{n^2 - \sin^2\alpha}.$$

This is the last missing explanation to what has been told in Chap. 6. Now, all the parameters got their analytical expressions, and there are no obstacles any more to complete simulation of the interference pattern that is described by the set of formulas from Chap. 6:

$$\sin^2\left\{\frac{1}{2}\left[k\,(z - z' - \Delta) + \Phi(x,y,z) - \Phi(x',y',z' + \Delta)\right]\right\};$$

$$\begin{cases} x' = x - S \\ y' = y + \theta z \\ z' = -\theta y + z \end{cases};$$

$$\Phi(x,y,z) = \frac{k}{2R}\,r^2 + c\cdot r^4.$$

Here $k = 2\pi/\lambda$ is the wavenumber and c is the coefficient, describing spherical aberration. The Fortran code listed below computes two-dimensional map in the xy plane with fixed $z = const$—distance from the center of the wedge to the screen.

```
c This program computes fringes in shear-plate interferometer


      program shear
      parameter (M=200)          !good graphics quality in Surfer requires M=200.
      real k, lambda, n
      common k, R, spheric
      pi=3.1415926
      pi2=2*pi
      n=1.5           !refractive index
      c=+0.81         !spherical aberration coefficient in wavelength per aperture
      size=20.        !mm side of the picture
      size05=0.5*size            !half of the side
      spheric=pi2*c/size05**4    !spherical aberration parameter
      z=30.                      !mm - image plane distance from the wedge
      alfa=pi/4.                 !45 degrees
      delta=8.        !wedge, arc seconds; 8" gives the best fitting to experiment
      delta=delta*4.848E-6       !radians
      lambda=632.8               !nm
      lambda=lambda*1.E-06       !mm
      k=pi2/lambda               !wavenumber, mm^-1
      t=10.                      !plate thickness, mm
c R=40000. fits best experiment
      R=+40000.                  !mm - sphere radius on the wedge, fixed value
      B=sqrt(n*n-sin(alfa)**2)
      S=t*sin(2*alfa)/B
      teta=2*delta*B
      D=2*t*B                    !optical path difference


      OPEN(UNIT=1, FILE='fringes.dat')
c The 2D xy map starts here:
      par=size/M
      do 1 i=1,M
      x=-size05+par*(i-1)
      x1=x-s
      do 1 j=1,M
      y=-size05+par*(j-1)
      y1=y+teta*z
      z1=-teta*y+z
c Interference fringes function:
      u=sin(0.5*(k*(z-z1-D)+fi(x,y,z,c)-fi(x1,y1,z1+D,c)))**2
      write(1,100) x, y, u
    1 continue
      close(1)

  100 format(1x,3E12.3)
      end



      function fi(x,y,z,c)
      real k
      common k, R, spheric
      H=R+z
      ro2=x*x+y*y
      fi=k/(2.*H)*ro2+spheric*ro2*ro2
      return
      end
```

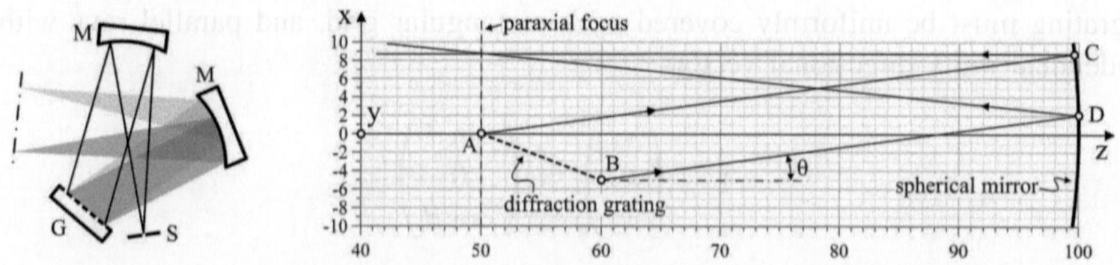

Fig. 12.9 In the crossed Czerny-Turner scheme, flat diffraction grating with grooves perpendicular to the drawing, sends monochromatic parallel bundles of rays to the spherical mirror. If we consider the *xz* plane as the horizontal one, then the *y* axis points vertically. The grating extends in vertical direction, making the problem three-dimensional. The points of the grating that do not belong in the *xz* plane send parallel off-axial rays to the mirror. The off-axial rays are supposed to produce bigger aberrations, which is the subject of our simulation

Another example with the reflection is the spectrometer in crossed Czerny-Turner geometry, analyzed in Chap. 9 (Fig. 12.9).

The rays reflected from the mirror must focus in as narrow spot as possible in order to fit single pixel of the CCD (charge coupled device, Chap. 9) and do not degrade spectral resolution. Thus, it is a very practical question how aberrations of a spherical mirror distort the focal spot. Assume that the front surface of a flat rectangular diffraction grating is confined in the *xz* plane between the points *A* and *B* with coordinates

$$A = \begin{pmatrix} a_x \\ 0 \\ a_z \end{pmatrix} \text{ and } B = \begin{pmatrix} b_x \\ 0 \\ b_z \end{pmatrix}$$

and vertically in between $\pm a_y$. In the particular case of Fig. 12.9, the origin resides at the center of the mirror curvature $R = 100$ mm;

$$A = \begin{pmatrix} 0 \\ 0 \\ 50 \end{pmatrix} \text{ and } B = \begin{pmatrix} -5 \\ 0 \\ 60 \end{pmatrix};$$

and $a_y = 5$ mm—roughly square grating area. All the dimensions are in millimeters.

Assume that the grating is installed in parallel rays. As such, monochromatic rays reflected at specific angles θ are also parallel at each spectral component. We choose any point at the grating and trace the ray, emerging at the angle θ, to the focal plane. Paraxial focal distance for the spherical mirror is

$$f = \frac{R}{2},$$

so that we have to analyze the cross section of the focused bundle somewhere around $z = 50$ mm. In order to obtain a picture, the surface of the diffraction grating must be uniformly covered by a rectangular grid, and parallel rays with identical unity directional vectors

$$\vec{v} = \begin{pmatrix} v_x \\ v_y \\ v_z \end{pmatrix} = \begin{pmatrix} \sin\theta \\ 0 \\ \cos\theta \end{pmatrix}$$

and different positions

$$\begin{pmatrix} x_j \\ y_j \\ z_j \end{pmatrix}, \quad j = 1, 2, \ldots, N$$

launched from every point j of the grid. Thus, each ray has the parametric form

$$\begin{pmatrix} x \\ y \\ z \end{pmatrix} + \begin{pmatrix} v_x \\ v_y \\ v_z \end{pmatrix} \cdot t.$$

In our system of coordinates, equation for the spherical mirror is merely

$$x^2 + y^2 + z^2 = R^2.$$

At intersection point C,

$$t = -U + \sqrt{U^2 - W}; \; U = a_x v_x + a_z v_z; \; W = a_x^2 + a_y^2 + a_z^2 - R^2.$$

The choice of sign has already been explained in the previous section. Hence, the intersection point itself is

$$\begin{pmatrix} c_x \\ c_y \\ c_z \end{pmatrix} = \begin{pmatrix} a_x + v_x t \\ a_y \\ a_z + v_z t \end{pmatrix}.$$

Obviously, the unity normal vector $\vec{s}$ directed to the center of the mirror curvature is

$$\vec{s} = \frac{1}{R} \begin{pmatrix} c_x \\ c_y \\ c_z \end{pmatrix}.$$

For the direction of the reflected vector we have

$$\vec{v}' = \vec{v} - 2\vec{s} \cdot (\vec{v}, \vec{s}) = \begin{pmatrix} v_x - \frac{2q}{R^2} c_x \\ -\frac{2q}{R^2} c_y \\ v_z - \frac{2q}{R^2} c_z \end{pmatrix}, \quad q = v_x c_x + v_z c_z.$$

Then the reflected ray itself in parametric form is

$$\begin{pmatrix} x \\ y \\ z \end{pmatrix} = \begin{pmatrix} c_x \\ c_y \\ c_z \end{pmatrix} + \vec{v}' t.$$

We need to trace these rays only to the focal plane, which should be somewhere around $z_0 = 50$ mm. This determines t:

$$t = \frac{z_0 - c_z}{v'_z}.$$

The only thing that has to be added is how to construct the grid on the diffraction grating. Since every node is always on the horizontal line $A - B$, set equally spaced independent discrete parameters $0 \le t_1 \le 1$ and $0 \le t_2 \le 1$ to form the grid of different horizontal and vertical dimensions:

$$\begin{aligned} x &= a_x + (b_x - a_x)\, t_1 \\ y &= -a_x + 2a_y\, t_2 \\ z &= a_z + (b_z - a_z)\, t_1 \end{aligned}.$$

This procedure is implemented in the Fortran code below.

```
c This program traces rays in Czemy-Tumer spectrometer with spherical mirror

      program czt2D
      parameter (M1=200, M2=30)
      R=100.                     !mm
      teta=10.                   !degrees
      teta=teta/180.*3.1415      !transformed to radians
      z0=50.4                    !mm, anticipated focal position

c Point A of the diffraction grating:
      ax=0.
      ay=5.
      az=50.
c Point B of the diffraction grating:
      bx=-5.
c The "y" coordinate is not needed because it is the same as for the point A.
      bz=60.

c Open the file to write the 2D data:
      OPEN(UNIT=1, FILE='2D.dat')
c Initial directional vector:
      vx=sin(teta)
      vy=0.
      vz=cos(teta)

c Gridding and ray-tracing routine starts here:
      pr1=1./M1
      pr2=1./M2
      par1=bx-ax
      par2=bz-az
      do 1 i=1, M1               !M1 is the number of nodes in "x" axis
      t1=pr1*i                   !first discrete parameter
      do 1 j=1, M2               !M2 is the number of nodes in "y" axis
      t2=pr2*j                   !second discrete parameter
c Coordinates of the nodes are sequentially  assigned:
      x=ax+par1*t1
      y=ay*(-1.+2.*t2)
      z=az+par2*t1

c Intersection with the mirror:
      U=x*vx+z*vz
      W=x*x+y*y+z*z-R*R
      t=-U+sqrt(U*U-W)
c Point "C":
      cx=x+vx*t
      cy=y
      cz=z+vz*t

c Normal vector:
      sx=-cx/R
      sy=-cy/R
      sz=-cz/R

c Reflected directional vector:
      q=vx*cx+vz*cz
      R2=R*R
      v1x=vx-2*cx*q/R2
      v1y=-2*cy*q/R2
      v1z=vz-2*cz*q/R2

c Trace the rays to anticipated focal plane:
      t=(z0-cz)/v1z
      x=cx+v1x*t
      y=cy+v1y*t
      z=cz+v1z*t

c Write coordinates of each ray in the bundle:
      write(1,100) x, y
    1 continue
      close(1)


  100 format(1x, 3E13.5)
      end
```

Figure 12.10 portrays several sequential cross-sectional views of the focused bundle of rays as the observation plane moves through the paraxial focus $z = f = 50$ mm from right to left.

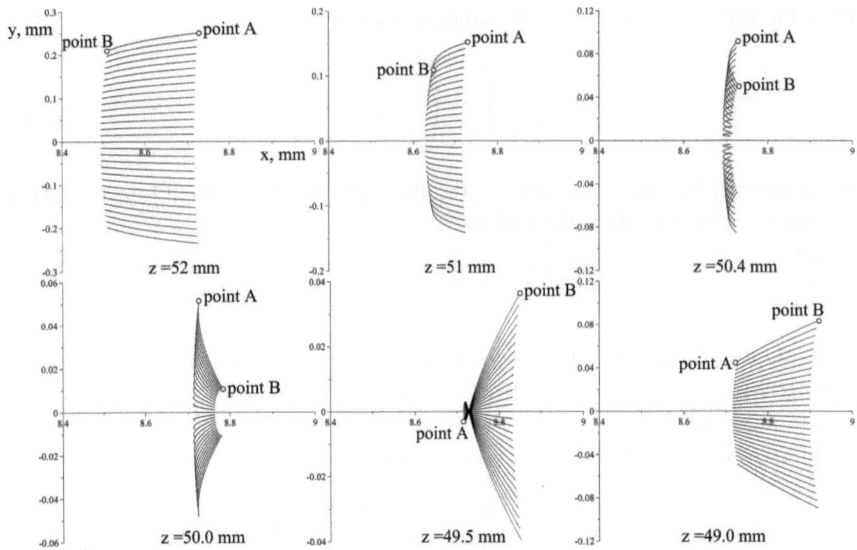

Fig. 12.10 Series of cross-sectional views of the focused beam. The optimal focusing may be considered at $z = 50.4$ mm: the width of the spot in x direction—the key factor for focusing onto narrow pixel of a linear CCD—is only 40 μ. In y direction, the smallest spot size of about 80 μ is at $z = 49.5$ mm

12.4 Refraction at Birefringent Interfaces

From Chap. 5, we have a debt of detailed mathematics of computing refraction at birefringent interfaces. The problem was clearly identified: for the extraordinary wave refractive index depends on the propagation direction, and this direction, in turn, depends on the refractive index. This implies solving a non-linear equation, which can be done only numerically. Consider detailed geometry of this problem (Fig. 12.11).

To begin with, recall the basics of birefringence. In uniaxial crystals, ellipsoid of velocities of light is axially symmetrical

$$\frac{v_x^2}{v_e^2} + \frac{v_y^2}{v_o^2} + \frac{v_z^2}{v_o^2} = 1$$

so that only xy plane may be considered:

$$\frac{v_x^2}{v_e^2} + \frac{v_y^2}{v_o^2} = 1$$

with v_o and v_e being velocities of the ordinary and extraordinary waves. Since the speed of light in the medium with refractive index n is ratio to the speed of light in vacuum c

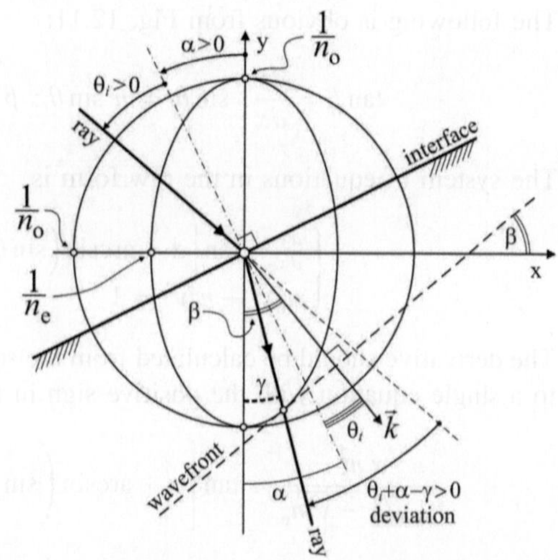

Fig. 12.11 Refraction of the extraordinary ray. The *dashed wavefront line* is the tangent to the extraordinary *ellipsoid*. The wavefront makes the right angle with the wave vector $\vec{k} = 2\pi/\lambda$. Angles of incidence θ_i and refraction θ_t satisfy the Snell law. Refracted extraordinary ray propagates through the tangent point, but not along the $\vec{k}$

$$v = \frac{c}{n},$$

we have

$$v_{\rm e,o} = \frac{c}{n_{\rm e,o}}; \quad n_{\rm e}^2 \frac{v_x^2}{c^2} + n_{\rm o}^2 \frac{v_y^2}{c^2} = 1;$$

where $n_{\rm o}$ and $n_{\rm e}$ are the refractive indices of the ordinary and extraordinary waves. In this system of coordinates, new variables x and y along the axes under the same names compose the extraordinary ellipse shown in Fig. 12.11:

$$n_{\rm e}^2 x^2 + n_{\rm o}^2 y^2 = 1; \quad r = \sqrt{x^2 + y^2}; \; r = \frac{1}{n}.$$

Thus, the refractive index n of the ray, propagating in the particular direction, is the reciprocal of the radius-vector r of the ellipse in this direction. The ordinary circle is

$$n_{\rm o}^2 x^2 + n_{\rm o}^2 y^2 = 1.$$

Propagation direction of the extraordinary ray, i.e. the angle γ, determines the speed of light for this ray, i.e. the extraordinary refractive index $n_{\rm e}$. In turn, this refractive index satisfies the Snell law for the wavefront of this wave. According to the Huygens principle, this wavefront is the tangent to the ellipse. These statements make a system of two equations to determine both the γ and $n_{\rm e}$. Hence, we only have to compose these equations.

The following is obvious from Fig. 12.11:

$$\tan\beta = \frac{dy}{dx};\ \sin\theta_i = n\,\sin\theta_t;\ \beta = \alpha + \theta_t;\ \tan\gamma = \frac{x}{y}.$$

The system of equations in the raw form is

$$\begin{cases} \frac{dy}{dx} = \tan\left[\alpha + \arcsin\left(\sin\theta_i \cdot \sqrt{x^2+y^2}\right)\right] \\ n_o^2 x^2 + n_o^2 y^2 = 1 \end{cases}.$$

The derivative should be calculated from the second equation to reduce the system to a single equation with the positive sign in the left-hand side:

$$\frac{x\,n_e^2}{n_o\sqrt{1-x^2 n_e^2}} = \tan\left[\alpha + \arcsin\left(\sin\theta_i \cdot \frac{\sqrt{1+x^2\left(n_o^2-n_e^2\right)}}{n_o}\right)\right].$$

This form of the equation is unstable for numerical solutions because of singularity in the denominator in the left-hand side. The trick is to rewrite it in a stable form against the new variable A:

$$A \equiv \tan\gamma = \frac{x\,n_o}{\sqrt{1-x^2 n_e^2}}.$$

Then the equation rewrites to a stable form

$$\left(\frac{n_e}{n_o}\right)^2 A = \tan\left[\alpha + \arcsin\left(\sin\theta_i \cdot \sqrt{\frac{1+A^2}{n_o^2+n_e^2 A^2}}\right)\right].$$

Note that the square root is always less than unity because both refractive indices are greater than unity, therefore arcsine does always exist.

Every numerical routine for solving equations requires starting point to begin from, commonly called the initial guess. The above equation provides very efficient initial guess, using closeness of the ordinary and extraordinary refractive indices:

$$p = \frac{n_e}{n_o} \approx 1.$$

Indeed, even for calcite with its exceptionally big difference—$n_o = 1.6584$, and $n_e = 1.4864$—$p \approx 0.9$. As such, the initial guess may be written as

$$A_0 = \frac{1}{p^2}\tan\left[\alpha + \arcsin\left(\frac{\sin\theta_i}{n_o}\right)\right].$$

Next, we have to prove that our initial guess lies within domain of the equation. It means that the square root must not exceed unity:

$$\sqrt{\frac{1+A_0^2}{n_o^2+n_e^2A_0^2}} \leq 1.$$

This holds true for any A_0 for the same reasons as before. We are ready to begin computations.

Fortran is supported by IMSL—the famous International Mathematical Scientific Library developed by Visual Numerics. This package offers the routine ZREAL for solving nonlinear equations. Do not forget to link the IMSL library during compilation. Transcript of the Fortran code is listed below. This program computes deviation angle in degrees for an arbitrary direction of the optical axis, and prints out solution for equation x, actual number of iterations *itnumber*, actual function value F in the end of iterations, values of α and deviation angle in degrees. Notations of the code are consistent with Fig. 12.11 and Fig. 5.7 of Chap. 5.

```
c This program computes refractions in calcite in geometry of the Nicol prism

      program nicolplus
      external F
      common no, p2, alfa, tetai
      real ne,no !refractive indices
      real x(1), xguess(1)
      dimension info(1)
      pi=3.1415926535
      alfa=48.25                 !degrees; for Nicol 48.25 degrees axis.
      tetai=22.                  !degrees; for Nicol delta=68 degrees.
      ne=1.4864                  !extraordinary refractive index: calcite
      no=1.6584                  !ordinary refractive index: calcite
      p=ne/no
      p2=p**2
      alfa=alfa/180.*pi
      tetai=tetai/180.*pi

c Initial settings for ZREAL:
      errabs=1.E-4
      errrel=1.E-4
      eps=1.E-4
      eta=1.E-4
      Nroot=1
      itmax=1000

c Initial guess:
      A0=tan(alfa+asin(sin(tetai)/no))/p2
      xguess(1)=A0

      call ZREAL(F, errabs, errrel, eps, eta, nroot, itmax, xguess, x, info)
      print*,'x:', x(1), 'itnumber:', info(1), 'F:', F(x(1))
      gamma=atan(x(1))
      deviation=(tetai+alfa-gamma)/pi*180.
      print*, 'alfa:', alfa/pi*180., 'deviation:', deviation


  100 format(1x,2E12.3)
      end


c This function is designed for ZREAL:
      function F(x)
      real no, alfa, tetai, x(1)
      common no, p2, alfa, tetai
      A=x(1)
      A2=A*A
      F=p2*A-tan(alfa+asin(sin(tetai)*sqrt((1+A2)/(1+A2*p2))/no))
      return
      end
```

Routine ZREAL is supposed to solve multidimensional systems of nonlinear equations. Therefore, the code explicitly shows one-dimensional arrays of the argument $x(1)$ and initial guess $xguess(1)$. Screenshot of the result is shown in Fig. 12.12.

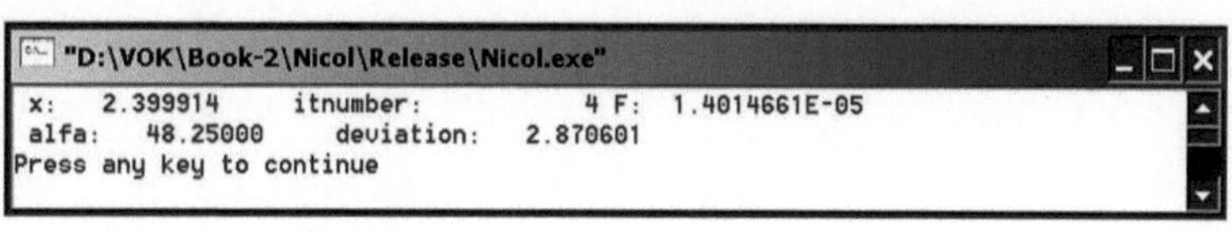

Fig. 12.12 Computing deviation angle in the Nicol prism. Only four iterations were required to reach 10^{-5} error of solving the equation

List of Common Mistakes

- The waist and focus do not coincide on Gaussian beams.

Further Reading

M. Born, E. Wolf, Principles of Optics, Pergamon Press; 4th edition (1968).
A. Yariv, Quantum Electronics, Wiley, 3rd edition (1989).
E. O'Neil, Introduction to Statistical Optics, Dover Publications; 4th edition (2004).

Index

V. Protopopov, *Practical Opto-Electronics*, Springer Series in Optical Sciences 184,
DOI: 10.1007/978-3-319-04513-9,

MIX
Papier aus verantwortungsvollen Quellen
Paper from responsible sources
FSC® C105338

If you have any concerns about our products, you can contact us on
ProductSafety@springernature.com

In case Publisher is established outside the EU, the EU authorized representative is:
Springer Nature Customer Service Center GmbH
Europaplatz 3, 69115 Heidelberg, Germany

Printed by Libri Plureos GmbH
in Hamburg, Germany